Concepts and Applications of Protein Structure

Concepts and Applications of Protein Structure

Edited by **Steven Tiff**

Callisto Reference

New York

Published by Callisto Reference,
106 Park Avenue, Suite 200,
New York, NY 10016, USA
www.callistoreference.com

Concepts and Applications of Protein Structure
Edited by Steven Tiff

International Standard Book Number: 978-1-63239-122-3 (Hardback)

Printed in China.

Contents

	Preface	**IX**
Section 1	**Introduction**	**1**
Chapter 1	**An Evolutionary Biology Approach to Understanding Neurological Disorders** Azhari Aziz, Jasmina Ilievska, Paul R. Fisher and Naomi E. Bishop	**3**
Chapter 2	**Anhydrous and Hydrated Protein Models Derived from High-Resolution and Low-Resolution Techniques** Helmut Durchschlag and Peter Zipper	**43**
Chapter 3	**Structure and Dynamics of Proteins from Nuclear Magnetic Resonance Spectroscopy** Homayoun Valafar and Stephanie J. Irausquin	**67**
Section 2	**Structure Prediction**	**93**
Chapter 4	**An Exhaustive Shape-Based Approach for Proteins' Secondary, Tertiary and Quaternary Structures Indexing, Retrieval and Docking** Eric Paquet and Herna L. Viktor	**95**
Chapter 5	**Refinement of Protein Tertiary Structure by Using Spin-Spin Coupling Constants from Nuclear Magnetic Resonance Measurements** Jürgen M. Schmidt and Frank Löhr	**107**
Chapter 6	**Protein Structure Alphabetic Alignment** Jiaan Yang and Wei-Hua Lee	**133**

Section 3 Energy and Thermodynamics 157

Chapter 7 **Theoretical Analyses of Photoinduced Electron Transfer from Aromatic Amino Acids to the Excited Flavins in Some Flavoproteins** 159
Kiattisak Lugsanangarm, Nadtanet Nunthaboot, Somsak Pianwanit, Sirirat Kokpol and Fumio Tanaka

Chapter 8 **On the Relationship Between Residue Solvent Exposure and Thermal Fluctuations in Proteins** 191
Yu-Tung Chien, Jenn-Kang Hwang and Shao-Wei Huang

Chapter 9 **Estimating Hydrogen Bond Energy in Integral Membrane Chromoproteins by High Hydrostatic Pressure Optical Spectroscopy** 207
Liina Kangur, John D. Olsen, C. Neil Hunter and Arvi Freiberg

Chapter 10 **Preserving Proteins Under High Pressure and Low Temperature** 229
Takahiro Takekiyo, Minoru Kato and Yukihiro Yoshimura

Chapter 11 **A Stable Protein – CutA1** 249
Azumi Hirata, Aya Sato, Takashi Tadokoro, Yuichi Koga, Shigenori Kanaya and Kazufumi Takano

Section 4 Function and Interaction 263

Chapter 12 **Functional Difference Between Deuterated and Protonated Macromolecules** 265
Takashi Sugiyama and Tohru Yoshioka

Chapter 13 **Ligand-Binding Proteins: Structure, Stability and Practical Application** 283
Olga Stepanenko, Alexander Fonin, Olesya Stepanenko, Irina Kuznetsova and Konstantin Turoverov

Chapter 14 **Slit/Robo Signaling: Inhibition of Directional Leukocyte Migration** 309
Ilya M. Mukovozov and Lisa A. Robinson

Section 5 Applications 335

Chapter 15 **Fibrinolytic Enzymes from Medicinal Mushrooms** 337
Chung-Lun Lu and Shiu-Nan Chen

Chapter 16 **Prediction and Rational Design of Antimicrobial Peptides** 363
William F. Porto, Osmar N. Silva and Octávio L. Franco

Chapter 17 **Phospholipases A_2 Protein Structure and Natural Products Interactions in Development of New Pharmaceuticals** 383
Marcos Toyama, Selma D. Rodrigues, Daneila O. Toyama, Veronica C.G. Soares, Camila Ap Cotrim, Rafael Ximenes and Marcelo L. Santos

Permissions

List of Contributors

Preface

Over the recent decade, advancements and applications have progressed exponentially. This has led to the increased interest in this field and projects are being conducted to enhance knowledge. The main objective of this book is to present some of the critical challenges and provide insights into possible solutions. This book will answer the varied questions that arise in the field and also provide an increased scope for furthering studies.

This book compiles the latest studies and advancements in understanding protein structure. Due to several advancements in the field of science and technology, experts today are able to explain various protein structures thoroughly. This book is a collection of several aspects and researches related to protein structures. Several experts have given their contribution in the compilation of data enclosed in this book. This book is a useful reference of knowledge for both students and experts dealing with protein structure.

I hope that this book, with its visionary approach, will be a valuable addition and will promote interest among readers. Each of the authors has provided their extraordinary competence in their specific fields by providing different perspectives as they come from diverse nations and regions. I thank them for their contributions.

Editor

Section 1

Introduction

1

An Evolutionary Biology Approach to Understanding Neurological Disorders

Azhari Aziz[1], Jasmina Ilievska[1,2],
Paul R. Fisher[2] and Naomi E. Bishop[1,3]
[1]*Cell Biology and Molecular Pathogenesis Laboratory,*
[2]*Molecular Cell Biology Laboratory,*
Department of Microbiology,
[3]*Olga Tennison Autism Research Centre,*
La Trobe University, Melbourne, Victoria,
Australia

1. Introduction

Many common human neurological disorders, including epilepsy, Alzheimer's disease, Parkinson's disease, autism spectrum disorders, and schizophrenia show complex heritability and genetics. While studies of single-gene diseases typically provide a more straightforward opportunity to understand the underlying molecular mechanisms of disease, complex diseases are more common and inherently more difficult to study. Nonetheless, researchers have begun to make dramatic inroads into the study of complex human diseases, including many neurological disorders, in the post-human genome sequence era. This is largely due to new technologies and resources that are promoting our understanding of protein structure and function, thereby facilitating the association of disease phenotypes with genetic loci. The online Mendelian inheritance in man (OMIM) database lists those genes implicated in human disease, and this highlights progress made in this field, where around 10% of human genes have a known disease-association (Amberger et al., 2009).

In the first few sections of this paper, we highlight the differences and similarities between simple and complex human genetic disorders, and key methods to study these disorders. We emphasize the key role comparative and evolutionary biology techniques play in increasing our understanding of the pathophysiology of complex human disorders, including in the assessment of the functional traits of gene products implicated in human disease. Several human neurological disorders are used to illustrate the power of this methodology. In the last sections of this paper, the significance and implications of comparative and evolutionary biology data are highlighted using schizophrenia, and autism as specific examples. The surprising recent links between neurological disorders and cancer are discussed in the final section. We conclude that exploration of the evolutionary history of human genes, and comparison of protein structure, helps us understand how and why human neurological disorders originated, influences the choice of appropriate animal models for human disorders, and informs our interpretation of data from model organisms, including the evaluation of novel therapeutics. We conclude that comparative and evolutionary biology, including techniques facilitating the prediction of protein function, has a major role in facilitating further understanding of human neurological disorders and in the development of therapeutic interventions.

2. Classification of human genetic disorders

A genetic disorder is a disease caused by an abnormality, or abnormalities, in genetic material or genome architecture. Genetic disorders are traditionally subdivided into four types: (i) single-gene disorders (often referred to as Mendelian or monogenic diseases), (ii) mitochondrial genome disorders; (iii) chromosomal disorders (where there are gross changes in chromosome structure, such as loss, duplication and/or translocation) diseases, and (iv) multigenic or complex diseases.

Much progress has been made in the last few decades in identifying the molecular cause of many rare genetic diseases (Amberger et al., 2009). Most of these are highly-penetrant traits due to single-gene mutations, and therefore follow classical Mendelian inheritance patterns (Antonarakis & Beckmann, 2006). Good progress has also been made in understanding mitochondrial genome disorders, particularly in the 30 years since the publication of the reference sequence for human mitochondrial DNA (Anderson et al., 1981; Kumar, 2008; Tuppen et al., 2010). Furthermore, a wide variety of chromosomal disorders have been characterized, where defects can be visualised microscopically (Theisen A & Shaffer LG, 2010). These three classes of genetic disorders are individually rare, although chromosomal disorders are being reported slightly more frequently in recent decades, due to factors such as increased parental age (Jones, 2008; Fonseka & Griffin, 2011), and technological advances facilitating detection of smaller deletions and duplications (Berg et al., 2010; Shaffer et al., 2007; Slavotinek, 2008). Despite these broad categories frequently being used to classify human genetic disorders, it must be borne in mind that the phenotypic expression of genetic mutations varies, and this is discussed next, before we focus our attention on complex genetic disorders.

2.1 The complexity of single-gene disorders

It is now well-known, even for well-characterized Mendelian genetic disorders, that individuals with a specific mutation can display phenotypic differences. This includes variation in the age of onset, in the severity of disease symptoms, and/or in phenotypic characteristics. Indeed, phenotypic pleiotropy is the rule, rather than the exception, even for single-gene disorders (Nadeau, 2001). The variation in individual phenotype can be affected by environmental factors, allelic variation and/or 'modifier' genes. Modifier genes can affect transcription and levels of gene expression directly, or affect phenotype at the cellular, tissue, or organism-level (Nadeau, 2001). While increasing numbers of human modifier genes are being identified, most of the progress in understanding genetic modifiers is dependent on model organisms, such as mice, where gene targeting experiments can be carried out using inbred strains (Nadeau, 2001, 2003). One particularly relevant class of modifiers are referred to as protective alleles, as their presence prevents disease from occurring (Nadeau, 2001, 2003). These findings provide insights relevant to the development of novel therapeutics, as therapeutics could be based on mimicking and/or possibly enhancing the effects of these protective alleles. This class of gene also means that the same genetic mutation can lead to a different disease phenotype depending on the genetic background (Lobo, 2008). Therefore, while genetic disorders are frequently classified as monogenic or complex (see below), the distinction between the two types is becoming increasingly blurred.

Other emerging aspects of monogenic disorders overlapping with those of complex disorders, are those due to the multi-functional nature of many genes. This multi-functionality can make it difficult to predict phenotype from genotype. This is illustrated by metabolic genes, where some gene products are referred to as 'moonlighting proteins' due to the multiple phenotypic effects of mutations (Jeffery, 2009; Sriram et al., 2005). Such multi-functionality also contributes to the observation that distinct phenotypes can be associated with different mutations in the same gene. For example, the *LMNA* gene encodes two proteins and is linked to five diseases (Vigouroux & Bonne, 2002), while mutation of the *ERCC2* gene may cause xeroderma pigmentosum (XP), Cockayne syndrome with XP, or trichothiodystrophy, three phenotypically different disorders (Lehmann et al., 2001). In other cases, different mutations in a single gene can cause different diseases via mechanistically different pathways. For example, the *FMR1* gene is considered 'a gene with three faces' (Oostra & Willemsen, 2009). Mutation of *FMR1* is best characterised as the cause of fragile X syndrome mental retardation, which is inherited in an X-linked dominant pattern, and is due to a lack of *FMR1* mRNA and protein expression. However, mutations leading to high levels of *FMR1* mRNA are linked to tremor/ataxia syndrome via mRNA 'toxicity', while the gene is also linked to premature ovarian insufficiency via a third uncharacterized molecular pathway, possibly affecting the production of *FMR1* mRNA isoforms (Oostra & Willemsen, 2009; Tassone et al., 2011).

Another factor complicating the phenotype of Mendelian disorders is the finding that heterozygotes for some recessively inherited Mendelian disorders, whom show no symptoms of the homozygotic phenotype, are at risk of an apparently unrelated disorder (Sidransky, 2006; Sriram et al., 2005). For example, patients who are heterozygous for the gene deficient in Gaucher disease are at an increased risk of neurodegenerative synucleinopathies, such as Parkinson disease (Sidransky, 2006). An additional complication arises in patients who show clinical symptoms consistent with a single-gene defect in a metabolic pathway, but do not have a complete deficiency in any one enzyme, but rather have multiple partial defects. This phenomenon is referred to as synergistic heterozygosity (Vockley et al., 2000).

Finally, while some genetic disorders are largely polygenic and complex in nature, a subset is inherited in a classical Mendelian manner (see Fig. 1). For example, with Alzheimer disease (ALZ) and Parkinson disease (PKD), a subset of the diseases (prefixed by the term 'familial') are inherited in a Mendelian manner. With PKD, around 5% of cases are due to mutations in one of several specific genes with either autosomal dominant or recessive inheritance patterns (Gasser, 2009; Lesage & Brice, 2009; Shulman et al., 2011), but PKD-associated genes with a more modest penetrance are now beginning to be identified (International Parkinson Disease Genomics Consortium, 2011; Liu et al., 2011; Shulman et al., 2011). With ALZ, around 0.1% of cases are inherited in an autosomal dominant manner, while one *APOE* allele, present in 2% of Caucasian populations, has recently been reclassified from 'risk gene' status to being considered moderately penetrant with semi-dominant inheritance (Blennow et al. 2006; Genin et al., 2011). Nonetheless, ALZ in most patients is influenced by a combination of multiple genetic risk factors and protective alleles (Sherva & Farrer, 201; Waring & Rosenberg, 2008).

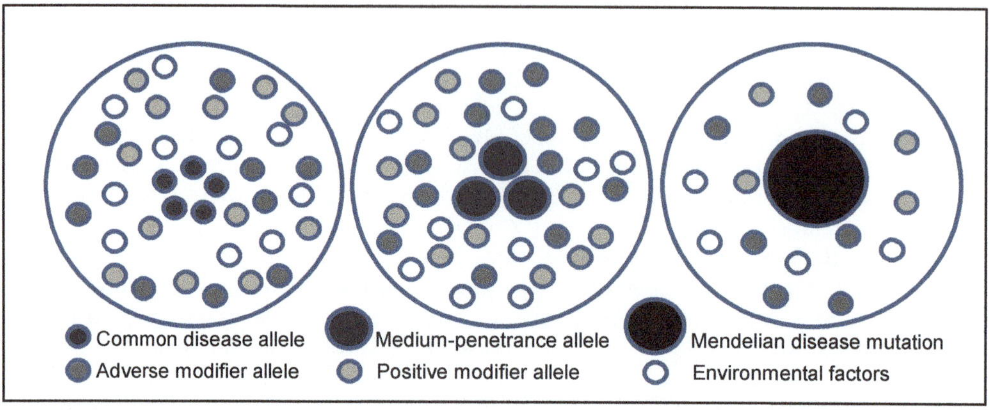

Fig. 1. Multiple genetic and evironmental variants in different combinations affect phenotypes. Common variants (left) or rare mutations (right), or alleles of medium penetrance (middle) can all cause human genetic disorders. Penetrance of a disease allele can be affected by so-called modifier alleles.

Therefore, Mendelian disorders have more in common with multi-factorial diseases than originally thought, and both are affected by genetic background and environmental conditions. Furthermore, the rare Mendelian forms of common complex disorders are providing key insights about the pathogenesis of many complex diseases by highlighting cellular pathways perturbed in the disease state (discussed further in Section 4.4) and this is leading to testable hypothese about disease etiology (Peltonen et al., 2006). Complex genetic disorders are discussed next, emphasizing the importance of evolutionary and comparative biology, while the relevance of these areas of research to multifunctional genes will be discussed further in Section 4.

2.2 Complex genetic disorders

While most Mendelian disorders are rare, there are over 7000 such disorders, and so they collectively affect hundreds of millions of people worldwide (Amberger et al., 2009). By contrast, most of the common disorders of children and adults are complex diseases, and a single highly-penetrant gene is not causative of the disease phenotype (see Fig. 1). Indeed, the causes of such disorders are usually heterogeneous, and a combination of effects from more than one gene, combined with non-genetic factors (environment), play a role in disease development (Davey Smith et al., 2005). Such disorders in children include mental retardation, autism spectrum disorders, attention deficit/hyperactivity disorder, and cancer. In adults, common complex disorders include schizophrenia, bipolar disorder, diabetes, coronary heart disease, hypertension, obesity, and cancer. The complex, multigenic, nature of these diseases has made them inherently more difficult to study. However, in the next section of this paper, we will discuss the key methods used to determine the genetic underpinnings of common complex disorders. Understanding the etiology of these multifactorial diseases is essential for the development of effective means of treatment and/or prevention.

3. Studying complex human genetic disorders

Complex disorders often cluster in families without clearly demonstrating Mendelian inheritance patterns. This makes it difficult to determine the genetic versus non-genetic contribution to the disease phenotype, and to calculate the heritable component of the disorder. Below we will discuss methods used to establish the heritability of human complex disorders, generation of the genetic variation that underpins these disorders, and discuss how to establish which genes are responsible for complex human disease.

3.1 Heritability of complex human diseases

Heritability is usually defined as the proportion of total phenotypic variation that can be attributed to genetic variability (Lee et al., 2011; Visscher et al., 2008). While the interaction of environment on phenotype makes heritability difficult to measure accurately in some cases (Ober & Vercelli, 2011), methods of obtaining unbiased estimates of heritability from various types of pedigree data are well established for both continuous phenotypes and complex human disorders (Lee et al., 2011; Visscher et al., 2008). Furthermore, animal models are invaluable in dissecting aspects of genetic and environmental interactions that are more difficult to assess in human studies (Complex Trait Consortium, 2004) and is discussed further in Section 4.5.

For many human diseases, recent data suggest that the heritable component of many has previously been underestimated (Lee et al., 2011). This has been due to limitations of the methodology employed, as well as other factors, such as evidence demonstrating monozygotic twins are less genetically similar than once thought (Zwijnenburg et al., 2010). A further example is that of PKD, which was long considered a non-hereditary disorder (Shulman et al., 2011; Westerlund et al., 2010). Despite extensive efforts to find environmental risk factors for the disease, genetic variants now stand out as the major causative factor (Shulman et al., 2011; Westerlund et al., 2010; Wirdefeldt et al., 2011). This shift of focus away from environmental toxins, towards genetic contributions, is now leading to rapid progress in understanding PKD and in guiding the development of the next generation of therapeutics (Shulman et al., 2011).

While it is clear that genetics underpins the pathophysiology of complex human disorders, the genetic alleles contributing to the disease phenotype are not always inherited. *De novo* mutations are increasingly being implicated in human disease and, by definition, these mutations are not present in the biological parents of the affected individual. Nonetheless, depending on the severity of the phenotype, and on any effects on fitness, these novel mutations may be transmitted to subsequent generations. Indeed, the rise of techniques such as intracytoplasmic sperm injection (ICSI), can facilitate transmission of *de novo* mutations even if they lead to infertility (Jiang et al., 1999). There is a wide variety in *de novo* human germline mutations, and these can include duplications or deletions of various size, as well as alterations in the number of chromosomes (Arnheim & Calabrese, 2009). The frequency of *de novo* mutations in germ-line cells increases with parental age, and can also be caused by environmental factors such as radiation exposure (Sasaki, 2006), genotoxic chemicals (Phillips & Arlt, 2009), or congenital viral infections (Ansari & Mason, 1977; Fortunato & Spector, 2003; Nusbacher et al., 1967; Vijaya-Lakshmi et al., 1999).

While *de novo* mutations typically occur in gametes, and are often defined as such, new mutations can also occur in the precursors of germ cells, leading to germline mosaicism (Arnheim & Calabrese, 2009), or occur post-fertilization, during embryonic/foetal development and somatic cells (Lupski, 2010). Indeed, new mutations can develop at any time, with cancer being the best-known example of a genetic disorder caused by somatic cell mutagenesis, while Proteus syndrome is a recently-identified disease linked to somatic mosaicism (Lindhurst et al., 2011). While mutations in the genome of somatic cells cannot be passed on to future generations, they may have detrimental effects. While *de novo* mutations are best-studied in cancer (see Section 6.2.3), a contributing role of somatic *de novo* mutations, such as those occurring during brain development, to neurological disorders (see Section 6.1.2) has not been explored.

Mutation frequencies also vary widely across the genome, and often concentrate at certain positions or 'hotspots' (see also Sections 3.2, 6.1.2 & 6.2.3), which have structural and functional features affecting mutagenesis (Ananda et al., 2011; Arnheim & Calabrese, 2009; Carvalho et al., 2010; Rogozin & Pavlov, 2003). For example, CpG context elevates the mutation rate by an order of magnitude (Schmidt et al., 2008), non-B DNA structures induced by palindromic AT-rich repeats facilitate recurrent translocations on chromosomes 11 and 22 at positions 11q23 and 22q11 (Kurahashi et al., 2006), while interspersed repetitive elements such as *Alu*, LINE, long-terminal repeats, and simple tandem repeats are frequently observed at breakpoints in the 9q34.3 subtelomere region (Yatsenko et al., 2009). However, at any given point, multiple mechanisms are acting, making prediction of mutational site and frequency difficult (Arnheim & Calabrese, 2009).

Despite *de novo* mutations most typically being deleterious, rather than neutral or advantageous, their very existence is evidence for ongoing adaptive evolution. Therefore, genetic disorders can be considered 'side-effects' or manifestations of the fundamental mechanisms that provide the genetic variation necessary for evolution to occur. The interaction between the evolutionary past of the human genome and human genetic disease is discussed next.

3.2 Human evolution and genetic disorders

Duplicated regions of DNA play a key role in the evolution of novel gene functions (Conant & Wolfe, 2008; Lynch & Conery, 2000; Ohno, 1970), but are also a source of genetic instability, leading to mutations implicated in both rare and common human genetic disorders (Marques-Bonet & Eichler, 2009). It is therefore relevant to explore the origins of human duplicated sequences. Current evidence indicates that many segmental duplications occurred the hominid lineage and, more specifically, in the common ancestor of African great apes (chimpanzee, gorilla, humans) after divergence from the Ponginae or Asian great ape (orangutan) lineage (Bailey & Eichler, 2006; Carvalho et al., 2010; Koszul & Fischer, 2009; Marques-Bonet & Eichler, 2009). A considerable portion of duplicated human sequences have also been found to correspond to expanded gene families, some of which show signatures of positive selection (Marques-Bonet & Eichler, 2009). Duplications specific to the *Homo sapiens* lineage have also been detected, and include duplications in gene families implicated in neurotransmission, and these may play a role in higher-order brain function in humans (Han et al., 2009).

Different classes of repetitive DNA sequence have been identified (Bao & Eddy, 2002). Some, such as LINES (e.g. L1 family) or SINES (e.g. *Alu* family), are long and short retrotransposable elements, found interspersed throughout the genome. Others are concentrated in certain regions, such as centromeres and telomere-adjacent sequences. These latter regions are also sites of increased genomic instability, which are associated with disease-causing chromosomal breakpoints (Stankiewicz & Lupski, 2002). *Alu* elements have propagated to more than one million copies in primate genomes, and likewise contribute to human genomic diversity (Batzer & Deininger, 2002). Indeed, one in 50 individuals will carry a *de novo* L1 insertion, and one in 20 individuals a *de novo Alu* insertion (Collier & Largaespada, 2007). The active nature of many human retrotransposons is therefore linked to disease-causing somatic and germline mutations (Collier & Largaespada, 2007; Wallace et al., 1991; Oldridge et al., 1999; Claverie-Martin et al., 2003). Repeats may also contribute to DNA secondary structures that are more prone to breakage (Yatsenko et al., 2009). One novel aspect of our increased understanding of the role of repetitive DNA sequence in *de novo* mutations, and our ability to detect such sequences, is that this information can now be used to predict rearrangements that will contribute to genomic disorders (Carvalho et al., 2010; Ou et al., 2011; Sharp et al., 2006).

Therefore, while duplicated sequences in primate genomes predispose apes and humans to extensive genetic diversity and biological innovation, the downside is that many *de novo* genomic changes are mediated by recombination events between these duplications. This characteristic of hominids, and *Homo sapiens* in particular, makes humans particularly susceptible to genomic rearrangements. These rearrangements, in turn, then play a major role in human genetic disease pathogenesis (Inoue & Lupski, 2002; Marques-Bonet & Eichler, 2009). The evolutionary history of some specific genomic rearrangements is discussed next.

3.2.1 Evolutionary history of specific human disease mutations

Using an evolutionary perspective, we can use comparative genomic analyses to calculate the age of appearance of segmental duplications mediating specific disease-causing mutations. Such analyses have revealed that the segmental duplication flanking the Charcot–Marie–Tooth disease region on chromosome 17 (at position 17p12) has an origin in the hominoid ancestor after the divergence of chimpanzees and humans, those flanking the DiGeorge syndrome region on chromosome 22 (22q11.2) expanded after the divergence of hominoids from Old World monkeys, the duplications flanking the Angelman/Prader–Willi region on chromosome 15 (15q11–q13) began to expand before the divergence of the Old World monkeys, while the Smith-Magenis syndrome segmental duplications (17p11.2) date back to after the divergence of New World monkeys (Marques-Bonet & Eichler, 2009). These, and other similar data, have demonstrated that the predisposing genomic features contributing to many genomic disorders have emerged within the last 25 million years (Marques-Bonet & Eichler, 2009).

Using similar methodology, the evolutionary history of human genetic disorders where *Alu* elements are implicated, have also been dated. For example, *Alu* elements mediating lipoprotein lipase deficiency (*LPL* gene) are found in human, ape, and monkey genomes; those *Alu*-elements implicated in Lesch-Nyhan syndrome mutations (*HPRT* gene) are common to human, chimpanzee, and gorilla; while those implicated in ApoB deficiency are restricted to human and great ape genomes (Martinez et al., 2001). Of note, while *Alu* elements acted as 'selfish DNA' when they first inserted into primate genomes, many have subsequently gained regulatory function, a process known as exaptation (Hasler & Strub, 2006). Therefore, despite contributing to the pathogenesis of human genetic disorders, *Alu* elements are thought to have played a role in the divergence of primates, and to have contributed to the regulatory and developmental complexity in primate lineages (Hasler & Strub, 2006). Therefore, *Alu*-depending human genetic diseases also date to primate lineages. Methods used to predict the different phylogenetic ages of genetic disorders instigated by DNA repeats and duplications can therefore being used to both predict and explain the different susceptibility of various primate species to genetic diseases (Martinez et al., 2001; Marques-Bonet & Eichler, 2009).

3.3 Identifying and characterizing genetic determinants of human disease

In the last few decades there has been rapid progress in human disease gene identification, due to recombinant DNA technologies, genome sequencing and analysis methods (Strachan & Read, 2010). With the vast range of resources now available, identification of novel disease genes is currently occurring on a weekly, if not daily, basis. There is no standard procedure for gene identification, however, and identifying the genes responsible for human disease requires information about both gene position and biological function. Functional data is proving a bottleneck for progress in understanding complex diseases in particular and, as outlined below, our understanding of evolutionary biology is of great benefit to studies aimed at identifying and characterizing the genetic determinants of human disease phenotypes.

Genetic linkage- and association-based analyses have been very successful in identifying rare genetic variants with highly penetrant effects, such as those causing Mendelian diseases, and have also been used to investigate the genetics of complex disorders (Altshuler et al., 2008; Jordi, 2000; Ku et al., 2011). More recently, techniques such as genotyping arrays and next-generation DNA sequencing are facilitating the identification of mutations causative of the many as-yet-uncharacterised Mendelian disorders, and of the genetic variation contributing to complex genetic disorders (Kingsley, 2011; Kuhlenbäumer et al., 2011; Roberts et al., 2010).

Two key issues have been emerging from these recent genetic studies. The first is the finding that genome-wide association studies (GWAS) have not been particularly effective in identifying complex genetic disorder risk genes, and the numbers and impact of identified genetic risk factors has been 'disappointing' (Davey Smith et al., 2005; Manolio et al., 2010; Gandhi & Wood, 2010). If sample numbers are sufficiently great, however, GWAS may be better placed to unambiguously identify risk or protective loci for complex diseases (Sullivan, 2001; Wray et al., 2008). Secondly, there are great ongoing difficulties in differentiating disease-causing mutations from rare benign variants (Kuhlenbäumer et al., 2011). These difficulties not only highlight the need for greater 'power' in GWAS, but also the need for follow-through on genetic findings and the application of many aspects of what is referred to as 'integrative genomics' (Giallourakis et al., 2005). Below, we outline some of the varied ways in which our understanding of evolutionary biology is of vital importance to leveraging information obtained from genetic studies. The resulting information can provide key insights into the biological function of uncharacterised genes, can be used to predict candidate disease genes, to predict detrimental mutations, and can provide valuable information about biological pathways altered in different disease states, which may lead to novel therapeutics.

3.3.1 Validating candidate disease genes

GWAS and linkage studies generate large sets of potential disease genes. However, it remains difficult to identify the most likely disease-related genes. Various computational methods for disease gene identification have been described (Oti et al., 2011; Tiffin et al., 2006), and many of these have as their basis data from the field of evolutionary biology. Many software tools apply some of this type of information to genetic datasets, and can be used to determined which genes are the most likely to be involved in the disease in question. The Gene Prioritization Portal website provides an up-to-date summary of web-based candidate disease gene prioritization and prediction tools (Tranchevent et al., 2010). However, most gene prediction tools were designed to study Mendelian disorders, and not for the analysis of complex genetic disorders. The exceptions are the web-based tool, CANDID, specifically designed to prioritize genes implicated in complex human genetic traits (Hutz et al., 2008), and CAESAR, which is not web-based (Gaulton et al., 2007). Tools applicable to complex disorders can be expected to expand over the coming years, due to the large amount of data that will be generated using new genome analysis methods.

A major drawback for the prediction and prioritization of disease genes is that most candidate gene identification tools are reliant on how well-characterized each human gene is, and whether its molecular and cellular function are known. Bear in mind, then, that: (i) over 98% of all gene ontology (GO) annotations are computationally inferred, have not been curated, and are considered by the GO consortium to be potentially unreliable (du Plessis L et al., 2011); (ii) errors in the sequence databases affect at least 1 in 6 sequences (Lagerstrom et al., 2006; Slater & Bishop, 2006; Haitina et al., 2009; Bishop, unpublished data), and this may affect the output from tools such as PROSPECTR (Adie et al., 2005) that use sequence features to rank genes in order of their likelihood of involvement in disease; (iii) while high-throughput protein-protein interaction detection studies have great potential for increasing our understanding of complex genetic disorders, the paucity and unreliability of available data currently limit the power of these approaches (Chen et al., 2008; Chua & Wong, 2008; Kuchaiev et al., 2009); and (iv) many domains in proteins are of unknown function. Tools, such as SUSPECTS (Adie et al., 2006), which rely on detecting shared domains, annotation, and patterns of expression, are clearly limited by the incompleteness and inaccuracy of these data. Therefore the gaps in our knowledge about gene product function are greatly hampering our understanding of complex genetic disease. Computational techniques, many anchored by evolutionary understanding, are helping direct and accelerate our understanding of the biological function of human genes, and the effect of human gene mutation and variation. These techniques are discussed next.

3.3.2 Gene age-based candidate gene prioritization

The first systematic study comparing the sequence characteristics of human disease genes (listed in OMIM) with genes not known to be involved in disease, found a subset of sequence-based features to be significantly different between the two sets of genes (Adie et al., 2005). These researchers created a web-based tool called PROSPECTR based on those features, which enriches lists for disease genes (Adie et al., 2005). Relevant DNA characteristics, more common in disease genes, include larger gene length and the presence of a mouse homologue (Adie et al., 2005). However, one limitation of this, and other, disease gene prediction algorithms is that they are developed on the basis of known disease genes, and many disease genes remain unidentified (reviewed by Ropers, 2007).

While PROSPECTR examines whether murid orthologues of a gene exist (Adie et al., 2005), the evolutionary history of human 'disease genes' has subsequently been explored in greater depth. A comprehensive study on the evolutionary 'age' of genes mutated in human diseases compared to those not implicated, revealed that human disease genes are more likely to be 'old' genes (Domazet-Loso & Tautz, 2008). These so-called 'old genes' are classified on the basis that they have orthologues in urochordates and/or more anciently-diverging lineages, and contrast with 'recent genes' where orthologues are restricted to chordate lineages. The over-representation of human disease genes among old genes is even more pronounced among those genes with tissue-specific expression profiles (Nagaraj et al., 2010). The evolutionary history of Mendelian disease genes, compared to genes implicated in complex disease, has also been examined, and more recently-evolving genes also divided into 'middle-aged' and 'young' gene categories (Cai et al., 2009). This study found that Mendelian disease genes tend to be older than non-disease genes, while complex disease genes are typically middle-aged (Cai et al., 2009). Therefore, despite not being evolutionarily ancient, most complex disease genes originated during the emergence of vertebrates, and are not human- or primate-specific.

4. Determining human disease gene function

The function of many human gene products is unknown or very poorly understood, and this greatly hampers progress in all studies on complex human genetic disorders. Below we discuss key methods used to predict and understand gene function, where a thorough understanding of evolutionary biology is imperative.

4.1 Sequence-based approaches to predicting gene function

The main method for predicting the function of a gene product in the absence of experimental data is termed 'homology-based transfer' (Friedberg, 2006; Sleator & Walsh, 2010). This approach is based on the detection of significant amino acid sequence similarity to a protein(s) of known function using programmes such as BLAST (Altschul et al., 1997). As sequence similarity suggests a common evolutionary origin, the function of the known protein is then transferred to the query protein. This method is not foolproof, however, and exceptions have been described at both ends of the similarity scale (reviewed by Sleator & Walsh, 2010). Understanding which residues are essential for protein function can be important in evaluating the relevance of similarities detected between proteins, such as those in conserved motifs. Furthermore, there are many proteins where homology-based prediction cannot be used. Therefore, more recently, non-homology based computational approaches have begun to emerge (reviewed by Sleator & Walsh, 2010). These methods are based on a combination of sequence, structural prediction methods, evolutionary history, biochemical properties, and genetic and genomic knowledge.

4.2 Predicting the effect of mutation on gene function

A major problem in the search for disease-causing mutations is the fact that some of them are difficult to recognize. Tools to evaluate the functional impact of mis-sense mutations is limited by the few solved protein tertiary structures, and on the limitations of software to predict effects of mutations on protein domain function and/or protein conformation, much of which incorporates evolutionary conservation data (reviewed by Ropers, 2007). Complicating matters further, even silent mutations have been found to be pathogenically relevant (Kimchi-Sarfaty et al., 2006; Pagani et al., 2005). Furthermore, non-coding mutations may not be examined or detected and, even if they are, it is currently even more challenging to predict whether these have functional effects. For example, intronic changes may alter the splicing pattern (Richards et al,. 2007; Lenski et al., 2007) and promoter mutations may affect gene expression levels (Almeida et al., 2006; Borck et al., 2006). Unlike the situation found with Mendelian disease loci, an increasing proportion of the loci being associated with complex disorders are being found outside protein-coding regions of the genome (Pomerantz et al., 2010). The thousand genomes project will sequence the complete genome sequence of more than 1,000 humans, and will provide valuable information about variants normally present in the human population (Marth et al., 2011).

Genetic mutations implicated in human disease are often mis-sense/nonsense mutations, or involve small duplications/deletions in coding regions. However, recent sequencing of multiple human exomes (coding regions of the genome) suggest that these types of mutation are actually quite common and such changes are frequently benign (Ng et al., 2008; Ng et al,. 2009). One way of distinguishing disease-causing mutations, at least in Mendelian disease genes, is because they occur more frequently in evolutionarily well-conserved amino acid residues, than in non-conserved ones, and these changes are expected to have a more severe impact on the function of the resulting protein. By contrast, the distribution of mutations, such as non-synonymous single nucleotide polymorphisms, (nsSNPs or cSNPs) contributing to complex human diseases, are often difficult to distinguish from the distribution for "normal" human variation (Thomas & Kejariwal, 2004). These results indicate that individual SNPs implicated in complex genetic disorders will have more subtle effects on function in isolation. This observation further suggests a disease architecture involving the concerted contribution of multiple genetic loci, each with a small individual effect. Indeed, this explanation is suggested to be the basis of a large proportion of common complex disorders, and is known as the common-variant common-disease (CVCD) model of complex genetic disease (Grady et al., 2003; Visscher et al., 2011). However, as discussed in Section 2.1 above, rare genetic changes can also be causative of a subset of common disorders, and is described as the RVCD (rare-variant common-disease) model. However, there is no cut-off point between these two models (see Fig. 1), and there is a broad variety in both the frequency of disease gene variation and in the penetrance of a given genetic change, which together combine to cause a given disease phenotype (Grady et al., 2003; Visscher et al., 2011).

Computational prediction of the effects of genetic variation utilizes evolutionary conservation of the resultant gene product, in combination with predictions of the changes to the physicochemical properties (Mooney, 2005; Ng & Henikoff, 2006; Ng et al., 2008; Tarpey et al., 2009). Computer-based tools are also used to predict conserved domains and motifs, and can be used to determine whether nsSNPs or other genetic changes are likely to contribute affect protein function. Examples of such databases include PROSITE (Hulo et al., 2008), BLOCKS (Henikoff et al., 2000), and PRINTS (Attwood et al., 2003). Variation in the sequence of orthologues with conserved function can also be used to indicate the amino acid variation possible in a domain, or motif, which is still predicted to maintain some degree of functionality.

Nonetheless, while sequence-based approaches provide a good basis for predicting the function of genes of unknown function, in many cases there may be little or no sequence similarity between an unknown gene product and any characterized gene product. Fortunately, due to evolutionary constraints, there is often still significant structural similarity between an uncharacterized protein and a characterized protein, and this can be a useful indicator of function (Shatsky et al., 2008; Todd et al., 2001; Watson et al., 2005). Therefore, recent developments are aimed at combining both sequence and structural information to increase the likelihood of a functional prediction (Laskowski et al., 2005; Pierri et al., 2010; Skolnick & Brylinski, 2009). However, there is much room for improvement to the currently available approaches for the prediction of protein function, as only ~1% of proteins on the UniProt database have experimentally-supported function, ~65% have some functional annotation, and over one-third are uncharacterized or have no predicted function (Barrell et al., 2009; Erdin et al., 2011; Goldsmith-Fischman & Honig, 2003; Laskowski et al., 2003; Magrane & Consortium, 2011). This may be particularly important for disease genes, as many intrinsically-disordered proteins, lacking stable secondary and tertiary structures, are being found associated with many complex human diseases, including cancer, diabetes, neurodegenerative diseases, and cardiovascular disease (Midic et al., 2009; Uversky, 2009; Uversky et al., 2008, 2009; Wang et al., 2011).

4.3 Evolutionary pedigree and co-expression to leverage functional prediction

Expression data can also provide information relevant to genetic disorders. For example, it is useful to know whether candidate genes are expressed in the tissue affected by the disease. Furthermore, genes involved in similar cellular processes are also more likely to be co-transcribed, hence the 'guilt-by-association' algorithm (Walker et al., 1999; Oliver, 2000). Not only can 'power' be added to the analysis of co-expression datasets, a better understanding of the evolutionary history of human genes may lead to novel ways to interpret sequence data and predict protein function (Eng et al., 2009; Thornton & DeSalle, 2000). For example, proteins with a similar evolutionary pedigree, and emerging at the same 'point' in the evolutionary history of the Eukaryota, are assumed to have evolved in parallel. This, in turn, indicates they are more likely to have a common function (Eisenberg et al., 2000). The 'guilt-by-association' principle greatly informs our understanding of protein-protein interaction networks, as it is known that most cellular functions are carried out by networks of interacting proteins (Qiu & Noble, 2008). Understanding protein-protein interaction networks is also informing computational approaches aimed at understanding the role of pathways affected in complex genetic diseases, and this is discussed next.

4.4 Network studies to leverage functional prediction

Network-based models incorporating protein-protein interaction (PPI) data are a relatively new way for studying disease-related genes. Nonetheless, this approach has already been proved to be effective for the identification of complex disease genes, including those involved in colon cancer (Nibbe et al., 2009). Therefore, analyzing the functional networks of human genes is providing a key framework for prioritizing candidate disease genes. Such analyses may also lead to the identification of key cellular pathways common to complex diseases that may be amenable to therapeutic intervention.

Based on large- and small-scale PPI studies, preliminary protein-protein interaction networks are being created (e.g. Fig. 2). These can be accessed and viewed using a number of web tools, such as STRING and BioGRID (Han, 2008). Proteins sharing a particular functional category cluster in the same location of PPI networks, and are referred to as functional modules, and placement of proteins in PPI networks can be used to inform protein function prediction studies (Dziembowski & Seraphin, 2004; Yook et al., 2004; Makino & Gojobori, 2006). Understanding networks of PPIs can be used to predict additional genes that, when mutated, may cause the same disease as that associated with mutations in interacting partners (McGary et al., 2010), and also explains why so many different mutated genes can cause the same or similar complex disease (Bill & Geschwind, 2009; Bourgeron, 2009; Crespi et al., 2010; Gilman et al., 2011; Guilmatre et al., 2009).

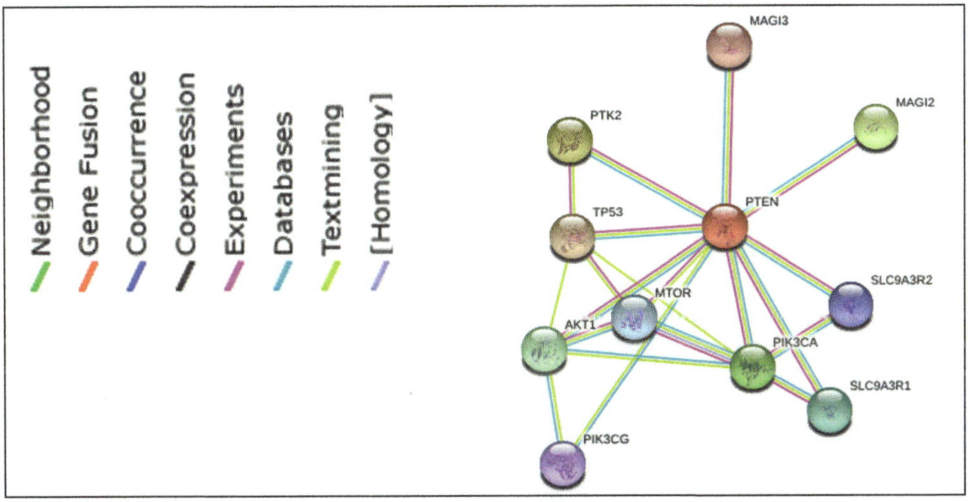

Fig. 2. Protein-protein interaction network example. Multiple lines of evidence from the STRING database (http://string-db.org/) demonstrate that the PTEN cancer gene product (red) is a hub-protein.

Relating to the network-based nature of many gene products is the concept that some proteins interact with multiple other proteins, referred to as 'hub' genes (Fig. 2). Multi-functional genes may impact on numerous pathways, while different mutations may cause different diseases (Gillis & Pavlidis, 2011). The encoded hub proteins have multi-functional cellular roles, and are typically annotated with multiple GO categories (Gillis & Pavlidis,

2011). The multi-functional nature of many proteins also contributes to phenotypic pleiotropy, where mutation(s) in a single gene can affect multiple phenotypic traits. Genes leading to pleiotropic phenotypes tend to be more evolutionarily conserved and are more likely to have essential functions (Eisenberg & Levanon, 2003; Feldman et al., 2008; Fraser et al., 2002; Gandhi et al., 2006; Goh et al., 2007; Jeong et al., 2001; Saeed & Deane, 2006). Overall, human disease-causing genes have an intermediate essentiality, being less than that of housekeeping genes, but greater than that of non-disease genes (Feldman I et al., 2008; Goh et al., 2007; Liao & Zhang, 2008; Tu et al., 2006). These findings are, to some extent, commonsense, as mutation of a housekeeping gene would be expected to lead to embryonic lethality.

As gene products that are peripheral in protein-protein interacting networks are known to have a higher evolutionary rate than hub proteins, these data also suggest human disease genes differ in rates of evolutionary change compared to non-disease genes. However, current results on this topic are inconsistent, with some studies indicating disease genes are evolving more slowly than non-disease genes (Blekhman et al., 2008; Tu et al., 2006), and other studies suggesting disease genes are evolving faster than non-disease genes (Huang et al., 2004; Smith & Eyre-Walker, 2003). There are a number of confounding factors contributing to these inconsistencies in establishing the rate of evolution of human disease genes. Comparing proteins based on PPI network interactions, genes encoding proteins that are part of 'modules' tend to be more conserved, evolutionarily old, and ubiquitously expressed. By contrast, genes encoding proteins outside modules are less well-conserved, evolutionarily younger, and enriched with at least some degree of tissue-specific expression (Dezso et al., 2008). Therefore, there appear to be different classes of disease genes, with different evolutionary patterns and different PPI patterns, and this may relate to the evolutionary 'age' of the disease gene (Nagaraj et al., 2010). Characterization the differing traits or classes of disease genes may contribute to our understanding of the different functional 'style' of disease genes. It also appears that Mendelian disease genes and complex disease genes have different evolutionary profiles (see Section 3.2.3) and, of course, these studies are limited by our current knowledge of disease genes, which remains incomplete.

Networks of PPIs are also informing efforts aimed at understanding whether human evolution is heading towards, or away from, susceptibility to a particular disease. One crucial aspect of this discussion, is the presence of alleles that increase the risk of one disorder, while simultaneously decreasing the risk for another. For example, multiple sclerosis and rheumatoid arthritis, or ankylosing spondylitis and multiple sclerosis, are negatively correlated (Sirota et al., 2009). By contrast, other alleles can increase the susceptibility to more than one complex disorder. This leads to multiple sclerosis and autoimmune thyroid disease, or type 1 diabetes and coeliac disease, commonly co-occurring (reviewed by Sirota et al., 2009). While these studies indicate the complexities of disease susceptibility, tools are being developed to leverage PPI data to predict such disease interactions (Chen et al., 2008; Chua & Wong L, 2008).

Overall PPI networks are crucial for the development of testable hypotheses regarding the underlying pathogenic mechanisms of many complex genetic disorders. However, PPI networks are incomplete and include both false-positive and false-negative interactions (Chen et al., 2008; Gandhi et al., 2006; Kuchaiev et al., 2009). Development of a reliable and complete human PPI network will provide an invaluable framework to study the contribution of multiple genes to complex genetic disorders, and will require both computational and experimental data to achieve accuracy and completeness.

4.5 Role of animal models

Animal research, including animal models of disease, has been responsible, at least in part, for every major medical advance made during the last century (Müller & Grossniklaus, 2010). The reason why model organisms are able to contribute so effectively to our understanding of human diseases lies in the high degree of molecular conservation found between metazoan species, and in the conserved nature of protein-protein, and other, networks (Gandhi et al., 2006). Indeed, even bacteria, plants, protists, and fungi are being exploited to explore differing aspects of biology relevant to human disease (Annesley & Fisher, 2009; Ilievska et al., 2011; McGary et al., 2010; Spradling et al., 2006).

As discussed above, a huge number of susceptibility alleles for a range of complex human genetic disorders have now been identified, but the function of many of these genes is poorly understood. Therefore, while risk-associated loci are being successfully identified (Easton et al., 2007; Hindorff et al., 2009; Jia et al., 2009), these findings are rarely followed up and the contribution of the allele to the molecular basis of disease rarely evaluated (McCarthy & Hirschhorn, 2008). That this failure is leading to a bottleneck in our understanding of complex disorders was highlighted by a recent *Nature Genetics* editorial, which suggested that significant investment in functional characterization of risk loci is needed (Axton, 2010). There are a number of ways to investigate the molecular and cellular function of a gene and its alleles, and these include *in vitro* studies, cell culture systems, and the use of whole animal models. These tools can be used to test hypotheses gained from thorough *in silico* studies. In this section we will discuss the role of animal models, as this is of greatest relevance to the topic of this paper. Although *in vitro* and cell culture studies can be of great benefit, ultimately good animal models provide the best biological models for complex disease.

High-throughput phenotypic screens of RNAi knockdowns in *Caenorhabditis elegans* and *Drosophila melanogaster* often provide the first inkling of the biological function of an uncharacterized human gene (Buckingham et al., 2004). Efforts are also underway to systematically knockout all the genes in the mouse genome to facilitate phenotypic and functional screening (Guan et al., 2010). Animal models also provide a vital system amenable for dissection of the contributions of genetic, environmental and developmental components to the etiology of complex human disorders, and for evaluation of novel therapeutics, which can be achieved in no other way (Complex Trait Consortium, 2004; Iwata et al., 2010).

Many recent advances relevant to studying complex disorders have also been made. This includes the development of a well-defined collection of recombinant inbred mice with different genetic backgrounds (Complex Trait Consortium, 2004). Studies from many organisms indicate that the phenotype of some gene knockouts only becomes apparent upon inactivation of another gene (Barbaric et al., 2007), indicating genetic background can be very important. An alternative, but related, approach to investigating human disease mechanisms, is to study animals that already have a disease-related phenotype of interest. These orthologous phenotypes, or phenologs, can be used to predict novel genes associated with a disease. This approach has been used to predict genes for angiogenesis, breast cancer, autism spectrum disorder, and Waardenburg syndrome, among others, in many diverse model organisms (Gilby, 2008; McGary et al., 2010; Pearson et al., 2011). Finally, while models were previously limited to studying one variant/gene at a time, efforts are now being made to investigate the cooperative interactions of multiple genes. For example, quintuple knockout mice have been used to study the role of multiple immune system genes in asthma (Dahlin et al., 2011). Therefore, new resources and tools are being developed to study complex diseases more effectively in model organisms.

However, care must be taken in extrapolating data from animal models to the human situation, as no model organism can exactly reproduce the disease of another. It is as important to understand the differences, as it is to highlight the similarities, between the animal model and the human disease. The ability to gain a thorough understanding of the key differences and similarities between species will minimize misinterpretation of data gleaned from model organisms, and lead to the improved use of animal models to both understand and develop treatments for human genetic disorders. The importance in understanding the differences between human and animal gene expression and physiology will be discussed below, using autism spectrum disorder as an example.

5. Human evolution and disease susceptibility

Evolutionary analysis has been applied to many aspects of human disease. As discussed in Section 3.3.2 above, one of the earliest findings revealed that Mendelian disease-associated nsSNPs are more frequently found in conserved amino acid positions, and these positions can be conserved even in more distantly-related proteins, while complex disease-associated SNPs are frequently not (Miller & Kumar, 2001; Ng & Henikoff, 2002; Thomas et al., 2003; Thomas & Kejariwal, 2004). Some other aspects of evolutionary biology, increasing our understanding of human genetic disease, are discussed next.

Over the last 100,000 years, humans have adapted to many changes in environment as they moved out of Africa and modified both diet and lifestyle, factors which influence the incidence of common genetic variants by positive selection of those alleles that prove advantageous (Sabeti et al., 2006). This theory is supported by the finding that complex disease-associated gene variants show heterogeneity in allelic frequency among different human populations, leading to a non-homogeneous world-wide distribution of disease alleles (Ioannidis et al., 2004). These findings have implications for GWAS, as disease variants differ between human populations (e.g. between Hispanics and those of African descent, Asian descent or European descent), and will affect the reproducibility of results from genetic studies of complex disease depending on the ethnic mix studied (Marigorta et al., 2011). Therefore, the recent evolutionary history of humankind affects the present global patterns of susceptibility to disease. The classic example of this is the mutation in the *Hemoglobin B* gene, undergoing positive selection in African populations as it promotes malaria resistance, while simultaneously being causative of sickle cell anaemia (Currat et al., 2002; Williams TN, 2006). Another example is a mutation in a regulatory region of the lactase gene (*LCT*) that mediates adult tolerance to lactose. Evidence suggests this particular variant was selected in parts of Europe after the domestication of cattle (reviewed by Sabeti et al., 2006). However, identifying and understanding traits that have been targets of selection is a challenging task. It took forty years of effort, by a succession of researchers, to unravel the association between malaria and the sickle cell mutation and, even now, there is still work to be done on understanding exactly how the sickle-cell state inhibits malaria pathogenesis (Sabeti et al., 2006).

A recent study examined Mendelian-disease genes and found these genes are under widespread negative ('purifying') selection (Blekhman et al., 2008). By contrast, in this study (Blekhman et al., 2008), genes contributing to an increased risk of complex genetic disease showed little signs of evolutionary conservation, and may be targets of both positive and purifying selection. This latter conclusion was supported by a subsequent study by Corona and colleagues (2010). In their study of genes contributing to seven complex genetic diseases, only genes affecting three diseases showed signs of recent positive selection; those increasing susceptibility to Crohn disease, rheumatoid arthritis, and diabetes (Corona et al., 2010). Of note, alleles decreasing susceptibility to Crohn disease also showed signs of positive selection (Corona et al., 2010). Overall, they found evidence for an evolutionary trajectory towards a decreasing risk of Crohn disease, but an increasing risk of type 1 diabetes.

Therefore, we may need to 'think outside the box' about why some complex disorders occur in the human population, and look not only for the disadvantages of a disease-causing mutation, but also consider unthought-of selective advantages. Furthermore, an allele increasing susceptibility to disease in the modern era, may have increased fitness in an earlier human environment. For example, rheumatoid arthritis susceptibility alleles are thought to enhance resistance to tuberculosis (Mobley, 2004; Rothschild et al., 1992), while the type 1 diabetes risk gene *IFIH1*, helps protect against enterovirus infection (Nejentsev et al., 2009). Therefore, for polygenic disorders, not only can the same polymorphisms contribute to more than one disease, some alleles may increase the risk for one disorder while simultaneously decreasing the risk for another (Sirota et al., 2009). Progress in our understanding of these complexities will be facilitated by an increased understanding of the function of disease genes and risk alleles, and rigorous studies examining natural selection in human disease-risk genes.

6. Applications of evolutionary biology to the study of human neurological disorders

Below, how knowledge from evolutionary biology-based analyses are providing us with crucial information impacting greatly on our understanding of neurodevelopmental disorders, and a surprising link to cancer, will be illustrated.

6.1 Neurodevelopmental disorders

Both schizophrenia (SCZ) and autism spectrum disorder (ASD) are considered common neurodevelopmental disorders (~1% of the population affected), which are typically diagnosed in early adulthood (SCZ) or childhood (ASD) (Bale et al., 2010; Costa e Silva, 2008; Lewis & Levitt, 2002; Owen et al., 2005). Both ASD and SCZ are behaviourally-based diagnoses, with the diagnostic criteria outlined in DSM-IV (American Psychiatric Association, 1994) and ICD-10 (World Health Organization, 1993). Briefly, SCZ is diagnosed on the presence of a collection of positive symptoms, negative symptoms, and cognitive deficits, while ASD is diagnosed on the basis of a triad of behavioural manifestations: social deficits, impaired communication skills, together with repetitive behaviours and/or restricted interests (see Fig. 3).

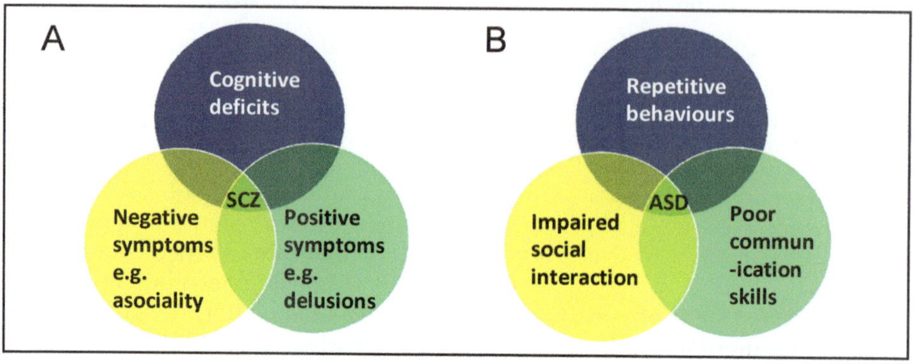

Fig. 3. Diagnostic criteria for (A) schizophrenia and (B) autism spectrum disorder.

Both ASD and SCZ are considered complex genetic disorders, with heritability estimates of around 80% for both (Ronald & Hoekstra, 2011; Sullivan et al., 2003). While some genetic disorders with Mendelian inheritance lead to syndromic forms of ASD (i.e. a phenotype of which ASD is typically one part), and while some alleles of intermediate penetrance appear to contribute, a large proportion of ASD cases fit the CVCD model (Eapen, 2011). As predicted by the CVCD model, parents of children with ASD share a subset of phenotypic traits, without having the 'full' ASD phenotype (Bernier et al., 2011; Robinson et al., 2011). A further 5-10% of ASD is caused by *de novo* mutations (reviewed by Eapen V, 2011), and this is discussed further below. For SCZ, the CVCD model of inheritance also dominates, as individuals with SCZ are less likely to pass on their genes to the next generation (Crow, 2011). Both ASD and SCZ are large, active areas of scientific research, and below we highlight those aspects where evolutionary biology is relevant.

6.1.2 Recurrent rearrangements in neurodevelopmental disorders

Recurrent microdeletions at 1q21.1, 15q11.2, 16p11.2, 16p13.11 and 22q11.2 are found in patients with a broad spectrum of neurospsychiatric conditions, including developmental delay, SCZ, psychotic disorder, ASD, or mental retardation (Brunetti-Pierri et al., 2008; Kumar et al., 2008; Sahoo et al., 2011; Stankiewicz & Lupski, 2010). Of note, recurrent rearrangements in synaptic and neuro-developmental genes affect shared biologic pathways contributing to the risk of developing several disorders, including SCZ, ASD, and mental retardation (Guilmatre et al., 2009; Sahoo et al., 2011).

Certain structural features make some chromosomal regions more prone to rearrangements, such as deletion, duplication or translocation (Smith et al., 2010), and this is discussed further in Section 6.2.3 below. For example, the 16p11.2 microdeletion linked to ASD is mediated by segmental duplications (Kumar et al., 2007), while deletions at 15q11-13 occur during meiosis and are caused by a number of repeated DNA elements in this chromosomal region. The 15q duplication syndrome is the result repetitive sequences mediating unequal but homologous recombination and, of note, also causes ASD (Chamberlain & Lalande, 2010). This indicates that duplication or deletion of some genes can cause an overlapping phenotype. However, this is not a universal phenomenon. Deletion of 16p11.2 leads to a phenotype of ASD with macrocephaly, while duplication to leads to SCZ and microcephaly (Brunetti-Pierri et al., 2008; McCarthy et al., 2009; Shinawi et al., 2010). A further example is duplication of 1q21.1, which is implicated in ASD with macrocephaly. Deletion of this region causes SCZ and microcephaly (Dumas & Sikela, 2009; Crespi et al., 2010). This is not restricted to neurodevelopmental disorders, as the most common locus affected in Charcot-Marie-Tooth (CMT) neuropathy, *PMP22*, when deleted causes a different neurological phenotype to that associated with gene gain (Chance, 2006). Duplication of *PMP22* leads to CMT type 1A, while *PMP22* deletion leads to a disease known as hereditary neuropathy with liability to pressure palsies (HNPP). Therefore, CNV gain, versus CNV loss, at identical loci can mediate similar or distinct phenotypes.

6.1.3 Phylostratigraphy of neurodevelopmental genes

As might be expected for neurodevelopmental disorders, many ASD- and SCZ-implicated gene products participate in protein-protein interaction networks implicated in neuron function (Bourgeron, 2009; Bill & Geschwind, 2009; Gilman et al., 2011; Sun et al., 2010; Torkamani et al., 2010; Voineagu et al., 2011). These genetic and PPI studies are supported by pathological findings, as structural alterations of dendritic spines are associated with both SCZ and ASD (reviewed by Penzes et al., 2011).

In addition to furthering our understanding of the evolution of human disease, model organisms play a major part in developing an understanding of the etiology of human genetic disorders. As discussed above, model organisms play a key role in characterisation of the normal and abnormal functions of risk genes (Aitman et al., 2011). Therefore, it is important to determine whether the genes implicated in SCZ and ASD are conserved in model organisms. Phylostratigraphy is a term applied to the application of phylogenetic methods to evaluate the evolutionary origin of disease genes and/or the origin of genes contributing to major evolutionary adaptations (Domazet-Loso et al., 2007; Domazet-Loso & Tautz, 2008). Such an approach has not yet been applied to any neurodevelopmental disorder. However, some data are available relating to the conservation of key PPI networks implicated in SCZ and ASD. Of relevance to SCZ and ASD, as well as other neurological disorders, is evidence that the core components of the nervous system and immune system are conserved in vertebrates. Indeed, the core components of the synapse are found in cnidarians, which form primitive nerve networks, and evolved around 680 million years ago (Galliot et al., 2009; Grimmelikhuijzen & Westfall, 1995). Furthermore, many synaptic genes are found in sponges, the oldest-surviving metazoan phyletic lineage, which actually lack synapses (Kosik et al., 2008; Srivastava et al., 2010). These data suggest many of the genes involved in neurological disorders have a more ancient evolutionary origin than previously thought.

A contributing role of the adaptive and innate immune systems, to ASD or SCZ etiology has also been suggested by some studies (Ashwood et al., 2006; Muller et al., 2000; Sun et al., 2010; Voineagu et al., 2011). This role would need to be studied in vertebrate models (possessing adaptive and innate immune systems), as non-vertebrate species only have an innate immune system. Nonetheless, a comparison of phenotypes between vertebrates and non-vertebrate may provide vital information about the relative contributions of immune dysfunction to the ASD or SCZ phenotype. A similar logic has been used to explore novel cell-death pathways, where regulated cell death processes are examined in species lacking classical caspase enzymes (Degterev & Yuan, 2008; Guisti et al., 2010; Smirlis & Soteriadou, 2011). A thorough, systematic analysis of differences in key disease networks is also required to gain the best insights into the strengths and weaknesses of each specific model organism. As the genomes of all currently used model organisms have been sequenced, accurate network analyses and disease network analyses are now imperative if we are to understand disease evolution and the limitations and/or benefits of various model organisms for disease.

6.1.4 Evolution of common neurodevelopmental disorders

A number of hypotheses regarding the evolution of autistic and schizotypal traits have been proposed, promoting controversy and debate. For example, it has been proposed that the intense focus and repetitive behaviours of ASD may have been beneficial to hunters and gatherers (Reser, 2007). This hypothesis suggests that, subsequent to the ascendance of agriculture and the development of complex community-based lifestyles, these alleles have become increasingly disadvantageous, and the common alleles responsible for ASD traits only remain in the human gene pool because they previously had adaptive function (Reser, 2011). Alleles contributing to SCZ have likewise been hypothesized to have prior adaptive function, and contributed to a fitness advantage in the ancestral human environment, such as providing physiological and behavioural characteristics increasing survival in conditions of nutritional paucity and stress (Reser, 2007). The clinical ASD or SCZ diagnosis would then ensue when individuals with subclinical phenotypes mate (Del Giudice et al., 2010), following the CVCD model.

Potential evolutionary advantages afforded by sub-clinical phenotypes of other psychiatric disorders have likewise been proposed, based on data indicating that a large proportion of these disorders are due to the inheritance of multiple copies of low-risk gene variants, which are present in the parents and siblings of affected individuals (Bernier et al., 2011; Hoffman & Stat, 2010; Robinson et al., 2011). Providing more direct evidence for these hypotheses is difficult. Indeed, the debate over the amount of adaptive evolution occurring in the genome itself is far from resolved (Amos & Bryant, 2011; Eyre-Walker, 2006). Furthermore, little evidence supports adaptive evolution in genes linked to brain development (Voight et al., 2006). Finally, as the genetic and molecular pathways underpinning both ASD and SCZ remain very poorly understood, this adds to these difficulties.

Inverse comorbidity, a concept introduced in Section 5, may also be relevant to ASD and SCZ, and the example of sickle cell anaemia gene selection occurring due to protection of carriers from malaria (Currat et al., 2002; Williams TN, 2006), was provided. Recent data supporting immune system differences in individuals with ASD or SCZ (Cohly & Panja, 2005; Müller & Schwarz, 2010), suggests that other selective pressures may be involved for risk genes for these disorders. Of relevance, there is much stronger evidence for adaptive evolution in genes with known immune-system function (reviewed by Eyre-Walker, 2006). Another possible selective pressure is found in recent studies supporting a lower than expected occurrence of cancer in patients with SCZ (Tabares-Seisdedos et al., 2011), and cancer will be discussed further later in this paper. Understanding the molecular pathways underpinning these disorders will help us clarify these issues further. However, these emerging findings suggest the selective pressure on ASD or SCZ disease alleles may not be related to the behavioural phenotype.

6.1.5 Model organisms for neurodevelopmental disease

Many professionals were initially sceptical whether phylogenetically 'lower' species could be successfully used to study the molecular and cellular mechanisms of human brain disorders. Surprisingly, many animal species are emerging as excellent model systems for such disorders, and their tractable nature, plus the ability to control for environment and genetic background, has already led to far-reaching advances in our understanding of many neurological diseases (Chesselet, 2005; Shah et al., 2010; Tayebati, 2006). For example, many recent reviews discuss how animal models are proving their usefulness in improving our understanding of the pathophysiology of SCZ and in the development of novel therapeutic strategies (Arguello et al., 2010; Feifel & Shilling, 2010; Lazar et al., 2011; Powell, 2010; Young et al., 2010). A greater understanding of the genetics underlying SCZ will also inform the development of future animal models. Likewise, for ASD, there are a wide variety of animal models available, which are becoming increasingly well characterized (Patterson, 2011; Tordjman et al., 2007). These include naturally-bred rodents (Gilby, 2008; Pearson et al., 2010) and transgenic mouse models (Minshew & McFadden, 2011; Robertson & Feng, 2011). Some wide-reaching findings have already been made based on results from these model systems. For example, multiple studies indicate neurodevelopmental disorders may be treatable, even in adults (Ey et al., 2011; Silva & Ehninger, 2009). Furthermore, rodent models are replicating some of the co-morbidities associated with an ASD diagnosis, such as immune abnormalities (Heo et al., 2011) and epilepsy (Gilby, 2008; Peñagarikano et al., 2001).

Despite the progress and potential that animal models of disease provide, caution must be exercised with the interpretation of data from such model systems, particularly when considering disorders affecting the central nervous system (CNS). Factors to consider when evaluating animal model data include the developmental trajectories unique to humans. The most dramatic of these is in the timing of maturation and pruning of the CNS during childhood (reviewed by Dean, 2009). Regulation of gene expression differs between animal species, with differences in microRNAs (Berezikov et al., 2006), DNA methylation patterns (Enard et al., 2004) and, as discussed next, mRNA splicing, also being detected.

Understanding the role of alternative isoforms is vitally important, as aberrant gene splicing is emerging as a key contributor to a variety of neurological diseases (Anthony & Gallo, 2010) and cancers (Ward & Cooper, 2010). Alternative splicing is considered a major mechanism underpinning metazoan biological complexity, including the increasingly-complex brain function of metazoa, with neurons having specific systems for regulating mRNA splicing and generating brain-specific isoforms (Ule & Darnell, 2007). Indeed, increasing numbers of splice site mutations are being implicated in the etiology of ASD and SCZ (Glatt et al., 2011). However, differences in splicing occur, even between closely-related species such as humans and chimpanzees (Calarco et al., 2007; Blekhman et al., 2010; Lin et al., 2010). Therefore, conservation of splice variation may be relevant to disease etiology, and different profiles should be assessed in animal models of disease. Current best-practice guidelines for preclinical studies for neurological disease in animal models, have recently been published (Shineman et al., 2011).

6.1.6 Evolution of the *Deleted In Autism 1* gene

DIA1 (*Deleted In Autism 1*) is implicated in an autosomal recessive form of ASD (Morrow et al., 2008; Aziz et al., 2011a). An evolutionary biology-based approach to understanding the role of this gene in ASD has illustrated the importance of many of the principles outlined above. While *DIA1* is conserved from cnidaria to humans, it is not detected in nematodes, suggesting *C. elegans* is not a suitable model in which to study the cellular role of this gene. Strikingly, a related gene was found in humans using phylogenetic-based analyses, *DIA1R*, which localizes to the X chromosome (Aziz et al., 2011b). *DIA1R* is vertebrate-specific and, as with *DIA1*, is implicated in ASD (Aziz et al., 2011a, 2011b). Of possible relevance to the ASD phenotype, *DIA1R* had been 'lost' in fish of a solitary nature, while those retaining the gene are 'social' schooling fish (Aziz et al., 2011b). Of further relevance to the use of animal models, *DIA1R* was found to be X-inactivated in mouse, but not in humans (Aziz et al., 2011b; Yang et al., 2010), and splicing may be species-specific. Indeed, we have preliminary evidence for brain-specific splicing of *DIA1* that is primate-specific, and not found in other vertebrate lineages (Aziz & Bishop, unpublished data). This type of evolutionary-based evidence facilitates an educated approach to the choice of model organism for functional studies, and highlights issues that may arise from studies in mice or fish.

6.2 Cancer

Natural selection and evolution is dependent on genetic variability and the occurrence of new mutations in the germ-line. However, mutation occurs in both germinal and somatic cell lineages and, over a human lifespan, somatic mutations accumulate and may lead to cancer (Greaves, 2007). Therefore, as with other genetic disorders, carcinogenesis is another manifestation of the biological processes on which evolution depends. Cancer is considered a probabilistic disease, and is inevitable in long-lived organisms such as humans, where the lifetime risk is around one in three (Greaves, 2007; Simpson & Camargo, 1998). In most, but not all, cases this is due to multiple genetic changes accumulating in cells (Maffini et al., 2004; Stratton et al., 2009; Touw & Erkeland, 2007).

Genes contributing to cancer are often divided into two broad groups: caretakers and gatekeepers (Kinzler & Vogelstein, 1997; Macleod, 2000; Michor et al., 2004; Russo et al., 2006). Mutations affecting caretaker genes promote neoplasia indirectly, increasing genetic instability and mutation rates, which leads to defects in many genes including gatekeeper genes. Mutations affecting gatekeeper genes directly promote cancer progression, and have roles in cell differentiation, growth and/or death. Gatekeeper genes may be further divided into oncogenes and tumour-suppressor genes. Well-known caretaker genes include *BRCA1*, *BRCA2*, *ATM* and *FANCA*; gatekeeper genes include *TSC1*, *TSC2*, *Rb*, *NF1*, *NF2* and *PTEN* (see Fig. 3); while some genes are multifunctional and can act as gatekeepers and/or caretakers, including *p53* and *ARF* (Dominguez-Brauer et al., 2010; Rubbi & Milner, 2005; Russo et al., 2006).

6.2.1 Cancer and neurological disorders

One surprising recent finding is the emerging link between a number of human disorders and cancer. Recent data indicate a lower than expected occurrence of cancer in patients with SCZ, Down syndrome, Parkinson disease, Alzheimer disease and multiple sclerosis (Tabares-Seisdedos et al., 2011). What is striking, is that most of the disorders found to protect against cancer, and which lead to an inverse cancer morbidity in humans, are neurological disorders. This is thought to occur due to the genetic and molecular connections between cancer and these complex human diseases (Tabares-Seisdedos et al., 2011). Of greater concern, is a higher incidence of certain cancers in patients with ASD (Crespi, 2011). This is likewise due to shared molecular pathways (Crespi, 2011), and ASD also shares genetic connections with SCZ and Alzheimer disease (Crespi et al., 2010; Sokol et al., 2011). Indeed, some lines of evidence suggest that ASD and SCZ involve diametric etiology (Crespi et al., 2010). Exploring these links will have far-reaching implications.

6.2.2 Evolution of cancer genes

A link between cancer formation and the evolution of multicellularity was predicted, and this hypothesis was recently explored (Domazet-Loso & Tautz, 2010). Using a phylostratigraphic strategy, Domazet-Loso and Tautz (2010) found different evolutionary origins for gatekeeper, compared to caretaker, human cancer genes. Genes with a caretaker function have an evolutionary origin in the first cellular organisms, and are also found in bacteria and/or archaea (Domazet-Loso & Tautz, 2010). By contrast, only genes with gatekeeper functionality correspond to the origin of metazoa, and are detected in Porifera and/or Cnidaria and subsequently diverging phyla (see Fig. 4). For example, genes encoding the ESCRT (endosomal sorting complex required for transport) proteins, are essential for the downregulation of many cell-surface signalling molecules (reviewed by Ilievska et al., 2011). Not only do many of the genes encoding ESCRT subunits appear to have caretaker functions, with mutations being linked to both cancer and neurological disorders, these genes also have an origin pre-dating the metazoan lineage (reviewed by Ilievska et al., 2011). Model organisms, such as the amoeba, *Dictyostelium discoideum* (Annesley & Fisher, 2009; Williams RS et al., 2006), can therefore be used to study the fundamental molecular roles of these proteins (Blanc et al., 2009; Mattei et al., 2006). Overall, the evolutionary age of cancer genes parallels their role in human cancer etiology (Fig. 4).

6.2.3 Heritability of cancer and co-heritability with neurological disease

Most tumour gene mutations can be inherited as well as acquired, although inherited cancer syndromes are quite rare. The cancer syndromes are inherited in a dominant Mendelian manner and are often associated with developmental defects and benign tumours (Knudson, 1971, 1993; Ponder, 2001). Single genes and deletions of multiple genes can both cause syndromes of which cancer is one part. For example, deletions of multiple genes at position 11p13, is clinically associated with WAGR syndrome and patients have a collection of symptoms including mental retardation, aniridia and kidney tumours (Fischbach et al., 2005). Mutations in *TSC1* or *TSC2* cause tuberous sclerosis, a disease characterised by non-malignant hamartoma formation in many organs, with epilepsy, ASD and/or mental retardation being frequent comorbidities (de Vries, 2010), while *PTEN* mutations are identified in human cancers, and also in the germline of patients with hamartoma tumour-related syndromes (PHTSs). In addition to harmatomas, ASD is a common comorbidity in individuals with *PTEN* mutations (Rodríguez-Escudero et al., 2011). Molecular network analyses (e.g. Fig. 2) are being used to establish the molecular pathways leading to the multiple phenotypes caused by mutation a single 'cancer gene' (de Vries, 2010; Rodríguez-Escudero et al., 2011). Genetic and functional studies are also being used, with current data indicating that different mutations in the *PTEN* gene can cause cancer alone, or cancer with ASD (Rodríguez-Escudero et al., 2011).

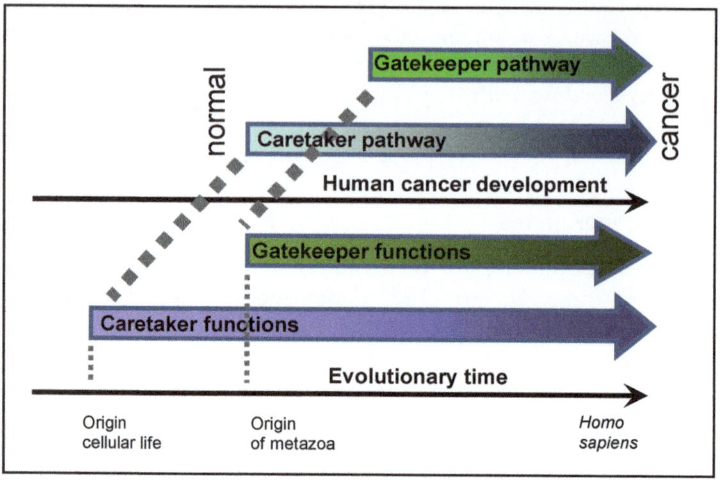

Fig. 4. The evolutionary age of cancer genes parallels their role in carcinogenesis (adapted from Michor & Tautz, 2010).

6.2.4 Recurrent genome rearrangements and cancer

Both somatic-cell chromosome rearrangements and germ-line rearrangements are implicated in both human variation and disease, including the development of cancers (Hanahan & Weinberg, 2000; Kidd et al., 2008; Inoue & Lupski, 2002; Stankiewicz & Lupski, 2002). On an evolutionary timescale, chromosome rearrangements have also played a key role in the divergence of species and differences in chromosomal arrangements between species (Drosophila 12 Genomes Consortium, 2007; Kehrer-Sawatzki & Cooper, 2007; Pevzner & Tesler, 2003; Peng et al., 2006). Furthermore, the evolutionary chromosome breakpoints and chromosome breakpoints found in diseases, including cancer and developmental disorders, overlap (Darai-Ramqvist et al., 2008; Lindsay et al., 2006; Murphy et al., 2005). Therefore, understanding the mechanisms of chromosome breakage and rearrangement is an active field of biological research.

Initially, genomic rearrangements were thought to occur randomly at non-specific, independent sites (Nadeau & Taylor, 1984; Ohno, 1970). However, subsequent studies, including those on cancer breakpoints, found non-random patterns (Cohen et al., 1996; Larkin et al., 2003; Pevzner & Tesler G, 2003; Sankoff et al., 2002), leading to the fragile-site model, where specific regions in the genome are 'hot spots' for genome rearrangements. Since that time, the most highly-fragile spots in the human genome have been mapped (Smith CL et al., 2010). The fragile-site model leads to various testable hypotheses as to the nature of these hot spots, and factors such as DNA sequence and chromosome structure were proposed. Since then, evolutionary breakpoints have been shown, for example, to be gene-rich regions with significantly more segmental duplications and/or repetitive elements than expected, which may facilitate homologous recombination (Bailey et al., 2004; Bulazel et al., 2007; Everts-van der Wind et al., 2004, 2005; Kehrer-Sawatzki & Cooper, 2008; Murphy et al., 2005; Schibler et al., 2006). These regions also encompass genes with higher densities of copy number variation and SNPs (Larkin et al., 2009). Understanding chromosomal 'hotspots' will not only inform our understanding of evolutionary biology, but such studies will also increase our understanding of the etiology of human genetic disorders caused by chromosomal rearrangements.

7. Conclusion

Genetic disorders are an inescapable component of evolution. However, comparative and evolutionary biology methodologies also play an important role in developing and improving our comprehension of many aspects of complex human genetic disorders including schizophrenia and autism spectrum disorder. Thorough *in silico* analyses of disease genes, the encoded proteins, their structure and function, are fundamental tools for evaluating differences and similarities between human genes and genetic pathways, and the comparison of these with equivalents in animal models. Such explorations of the evolutionary conservation of human genes, proteins, protein structures, and resulting cellular networks enables us to: (i) understand how and why human diseases originated; (ii) predict disease genes; (iii) best predict the impact of genetic variation on cellular function; (iv) choose appropriate animal models for human diseases; (v) better interpret data obtained in studies using model organisms; and (vi) evaluate more accurately the validity of therapeutics. Therefore, comparative and evolutionary biology is of major relevance to our understanding of, and in the development of treatments for, many human neurological disorders.

8. Acknowledgment

Thanks to all members of the NEB and PRF laboratories for helpful discussions. We apologize to those authors whose work we have be unable to cite directly, due to space limitations. AA was supported by the Malaysian Ministry of Higher Education and the Islamic Science University of Malaysia. JI was supported by an Australian postgraduate award. This work was supported by grants from the Australian Research Council and the Thyne Reid Memorial Trust.

9. References

Adie EA, Adams RR, Evans KL, Porteous DJ, Pickard BS. (2005). Speeding disease gene discovery by sequence based candidate prioritization. *BMC Bioinformatics,* 6:55.

Adie EA, Adams RR, Evans KL, Porteous DJ, Pickard BS. (2006). SUSPECTS: enabling fast and effective prioritization of positional candidates. *Bioinformatics,* 22(6):773-4.

Aitman TJ, Boone C, Churchill GA, Hengartner MO, Mackay TF, Stemple DL. (2011). The future of model organisms in human disease research. *Nat Rev Genet,* 12(8):575-82.

Almeida AM, Murakami Y, Layton DM, Hillmen P, Sellick GS, et al. (2006). Hypomorphic promoter mutation in PIGM causes inherited glycosylphosphatidylinositol deficiency. *Nat Med,* 12(7):846-51.

Altschul SF, Madden TL, Schäffer AA, Zhang J, Zhang Z, et al. (1997). Gapped BLAST and PSI-BLAST: a new generation of protein database search programs. *Nucleic Acids Res,* 25(17):3389-402.

Altshuler D, Daly MJ, Lander ES. (2008). Genetic mapping in human disease. *Science,* 322(5903):881-8.

Amberger J, Bocchini CA, Scott AF, Hamosh A. (2009). McKusick's online mendelian inheritance in man (OMIM). *Nucleic Acids Res,* 37:D793-6.

American Psychiatric Association (1994). *Diagnostic & Statistical Manual of Mental Disorders 4th Edition,* American Psychiatric Association, Washington DC, USA.

Amos W, Bryant C. (2011). Using human demographic history to infer natural selection reveals contrasting patterns on different families of immune genes. *Proc Biol Sci,* 278(1711):1587-94.

Ananda G, Chiaromonte F, Makova KD. (2011). A genome-wide view of mutation rate co-variation using multivariate analyses. *Genome Biol,* 12(3):R27.

Anderson S, Bankier AT, Barrell BG, de Bruijn MH, Coulson AR, et al. (1981). Sequence and organization of the human mitochondrial genome. *Nature,* 290(5806):457-65.

Annesley SJ, Fisher PR. (2009). *Dictyostelium discoideum* - a model for many reasons. *Mol Cell Biochem,* 329(1-2):73-91.

Ansari BM, Mason MK. (1977). Chromosomal abnormality in congenital rubella. *Pediatrics,* 59(1):13-5.

Anthony K, Gallo JM. (2010). Aberrant RNA processing events in neurological disorders. *Brain Res,* 1338:67-77.

Antonarakis SE, Beckmann JS. (2006). Mendelian disorders deserve more attention. *Nat Rev Genet,* 7(4):277-82.

Arguello PA, Markx S, Gogos JA, Karayiorgou M. (2010). Development of animal models for schizophrenia. *Dis Model Mech,* 3(1-2):22-6.

Arnheim N, Calabrese P. (2009). Understanding what determines the frequency and pattern of human germline mutations. *Nat Rev Genet,* 10(7):478-88.

Ashwood P, Wills S, Van de Water J. (2006). The immune response in autism: a new frontier for autism research. *J Leukoc Biol,* 80(1):1-15.

Attwood TK, Bradley P, Flower DR, Gaulton A, Maudling N, et al. (2003). PRINTS and its automatic supplement, preprints. *Nucleic Acids Res,* 31(1):400-2.

Axton M. (2010). Editorial: On beyond GWAS. *Nat Genet,* 42(7):551.

Aziz A, Harrop SP, Bishop NE. (2011a). DIA1R is an X-linked gene related to Deleted In Autism-1. *PLoS One,* 6(1): e14534.

Aziz A, Harrop SP, Bishop NE. (2011b). Characterization of the Deleted In Autism 1 protein family: implications for studying cognitive disorders. *PLoS One,* 6(1):e14547.

Bale TL, Baram TZ, Brown AS, Goldstein JM, Insel TR, et al. (2010). Early life programming and neurodevelopmental disorders. *Biol Psychiatry,* 68(4):314-9.

Bao Z, Eddy SR. (2002). Automated de novo identification of repeat sequence families in sequenced genomes. *Genome Res,* 12(8):1269-76.

Bailey JA, Eichler EE. (2006). Primate segmental duplications: crucibles of evolution, diversity and disease. *Nat Rev Genet,* 7(7):552-64.

Bailey JA, Baertsch R, Kent WJ, Haussler D, Eichler EE. (2004). Hotspots of mammalian chromosomal evolution. *Genome Biol,* 5(4):R23.

Barbaric I, Miller G, Dear TN. (2007). Appearances can be deceiving: phenotypes of knockout mice. *Brief Funct Genomic Proteomic,* 6(2):91-103.

Barrell D, Dimmer E, Huntley RP, Binns D, O'Donovan C, et al. (2009). The GOA database in 2009-an integrated gene ontology annotation resource. *Nucleic Acids Res,* 37:D396-403.

Batzer MA, Deininger PL. (2002). Alu repeats and human genomic diversity. *Nat Rev Genet,* 3(5):370-9.

Berezikov E, Thuemmler F, van Laake LW, Kondova I, Bontrop R, et al. (2006). Diversity of microRNAs in human and chimpanzee brain. *Nat Genet,* 38(12):1375-7.

Berg JS, Potocki L, Bacino CA. (2010). Common recurrent microduplication syndromes: diagnosis and management in clinical practice. *Am J Med Genet A,* 152A(5):1066-78.

Bernier R, Gerdts J, Munson J, Dawson G, Estes A. (2011). Evidence for broader autism phenotype characteristics in parents from multiple-incidence autism families. *Autism Res,* (Sep 8) in press.

Bill BR, Geschwind DH. (2009). Genetic advances in autism: heterogeneity and convergence on shared pathways. *Curr Opin Genet Dev,* 19(3):271-8.

Blanc C, Charette SJ, Mattei S, Aubry L, Smith EW, et al. (2009). Dictyostelium Tom1 participates to an ancestral ESCRT-0 complex. *Traffic,* 10(2):161-71.

Blekhman R, Man O, Herrmann L, Boyko AR, Indap A, et al. (2008). Natural selection on genes that underlie human disease susceptibility. *Curr Biol,* 18(12):883-9.

Blekhman R, Marioni JC, Zumbo P, Stephens M, Gilad Y. (2010). Sex-specific and lineage-specific alternative splicing in primates. *Genome Res,* 20(2):180-9.

Blennow K, de Leon MJ, Zetterberg H. (2006). Alzheimer's disease. *Lancet,* 368(9533):387-403.

Bourgeron T. (2009). A synaptic trek to autism. *Curr Opin Neurobiol,* 19(2):231-4.

Borck G, Zarhrate M, Cluzeau C, Bal E, Bonnefont JP, et al. (2006). Father-to-daughter transmission of Cornelia de Lange syndrome caused by a mutation in the 5' untranslated region of the NIPBL gene. *Hum Mutat,* 27(8):731-5.

Brunetti-Pierri N, Berg JS, Scaglia F, Belmont J, Bacino CA, et al. (2008). Recurrent reciprocal 1q21.1 deletions and duplications associated with microcephaly or macrocephaly and developmental and behavioral abnormalities. *Nat Genet,* 40(12):1466-71.

Buckingham SD, Esmaeili B, Wood M, Sattelle DB. (2004). RNA interference: from model organisms towards therapy for neural and neuromuscular disorders. *Hum Mol Genet,* 13(R2):R275-88.

Bulazel KV, Ferreri GC, Eldridge MD, O'Neill RJ. (2007). Species-specific shifts in centromere sequence composition are coincident with breakpoint reuse in karyotypically divergent lineages. *Genome Biol,* 8(8):R170.

Cai JJ, Borenstein E, Chen R, Petrov DA. (2009). Similarly strong purifying selection acts on human disease genes of all evolutionary ages. *Genome Biol Evol,* 1:131-44.

Calarco JA, Xing Y, Cáceres M, Calarco JP, Xiao X, et al. (2007). Global analysis of alternative splicing differences between humans and chimpanzees. *Genes Dev,* 21(22):2963-75.

Carvalho CM, Zhang F, Lupski JR. (2010). Evolution in health and medicine Sackler colloquium: Genomic disorders: a window into human gene and genome evolution. *Proc Natl Acad Sci USA,* 107 (Suppl 1):1765-71.

Chamberlain SJ, Lalande M. (2010). Neurodevelopmental disorders involving genomic imprinting at human chromosome 15q11-q13. *Neurobiol Dis,* 39(1):13-20.

Chance PF. (2006). Inherited focal, episodic neuropathies: hereditary neuropathy with liability to pressure palsies and hereditary neuralgic amyotrophy. *Neuromolecular Med,* 8(1-2):159-74.

Chen PY, Deane CM, Reinert G. (2008). Predicting and validating protein interactions using network structure. *PLoS Comput Biol,* 4(7):e1000118.

Chesselet MF. (2005). Animal models of neurological disorders. *NeuroRx,* 2(3):395.

Chua HN, Wong L. (2008). Increasing the reliability of protein interactomes. *Drug Discov Today,* 13(15-16):652-8.

Claverie-Martin F, González-Acosta H, Flores C, Antón-Gamero M, García-Nieto V. (2003). De novo insertion of an Alu sequence in the coding region of the CLCN5 gene results in Dent's disease. *Hum Genet,* 113(6):480-5.

Cohen O, Cans C, Cuillel M, Gilardi JL, Roth H, et al. (1996). Cartographic study: breakpoints in 1574 families carrying human reciprocal translocations. *Hum Genet,* 97(5):659-67.

Cohly HH, Panja A. (2005). Immunological findings in autism. *Int Rev Neurobiol,* 71:317-41.

Collier LS, Largaespada DA. (2007). Transposable elements and the dynamic somatic genome. *Genome Biol,* 8 (Suppl 1):S5.

Complex Trait Consortium. (2004). The Collaborative Cross, a community resource for the genetic analysis of complex traits. *Nat Genet,* 36(11):1133-7.

Conant GC, Wolfe KH. (2008). Turning a hobby into a job: how duplicated genes find new functions. *Nat Rev Genet,* 9(12):938-50.

Corona E, Dudley JT, Butte AJ. (2010). Extreme evolutionary disparities seen in positive selection across seven complex diseases. *PLoS One,* 5(8):e12236.

Costa e Silva JA. (2008). Autism, a brain developmental disorder: some new pathophysiologic and genetics findings. *Metabolism,* 57(Suppl 2):S40-3.

Crespi B. (2011). Autism and cancer risk. *Autism Res,* 4:301-10.

Crespi B, Stead P, Elliot M. (2010). Evolution in health and medicine Sackler colloquium: Comparative genomics of autism and schizophrenia. *Proc Natl Acad Sci USA,* 107(Suppl 1):1736-41.

Crow TJ. (2011). "Just the facts" of schizophrenia in the context of human evolution: commentary. *Schizophr Res,* 129(2-3):205-7.

Currat M, Trabuchet G, Rees D, Perrin P, Harding RM, et al. (2002). Molecular analysis of the beta-globin gene cluster in the Niokholo Mandenka population reveals a recent origin of the beta(S) mutation. *Am J Hum Genet,* 70(1):207-23.

Dahlin JS, Ivarsson MA, Heyman B, Hallgren J. (2011). IgE immune complexes stimulate an increase in lung mast cell progenitors in a mouse model of allergic airway inflammation. *PLoS One,* 6(5):e20261.

Darai-Ramqvist E, Sandlund A, Müller S, Klein G, Imreh S, et al. (2008). Segmental duplications and evolutionary plasticity at tumor chromosome break-prone regions. *Genome Res,* 18(3):370-9.

Davey Smith G, Ebrahim S, Lewis S, Hansell AL, Palmer LJ, Burton PR. (2005). Genetic epidemiology and public health: hope, hype, and future prospects. *Lancet,* 366(9495):1484-98.

Dean B. (2009). Is schizophrenia the price of human central nervous system complexity? *Aust NZ J Psychiatry,* 43(1): 13-24.

Degterev A, Yuan J. (2008). Expansion and evolution of cell death programmes. *Nat Rev Mol Cell Biol,* 9(5):378-90.

Del Giudice M, Angeleri R, Brizio A, Elena MR. (2010). The evolution of autistic-like and schizotypal traits: a sexual selection hypothesis. *Front Psychol,* 1:41.

de Vries PJ. (2010). Targeted treatments for cognitive and neurodevelopmental disorders in tuberous sclerosis complex. *Neurotherapeutics,* 7(3):275-82.

Dezso Z, Nikolsky Y, Sviridov E, Shi W, Serebriyskaya T, et al. (2008). A comprehensive functional analysis of tissue specificity of human gene expression. *BMC Biol,* 6:49.

Domazet-Loso T, Tautz D. (2008). An ancient evolutionary origin of genes associated with human genetic diseases. *Mol Biol Evol,* 25(12):2699-707.

Domazet-Loso T, Brajkovi J, Tautz D. (2007). A phylostratigraphy approach to uncover the genomic history of major adaptations in metazoan lineages. *Trends Genet,* 23(11):533-9.

Dominguez-Brauer C, Brauer PM, Chen YJ, Pimkina J, Raychaudhuri P. (2010). Tumor suppression by ARF: gatekeeper and caretaker. *Cell Cycle,* 9(1):86-9.

Drosophila 12 Genomes Consortium. (2007). Evolution of genes and genomes on the Drosophila phylogeny. *Nature,* 450(7167):203-18.

Dumas L, Sikela JM. (2009). DUF1220 domains, cognitive disease, and human brain evolution. *Cold Spring Harb Symp Quant Biol,* 74:375-82.

du Plessis L, Skunca N, Dessimoz C. (2011). The what, where, how and why of gene ontology-a primer for bioinformaticians. *Brief Bioinform,* 12(6):723-35.

Dziembowski A, Séraphin B. (2004). Recent developments in the analysis of protein complexes. *FEBS Lett,* 556(1-3):1-6.

Eapen V. (2011). Genetic basis of autism: is there a way forward? *Curr Opin Psychiatry,* 24(3):226-36.

Easton DF, Pooley KA, Dunning AM, Pharoah PD, Thompson D, et al. (2007). Genome-wide association study identifies novel breast cancer susceptibility loci. *Nature,* 447(7148):1087-93.

Eisenberg E, Levanon EY. (2003). Preferential attachment in the protein network evolution. *Phys Rev Lett,* 91(13):138701.

Eisenberg D, Marcotte EM, Xenarios I, Yeates TO. (2000). Protein function in the post-genomic era. *Nature,* 405(6788):823-826.

Enard W, Fassbender A, Model F, Adorján P, Pääbo S, et al. (2004). Differences in DNA methylation patterns between humans and chimpanzees. *Curr Biol,* 14(4):R148-9.

Eng KH, Bravo HC, Kele S. (2009). A phylogenetic mixture model for the evolution of gene expression. *Mol Biol Evol,* 26(10):2363-72.

Erdin S, Lisewski AM, Lichtarge O. (2011). Protein function prediction: towards integration of similarity metrics. *Curr Opin Struct Biol,* 21(2):180-8.

Everts-van der Wind A, Kata SR, Band MR, Rebeiz M, Larkin DM, et al. (2004). A 1463 gene cattle-human comparative map with anchor points defined by human genome sequence coordinates. *Genome Res,* 14(7):1424-37.

Everts-van der Wind A, Larkin DM, Green CA, Elliott JS, Olmstead CA, et al. (2005). A high-resolution whole-genome cattle-human comparative map reveals details of mammalian chromosome evolution. *Proc Natl Acad Sci USA,* 102(51):18526-31.

Ey E, Leblond CS, Bourgeron T. (2011). Behavioral profiles of mouse models for autism spectrum disorders. *Autism Res,* 4(1):5-16.

Eyre-Walker A. (2006). The genomic rate of adaptive evolution. *Trends Ecol Evol,* 21(10):569-75.

Feifel D, Shilling PD. (2010). Promise and pitfalls of animal models of schizophrenia. *Curr Psychiatry Rep,* 12 (4): 327-334.

Feldman I, Rzhetsky A, Vitkup D. (2008). Network properties of genes harboring inherited disease mutations. *Proc Natl Acad Sci USA,* 105(11):4323-8.

Fischbach BV, Trout KL, Lewis J, Luis CA, Sika M. (2005). WAGR syndrome: a clinical review of 54 cases. *Pediatrics,* 116(4):984-8.

Fonseka KG, Griffin DK. (2011). Is there a paternal age effect for aneuploidy? *Cytogenet Genome Res,* 133(2-4):280-91.

Fortunato EA, Spector DH. (2003). Viral induction of site-specific chromosome damage. *Rev Med Virol,* 13(1):21-37.

Fraser HB, Hirsh AE, Steinmetz LM, Scharfe C, Feldman MW. (2002). Evolutionary rate in the protein interaction network. *Science,* 296(5568):750-2.

Friedberg I. (2006). Automated protein function prediction- the genomic challenge. *Brief Bioinform,* 7(3):225-42.

Galliot B, Quiquand M, Ghila L, de Rosa R, et al. (2009). Origins of neurogenesis, a cnidarian view. *Dev Biol,* 332(1):2-24.

Gandhi S, Wood NW. (2010). Genome-wide association studies: the key to unlocking neurodegeneration? *Nat Neurosci,* 13(7):789-94.

Gandhi TK, Zhong J, Mathivanan S, Karthick L, Chandrika KN, et al. (2006). Analysis of the human protein interactome and comparison with yeast, worm and fly interaction datasets. *Nat Genet,* 38(3):285-93.

Gasser T. (2009). Mendelian forms of Parkinson's disease. *Biochim Biophys Acta,* 1792(7):587-96.

Gaulton KJ, Mohlke KL, Vision TJ. (2007). A computational system to select candidate genes for complex human traits. *Bioinformatics,* 23(9):1132-40.

Genin E, Hannequin D, Wallon D, Sleegers K, Hiltunen M, et al. (2011). APOE and Alzheimer disease: a major gene with semi-dominant inheritance. *Mol Psychiatry,* 16(9):903-7.

Giallourakis C, Henson C, Reich M, Xie X, Mootha VK. (2005). Disease gene discovery through integrative genomics. *Annu Rev Genomics Hum Genet,* 6:381-406.

Gilby KL. (2008). A new rat model for vulnerability to epilepsy and ASDs. *Epilepsia,* 49(Suppl 8):108-10.

Gillis J, Pavlidis P. (2011). The impact of multifunctional genes on "guilt by association" analysis. *PLoS One,* 6 (2): e17258.

Gilman SR, Iossifov I, Levy D, Ronemus M, Wigler M, et al. (2011). Rare de novo variants associated with autism implicate a large functional network of genes involved in formation and function of synapses. *Neuron,* 70(5):898-907.

Giusti C, Luciani MF, Golstein P. (2010). A second signal for autophagic cell death? *Autophagy,* 6(6):823-4.

Glatt SJ, Cohen OS, Faraone SV, Tsuang MT. (2011). Dysfunctional gene splicing as a potential contributor to neuropsychiatric disorders. *Am J Med Genet B Neuropsychiatr Genet,* 156B (4):382-92.

Goh KI, Cusick ME, Valle D, Childs B, Vidal M, et al. (2007). The human disease network. *Proc Natl Acad Sci USA,* 104(21):8685-90.

Goldsmith-Fischman S, Honig B. (2003). Structural genomics: computational methods for structure analysis. *Protein Sci,* 12(9):1813-21.

Grady DL, Chi HC, Ding YC, Smith M, Wang E, et al. (2003). High prevalence of rare dopamine receptor D4 alleles in children diagnosed with attention-deficit hyperactivity disorder. *Mol Psychiatry,* 8(5):536-45.

Greaves M. (2007). Darwinian medicine: a case for cancer. *Nat Rev Cancer,* 7(3):213-21.

Grimmelikhuijzen CJ, Westfall JA. (1995). The nervous systems of cnidarians. *EXS,* 72:7-24.

Guan C, Ye C, Yang X, Gao J. (2010). A review of current large-scale mouse knockout efforts. *Genesis,* 48(2):73-85.

Guilmatre A, Dubourg C, Mosca AL, Legallic S, Goldenberg A, et al. (2009). Recurrent rearrangements in synaptic and neurodevelopmental genes and shared biologic pathways in schizophrenia, autism, and mental retardation. *Arch Gen Psychiatry,* 66(9):947-56.

Han JD. (2008). Understanding biological functions through molecular networks. *Cell Res,* 18(2):224-37.

Han MV, Demuth JP, McGrath CL, Casola C, Hahn MW. (2009). Adaptive evolution of young gene duplicates in mammals. *Genome Res,* 19(5):859-67.

Haitina T, Fredriksson R, Foord SM, Schiöth HB, Gloriam DE. (2009). The G protein-coupled receptor subset of the dog genome is more similar to that in humans than rodents. *BMC Genomics,* 10:24.

Hanahan D, Weinberg RA. (2000). The hallmarks of cancer. *Cell,* 100(1):57-70.

Häsler J, Strub K. (2006). Alu elements as regulators of gene expression. *Nucleic Acids Res,* 34(19):5491-7.

Henikoff JG, Pietrokovski S, McCallum CM, Henikoff S. (2000). Blocks-based methods for detecting protein homology. *Electrophoresis,* 21(9):1700-6.

Heo Y, Zhang Y, Gao D, Miller VM, Lawrence DA. (2011). Aberrant immune responses in a mouse with behavioral disorders. *PLoS One,* 6(7):e20912.

Hindorff LA, Sethupathy P, Junkins HA, Ramos EM, Mehta JP, et al. (2009). Potential etiologic and functional implications of genome-wide association loci for human diseases and traits. *Proc Natl Acad Sci USA,* 106(23):9362-9367.

Hoffman EJ, State MW. (2010). Progress in cytogenetics: implications for child psychopathology. *J Am Acad Child Adolesc Psychiatry,* 49(8):736-751.

Huang H, Winter EE, Wang H, Weinstock KG, Xing H, et al. (2004). Evolutionary conservation and selection of human disease gene orthologs in the rat and mouse genomes. *Genome Biol,* 5(7):R47.

Hulo N, Bairoch A, Bulliard V, Cerutti L, Cuche BA, et al. (2008). The 20 years of PROSITE. *Nucleic Acids Res,* 36:D245-9.

Hutz JE, Kraja AT, McLeod HL, Province MA. (2008). CANDID: a flexible method for prioritizing candidate genes for complex human traits. *Genet Epidemiol,* 32(8):779-90.

Ilievska J, Bishop NE, Annesley SJ, Fisher PR, (2011). The roles of ESCRT proteins in healthy cells and in disease, In: *Cell Biology,* S. Najman, (Ed.), InTech, Rijeka, Croatia.

Inoue K, Lupski JR. (2002). Molecular mechanisms for genomic disorders. *Annu Rev Genomics Hum Genet,* 3:199-242.

International Parkinson Disease Genomics Consortium (2011). Imputation of sequence variants for identification of genetic risks for Parkinson's disease: a meta-analysis of genome-wide association studies. *Lancet,* 377(9766):641-9.

Ioannidis JP, Ntzani EE, Trikalinos TA. (2004). 'Racial' differences in genetic effects for complex diseases. *Nat Genet,* 36(12):1312-8.

Iwata K, Matsuzaki H, Takei N, Manabe T, Mori N. (2010). Animal models of autism: An epigenetic and environmental viewpoint. *J Central Nervous Syst Dis,* 2:37-44. Retrieved from www.la-press.com

Jeffery CJ. (2009). Moonlighting proteins- an update. *Mol Biosyst,* 5(4):345-50.

Jeong H, Mason SP, Barabási AL, Oltvai ZN. (2001). Lethality and centrality in protein networks. *Nature,* 411(6833):41-2.

Jia L, Landan G, Pomerantz M, Jaschek R, Herman P, et al. (2009). Functional enhancers at the gene-poor 8q24 cancer-linked locus. *PLoS Genet,* 5(8):e1000597.

Jiang MC, Lien YR, Chen SU, Ko TM, Ho HN, et al. (1999). Transmission of de novo mutations of the deleted in azoospermia genes from a severely oligozoospermic male to a son via intracytoplasmic sperm injection. *Fertil Steril,* 71(6):1029-32.

Jones KT. (2008). Meiosis in oocytes: predisposition to aneuploidy and its increased incidence with age. *Hum Reprod Update,* 14(2):143-58.

Jorde LB. (2000). Linkage disequilibrium and the search for complex disease genes. *Genome Res,* 10(10):1435-44.

Kehrer-Sawatzki H, Cooper DN. (2007). Understanding the recent evolution of the human genome: insights from human-chimpanzee genome comparisons. *Hum Mutat,* 28(2):99-130.

Kehrer-Sawatzki H, Cooper DN. (2008). Molecular mechanisms of chromosomal rearrangement during primate evolution. *Chromosome Res,* 16(1):41-56.

Kidd JM, Cooper GM, Donahue WF, Hayden HS, Sampas N, et al. (2008). Mapping and sequencing of structural variation from eight human genomes. *Nature,* 453(7191):56-64.

Kimchi-Sarfaty C, Oh JM, Kim IW, Sauna ZE, Calcagno AM, et al. (2007). A "silent" polymorphism in the MDR1 gene changes substrate specificity. *Science,* 315(5811):525-8.

Kingsley CB. (2011). Identification of causal sequence variants of disease in the next generation sequencing era. *Methods Mol Biol,* 700:37-46.

Kinzler KW, Vogelstein B. (1997). Cancer-susceptibility genes -- Gatekeepers and caretakers. *Nature,* 386 (6627): 761- 763.

Knudson AG. (1993). All in the (cancer) family. *Nat Genet,* 5(2):103-4.

Kosik KS. (2009). Exploring the early origins of the synapse by comparative genomics. *Biol Lett,* 5(1):108-11.

Koszul R, Fischer G. (2009). A prominent role for segmental duplications in modeling eukaryotic genomes. *C R Biol,* 332(2-3):254-66.

Ku CS, Naidoo N, Pawitan Y. (2011). Revisiting Mendelian disorders through exome sequencing. *Hum Genet,* 129(4):351-70.

Kuchaiev O, Wang PT, Nenadic Z, Przulj N. (2009). Structure of brain functional networks. *Conf Proc IEEE Eng Med Biol Soc,* 2009:4166-70.

Kuchaiev O, Rasajski M, Higham DJ, Przulj N. (2009). Geometric de-noising of protein-protein interaction networks. *PLoS Comput Biol,* 5(8):e1000454.

Kuhlenbäumer G, Hullmann J, Appenzeller S. (2011). Novel genomic techniques open new avenues in the analysis of monogenic disorders. *Hum Mutat,* 32(2):144-51.

Kumar D. (2008). Disorders of the genome architecture: a review. *Genomic Med,* 2(3-4):69-76.

Kumar RA, KaraMohamed S, Sudi J, Conrad DF, Brune C. et al. (2008). Recurrent 16p11.2 microdeletions in autism. *Hum Mol Genet,* 17(4):628-38.

Kurahashi H, Inagaki H, Ohye T, Kogo H, Kato T, Emanuel BS. (2006). Chromosomal translocations mediated by palindromic DNA. *Cell Cycle,* 5(12):1297-303.

Lagerström MC, Hellström AR, Gloriam DE, Larsson TP, Schiöth HB, et al (2006). The G protein-coupled receptor subset of the chicken genome. *PLoS Comput Biol,* 2(6):e54.

Larkin DM, Everts-van der Wind A, Rebeiz M, Schweitzer PA, Bachman S, et al. (2003). A cattle-human comparative map built with cattle BAC-ends and human genome sequence. *Genome Res,* 13(8):1966-72.

Larkin DM, Pape G, Donthu R, Auvil L, Welge M, Lewin HA. (2009). Breakpoint regions and homologous synteny blocks in chromosomes have different evolutionary histories. *Genome Res,* 19(5):770-7.

Laskowski RA, Watson JD, Thornton JM. (2003). From protein structure to biochemical function? *J Struct Funct Genomics,* 4(2-3):167-77.

Lazar NL, Neufeld RW, Cain DP. (2011). Contribution of nonprimate animal models in understanding the etiology of schizophrenia. *J Psychiatry Neurosci,* 36(4):E5-29.

Lee SH, Wray NR, Goddard ME, Visscher PM. (2011). Estimating missing heritability for disease from genome-wide association studies. *Am J Hum Genet,* 88(3):294-305.

Lehmann AR. (2001). The xeroderma pigmentosum group D (XPD) gene: one gene, two functions, three diseases. *Genes Dev,* 15(1):15-23.

Lenski C, Kooy RF, Reyniers E, Loessner D, Wanders RJ, et al. (2007). The reduced expression of the HADH2 protein causes X-linked mental retardation, choreoathetosis, and abnormal behavior. *Am J Hum Genet,* 80(2):372-7.

Lesage S, Brice A. (2009). Parkinson's disease: from monogenic forms to genetic susceptibility factors. *Hum Mol Genet,* 18(R1):R48-59.

Lewis DA, Levitt P. (2002). Schizophrenia as a disorder of neurodevelopment. *Annu Rev Neurosci,* 25:409-32.

Liao BY, Zhang J. (2008). Null mutations in human and mouse orthologs frequently result in different phenotypes. *Proc Natl Acad Sci USA,* 105(19):6987-92.

Lin L, Shen S, Jiang P, Sato S, Davidson BL, et al. (2010). Evolution of alternative splicing in primate brain transcriptomes. *Hum Mol Genet,* 19(15):2958-73.

Lindhurst MJ, Sapp JC, Teer JK, Johnston JJ, Finn EM, et al. (2011). A mosaic activating mutation in AKT1 associated with the Proteus syndrome. *N Engl J Med,* 365(7):611-9.

Lindsay SJ, Khajavi M, Lupski JR, Hurles ME. (2006). A chromosomal rearrangement hotspot can be identified from population genetic variation and is coincident with a hotspot for allelic recombination. *Am J Hum Genet,* 79(5):890-902.

Liu X, Cheng R, Verbitsky M, Kisselev S, Browne A, et al. (2011). Genome-Wide association study identifies candidate genes for Parkinson's disease in an Ashkenazi Jewish population. *BMC Med Genet,* 12:104.

Lobo I. (2008) Same genetic mutation, different genetic disease phenotype. *Nature Education,* 1(1), Retrieved from http://www.nature.com/scitable/topicpage/same-genetic-mutation-different-genetic-disease-phenotype-938

Lupski JR. (2010). New mutations and intellectual function. *Nat Genet,* 42(12):1036-8.

Lynch M, Conery JS. (2000). The evolutionary fate and consequences of duplicate genes. *Science,* 290 (5494):1151-1155.

Macleod K (2000). Tumor suppressor genes. *Curr Opin Genet Dev,* 10(1):81-93.

Maffini MV, Soto AM, Calabro JM, Ucci AA, Sonnenschein C. (2004). The stroma as a crucial target in rat mammary gland carcinogenesis. *J Cell Sci,* 117(Pt 8):1495-502.

Magrane M, Consortium U. (2011). UniProt Knowledgebase: a hub of integrated protein data. *Database (Oxford),* 2011:bar009.

Makino T, Gojobori T. (2006). The evolutionary rate of a protein is influenced by features of the interacting partners. *Mol Biol Evol,* 23(4):784-9.

Manolio TA. (2010). Genomewide association studies and assessment of the risk of disease. *N Engl J Med,* 363(2):166-76.

Marigorta UM, Lao O, Casals F, Calafell F, Morcillo-Suárez C, et al. (2011). Recent human evolution has shaped geographical differences in susceptibility to disease. *BMC Genomics,* 12:55.

Marques-Bonet T, Girirajan S, Eichler EE. (2009). The origins and impact of primate segmental duplications. *Trends Genet,* 25(10):443-54.

Marth GT, Yu F, Indap AR, Garimella K, Gravel S, et al. (2011). The functional spectrum of low-frequency coding variation. *Genome Biol,* 12(9):R84.

Martinez J, Dugaiczyk LJ, Zielinski R, Dugaiczyk A. (2001). Human genetic disorders, a phylogenetic perspective. *J Mol Biol,* 308(4):587-96.

Mattei S, Klein G, Satre M, Aubry L. (2006). Trafficking and developmental signaling: Alix at the crossroads. *Eur J Cell Biol,* 85(9-10):925-36.

McCarthy SE, Makarov V, Kirov G, Addington AM, McClellan J, et al. (2009). Microduplications of 16p11.2 are associated with schizophrenia. *Nat Genet,* 41(11):1223-7.

McCarthy MI, Hirschhorn JN. (2008). Genome-wide association studies: potential next steps on a genetic journey. *Hum Mol Genet,* 17(R2):R156-65.

McGary KL, Park TJ, Woods JO, Cha HJ, Wallingford JB, et al. (2010). Systematic discovery of nonobvious human disease models through orthologous phenotypes. *Proc Natl Acad Sci USA,* 107(14):6544-9.

Michor F, Iwasa Y, Nowak MA. (2004). Dynamics of cancer progression. *Nat Rev Cancer,* 4(3):197-205.

Midic U, Oldfield CJ, Dunker AK, Obradovic Z, Uversky VN. (2009). Protein disorder in the human diseasome: unfoldomics of human genetic diseases. *BMC Genomics,* 10(Suppl 1):S12.

Miller MP, Kumar S. (2001). Understanding human disease mutations through the use of interspecific genetic variation. *Hum Mol Genet,* 10(21):2319-28.

Minshew N, McFadden K. Commentary for special issue of autism research on mouse models in ASD: a clinical perspective. *Autism Res,* 4(1):1-4.

Mobley JL. (2004). Is rheumatoid arthritis a consequence of natural selection for enhanced tuberculosis resistance? *Med Hypotheses,* 62(5):839-43.

Mooney S. (2005). Bioinformatics approaches and resources for single nucleotide polymorphism functional analysis. *Brief Bioinform,* 6(1):44-56.

Morrow EM, Yoo SY, Flavell SW, Kim TK, Lin Y, et al. (2008). Identifying autism loci and genes by tracing recent shared ancestry. *Science,* 321(5886):218-23.

Müller B, Grossniklaus U. (2010). Model organisms- a historical perspective. *J Proteomics,* 73(11):2054-63.

Müller N, Schwarz MJ. (2010). Immune system and schizophrenia. *Curr Immunol Rev,* 6(3):213-220.

Müller N, Riedel M, Gruber R, Ackenheil M, Schwarz MJ. (2000). The immune system and schizophrenia. An integrative view. *Ann NY Acad Sci,* 917:456-67.

Murphy WJ, Larkin DM, Everts-van der Wind A, Bourque G, Tesler G. et al. (2005). Dynamics of mammalian chromosome evolution inferred from multispecies comparative maps. *Science,* 309(5734):613-7.

Nadeau JH, Taylor BA. (1984). Lengths of chromosomal segments conserved since divergence of man and mouse. *Proc Natl Acad Sci USA,* 81(3):814-8.

Nadeau JH. (2001). Modifier genes in mice and humans. *Nat Rev Genet,* 2(3):165-74.

Nadeau JH. (2003). Modifier genes and protective alleles in humans and mice. *Curr Opin Genet Dev,* 13(3):290-5.

Nagaraj SH, Ingham A, Reverter A. (2010). The interplay between evolution, regulation and tissue specificity in the human hereditary diseasome. *BMC Genomics,* 11(Suppl 4):S23.

Nejentsev S, Walker N, Riches D, Egholm M, Todd JA. (2009). Rare variants of IFIH1, a gene implicated in antiviral responses, protect against type 1 diabetes. *Science,* 324(5925):387-9.

Ng PC, Henikoff S. (2006). Predicting the effects of amino acid substitutions on protein function (2006). Annu Rev *Genomics Hum Genet,* 7:61-80.

Ng PC, Henikoff S. (2002). Accounting for human polymorphisms predicted to affect protein function. *Genome Res,* 12(3):436-46.

Ng PC, Levy S, Huang J, Stockwell TB, Walenz BP, et al. (2008). Genetic variation in an individual human exome. *PLoS Genet,* 4(8):e1000160.

Nibbe RK, Markowitz S, Myeroff L, Ewing R, Chance MR. (2009). Discovery and scoring of protein interaction subnetworks discriminative of late stage human colon cancer. *Mol Cell Proteomics,* 8(4):827-45.

Nusbacher J, Hirschhorn K, Cooper LZ. (1967). Chromosomal abnormalities in congenital rubella. *N Engl J Med,* 276(25):1409-13.

Ober C, Vercelli D. (2011). Gene-environment interactions in human disease: nuisance or opportunity? *Trends Genet,* 27(3):107-15.

Ohno S. (1970). *Evolution by gene duplication.* Springer Verlag, New York, NY, USA.

Oldridge M, Zackai EH, McDonald-McGinn DM, Iseki S, Morriss-Kay GM, et al. (1999). De novo Alu-element insertions in FGFR2 identify a distinct pathological basis for Apert syndrome. *Am J Hum Genet,* 64(2):446-61.

Oliver S. (2000). Guilt-by-association goes global. *Nature,* 403(6770):601-3.

Oostra BA, Willemsen R. (2009). FMR1: a gene with three faces. *Biochim Biophys Acta,* 1790(6):467-77.

Oti M, Ballouz S, Wouters MA. (2011). Web tools for the prioritization of candidate disease genes. *Methods Mol Biol,* 760:189-206.

Ou Z, Stankiewicz P, Xia Z, Breman AM, Dawson B, et al. (2011). Observation and prediction of recurrent human translocations mediated by NAHR between nonhomologous chromosomes. *Genome Res,* 21(1):33-46.

Owen MJ, Craddock N, O'Donovan MC. (2005). Schizophrenia: genes at last? *Trends Genet,* 21(9):518-25.

Pagani F, Raponi M, Baralle FE. (2005). Synonymous mutations in CFTR exon 12 affect splicing and are not neutral in evolution. *Proc Natl Acad Sci USA,* 102(18):6368-72.

Patterson PH. (2011). Modeling autistic features in animals. *Pediatr Res,* 69(5 Pt 2):34R-40R.

Pearson BL, Pobbe RL, Defensor EB, Oasay L, Bolivar VJ, et al. (2011). Motor and cognitive stereotypies in the BTBR T+tf/J mouse model of autism. *Genes Brain Behav,* 10(2):228-35.

Peng Q, Pevzner PA, Tesler G. (2006). The fragile breakage versus random breakage models of chromosome evolution. *PLoS Comput Biol,* 2(2):e14.

Peltonen L, Perola M, Naukkarinen J, Palotie A. (2006). Lessons from studying monogenic disease for common disease. *Hum Mol Genet,* 15(Spec No 1):R67-74.

Peñagarikano O, Abrahams BS, Herman EI, Winden KD, Gdalyahu A, et al. (2011). Absence of CNTNAP2 leads to epilepsy, neuronal migration abnormalities, and core autism-related deficits. *Cell,* 147(1):235-46.

Penzes P, Cahill ME, Jones KA, VanLeeuwen JE, Woolfrey KM. (2011). Dendritic spine pathology in neuropsychiatric disorders. *Nat Neurosci,* 14(3):285-93.

Pevzner P, Tesler G. (2003). Genome rearrangements in mammalian evolution: lessons from human and mouse genomes. *Genome Res,* 13(1):37-45.

Phillips DH, Arlt VM. (2009). Genotoxicity: damage to DNA and its consequences. *EXS,* 99:87-110.

Pierri CL, Parisi G, Porcelli V. (2010). Computational approaches for protein function prediction: a combined strategy from multiple sequence alignment to molecular docking-based virtual screening. *Biochim Biophys Acta,* 1804(9):1695-712.

Pocklington AJ, Cumiskey M, Armstrong JD, Grant SG. (2006). The proteomes of neurotransmitter receptor complexes form modular networks with distributed functionality underlying plasticity and behaviour. *Mol Syst Biol,* 2:2006.0023.

Pomerantz MM, Shrestha Y, Flavin RJ, Regan MM, Penney KL, et al. (2010). Analysis of the 10q11 cancer risk locus implicates MSMB and NCOA4 in human prostate tumorigenesis. *PLoS Genet,* 6(11):e1001204.

Ponder BA. (2001). Cancer genetics. *Nature,* 411(6835):336-41.

Powell SB. (2010). Models of neurodevelopmental abnormalities in schizophrenia. *Curr Top Behav Neurosci,* 4:435-81.

Qiu J, Noble WS. (2008). Predicting co-complexed protein pairs from heterogeneous data. *PLoS Comput Biol,* 4(4):e1000054.

Reser JE. (2011). Conceptualizing the autism spectrum in terms of natural selection and behavioral ecology: the solitary forager hypothesis. *Evol Psychol,* 9(2): 207-238, Retrieved from http://www.epjournal.net/ filestore/EP09207238.pdf

Richards AJ, Laidlaw M, Meredith SP, Shankar P, Poulson AV, et al. (2007). Missense and silent mutations in COL2A1 result in Stickler syndrome but via different molecular mechanisms. *Hum Mutat,* 28(6):639.

Roberts R, Wells GA, Stewart AF, Dandona S, Chen L. (2010). The genome-wide association study- a new era for common polygenic disorders. J Cardiovasc Transl Res, 3(3):173-82.

Robertson HR, Feng G. (2011). Annual Research Review: Transgenic mouse models of childhood-onset psychiatric disorders. *J Child Psychol Psychiatry,* 52(4):442-75.

Robinson EB, Koenen KC, McCormick MC, Munir K, Hallett V, et al. (2011). A multivariate twin study of autistic traits in 12-year-olds: Testing the fractionable autism triad hypothesis. *Behav Genet,* 42(2):245-55.

Rodríguez-Escudero I, Oliver MD, Andrés-Pons A, Molina M, Cid VJ, et al. (2011). A comprehensive functional analysis of PTEN mutations: implications in tumor- and autism-related syndromes. *Hum Mol Genet,* 20(21):4132-42.

Rogozin IB, Pavlov YI. (2003). Theoretical analysis of mutation hotspots and their DNA sequence context specificity. *Mutat Res,* 544(1):65-85.

Ronald A, Hoekstra RA. (2011). Autism spectrum disorders and autistic traits: a decade of new twin studies. *Am J Med Genet B Neuropsychiatr Genet,* 156B(3):255-74.

Ropers HH. (2007). New perspectives for the elucidation of genetic disorders. *Am J Hum Genet,* 81(2):199-207.

Rothschild BM, Woods RJ, Rothschild C, Sebes JI. (1992). Geographic distribution of rheumatoid arthritis in ancient North America: implications for pathogenesis. *Semin Arthritis Rheum,* 22(3):181-7.

Rubbi C, Milner J. (2005). p53: Gatekeeper, caretaker, or both? In: *25 Years of p53 Research,* P Hainaut & KG Wiman, (Eds.), Springer, London, UK.

Russo A, Migliavacca M, Zanna I, Macaluso M, Gebbia N, et al. (2006). Caretakers and Gatekeepers. In: *Encyclopedia of Life Sciences*, John Wiley & Sons, Retrieved from http://onlinelibrary.wiley.com/ doi/10.1038/npg.els.0006048/pdf

Sabeti PC, Schaffner SF, Fry B, Lohmueller J, Varilly P, et al. (2006). Positive natural selection in the human lineage. *Science*, 312(5780):1614-20.

Saeed R, Deane CM. (2006). Protein protein interactions, evolutionary rate, abundance and age. *BMC Bioinformatics*, 7:128.

Sahoo T, Theisen A, Rosenfeld JA, Lamb AN, Ravnan JB, et al. (2011). Copy number variants of schizophrenia susceptibility loci are associated with a spectrum of speech and developmental delays and behavior problems. *Genet Med*, 13(10):868-80.

Sankoff D, Deneault M, Turbis P, Allen C. (2002). Chromosomal distributions of breakpoints in cancer, infertility, and evolution. *Theor Popul Biol*, 61(4):497-501.

Sasaki MS. (2006). Delayed manifestation and transmission bias of de novo chromosome mutations: their relevance for radiation health effect. *J Radiat Res*, 47(Suppl B):B45-56.

Schibler L, Roig A, Mahe MF, Laurent P, Hayes H, et al. (2006). High-resolution comparative mapping among man, cattle and mouse suggests a role for repeat sequences in mammalian genome evolution. *BMC Genomics*, 7:194.

Schmidt S, Gerasimova A, Kondrashov FA, Adzhubei IA, Kondrashov AS, et al. (2008). Hypermutable non-synonymous sites are under stronger negative selection. *PLoS Genet*, 4(11):e1000281.

Shaffer LG, Bejjani BA, Torchia B, Kirkpatrick S, Coppinger J, et al. (2007). The identification of microdeletion syndromes and other chromosome abnormalities: cytogenetic methods of the past, new technologies for the future. *Am J Med Genet C Semin Med Genet*, 145C(4):335-45.

Shah A, Garzon-Muvdi T, Mahajan R, Duenas VJ, Quiñones-Hinojosa A. (2010). Animal models of neurological disease. *Adv Exp Med Biol*, 671:23-40.

Sharp AJ, Hansen S, Selzer RR, Cheng Z, Regan R, et al. (2006). Discovery of previously unidentified genomic disorders from the duplication architecture of the human genome. *Nat Genet*, 38(9):1038-42.

Shatsky M, Nussinov R, Wolfson HJ. (2008). Algorithms for multiple protein structure alignment and structure-derived multiple sequence alignment. *Methods Mol Biol*, 413:125-46.

Sherva R, Farrer LA. (2011). Power and pitfalls of the genome-wide association study approach to identify genes for Alzheimer's disease. *Curr Psychiatry Rep*, 13(2):138-46.

Shinawi M, Liu P, Kang SH, Shen J, Belmont JW, et al. (2010). Recurrent reciprocal 16p11.2 rearrangements associated with global developmental delay, behavioural problems, dysmorphism, epilepsy, and abnormal head size. *J Med Genet*, 47(5):332-41.

Shineman DW, Basi GS, Bizon JL, Colton CA, Greenberg BD, et al. (2011). Accelerating drug discovery for Alzheimer's disease: best practices for preclinical animal studies. *Alzheimers Res Ther*, 3(5):28.

Shulman JM, De Jager PL, Feany MB. (2011). Parkinson's disease: genetics and pathogenesis. *Annu Rev Pathol*, 6:193-222.

Sidransky E. (2006). Heterozygosity for a Mendelian disorder as a risk factor for complex disease. *Clin Genet*, 70(4):275-82.

Silva AJ, Ehninger D. (2009). Adult reversal of cognitive phenotypes in neurodevelopmental disorders. *J Neurodev Disord*, 1(2):150-7.

Simpson AJ, Camargo AA. (1998). Evolution and the inevitability of human cancer. *Semin Cancer Biol,* 8(6):439-45.

Sirota M, Schaub MA, Batzoglou S, Robinson WH, Butte AJ. (2009). Autoimmune disease classification by inverse association with SNP alleles. *PLoS Genet,* 5(12):e1000792.

Skolnick J, Brylinski M. (2009). FINDSITE: a combined evolution/structure-based approach to protein function prediction. *Brief Bioinform,* 10(4):378-91.

Slater R, Bishop NE. (2006). Genetic structure and evolution of the Vps25 family, a yeast ESCRT-II component. *BMC Evol Biol,* 6:59.

Slavotinek AM. (2008). Novel microdeletion syndromes detected by chromosome microarrays. *Hum Genet,* 124(1):1-17.

Sleator RD, Walsh P. (2010). An overview of in silico protein function prediction. *Arch Microbiol,* 192(3):151-5.

Smith NG, Eyre-Walker A. (2003). Human disease genes: patterns and predictions. *Gene,* 318:169-175.

Smith CL, Bolton A, Nguyen G. (2010). Genomic and epigenomic instability, fragile sites, schizophrenia and autism. *Curr Genomics,* 11(6):447-69.

Smirlis D, Soteriadou K. (2011). Trypanosomatid apoptosis: 'apoptosis' without the canonical regulators. *Virulence,* 2(3):253-6.

Spradling A, Ganetsky B, Hieter P, Johnston M, Olson M, et al. (2006). New roles for model genetic organisms in understanding and treating human disease: report from the 2006 Genetics Society of America meeting. *Genetics,* 172(4):2025-32.

Sriram G, Martinez JA, McCabe ER, Liao JC, Dipple KM. (2005). Single-gene disorders: what role could moonlighting enzymes play? *Am J Hum Genet,* 76(6):911-24.

Srivastava M, Simakov O, Chapman J, Fahey B, Gauthier ME, et al. (2010). The Amphimedon queenslandica genome and the evolution of animal complexity. *Nature,* 466(7307):720-6.

Stankiewicz P, Lupski JR. (2002). Molecular-evolutionary mechanisms for genomic disorders. *Curr Opin Genet Dev,* 12(3):312-9.

Stankiewicz P, Lupski JR. (2010). Structural variation in the human genome and its role in disease. *Annu Rev Med,* 61:437-55.

Strachan T, Read AP. (2010). *Human Molecular Genetics* (4th ed.), Garland Science, London, UK.

Stratton MR, Campbell PJ, Futreal PA. (2009). The cancer genome. *Nature,* 458(7239):719-24.

Sullivan PF, Kendler KS, Neale MC. (2003). Schizophrenia as a complex trait: evidence from a meta-analysis of twin studies. *Arch Gen Psychiatry,* 60(12):1187-92.

Sullivan P. (2011). Don't give up on GWAS. *Mol Psychiatry,* 17(1):203.

Sun J, Jia P, Fanous AH, van den Oord E, Chen X, et al. (2010). Schizophrenia gene networks and pathways and their applications for novel candidate gene selection. *PLoS One,* 5(6):e11351.

Tabarés-Seisdedos R, Dumont N, Baudot A, Valderas JM, Climent J, et al. (2011). No paradox, no progress: inverse cancer comorbidity in people with other complex diseases. *Lancet Oncol,* 12(6):604-8.

Tarpey PS, Smith R, Pleasance E, Whibley A, Edkins S, et al. (2009). A systematic, large-scale resequencing screen of X-chromosome coding exons in mental retardation. *Nat Genet,* 41(5):535-43.

Tassone F, De Rubeis S, Carosi C, La Fata G, Serpa G, et al. (2011). Differential usage of transcriptional start sites and polyadenylation sites in FMR1 premutation alleles. *Nucleic Acids Res,* 39(14):6172-85.

Tayebati SK. (2006). Animal models of cognitive dysfunction. *Mech Ageing Dev,* 127(2):100-8.

Theisen A, Schaffer LG. (2010) Disorders caused by chromosome abnormalities. *App Clin Genet* 3, 159-174.

Thomas PD, Kejariwal A, Campbell MJ, Mi H, Diemer K, et al. (2003). PANTHER: a browsable database of gene products. *Nucleic Acids Res,* 31(1):334-41.

Thomas PD, Kejariwal A. (2004). Coding single-nucleotide polymorphisms associated with complex vs. Mendelian disease: evolutionary evidence for differences in molecular effects. *Proc Natl Acad Sci USA,* 101(43):15398-403.

Thornton JW, DeSalle R. (2000). Gene family evolution and homology: genomics meets phylogenetics. *Annu Rev Genomics Hum Genet,* 1:41-73.

Tiffin N, Adie E, Turner F, Brunner HG, van Driel MA. (2006). Computational disease gene identification: a concert of methods prioritizes type 2 diabetes and obesity candidate genes. *Nucleic Acids Res,* 34(10):3067-81.

Todd AE, Orengo CA, Thornton JM. (2001). Evolution of function in protein superfamilies. *J Mol Biol,* 307(4):1113-43.

Tordjman S, Drapier D, Bonnot O, Graignic R, Fortes S, et al. (2007). Animal models relevant to schizophrenia and autism: validity and limitations. *Behav Genet,* 37(1):61-78.

Torkamani A, Dean B, Schork NJ, Thomas EA. (2010). Coexpression network analysis of neural tissue reveals perturbations in developmental processes in schizophrenia. *Genome Res,* 20(4):403-12.

Touw IP, Erkeland SJ. (2007). Retroviral insertion mutagenesis in mice as a comparative oncogenomics tool to identify disease genes in human leukemia. *Mol Ther,* 15(1):13-9.

Tranchevent LC, Capdevila FB, Nitsch D, De Moor B, De Causmaecker P, et al. (2011). A guide to web tools to prioritize candidate genes. *Brief Bioinform,* 12(1):22-32.

Tu Z, Wang L, Xu M, Zhou X, Chen T, et al. (2006). Further understanding human disease genes by comparing with housekeeping genes and other genes. *BMC Genomics,* 7:31.

Tuppen HA, Blakely EL, Turnbull DM, Taylor RW. (2010). Mitochondrial DNA mutations and human disease. *Biochim Biophys Acta,* 1797(2):113-28.

Ule J, Darnell RB. (2007). Functional and mechanistic insights from genome-wide studies of splicing regulation in the brain. *Adv Exp Med Biol,* 623:148-60.

Uversky VN. (2009). Intrinsic disorder in proteins associated with neurodegenerative diseases. *Front Biosci,* 14: 5188-5238.

Uversky VN, Oldfield CJ, Dunker AK. (2008). Intrinsically disordered proteins in human diseases: introducing the D2 concept. *Annu Rev Biophys,* 37:215-46.

Uversky VN, Oldfield CJ, Midic U, Xie H, Xue B, et al. (2009). Unfoldomics of human diseases: linking protein intrinsic disorder with diseases. *BMC Genomics,* 10(Suppl 1):S7.

Vigouroux C, Bonne G (2002). Laminopathies: one Gene, two proteins, five diseases, In: *Madame Curie Bioscience Database*. Landes Bioscience, Austin, Texas, USA. Retrieved from http://www.ncbi.nlm.nih.gov/books/NBK6151

Vijaya-Lakshmi AN, Ramana MV, Vijayashree B, Ahuja YR, Sharma G. (1999). Detection of influenza virus induced DNA damage by comet assay. *Mutat Res,* 442(1):53-8.

Visscher PM, Hill WG, Wray NR. (2008). Heritability in the genomics era. *Nat Rev Genet,* 9(4):255-66.

Visscher PM, Goddard ME, Derks EM, Wray NR. (2011). Evidence-based psychiatric genetics, AKA the false dichotomy between common and rare variant hypotheses. *Mol Psychiatry,* (Jun 14) in press.

Vockley J, Rinaldo P, Bennett MJ, Matern D, Vladutiu GD. (2000). Synergistic heterozygosity: disease resulting from multiple partial defects in one or more metabolic pathways. *Mol Genet Metab,* 71(1-2):10-8.

Voight BF, Kudaravalli S, Wen X, Pritchard JK. (2006). A map of recent positive selection in the human genome. *PLoS Biol,* 4(3):e72.

Voineagu I, Wang X, Johnston P, Lowe JK, Tian Y, et al. (2011). Transcriptomic analysis of autistic brain reveals convergent molecular pathology. *Nature,* 474(7351):380-4.

Walker MG, Volkmuth W, Klingler TM. (1999). Pharmaceutical target discovery using Guilt-by-Association: schizophrenia and Parkinson's disease genes. *Proc Int Conf Intell Syst Mol Biol,* 1999:282-6.

Wallace MR, Andersen LB, Saulino AM, Gregory PE, Glover TW, et al. (1991). A de novo Alu insertion results in neurofibromatosis type 1. *Nature,* 353(6347):864-6.

Wang J, Cao Z, Zhao L, Li S. (2011). Novel strategies for drug discovery based on intrinsically disordered proteins (IDPs). *Int J Mol Sci,* 12(5):3205-19.

Ward AJ, Cooper TA. (2010). The pathobiology of splicing. *J Pathol,* 220(2):152-63.

Waring SC, Rosenberg RN. (2008). Genome-wide association studies in Alzheimer disease. *Arch Neurol,* 65(3):329-34.

Watson JD, Laskowski RA, Thornton JM. (2005). Predicting protein function from sequence and structural data. *Curr Opin Struct Biol,* 15(3):275-84.

Westerlund M, Hoffer B, Olson L. (2010). Parkinson's disease: Exit toxins, enter genetics. *Prog Neurobiol,* 90(2):146-56.

Williams RS, Boeckeler K, Gräf R, Müller-Taubenberger A, Li Z, et al. (2006). Towards a molecular understanding of human diseases using *Dictyostelium discoideum*. *Trends Mol Med,* 12(9):415-24.

Williams TN. (2006). Red blood cell defects and malaria. *Mol Biochem Parasitol,* 149(2):121-7.

Wirdefeldt K, Gatz M, Reynolds CA, Prescott CA, Pedersen NL. (2011). Heritability of Parkinson disease in Swedish twins: a longitudinal study. *Neurobiol Aging,* 32(10):1923.e1-8.

World Health Organization (1993). *The ICD-10 Classification of Mental and Behavioural Disorders: Clinical Descriptions and Diagnostic Guidelines,* World Health Organization, Geneva, Switzerland.

Wray NR, Goddard ME, Visscher PM. (2008). Prediction of individual genetic risk of complex disease. *Curr Opin Genet Dev,* 18(3):257-63.

Yang F, Babak T, Shendure J, Disteche CM. (2010). Global survey of escape from X inactivation by RNA-sequencing in mouse. *Genome Res,* 20(5):614-22.

Yatsenko SA, Brundage EK, Roney EK, Cheung SW, Chinault AC, et al. (2009). Molecular mechanisms for subtelomeric rearrangements associated with the 9q34.3 microdeletion syndrome. *Hum Mol Genet,* 18(11):1924-36.

Yook SH, Oltvai ZN, Barabási AL. (2004). Functional and topological characterization of protein interaction networks. *Proteomics,* 4(4):928-42.

Young JW, Zhou X, Geyer MA. (2010). Animal models of schizophrenia. *Curr Top Behav Neurosci,* 4:391-433.

Zhu J, Xiao H, Shen X, Wang J, Zou J, et al. (2010). Viewing cancer genes from co-evolving gene modules. *Bioinformatics,* 26(7):919-24.

Zwijnenburg PJ, Meijers-Heijboer H, Boomsma DI. (2010). Identical but not the same: the value of discordant monozygotic twins in genetic research. *Am J Med Genet B Neuropsychiatr Genet,* 153B(6):1134-49.

2

Anhydrous and Hydrated Protein Models Derived from High-Resolution and Low-Resolution Techniques

Helmut Durchschlag[1] and Peter Zipper[2]
[1]*Institute of Biophysics and Physical Biochemistry, University of Regensburg, Regensburg,*
[2]*Physical Chemistry, Institute of Chemistry, University of Graz, Graz,*
[1]*Germany,*
[2]*Austria*

1. Introduction

High-resolution crystallography provides information on the precise 3D structure of proteins or other (bio)macromolecules, including ligands and several water molecules. In the case of small proteins, NMR techniques may be equally purposive. Reconstructions from cryo-electron microscopy may also supply precise anhydrous models. The most intimate structural details of proteins can be visualized from application of these high-resolution techniques (Creighton, 2010a, 2010b; Serdyuk et al., 2007).

In general, however, only a small fraction of preferentially bound water molecules is identified by crystallographic techniques, owing to insufficient resolution and hydrogen bond network building. Discrete waters should have hydrogen bonded contact(s) to other solvent molecules or to protein; poorly placed waters tend to drift away during refinement (Rupp, 2010). Further deficiencies/errors in crystallographic work are due to missing parts (amino acid (AA) residues) and the occurrence of various radiation damages (Ravelli & Garman, 2006).

By contrast, the data from low-resolution solution techniques, such as small-angle X-ray scattering (SAXS) and hydrodynamics (analytical ultracentrifugation, viscometry etc.), inherently contain hydration and yield hydrated protein models (Durchschlag et al., 2007; Zipper & Durchschlag, 2010b). For many reasons, knowledge of hydration is essential for understanding the behaviour of proteins in solution and manifold interactions.

Water is important for structure, stability, dynamics, and function of native proteins; it is also involved in guiding protein folding, and, consequently, needs to be involved in protein structure predictions and modelling of folding pathways (Levy & Onuchic, 2006; Papoian et al., 2004). Minimizing the number of hydrophobic side-chains exposed to aqueous solvent is a major driving force behind protein structure formation. In a typical protein, a tightly packed core contains more than 80 % of the non-polar side chains and water molecules are generally excluded from protein interiors. There is a clear correlation between the hydrophobicities of AA residues and their tendency to occur in the interior (Creighton, 2010b). Water soluble proteins have all ionized groups on the surface, exposed to the solvent. Polar groups tend to be paired in hydrogen bonds. Fixed, preferentially bound water molecules occur in positions where they can build hydrogen bonds to polar groups. Overall, to some extent both bound and free waters are of relevance for the structure of proteins and their formation. Only the existence of water leads to the distinction between hydrophobic and hydrophilic parts. Water is the 'lubricant of life'.

Biophysically relevant hydration details are mandatorily required in certain cases: understanding protein interactions as precondition for flexibility, dynamics and functionality (Creighton, 2010a, 2010b; Serdyuk et al., 2007); very precise comparisons of high-resolution 3D structures and molecular parameters with data from solution techniques (Durchschlag et al., 2007; Durchschlag & Zipper, 2008; Zipper & Durchschlag, 2010b); gaining insight into radiation damage events (Durchschlag et al., 2003; Durchschlag & Zipper, 2007); construction of tailor-made nano-compounds in context with drug-design projects and the development of functionalized surfaces and polymers by mimicking proteins (Durchschlag & Zipper, 2008); improvement/check of crystallographic/NMR data with regard to hydration waters (Durchschlag & Zipper, 2003).

Experimental determinations of volume, surface, and hydration properties of proteins by SAXS or other low-resolution techniques turned out to be rather inaccurate, whereas the hydration numbers for individual AA residues, derived from NMR spectroscopy or thermodynamic considerations, seem to be rather precise (Durchschlag & Zipper, 2001, 2008). For different pH values, of course, different hydration numbers for the AA residues have been found.

Unlike experimental data, calculations of volume and surface properties of simple and complex proteins proved to yield reliable results, if based on the properties of the molecular constituents and the coordinates (Durchschlag & Zipper, 2005, 2008). Modern analytical procedures and programs yield the anhydrous volume (van der Waals volume), summing up the contributions of the constituents. For analyzing the surface area of proteins, rolling-ball mechanisms were most effective, characterizing either an anhydrous or a hydrated protein. Prediction of the values for hydrated volume and surface requires assumptions/estimations/findings for the amount of water bound to the protein, e.g. application of a shell model, blowing up (surface) AA residues, and use of individual waters. For the precise prediction of structural and hydrodynamic parameters (volume, surface, radius of gyration, sedimentation and diffusion coefficients, intrinsic viscosity etc.) realistic assumptions or estimations are sufficient, whereas visualization of biophysically realistic hydrated protein models obviously necessitates more sophisticated, advanced modelling techniques.

A critical inspection of anhydrous and hydrated protein models obtained by crystallography with models derived from quite different experimental techniques and calculation approaches allows a scrutinized comparison of the models under analysis. Among a variety of problems, the amount of hydration and the position of the individual water molecules turned out to be the most crucial points. To meet this challenge, a variety of techniques and approaches were examined and both models and molecular parameters were analyzed: (i)

Conventional and *ab initio* modelling approaches signify satisfactory agreement between crystal- and SAXS-based protein models, provided hydration contributions and other precautions are taken into account (Durchschlag et al., 2007). (ii) Recourse to crystallographic or model data also allows scattering and hydrodynamic modelling; in the case of multibead structures novel modelling refinements (e.g., efficient bead reductions) have to be adopted (Zipper et al., 2005; Zipper & Durchschlag, 2007, 2010a, 2010b). (iii) The creation of hydrated models from cryo-electron microscopy data necessitates qualified assumptions regarding hydration, e.g. in terms of voxel densities (Zipper & Durchschlag, 2002b). (iv) Combining the exact surface topography (molecular dot surface; derived from atomic or amino acid coordinates of proteins or appropriate models) and our recent hydration algorithms (program HYDCRYST) allows the prediction of individual water molecules preferentially bound to certain amino acid residues (Durchschlag & Zipper, 2001, 2002a, 2002b, 2003, 2004, 2005, 2006, 2008; Zipper & Durchschlag, 2002a, 2002b).

2. Methods

2.1 Coarse-grained models

Sphere (S), prolate and oblate ellipsoids of revolution (PE, OE) of different axial ratios, and hollow spheres (HS) of different hollowness may serve as approximations for simple anhydrous protein structures. Selected structures were modelled by assemblies of equal-sized densely packed beads (Zipper & Durchschlag, 2010a). By variation of the bead radius (r_b), the number of beads (N_b), i.e. the extent of the reduction process, can be varied systematically. The coordinates of the beads were generated by an in-house program.

2.2 Proteins: Recourse to experimental results and data banks

SAXS experiments yield scattering intensity profiles, $I(h)$, and pair-distance distribution functions, $p(r)$, in addition to a variety of molecular parameters such as radius of gyration, R_G, hydrated volume, V, and maximum particle diameter, d_{max} (Glatter & Kratky, 1982). Shape reconstructions may be obtained by conventional modelling or advanced *ab initio* modelling approaches, preferably by programs based on simulated annealing or on a genetic algorithm (Zipper et al., 2005). Hydrodynamic properties such as sedimentation coefficient, s, translational diffusion coefficient, D, and intrinsic viscosity, $[\eta]$, are determined by analytical ultracentrifugation and viscometry, respectively. Atomic coordinates and masses of proteins were obtained from the PDB and SWISS-PROT data banks (Berman et al., 2000; Boeckmann et al., 2003). For hydration predictions by HYDCRYST, only the coordinates of the protein were taken, i.e. in this case all crystallographically found waters were discarded.

2.3 Model construction, data reduction, and prediction of structures and molecular parameters

Size, shape and properties of simple and complex proteins can be calculated by bead modelling procedures: in context with solution techniques, but also in the case of crystallographic and EM data (Zipper et al., 2005; Byron, 2008). Pilot tests applying whole-body approaches yield rough estimates of scattering and hydrodynamic molecular parameters and allow estimates of partial specific volume, molecular volume and overall hydration (Durchschlag & Zipper, 2005). Among the more advanced approaches applied, Debye modelling (Glatter & Kratky, 1982) and the program CRYSOL (Svergun et al., 1995) may be used for the calculation of scattering profiles from (atomic) coordinates. For hydrodynamic modelling, the HYDRO program suite (consisting of several programs and many modern adaptations; e.g., programs HYDRO++ and HYDROPRO) (García de la Torre et al., 1994, 2000, 2007, 2010; Ortega et al., 2011a, 2011b), and ZENO (Kang et al., 2004; Mansfield & Douglas, 2008) should be emphasized. Filling-model strategies (instead of shell models usually applied for hydrodynamic modelling) have to be used for analyzing both scattering and hydrodynamic quantities.

In a broad range of circumstances, models composed of a multitude of beads have to be handled, in particular when resorting to (atomic) coordinates of huge macromolecules or if the biophysically relevant fine structure of hydrated models is required. Various programs, e.g. AtoB (Byron, 1997), SOMO (Brookes et al., 2010a, 2010b; Rai et al., 2005), PDB2AT, PDB2AM, and MAP2GRID (Zipper & Durchschlag, 2007) have been developed to transform crystallographic/NMR data information to bead models (unreduced initial models based on atoms, atomic groups, or amino acid residues, and reduced models at certain reduction levels). The hydration program HYDCRYST allows the efficient prediction of individual water molecules preferentially bound to proteins provided the accessible surface area has been calculated previously by a surface calculation program. Several complementary tools may be required, e.g., for conversion between data formats, calculation of scattering functions, or for a correct visualization of reduced and/or hydrated structures by RASMOL (Sayle & Milner-White, 1995).

2.4 Calculation of volumes and surfaces

Volume and surface properties of protein molecules can be calculated using crystallographic data sets as those deposited in the PDB. The molecular volumes thus obtained are either anhydrous volumes or, because of scarce waters, poorly hydrated volumes (Durchschlag & Zipper, 2008). Surface characteristics of proteins are obtained by using analytical surface calculation programs such as SIMS based on the rolling-ball strategy (Vorobjev & Hermans, 1997), yielding molecular surface and solvent-accessible surface areas (i.e. anhydrous and hydrated surfaces), in addition to a smooth 'molecular dot surface' required for values for the solvent-excluded volume and appliance of advanced hydration modelling strategies. The program SIMS may be applied either to the atomic coordinates or newly-created coordinates (e.g., gravity centres of AA residues) (Durchschlag & Zipper, 2008).

2.5 Calculation of hydration

The water molecules bound preferentially to proteins have properties different from those of the bulk water (higher order, lower mobility, higher density) (Durchschlag & Zipper, 2003). For roughly estimating the overall hydration (Durchschlag & Zipper, 2004, 2008), the hydration numbers found for the individual AA residues (Kuntz, 1971) may be used.

Knowledge of the exact anhydrous surface topography in terms of dot surface points, as obtained by SIMS, however, enables usage of advanced hydration algorithms. The normal vectors at these anhydrous dot surface points allow the creation of a huge amount of hypothetical points for potential positions of water molecules; these waters are located at the normal distance above the anhydrous protein surface, i.e. at a distance usually corresponding to the water radius (probe radius). For the selection of preferential positions for bound water molecules, a new version of our in-house program HYDCRYST (now including also essentials of the related program HYDMODEL) is applied (creation of models based on the initial atomic coordinates of the protein or reduced models derived from the coordinates of AA residues) (Durchschlag & Zipper, 2001, 2002a, 2003). The selected waters, assigned to the accessible AA residues, are then attached to the dry protein models. Overall, the method is based on geometrical and energetic constraints, owing to the placing of definite hydration numbers to definite AA residues. The extent of hydration can be modulated step by step by several input parameters, in particular by variation of the dot density and a scaling factor acting on the hydration numbers used. Thereby different extents of water binding (minimum, medium, and maximum hydration) may be simulated.

2.6 Calculation of SAXS functions and of structural and hydrodynamic parameters

Radius of gyration, R_G, and the pair-distance distribution function, $p(r)$, of the protein models were calculated from the radii and coordinates of the constituent beads and the bead volume as statistical weight. The $p(r)$ function gives the relative number of distances of two points inside a particle as a function of the distance r. Calculation approaches and the prediction of structural and hydrodynamic parameters have been described in detail previously (Zipper et al., 2005).

3. Results

In the first instance, whole-body models of different shape were selected, to test the applicability of the reduction and calculation approaches applied. Based on these findings, anhydrous models for many proteins were constructed and their predicted structural and hydrodynamic properties were compared to experimental data. Because of obvious deficiencies of the anhydrous (dry) protein models, sophisticated hydration strategies had to be developed in the following, to account for the hydration contributions of the protein structures in solution. Application of our advanced hydration algorithms to various proteins revealed the effectiveness of the approaches applied.

3.1 Coarse-grained models

Models of quite different shape were selected, to prove the effectiveness of the approaches under consideration. Fig. 1 depicts a few illustrative examples: a sphere (S), a hollow sphere (HS), a prolate ellipsoid (PE) and an oblate ellipsoid (OE), composed of a multitude of beads, together with a set of the reduced models. As may be expected, the distance distribution functions, $p(r)$, of the different model structures vary considerably (Fig. 2), with respect to their form, position of maximum, and particle diameter. However, the profiles of unreduced and reduced models coincide nearly completely, proving the applicability of the model reduction process applied. As may be taken from the structural and hydrodynamic parameters presented in Table 1, the comparison of whole-body and multibead models indicates good accordance (data for R_G, D (and therefore also for s), and $[\eta]$). As shown previously (Zipper & Durchschlag, 2010a), choice of $[\eta]_{RVC}$ represents a reasonable approximation for the value of the intrinsic viscosity.

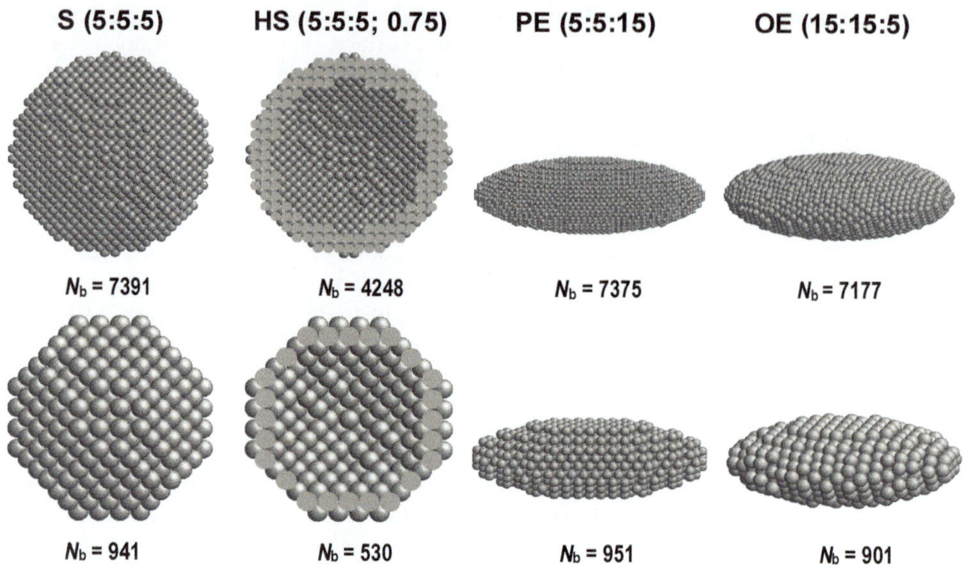

Fig. 1. Selected space-filling multibead models for sphere (S), hollow sphere (HS) with r_i/r_o = 0.75, and prolate and oblate ellipsoids of revolution (PE, OE). Each model is represented by a number of beads (N_b), depending on size, hollowness and axial ratio of the model under consideration. By increasing the bead radius of the model, N_b can be reduced considerably.

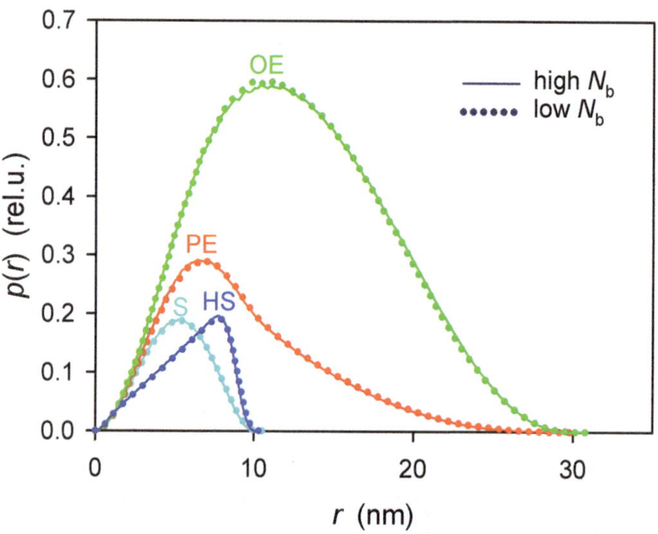

Fig. 2. Distance distribution functions $p(r)$ of the multibead models (S, HS, PE, OE) shown in Fig. 1. The profiles represent initial and reduced numbers of beads (high and low N_b); their integral values are proportional to the overall model volumes.

Shape	$a:b:c$ (nm)	N_b	r_b (nm)	$V \times 10^{-3}$ (nm³)	R_G (nm)	$D \times 10^7$ (cm²/s)	$[\eta]_{NVC}$ (cm³/g)	$[\eta]_{FVC}$ (cm³/g)	$[\eta]_{RVC}$ (cm³/g)	$[\eta]_{AVC}$[a] (cm³/g)
S	**5:5:5, WB**			0.5236	3.873	4.286	2.639[b]			
	–, MB	10777 – 65	0.20 – 1.00	0.357 ± 0.027	3.90 ± 0.10	4.46 ± 0.35[c]	2.58 ± 0.22[c]	4.38 ± 0.36[c]	2.81 ± 0.23[c]	
		7391 – 65	0.225 – 1.00			4.32 ± 0.15[d]	2.58 ± 0.23[d]	4.38 ± 0.37[d]	2.83 ± 0.23[d]	
		5577 – 65	0.25 – 1.00							2.87 ± 0.26
HS	**5:5:5, WB, r_i/r_o=0.75**			0.3027	4.449	4.286	4.564[e]			
	–, MB	6300 - 104	0.20 – 0.80	0.212 ± 0.009	4.49 ± 0.04	4.28 ± 0.04[d]	4.58 ± 0.13[d]	6.43 ±0.16[d]	4.81 ± 0.14[d]	4.87 ± 0.13
PE	**5:5:15, WB**			1.5708	7.416	2.671	3.889[f]			
	–, MB	9395 – 233	0.30 – 1.00	1.065 ± 0.030	7.42 ± 0.09	2.69 ± 0.03[d]	3.82 ± 0.12[d]	5.61 ± 0.17[d]	4.00 ± 0.10[d]	
		5949 – 233	0.35 – 1.00							4.69 ± 0.16
OE	**15:15:5, WB**			4.7124	9.747	1.865	3.620[g]			
	–, MB	10007 – 757	0.425 – 1.00	3.213 ± 0.040	9.77 ± 0.05	1.87 ± 0.01[d]	3.60 ± 0.06[d]	5.40 ± 0.08[d]	3.74 ± 0.05[d]	
		7177 – 757	0.475 – 1.00							4.08 ± 0.07

S: sphere, HS: hollow sphere, PE: prolate ellipsoid of revolution, OE: oblate ellipsoid of revolution, WB: whole-body, MB: multibead, N_b: number of beads, r_b: radius of beads, V: total volume, R_G: radius of gyration, D: translational diffusion coefficient, $[\eta]$: intrinsic viscosity, NVC: no volume correction, FVC: full volume correction, RVC: reduced volume correction, AVC: adjusted volume correction, M: molar mass.

[a] All values were computed by HYDRO++9beta.
[b] Based on M = 298.8 kg/mol.
[c] Obtained by executing HYDRO in single-precision mode.
[d] Obtained by executing HYDRO in double-precision mode.
[e] Based on M = 172.7 kg/mol.
[f] Based on M = 896.3 kg/mol.
[g] Based on M = 2688.8 kg/mol.

Table 1. Comparison of structural and hydrodynamic parameters of whole-body and multi-bead models shown in Fig. 1.

3.2 Anhydrous and hydrated protein models

Bovine pancreatic trypsin inhibitor (BPTI), a nonconjugated protein, was chosen as a small model protein, to demonstrate different types of anhydrous models and descriptions of reduction procedures to be applied in scattering and hydrodynamic modelling.

Fig. 3 compares the initial anhydrous model, the unreduced BPTI model based on atomic coordinates (A), with several approaches for reduced models (B-E). The most straightforward types of reduced models are based on AA residue coordinates, either on the entire residue or on parts of it (such as main chain and side chain, a concept adopted from SOMO). Procedures, such as SOMO and AtoB, are used in the program UltraScan II (Brookes et al., 2010a, 2010b; Demeler, 2005). An inspection of the $p(r)$ functions (Fig. 4) clearly proves that the straightforward procedures result in perfect agreement with the initial model, while the applied versions of SOMO, and in particular of AtoB, lead to remarkable discrepancies. This is partly due to the fast, but unprecise modus operandi for calculation of $p(r)$ in the present versions of UltraScan II (the errors caused by neglecting the bead radii increase with decreasing bead numbers). However, part of the disagreement of the $p(r)$ functions of SOMO and AtoB models with those of the initial and straightforward reduction procedures seems to be caused by the models themselves. The models created by SOMO and AtoB tend to be rather artificial and far from space-filling (Fig. 3) because of avoidance of overlapping beads. Our straightforward approaches, on the other hand, do not prevent the occurrence of bead overlaps. Even overlapping unequal-sized beads can be used for hydrodynamic modelling, either by applying a special expression for the interaction tensor in HYDRO (Zipper & Durchschlag, 1997, 1999) or by using ZENO. However, in principle, all types of anhydrous model constructions may be used as starting points for the creation of hydrated models. For obvious reasons, however, in the following we will restrict our considerations to our straightforward in-house approaches for the model reduction process.

The 30 S ribosomal subunit of *Thermus thermophilus* was used as a representative of a conjugated protein. It is a nucleoprotein composed of protein and ribonucleic acid moieties. As may be taken from Fig. 5, also in this case modelling may be achieved on the level of atomic coordinates (unreduced model: 51792 beads) and by applying AA and nucleotide residue coordinates (reduced model: 3917 beads) as well. Different approaches for the reduction process (running mean and grid reductions) and varying input parameters were applied, to test the applicability of the reduction procedures. The overall impression of the images obtained for all the models is satisfactory (Fig. 6). The comparison of the $p(r)$ functions, however, reveals that use of the cubic grid approach and equal-sized beads on lattice points (CLE) may fail (Fig. 7). And the same conclusion can be drawn from a meticulous inspection of the molecular parameters listed in Table 2. Again, the CLE variant rather leads to erroneous predictions of structural and hydrodynamic parameters.

Fig. 8 illustrates the steps of the hydration approach HYDRCRYST, again applied to the model protein BPTI. Starting from the anhydrous protein model (A), a myriad of dot surface points is created (B). The normal vectors at these points are used for the localization of potential water points (C). Selecting water molecules on the protein surface by HYDCRYST yields the water molecules to be tracked on the protein surface, eventually establishing a realistic hydrated protein model, revealing, however, no complete water shell (D). In comparison to this calculated model, the structure based on crystal data demonstrates only a few bound water molecules (E).

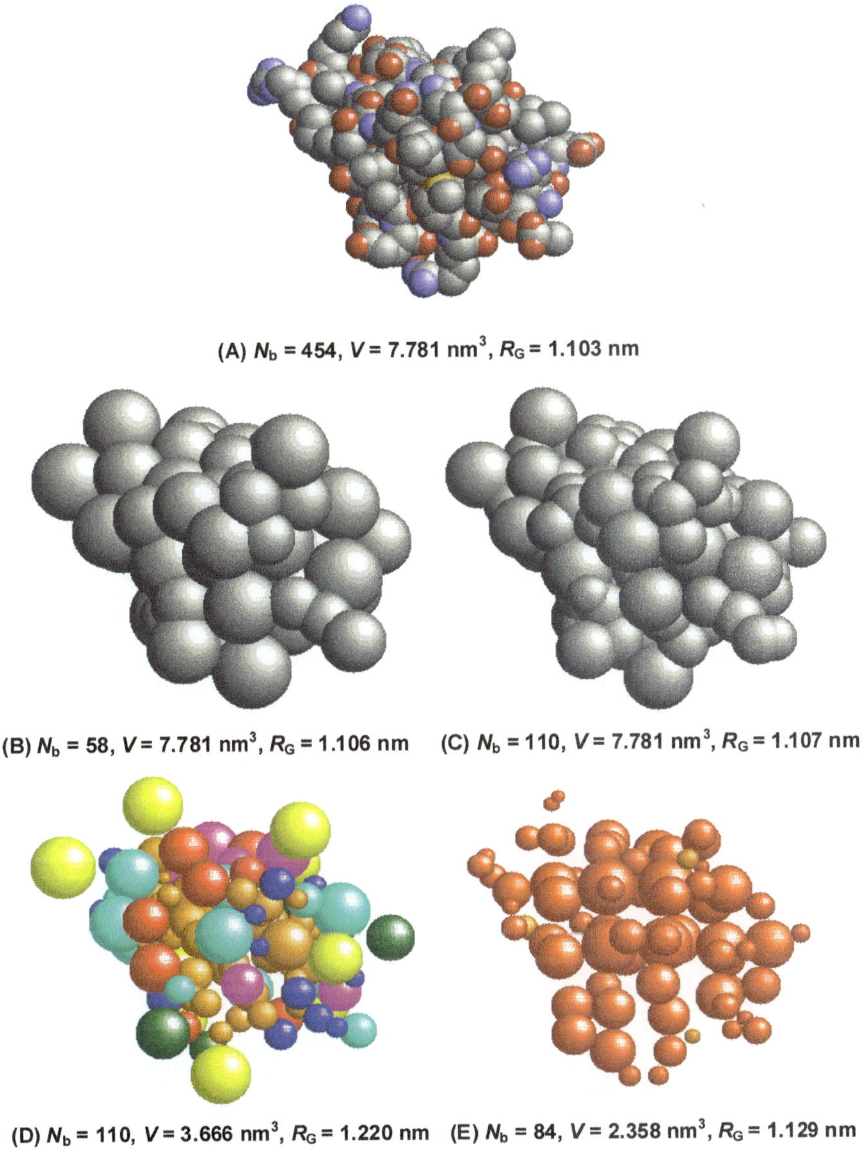

Fig. 3. Different types of anhydrous models for bovine pancreatic trypsin inhibitor (4PTI), created by different approaches: (A) Unreduced model for the protein based on atomic coordinates; the basic atoms are given in CPK colors, (B) Reduced model based on whole AA residues, (C) Reduced model based on AA residues split into a main-chain and a side-chain moiety, (D) SOMO model created by UltraScan II, (E) AtoB model created by UltraScan II. The given values for V and R_G were calculated from the coordinates and radii of beads.

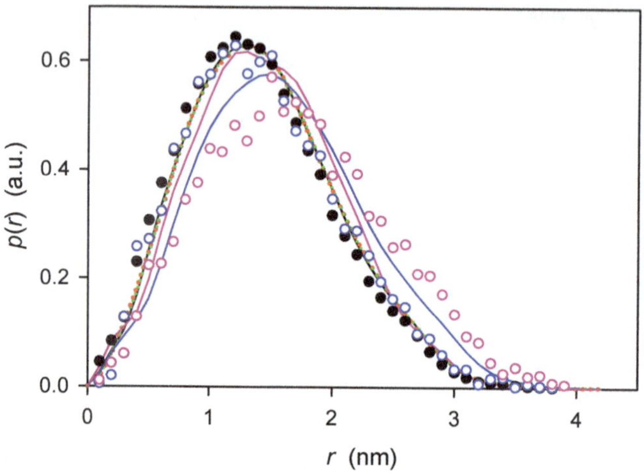

Fig. 4. Distance distribution functions $p(r)$ of the different types of anhydrous models for bovine pancreatic trypsin inhibitor (4PTI) shown in Fig. 3. The black line and the dotted red and green curves are the profiles obtained for the unreduced model and the models reduced to whole (red) or split (green) AA residues; the blue and pink lines represent the profiles calculated for the SOMO (blue) and AtoB (pink) models created by UltraScan II. These profiles were calculated by means of in-house programs from the coordinates and radii of the beads. The circles illustrate the $p(r)$ profiles of the unreduced model (black) and of the SOMO (blue) and AtoB (pink) model as provided directly by UltraScan II; these profiles were calculated from the coordinates of the beads only.

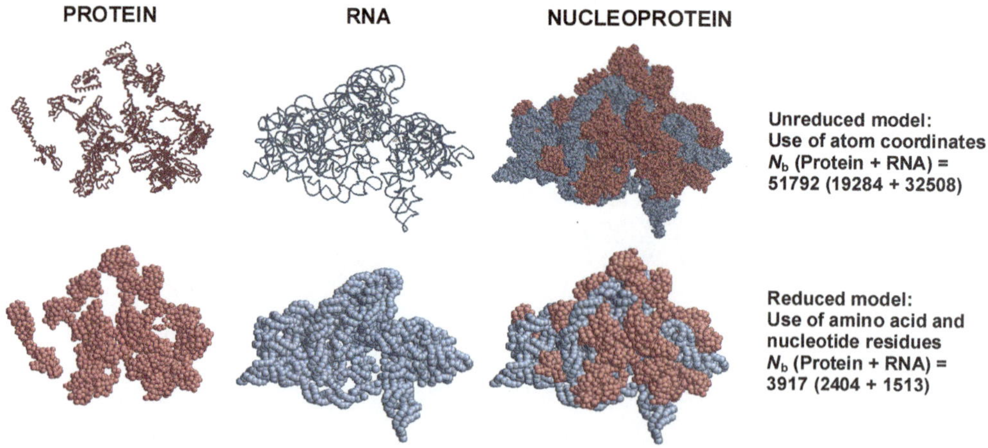

Fig. 5. Anhydrous models for the *Thermus thermophilus* 30 S ribosomal subunit (1FJG), a nucleoprotein composed of protein and RNA. The principle constituents (protein moiety in bluetint, and nucleic acid moiety in pinktint) are shown in backbone and space-filling formats, while the nucleoprotein is shown in space-filling format only. The images in the upper row represent the models based on atomic coordinates, while the lower row signifies the models based on the coordinates of AA and nucleotide residues.

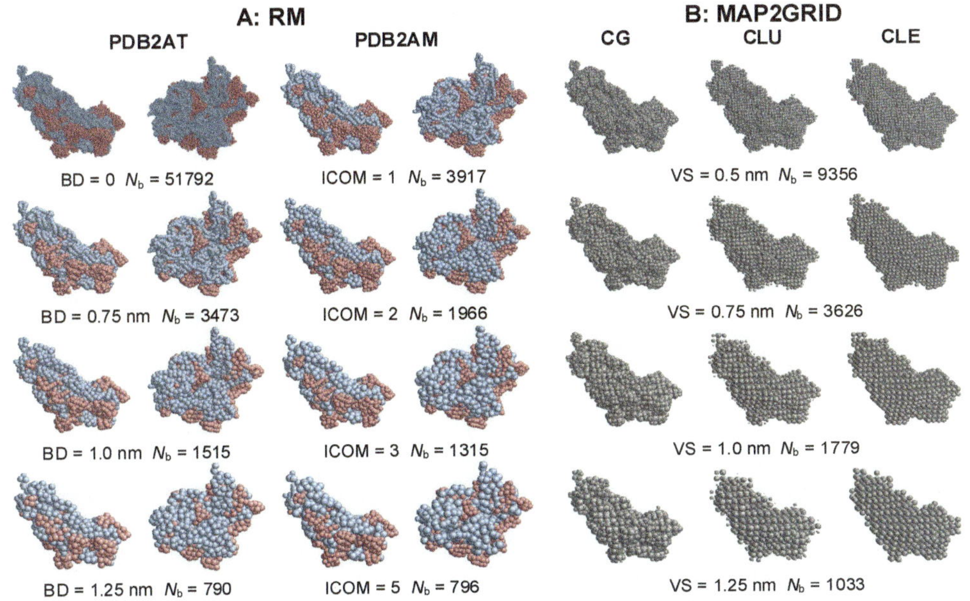

Fig. 6. Selected bead models for the *Thermus thermophilus* 30 S ribosomal subunit (1FJG), obtained by various approaches. (A) Running mean (RM) reductions to the crystal structure, applying different bead diameters BD (program PDB2AT) or compression indices ICOM (program PDB2AM) when transforming the original crystallographic information to various kinds of bead models. The models obtained with BD = 0 or ICOM = 1 correspond to the original crystal structure in atomic coordinates or residue coordinates, respectively. (B) Grid reductions (program MAP2GRID), mapping a given structure into a 3D cubic grid (C) of chosen voxel size (VS) and placing the unequal-sized (U) or equal-sized (E) beads at local centres of gravity (G) or on lattice points (L). Note that CG implicitly means CGU since the G approach is only applicable together with unequal-sized beads.

Model Method/mode	Parameter	N_b	V [a] (nm^3)	R_G [a] (nm)	χ [b]	$D \times 10^7$ [c] (cm^2/s)	$s \times 10^{13}$ [d] (s)	$[\eta]$ [e] (cm^3/g)
Crystal structure,	Initial	51792	800.49	6.652				
anhydrous			485.1	6.652		2.70	31.1	4.20
PDB2AT	BD (nm)							
	0.5	10484	800.49	6.650	0.0044			
			571.5	6.654		2.70	31.1	4.20
	1.0	1515	800.49	6.647	0.0079			
			692.3	6.650		2.71	31.3	4.15
	1.5	465	800.49	6.626	0.0374			
			676.5	6.633		2.75	31.7	3.99
PDB2AM	ICOM							
	1	3917	800.62	6.651	0.0050			
			664.5	6.661		2.70	31.1	4.21

	3	1315	800.62	6.646	0.0080			
			699.7	6.650		2.71	31.2	4.15
	5	796	800.62	6.638	0.0189			
			691.9	6.647		2.72	31.4	4.10
MAP2GRID	VS (nm)							
CG	0.5	9356	800.49	6.651	0.0032			
			664.0	6.649		2.69	31.0	4.24
	1.0	1779	800.49	6.650	0.0097			
			728.6	6.643		2.69	31.0	4.25
	1.5	678	800.49	6.640	0.0222			
			751.9	6.635		2.70	31.1	4.21
CLU	0.5	9356	800.49	6.656	0.0111			
			730.5	6.668		2.64	30.4	4.49
	1.0	1779	800.49	6.682	0.0388			
			776.9	6.685		2.58	29.8	4.77
	1.5	678	800.49	6.717	0.0767			
			790.4	6.721		2.52	29.1	5.16
CLE	0.5	9356	800.49	6.658	0.0753			
			779.7	6.744		2.60	30.0	4.71
	1.0	1779	800.49	6.679	0.286			
			799.3	7.013		2.50	28.8	5.27
	1.5	678	800.49	6.711	0.584			
			800.7	7.451		2.38	27.4	6.09

N_b: number of beads, V: total volume, R_G: radius of gyration, χ: goodness of the fit, D: translational diffusion coefficient, s: sedimentation coefficient, $[\eta]$: intrinsic viscosity; PDB2AT: program generating running-mean models by merging as many atoms in sequential order as fit into a bead of given diameter BD, PDB2AM: program generating running-mean models by merging a given number (compression index ICOM) of AA and nucleotide residues in sequential order to one bead, MAP2GRID: program generating bead models by mapping a given model onto a cubic or hexagonal grid of given edge length (voxel size) VS; CG, CLU, CLE: cubic grid models composed of unequal-sized beads placed on local centres of gravity (CG) or on lattice points (CLU) or of equal-sized beads placed on lattice points (CLE).

[a] The values given in the first line were obtained from the reduction program, the values given in the second line were obtained by the ZENO approach. Discrepancies in V are mainly due to the overlap of beads; the discrepancies in R_G for the models reduced with MAP2GRID (calculation mode CLE) result from the neglect of the special weights of the beads of these models by the ZENO approach.

[b] The values were obtained by comparing the calculated $p(r)$ function of the reduced models with the $p(r)$ function of the initial model. For computing the $p(r)$ functions weighting by volume was assumed throughout.

[c] The values are accurate to about $\pm$ 0.03 x 10^{-7} cm^2/s.

[d] The values are accurate to about $\pm$ 0.3 x 10^{-13} s.

[e] The values are accurate usually to about $\pm$ 0.06 cm^3/g, except for the models reduced by MAP2GRID where broader error bands are encountered (for mode CLU up to $\pm$ 0.08 cm^3/g, for mode CLE up to $\pm$ 0.09 cm^3/g).

Table 2. Comparison of structural and hydrodynamic parameters of anhydrous bead models for *Thermus thermophilus* 30 S ribosomal subunit (1FJG), generated by various reduction approaches.

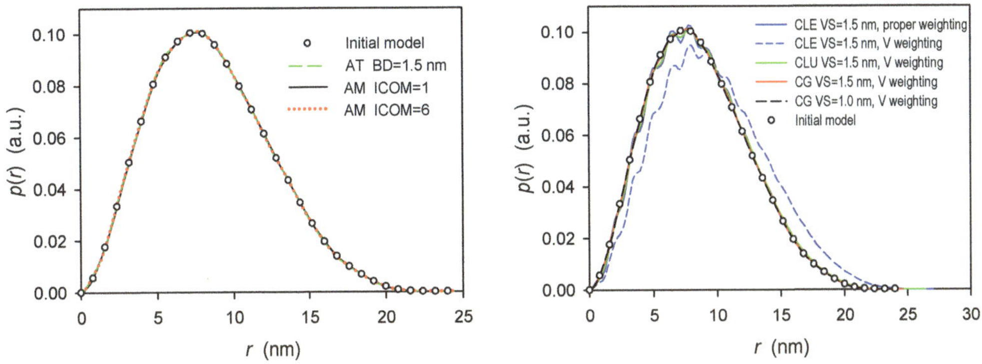

Fig. 7. Comparison of distance distribution functions $p(r)$ of selected unreduced and reduced bead models for the *Thermus thermophilus* 30 S ribosomal subunit (1FJG) as explained and shown in Fig. 6. The term 'V weighting' means that in the calculation of $p(r)$ the beads are weighted according to their volume. This kind of weighting is correct for grid models of type CG and CLU, but obviously not for type CLE .

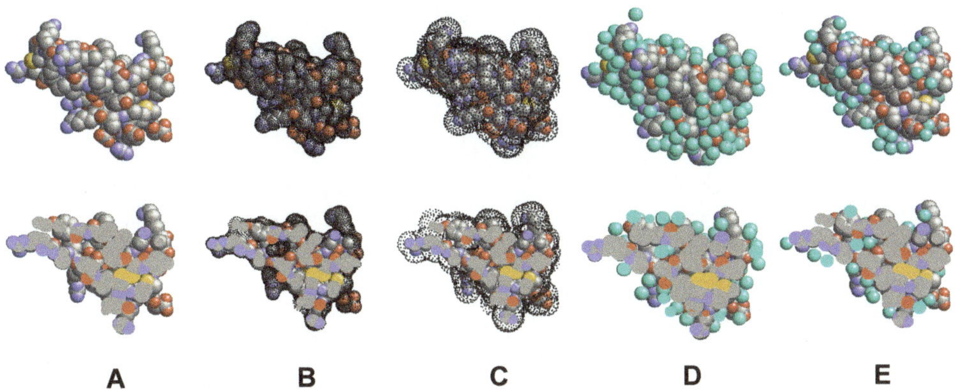

Fig. 8. Space-filling models for anhydrous and hydrated bovine pancreatic trypsin inhibitor (4PTI), together with central slabs: (A) Model for the anhydrous protein based on atomic coordinates; the basic atoms are given in CPK colors. (B) The anhydrous model and dot surface points (in black) created for the anhydrous contour. (C) The anhydrous model and surface points created by HYDCRYST for the contour of potential water points (in black) located at a certain distance from the initial surface points. (D) Model for the hydrated protein as obtained by HYDCRYST; bound waters are displayed in cyan. (E) Model for the hydrated protein as obtained by crystallography; waters are shown in cyan.

Some characteristics of BPTI and the model under discussion are outlined in Table 3, together with the properties of some other selected proteins discussed later on. In the following, our hydration approaches were applied to several proteins differing in size and complexity.

Carbonmonoxide myoglobin has been used as an example of a liganded protein (ligands haem and carbonmonoxide CMO). The hydration approach is demonstrated both for the unreduced model based on atomic coordinates and the reduced model utilizing AA residue coordinates (Fig. 9). Evidently, the approach works in both cases.

Protein	PDB entry	M (kg/mol)	N_{AA}	Type of coordinates (AT or AM)	N_b (anhydrous model)	N_w (HYDCRYST)	N_b (hydrated model)	N_w (crystal)
BPTI	4PTI	6.5 [a]	58	AT	454	125	579	60
Apoferritin	2W0O	464.7 [b]	4080	AM	4080	6519	10599	5352
Aquaporin 1	1J4N	26.2 [c]	249	AT	1852	345	2197	114
Aquaporin Z	2O9D	47.3 [d]	464	AT	3356	577	3933	96

Symbols and abbreviations: M: molar mass; N_{AA}: number of AA residues; N_b: number of beads used for modelling; N_w: number of water molecules (of radius r_w = 0.145 nm in the case of HYDCRYST); AT: atomic coordinates; AM: AA residue coordinates.

[a] Composed of one chain.

[b] Contains 24 identical chains.

[c] Asymmetric unit containing one chain.

[d] Asymmetric unit containing two nearly identical chains.

Table 3. Properties of selected proteins and models used for hydration calculations.

A representative of large proteins, apoferritin, is shown in Fig. 10, highlighting some further aspects of particular interest. Water molecules are bound both to the outer and inner surface of this hollow protein; in addition, a few waters seem to occur also in some channels between interior and outside. Again, the number of waters predicted by HYDCRYST exceeds the number found by crystallography (Table 3).

A comparison of structural and hydrodynamic parameters of anhydrous and hydrated models for apoferritin (Table 4) reveals that the properties of the hydrated protein (V, R_G, D, s, $[\eta]$) considerably deviate from those of the anhydrous one. It is quite obvious that a critical comparison of crystallographic data with the findings of solution techniques requires strict consideration of hydration contributions.

There is compelling evidence that water molecules can also be visualized in typical water channels of membrane proteins. The aquaporins shown in Fig. 11 are illustrative examples showing that the existence of a water channel as predicted by HYDCRYST is in full agreement with the crystallographic waters found in this case (Table 3). Some pilot tests also showed that the width of channels can be determined by variation of the probe radius.

Quantifying the results of anhydrous and hydrated proteins in terms of distance distribution functions $p(r)$ again demonstrates that the properties of hydrated models differ significantly from anhydrous models. For the proteins selected (BPTI, apoferritin, aquaporin 1, aquaporin Z), this is shown in Fig. 12. The profiles for the proteins plus the HYDCRYST waters deviate significantly from those of the anhydrous proteins. This effect is most pronounced in the case of small proteins (such as BPTI), whereas, due to the size ratio between protein and waters, the difference is rather insignificant with large molecules (such as apoferritin). For a given protein, owing to the scarce numbers of waters in the crystal data set, generally the functions obtained on the basis of the proteins plus crystal waters rather resemble the anhydrous proteins.

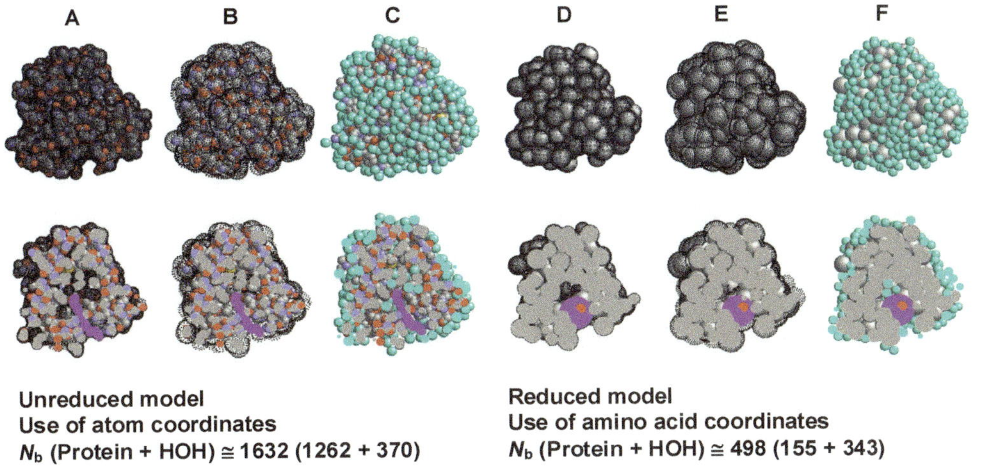

Fig. 9. Space-filling models for anhydrous and hydrated carbonmonoxide myglobin (1VXC), together with central slabs: (A-C) Unreduced model for the anhydrous protein based on atomic coordinates plus dot surface points (A) or surface points created by HYDCRYST (B) or for the hydrated protein as obtained by HYDCRYST (C). The basic atoms of the protein and the ligand CMO are given in CPK colors and the ligand haem in purple; dot surface points and potential water points are shown in black, and the bound waters are displayed in cyan. (D-E) Reduced model for the anhydrous protein based on AA residue coordinates plus dot surface points (D) or surface points created by HYDCRYST (E) or for the hydrated protein as obtained by HYDCRYST (F). AA residues are shown in gray, and the ligands haem and CMO in purple and red, respectively.

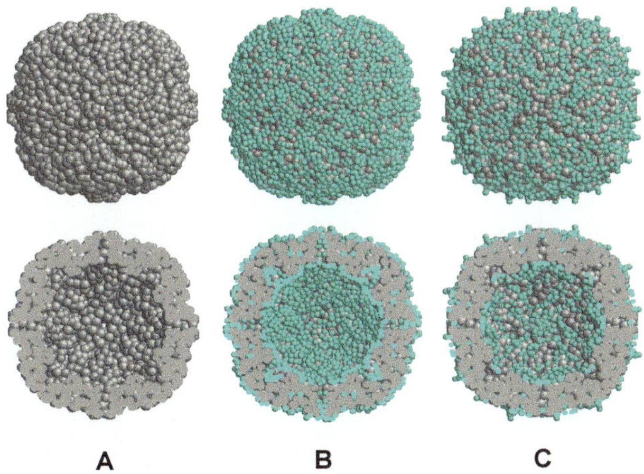

Fig. 10. Space-filling model for anhydrous and hydrated apoferritin (2W0O), together with illustrative central slabs. (A) Model for the anhydrous protein based on AA residue coordinates; AA residues are given in gray. (B) Model for the hydrated protein as obtained by HYDCRYST; bound waters are displayed in cyan. (C) Model for the hydrated protein as obtained by crystallography; waters are shown in cyan.

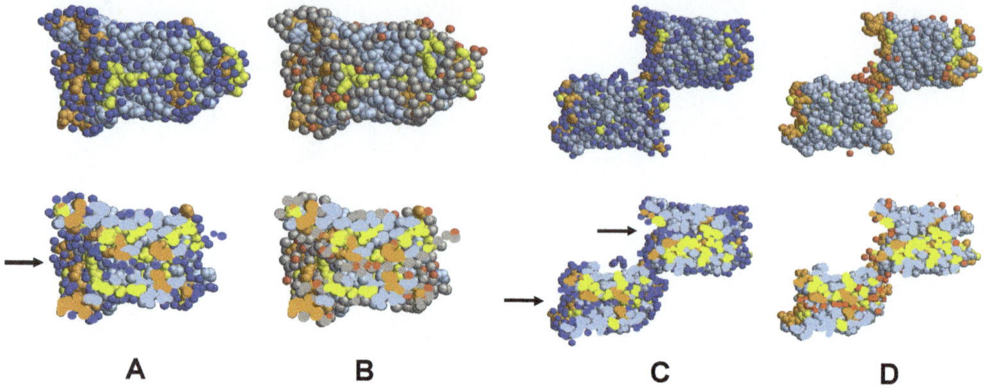

Fig. 11. Space-filling models and slabs for the asymmetric units of aquaporin 1 (1J4N; A, B) and aquaporin Z (2O9D; C, D), together with illustrative central slabs. Models for the anhydrous proteins are based on atomic coordinates; the atoms of hydrophobic AA residues are drawn in bluetint, those of polar residues in yellow and those of charged residues in orange. Models for the hydrated proteins refer to proteins plus waters predicted by HYDCRYST (A, C) or found by crystallography (B, D); bound waters are displayed in blue (A,C) and red (B,D), respectively. Water channels are indicated by arrows.

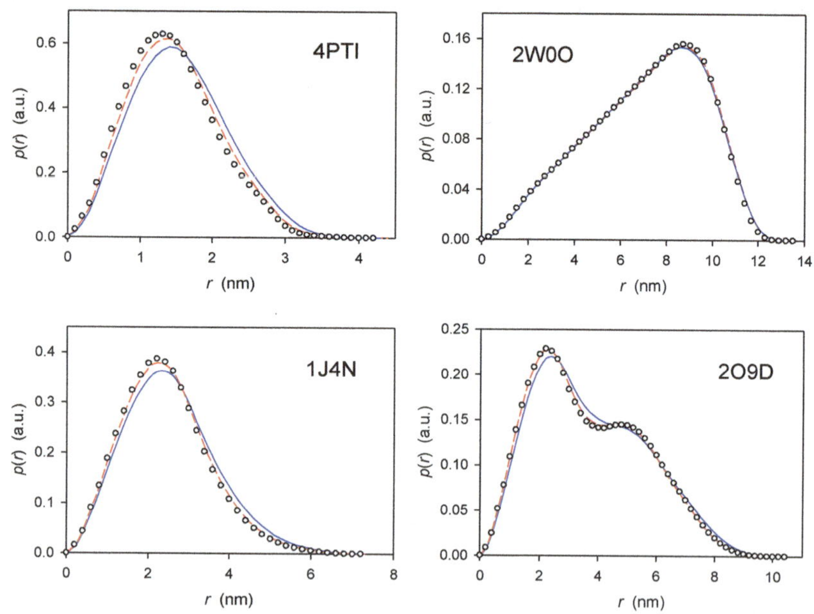

Fig. 12. Comparison of distance distribution functions $p(r)$ of space-filling multibead models for the selected anhydrous and hydrated proteins: BPTI (4PTI), apoferritin (2W0O), aquaporin 1 (1J4N), aquaporin Z (2O9D). ○: anhydrous model; ——: hydrated model: protein plus waters predicted by HYDCRYST; – – –: hydrated model: protein plus waters found by crystallography.

Model Method/mode	f_K	N_w	N_b	V (nm³)	R_G (nm)	$D \times 10^7$ [a] (cm²/s)	$s \times 10^{13}$ [a] (s)	$[\eta]$ [b] (cm³/g)
Crystal structure, anhydrous			32736 [c]	564.0	5.305	3.45 [d]	18.2 [d]	3.21 [d]
Anhydrous RM models:								
AT, BD = 1.0 nm			1056	564.0	5.289	3.52	18.64	3.23
AM, ICOM = 1			4080	564.0	5.292	3.49	18.48	3.25
						3.45 [d]	18.2 [d]	3.21 [d]
AM, ICOM = 10			408	564.0	5.282	3.54	18.72	3.26
Anhydrous CG models:								
AT, VS = 1.0 nm			1082	564.0	5.296	3.48	18.38	3.36
AM, VS = 1.0 nm			854	564.0	5.303	3.51	18.56	3.29
Hydrated models:								
AM, SC0 (low hydration)	1.0	6519	10599	723.0	5.323	3.33	17.64	3.71
						3.31 [d]	17.5 [d]	3.62 [d]
AM, SC0, CG, VS = 1.0 nm			1235	723.0	5.324	3.40	17.96	3.63
AM, EC0	1.0	[6519]	4080	723.0	5.318	3.42	18.09	3.48
						3.37 [d]	17.8 [d]	3.44 [d]
AM, SC9 (high hydration)	4.0	8347	12427 [c]	767.6	5.365	3.29 [d]	17.4 [d]	3.72 [d]
AM, SC9, CG, VS = 1.0 nm			1262	767.6	5.360	3.37	17.83	3.72
AM, EC9	4.0	[8347]	4080	767.6	5.361	3.39	17.92	3.59
						3.34 [d]	17.6 [d]	3.55 [d]

f_K: hydration factor acting on the hydration numbers, N_w: number of water molecules, N_b: number of beads, V: total volume, R_G: radius of gyration, D: translational diffusion coefficient, s: sedimentation coefficient, $[\eta]$: intrinsic viscosity; AT: models based on atoms or atomic groups, AM: models based on AA residues, BD: bead diameter, ICOM: compression index, VS: voxel size, SC: hydration is expressed by attachment of N_w discrete beads to the anhydrous model, EC: hydration is expressed by appropriately increased dimensions of the solvent-accessible beads of the anhydrous model, CG: models obtained by mapping a given model onto a cubic grid (edge length VS) and placing the resulting unequal-sized beads on local centres of gravity.

[a] If not stated otherwise the data were obtained by means of program HYDRO.

[b] If not stated otherwise the data represent $[\eta]_{RVC}$ obtained by means of program HYDRO applying the approach of reduced volume correction.

[c] Too many beads for prediction of hydrodynamic parameters by means of program HYDRO.

[d] The data were obtained by means of program ZENO; the limits of error typically amount to ± 0.04 x 10^{-7} cm²/s for D, ± 0.2 x 10^{-13} s for s, and ± 0.05 cm³/g for $[\eta]$.

Table 4. Comparison of structural and hydrodynamic parameters of anhydrous and hydrated models for apoferritin (2W0O).

The multidrug resistance transporter Sav1866 is another example of a membrane protein. In this case crystallography was able to identify only very few water molecules, while our calculative approach HYDCRYST reveals many waters (Fig. 13). In context of membrane proteins, it has to be mentioned, however, that the preferential water molecules identified only indicate that at these positions individual waters could exist in principle, provided that they have contact with water. On the other hand, the model also shows that waters are preferentially bound to charged and polar residues, while hydrophobic residues are avoided.

Finally, modelling of the giant protein *Lumbricus terrestris* haemoglobin calls for tough measures. Fig. 14 demonstrates SAXS-based experimental scattering intensities, $I(h)$, and distance distribution functions, $p(r)$, of the native HBL complex and of its dodecameric subunit, respectively. The two profiles render information of the molecules under investigation in the reciprocal space and real space, respectively. The two profiles also show impressively the inverse relation between particle size (real space) and the decay of scattering intensity (reciprocal space) by reflecting the different size of the subunit and the complex.

Modelling of the anhydrous and hydrated HBL complexes requires consideration of several measures and precautions regarding model reduction and careful attention of AA residues missing in the crystal structure. Additional, appropriately located beads may serve as substitutes for the missing residues. Figs. 15 and 16 demonstrate that all structural features are retained through the reduction process, and the possibility to generate hydrated objects in both cases.

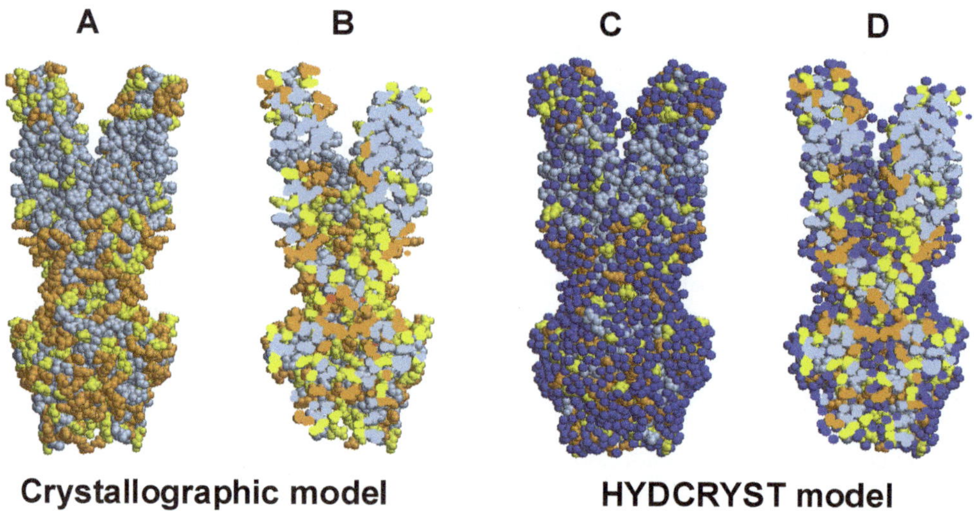

Fig. 13. Space-filling models and slabs for the hydrated multidrug resistance transporter Sav1866 (2ONJ), based on crystallographic data (A, B) and HYDCRYST modelling procedures (C, D). Groups of special AA residues are highlighted in bluetint (hydrophobic residues), yellow (polar residues), or orange (charged residues). Waters localized by crystallography are shown in red, and waters identified by HYDCRYST are displayed in blue.

Advanced 3D reconstructions of the HBL complex, obtained from cryoelectron microscopy (Krebs et al., 1998), yield a voxel density distribution that can be interpreted in terms of anhydrous and hydrated protein volumes (Fig. 17). Thus, the EM-based data allow construction of anhydrous protein models and hydrated models as well (Fig. 18). The resemblance between the crystallography-based and EM-based protein models is striking.

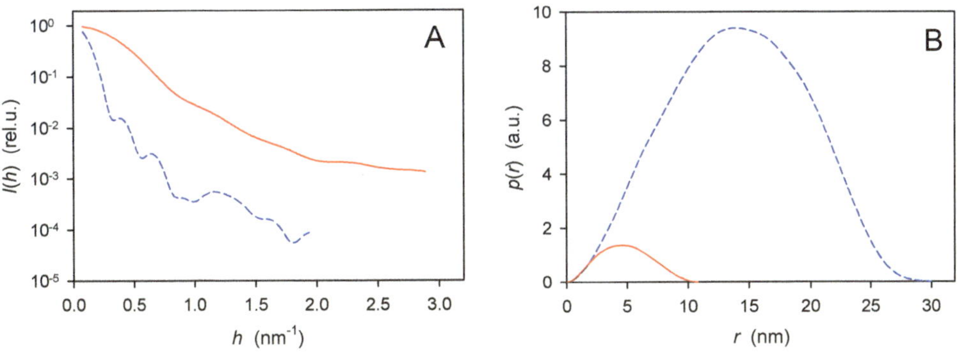

Fig. 14. Experimental scattering profiles of the HBL complex of *Lumbricus terrestris* haemoglobin (blue, dashed lines) and its dodecameric subunit (red, solid lines). (A) Scattering intensities, $I(h)$, normalized to $I(0)$=1; (B) distance distribution functions, $p(r)$, with areas under the $p(r)$ functions proportional to the particle masses.

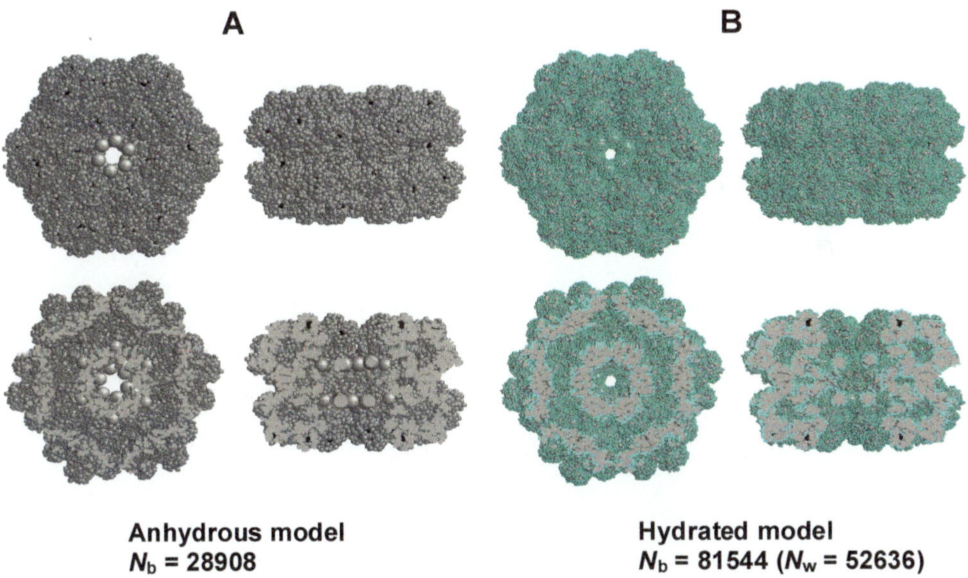

Fig. 15. Space-filling models for anhydrous and hydrated HBL complexes of *Lumbricus terrestris* haemoglobin (2GTL), together with illustrative central slabs. (A) Model for the anhydrous protein based on AA and substitute residue coordinates. AA residues are displayed in gray and haem groups in black; 24 additional large beads (in gray) are substitutes for the AA residues missing in the crystal structure at the N and C termini of the linker chains. (B) Model for the hydrated protein as obtained by HYDMODEL; bound waters are displayed in cyan.

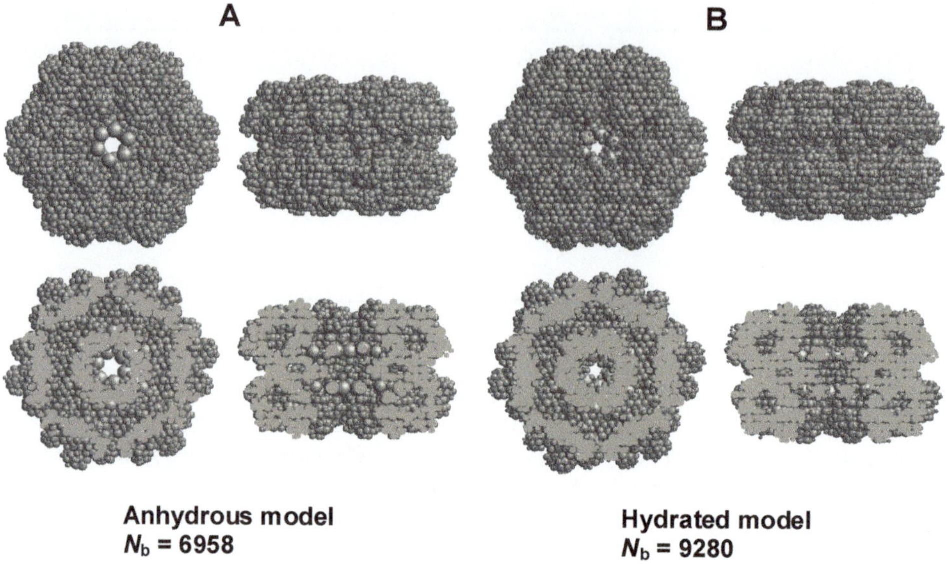

Fig. 16. Space-filling reduced models for anhydrous and hydrated HBL complexes of *Lumbricus terrestris* haemoglobin (2GTL), together with illustrative central slabs. The models were generated by means of MAP2GRID, by mapping the anhydrous and hydrated models shown in Fig. 15 onto hexagonal grids of edge length 1.05 nm and placing the resulting unequal-sized beads on local gravity centres. (A) Model for the anhydrous protein. (B) Model for the hydrated protein.

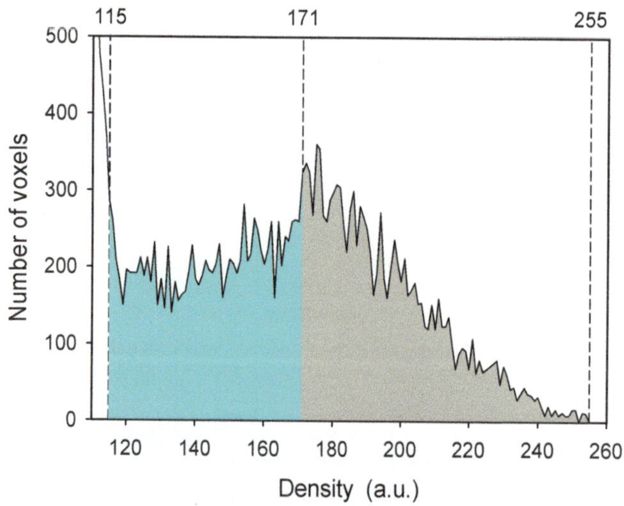

Fig. 17. Voxel density distribution of a 3D reconstruction of the HBL complex of *Lumbricus terrestris* haemoglobin as obtained from cryoelectron microscopy. The density peak between 115 and 255 is caused by the protein; the threshold at 171 is compatible with the anhydrous protein volume, while lower thresholds correspond to hydrated volumes.

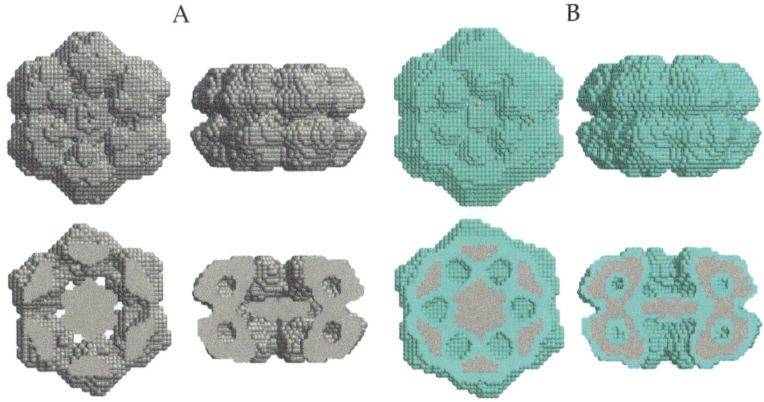

Fig. 18. Top and side views of EM models for the HBL complex of *Lumbricus terrestris* haemoglobin. The different colors represent the anhydrous and hydrated models (shown in gray and cyan, respectively) as given by the thresholds and color codes in Fig. 17.

4. Discussion

Several classes of coarse-grained models and biologically relevant proteins of different size and complexity have been used, to test the applied approaches concerning bead reduction and hydration algorithms, and to compare protein structures and molecular parameters obtained by quite different high-resolution and low-resolution techniques in the crystal, solid state or solution.

With all models and proteins, the bead reduction steps applied were successful, provided the reduction process was not too excessive and grid approaches with beads on lattice points are avoided. Reductions by a factor of 10 provided parameter predictions identical to the initial models, while reductions exceeding this factor can lead to slight deviations.

In the case of proteins, usage of precise anhydrous 3D models (derived from atomic or amino acid coordinates or appropriate models) along with computation of the exact surface topography (molecular dot surface) and our recent hydration approaches (program HYDCRYST) allow the prediction of discrete water molecules preferentially bound to particular residues. In this context, various approaches and procedural methods were tested: sequence of assignment to accessible residues, atomic vs. amino acid coordinates, original vs. coarse-grained models, fine-tuning of input parameters, variation of channel characteristics (e.g., width), rugosity effects. A critical comparison of the water sites on the surface, in active centres, ligand binding sites, interior, crevices, channels, contact areas etc. proves far-reaching identity of crystallographic data, if available, and our predictions. Examples presented include proteins ranging from simple to complex, multisubunit, liganded proteins and water-channels in membrane proteins (e.g., aquaporins) as well. Our hydration algorithms allow the prediction of the number and position of discrete water molecules, even in those cases where no or scant crystallographic waters or water channels have been identified. Our approaches may be used in the future as useful tools for improving crystal data.

The good agreement of the results found for hydrated models by crystallography, SAXS, and other techniques offers the possibility to complement different techniques and to predict details such as the localization of potential water sites - even in those cases where no or insufficient amounts of waters, water clusters or water channels have been identified by crystallography. Variation of input parameters (such as probe radius) should also allow the width and type of channels (other than water channels) to be established. Visualization of protein sites of special concern (charged, hydrophilic and hydrophobic residues and patches, radiosensitive groups, active centres of enzymes, ligand binding sites, docking sites and contact areas) together with individual waters provides the basis for a much deeper understanding of the mechanisms of biological action and effective biotechnological application.

Considering quite different proteins, it can be stated that the majority of waters is bound to polar AA residues, located primarily outside. Some hydrophobic patches on the protein surface prevent the formation of a perfect, uniform water shell. The hydrated models also reveal that some water molecules can be found in the protein interior (e.g. in crevices, clefts, cavities or between subunits).

Both structural and hydrodynamic parameters and scattering profiles proved good agreement between observed and predicted quantities. In conclusion, about two water molecules were found per AA residue, corresponding to about 0.35 g of water per g of protein. Anhydrous and hydrated models differ substantially in their volume-to-mass ratios (1.2 vs. 1.6-1.7). Detailed data on various proteins may be found in previous papers (Durchschlag et al., 2007; Durchschlag & Zipper, 2001, 2002a, 2002b, 2003, 2004, 2005, 2006, 2008; Zipper & Durchschlag, 2002a, 2002b, 2010b).

5. Acknowledgement

The authors are much obliged to several scientists for use of their computer programs: to Y.N. Vorobjev for SIMS, B. Demeler for UltraScan, to J. García de la Torre for the HYDRO suite, to M.L. Mansfield for ZENO, and R.A. Sayle for RASMOL.

6. References

Berman, H.M.; Westbrook, J.; Feng, Z.; Gilliland, G.; Bhat, T.N.; Weissig, H.; Shindyalov, I.N. & Bourne, P.E. (2000). The protein data bank. *Nucleic Acids Res*, 28, 235-242

Boeckmann, B.; Bairoch, A.; Apweiler, R.; Blatter, M.-C.; Estreicher, A.; Gasteiger, E.; Martin, M.J.; Michoud, K.; O'Donovan, C.; Phan, I.; Pilbout, S. & Schneider, M. (2003). The SWISS-PROT protein knowledgebase and its supplement TrEMBL in 2003. *Nucleic Acids Res*, 31, 365-370

Brookes, E.; Demeler, B. & Rocco, M. (2010a). Developments in the US-SOMO bead modeling suite: New Features in the direct residue-to-bead method, improved grid routines, and influence of accessibile surface area screening. *Macromol Biosci*, 10, 746-753

Brookes, E.; Demeler, B.; Rosano, C. & Rocco, M. (2010b). The implementation of SOMO (SOlution MOdeller) in the UltraScan analytical ultracentrifugation data analysis suite: enhanced capabilities allow the reliable hydrodynamic modeling of virtually any kind of biomacromolecule. *Eur Biophys J*, 39, 423-435

Byron, O. (1997). Construction of hydrodynamic bead models from high-resolution X-ray crystallographic or nuclear magnetic resonance data. *Biophys J*, 72, 408-415

Byron, O. (2008). Hydrodynamic modeling: the solution conformation of macromolecules and their complexes. *Methods Cell Biol*, 84, 327-373

Creighton, T.E. (2010a). *The Physical and Chemical Basis of Molecular Biology*. Helvetian Press, ISBN 978-0-9564781-0-8, www.HelvetianPress.com

Creighton, T.E. (2010b). *The Biophysical Chemistry of Nucleic Acids & Proteins*. Helvetian Press, ISBN 978-0-9564781-1-5, www.HelvetianPress.com

Demeler, B. (2005). UltraScan – A comprehensive data analysis software package for analytical ultracentrifugation experiments. In: *Analytical Ultracentrifugation: Techniques and Methods*. Scott, D.J., Harding, S.E. & Rowe, A.J. (Eds.), pp. 210-230, Royal Society of Chemistry, ISBN 0-85404-547-3, Cambridge, UK

Durchschlag, H.; Hefferle, T. & Zipper, P. (2003). Comparative investigations of the effects of X- and UV-irradiation on lysozyme in the absence or presence of additives. *Radiat Phys Chem*, 67, 479-486

Durchschlag, H. & Zipper, P. (2001). Comparative investigations of biopolymer hydration by physicochemical and modeling techniques. *Biophys Chem*, 93, 141-157

Durchschlag, H. & Zipper, P. (2002a). Modelling of protein hydration. *J Phys Condens Matter*, 14, 2439-2452

Durchschlag, H. & Zipper, P. (2002b). Modeling of protein hydration with respect to X-ray scattering and hydrodynamics. *Prog Colloid Polymer Sci*, 119, 131-140

Durchschlag, H. & Zipper, P. (2003). Modeling the hydration of proteins: prediction of structural and hydrodynamic parameters from X-ray diffraction and scattering data. *Eur Biophys J*, 32, 487-502

Durchschlag, H. & Zipper, P. (2004). Modeling the hydration of proteins at different pH values. *Prog Colloid Polymer Sci*, 127, 98-112

Durchschlag, H. & Zipper, P. (2005). Calculation of volume, surface, and hydration properties of biopolymers. In: *Analytical Ultracentrifugation: Techniques and Methods*. Scott, D.J., Harding, S.E. & Rowe, A.J. (Eds.), pp. 389-431, Royal Society of Chemistry, ISBN 0-85404-547-3, Cambridge, UK

Durchschlag, H. & Zipper, P. (2006). Tracking water molecules on protein surfaces. *Bussei Kenkyu*, 87, 68-69

Durchschlag, H. & Zipper, P. (2007). X-ray-based structural models for the *in situ* X-irradiation of a sulfur-containing enzyme. *Radiat Phys Chem*, 76, 1295-1301

Durchschlag, H. & Zipper, P. (2008). Volume, surface and hydration properties of proteins. *Prog Colloid Polymer Sci*, 134, 19-29

Durchschlag, H.; Zipper, P. & Krebs, A. (2007). A comparison of protein models obtained by small-angle X-ray scattering and crystallography. *J Appl Cryst*, 40, 1123-1134

García de la Torre, J.; Amorós, D. & Ortega, A. (2010). Intrinsic viscosity of bead models for macromolecules and nanoparticles. *Eur Biophys J*, 39, 381-388

García de la Torre, J.; del Rio Echenique, G. & Ortega, A. (2007). Improved calculation of rotational diffusion and intrinsic viscosity of bead models for macromolecules and nanoparticles. *J Phys Chem B*, 111, 955-961

García de la Torre, J.; Huertas, M.L. & Carrasco, B. (2000). Calculation of hydrodynamic properties of globular proteins from their atomic-level structure. *Biophys J*, 78, 719-730

García de la Torre, J.; Navarro, S.; López Martínez, M.C., Díaz, F.G. & López Cascales, J. (1994) HYDRO: A computer program for the prediction of hydrodynamic properties of macromolecules. *Biophys J*, 67, 530-531

Glatter, O. & Kratky, O. (Eds.). (1982). *Small Angle X-ray Scattering*. Academic Press, London, ISBN 0-12-286280-5

Kang, E.-H.; Mansfield, M.L. & Douglas, J.F. (2004). Numerical path integration technique for the calculation of transport properties of proteins. *Phys Rev E*, 69, 031918

Krebs, A.; Lamy, J.; Vinogradov, S.N. & Zipper, P. (1998). *Lumbricus terrestris* hemoglobin: a comparison of small-angle X-ray scattering and cryoelectron microscopy data. *Biopolymers*, 45, 289-298

Kuntz, I.D. (1971). Hydration of macromolecules. III. Hydration of polypeptides. *J Am Chem Soc*, 93, 514-516

Levy, Y. & Onuchic, J.N. (2006). Water mediation in protein folding and molecular recognition. *Annu Rev Biophys Biomol Struct*, 35, 389-415

Mansfield, M.L & Douglas, J.F. (2008). Improved path integration method for estimating the intrinsic viscosity of arbitrarily shaped particles. *Phys Rev E*, 78, 046712

Ortega, A.; Amorós, D. & García de la Torre, J. (2011a). Prediction of hydrodynamic and other solution properties of rigid proteins from atomic- and residue-level models. *Biophys J*, 101, 892-898

Ortega, A.; Amorós, D. & García de la Torre, J. (2011b). Global fit and structure optimization of flexible and rigid macromolecules and nanoparticles from analytical ultracentrifugation and other dilute solution properties. *Methods*, 54, 115-123

Papoian, G.A.; Ulander, J.; Eastwood, M.P., Luthey-Schulten, Z. & Wolynes, P.G. (2004). Water in protein structure prediction. *Proc Natl Acad Sci USA*, 101, 3352-3357

Rai, N.; Nöllmann, M.; Spotorno, B.; Tassara, G.; Byron, O. & Rocco, M. (2005). *SOMO* (*SO*lution *MO*deler): Differences between X-ray and NMR-derived bead models suggest a role for side chain flexibility in protein hydrodynamics. *Structure*, 13, 723-734

Ravelli, R.B.G. & Garman, E.F. (2006). Radiation damage in macromolecular cryo-crystallography. *Curr Opin Struct Biol*, 16, 624-629

Rupp, B. (2010). *Biomolecular Crystallography: Principles, Practice, and Application to Structural Biology*. Garland Science, ISBN 978-0-8153-4081-2, New York

Sayle, R.A. & Milner-White, E.J. (1995). RASMOL: biomolecular graphics for all. *Trends Biochem Sci*, 20, 374-376

Serdyuk, I.N.; Zaccai, N.R. & Zaccai, J. (2007). *Methods in Molecular Biophysics: Structure, Dynamics, Function.* Cambridge University Press, ISBN-13 978-0-521-81524-6, Cambridge, UK

Svergun, D.; Barberato C. & Koch, M.H.J. (1995). *CRYSOL* - a program to evaluate X-ray solution sattering of biological macromolecules from atomic coordinates. *J Appl Cryst*, 28, 768-773

Vorobjev, Y.N. & Hermans, J. (1997). SIMS: computation of a smooth invariant molecular surface. *Biophys J*, 73, 722-732

Zipper, P. & Durchschlag, H. (1997). Calculation of hydrodynamic parameters of proteins from crystallographic data using multibody approaches. *Prog Colloid Polym Sci*, 107, 58-71

Zipper, P. & Durchschlag, H. (1999). Prediction of hydrodynamic parameters from 3D structures. *Prog Colloid Polym Sci*, 113, 106-113

Zipper, P. & Durchschlag, H. (2002a). Prediction of structural and hydrodynamic parameters of hydrated proteins by computer modeling based on the results from high-resolution techniques. *Physica A*, 304, 283-293

Zipper, P. & Durchschlag, H. (2002b). Modeling of complex protein structures. *Physica A*, 314, 613-622

Zipper, P. & Durchschlag, H. (2007). Modeling complex biological macromolecules: reduction of multibead models. *J Biol Phys*, 33, 523-539

Zipper, P. & Durchschlag, H. (2010a). Hydrodynamic multibead modeling: problems, pitfalls and solutions. 1. Ellipsoid models. *Eur Biophys J*, 39, 437-447

Zipper, P. & Durchschlag, H. (2010b). Hydrodynamic multibead modeling: problems, pitfalls and solutions. 2. Proteins. *Eur Biophys J*, 39, 481-495

Zipper, P.; Durchschlag, H. & Krebs, A. (2005). Modelling of biopolymers. In: *Analytical Ultracentrifugation: Techniques and Methods.* Scott, D.J., Harding, S.E. & Rowe, A.J. (Eds.), pp. 320-371, Royal Society of Chemistry, ISBN 0-85404-547-3, Cambridge, UK

3

Structure and Dynamics of Proteins from Nuclear Magnetic Resonance Spectroscopy

Homayoun Valafar and Stephanie J. Irausquin
University of South Carolina, Columbia, SC, USA

1. Introduction

Nuclear Magnetic Resonance (NMR) spectroscopy has become an indispensable tool in the characterization of structure and dynamics of biological macromolecules such as proteins. In recent years, NMR spectroscopy has contributed to our understanding of protein biophysics, especially when considering time-averaged dynamical events spanning various time scales. It has also played an important role in structural genomics and protein structure initiatives. The main focus of this chapter is to introduce the NMR phenomenon by providing a broad description of NMR spectroscopy and its contribution to the characterization of structure and dynamics of proteins.

2. Traditional experimental approaches to structure determination of proteins

Traditional experimental approaches to structure determination include X-ray crystallography and solution-state NMR spectroscopy. Presently, X-ray crystallography continues to be the most applied technique for the structural characterization of proteins and protein complexes at atomic resolution (Sali *et al.*, 2003); as is evidenced by the Protein Data Bank (PDB), which currently reports approximately 87% of their structures as being acquired by this particular method (Berman *et al.*, 2000). Crystallography begins with the expression and purification of the protein or proteins of interest. The subsequent step is to produce crystals of sufficient quality (at least 2.5Å), in order to obtain high-resolution data for structure determination (Liu & Hsu, 2005; Sali *et al.*, 2003). Typical X-ray diffraction experiments require only a small single crystal sample (of a few micrometers) in order to physically interrupt the flow of X-rays from a source and cause them to scatter or diffract (Ooi, 2010). Diffracted X-rays are then identified by a detector (Ooi, 2010). The way in which X-rays are diffracted depends on the structure of the crystal; therefore the diffraction pattern that results is unique to each structure (Ooi, 2010). Data collected by the detector is then processed so as to create a visual image of the information. This allows for the direct inference of the types and arrangements of atoms, molecules and/or ions, as well as bond lengths and angles within the crystal (Ooi, 2010). Crystallization is often regarded as a slow and resource-intensive method (Liu & Hsu, 2005). What's more, since crystallization conditions cannot be predetermined, it is often necessary to screen a wide range of conditions related to pH, salt, protein concentration, and cofactors (Liu & Hsu, 2005; Sali *et al.*, 2003). Although recent technologies allowing for the use of smaller sample volumes has led to the automation of high-throughput crystallization, most structures will still require a great deal of time (sometimes as long as days or weeks) in order to produce a high-resolution structure (Abola *et al.*, 2000; Ooi, 2010; Sali *et al.*, 2003). This includes the iterative process of refinement, in which the molecular model is continually compared to the experimental data utilizing statistical methods (Ooi, 2010). Except for the exceptionally rare case of well-ordered crystals from rigid molecules, disorder (which is commonly modeled as part of the refinement process) is a common phenomenon that occurs when some of the atoms in the structure adopt different orientations within different unit cells in the crystal (Ooi, 2010). Disorder may take on the form of discrete conformational sub-states for side chains or surface loops, or even small changes in the orientation of entire molecules throughout the crystal (Adams *et al.*, 2003; Wilson & Brunger, 2000). In addition, crystal structures of multi-component systems and membrane proteins are still limited and refractory for structure determination (Liu & Hsu, 2005). Despite these limitations, X-ray crystallography represents a mature approach (Adams *et al.*, 2003) and continues to be regarded as the 'gold standard' for structure determination (Sali *et al.*, 2003).

The field of NMR, on the other hand, is still relatively young and constantly evolving (Markley *et al.*, 2003). In fact only about 12% of the total structures deposited at the PDB are determined by NMR (Berman *et al.*, 2000). Since NMR does not require crystals in order to produce a three-dimensional structure, samples appropriate for structure determination can be identified relatively quickly (Liu & Hsu, 2005). In addition, NMR experiments can be conducted in aqueous solutions under conditions that are physiologically similar to those in which the protein normally functions (Liu & Hsu, 2005; Montelione *et al.*, 2000). Protein NMR methods have advanced to the point that small to medium sized protein domain structures may be determined rather routinely (Markley *et al.*, 2003). Conventional protein NMR experiments first require successful expression and preparation (purification, isotopic labeling, sample concentration and stability) (Christendat *et al.*, 2000; Markley *et al.*, 2003). NMR methods resolve signals from ^{1}H, ^{15}N, and ^{13}C nuclei of a protein and assign them to specific nuclei in the structure of a molecule (Markley *et al.*, 2003). The assigned chemical shifts are then able to provide reliable information, which reveal the secondary structure of the protein (Markley *et al.*, 2003; Wishart & Nip, 1998; Wishart & Sykes, 1994; Wishart *et al.*, 1991, 1992). Similar to X-ray crystallography, refinement continues iteratively until a self-consistent set of experimental constraints produces a collection of structures that also satisfies standard covalent geometry and steric overlap considerations (Markley *et al.*, 2003). Additional structural restraints may be acquired by evaluating data from one or more different classes of NMR experiments. For example, NOE spectra provide ^{1}H-^{1}H distance constraints; three-bond J coupling experiments specify torsion angle restraints; and residual dipolar couplings from partially ordered proteins provide both distance and spatial constraints for pairs of coupled nuclei (Markley *et al.*, 2003). NMR spectroscopy is capable of performing structural studies of small proteins which display any one of the following characteristics: partial disorder, multiple stable conformations, weak interactions with important ligands or cofactors, or do not crystallize readily (Markley *et al.*, 2003). Moreover, it is a method that can also reveal critical information with regard to overall protein-folding, the existence of multiple-folded conformations, protein-ligand or protein-protein interactions, and even local dynamics (Markley *et al.*, 2003). In fact, the clear advantage of NMR methods is that they are able to deliver the timescale of transitions (from picoseconds to seconds) at atomic resolution in steady-state conditions (Henzler-Wildman & Kern, 2007).

3. A brief introduction to NMR spectroscopy

NMR spectroscopy is a technique that relies on the magnetic precession that is observed when the nuclei of certain atoms are immersed in a magnetic field (Berg *et al.*, 2002; Hornak, 1997). A limited number of isotopes display this property; the most biochemically common isotopes for experiments with proteins and nucleic acids include hydrogen-1 (^{1}H), carbon-13 (^{13}C), nitrogen-15 (^{15}N), and phosphorous-31 (^{31}P) (Berg *et al.*, 2002). Protons, electrons, and neutrons possess quantum spin, which are described in multiples of ½ and can be either positive or negative (Hornak, 1997). For simplicity's sake we discuss NMR concepts using the hydrogen nucleus, which contains one proton, as an example. The spinning of a proton produces a magnetic moment, which can take on one of two orientations or spin states (referred to as α and β) when a magnetic field is applied (Berg *et al.*, 2002). The energy difference between the two states is proportional to the strength of the magnetic field (Berg *et al.*, 2002). Because the α state is aligned with the field, it is slightly more populated and therefore has a slightly lower energy (Berg *et al.*, 2002). By providing a pulse of electromagnetic radiation that corresponds to the energy difference between the α and β states, a spinning proton in the α state can be raised to a β or excited state, allowing a resonance to be acquired (Berg *et al.*, 2002). A resonance spectrum can be obtained for any molecule either by altering the amount of electromagnetic radiation while the magnetic field stays constant, or by changing the magnetic field while the frequency of electromagnetic radiation remains constant (Berg *et al.*, 2002). These properties can then be used to analyze the chemical environment of the hydrogen nucleus (Berg *et al.*, 2002). The flow of electrons around a magnetic nucleus produces a small magnetic field which opposes that of the externally applied field (Berg *et al.*, 2002). The electron density around each nucleus in a molecule varies as a result of the type of nuclei and bonds in the molecule (Hornak, 1997). Nuclei in different environments will also resonate at slightly different field strengths or radiation frequencies (Berg *et al.*, 2002). Nuclei of a perturbed sample absorb electromagnetic radiation at a frequency which can be measured (Berg *et al.*, 2002). The difference between the resonance frequency of the nucleus and a standard, relative to the standard is referred to as a chemical shift and is reported in parts per million (ppm, symbolized by δ), usually with values between 0 and 9 (Berg *et al.*, 2002; Hornak, 1997). For example, when using the water-soluble derivative of tetramethylsilane (TMS) as the standard compound, the chemical shift of a CH_3 proton usually exhibits a chemical shift of 1ppm compared to that of an aromatic proton which is typically 7ppm (Berg *et al.*, 2002). The manner in which chemical shifts are calculated allows for NMR spectra, obtained using spectrometers at differing field strengths, to be compared (Hornak, 1997). Fig. 1 provides an example of a one-dimensional NMR spectrum of Galactose penta-acetate ($C_{16}H_{22}O_{11}$) with chemical shifts for the hydrogens clearly resolved. Nuclei experiencing the same chemical shift are referred to as equivalent, while those experiencing different environments or having different chemical shifts are considered nonequivalent (Hornak, 1997). Nuclei which are close to one another have an influence on each other's magnetic field; this effect, referred to as J coupling, is observable in the NMR spectrum when the nuclei are nonequivalent and their distance is less than or equal to three bond lengths (Hornak, 1997).

Utilizing the one-dimensional NMR technique, it is possible to resolve most protons for a few proteins; using the obtained information, we may then deduce changes to a particular chemical group under different conditions (Berg *et al.*, 2002). However, in instances where one-dimensional NMR spectra are far too complex for interpretation due to the overlapping of signals (refer to Fig. 2), the introduction of additional spectral dimensions is not only helpful but necessary for resolving individual resonances in larger proteins.

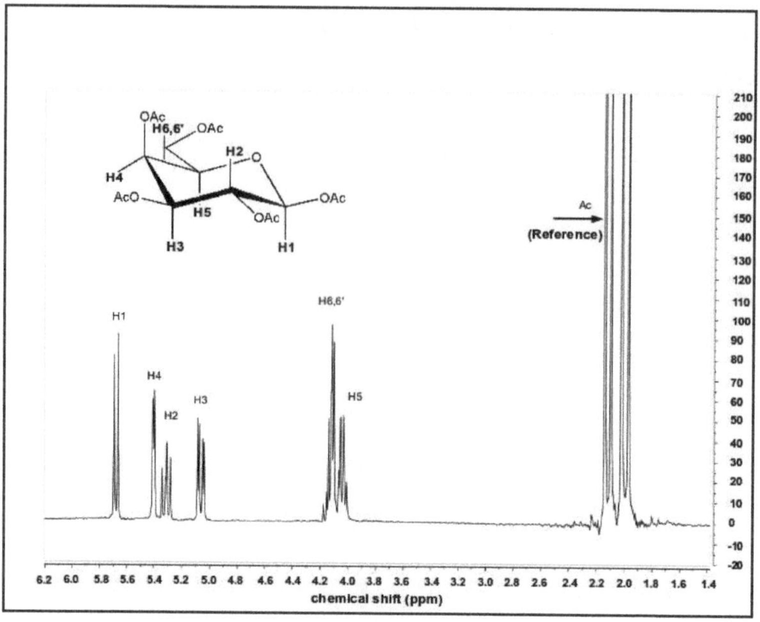

Fig. 1. Example of a one-dimensional NMR spectrum of Galactose penta-acetate with clearly resolved hydrogens. [Figure courtesy of John Glushka, Complex Carbohydrate Research Center, The University of Georgia.]

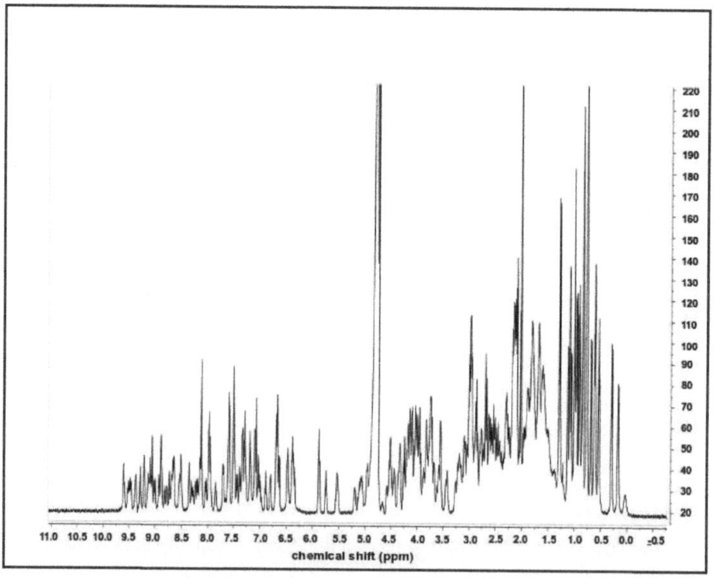

Fig. 2. Example of a one-dimensional NMR spectrum of a small protein (Rubredoxin) with overlapping proton resonances. [Figure courtesy of John Glushka, Complex Carbohydrate Research Center, The University of Georgia.]

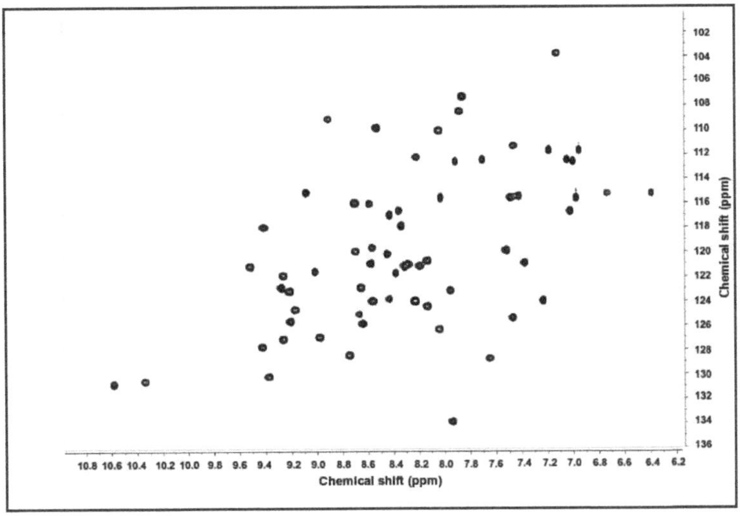

Fig. 3. Example of a two-dimensional HSQC spectrum of protein G. [Figure courtesy of John Glushka, Complex Carbohydrate Research Center, The University of Georgia.]

In order to analyze NMR data, it is important to establish which chemical shift corresponds to which atom. This task is often referred to as resonance assignment, and is dependent upon the protein being isotopically labeled. Standard methods usually begin with two-dimensional ^{1}H-^{15}N Heteronuclear Single Quantum Coherence (2D HSQC) experiments, serving as the initial reference spectrum for signal identification (Markley *et al.*, 2003). In this experiment, magnetization is transferred from the Hydrogen attached to ^{15}N via J coupling; the chemical shift is then evolved on the Nitrogen with the magnetization being transferred back to the Hydrogen for detection (Cavanagh *et al.*, 1995). This particular experiment, illustrated in Fig. 3, reveals all ^{1}H-^{15}N correlations, which are mainly backbone amide groups (Cavanagh *et al.*, 1995). From this experiment, one can determine whether other experiments would be useful before spending the time and resources required for their implementation. In cases where significant degeneracy is present in the 2D HSQC, three-dimensional spectrum (such as HNCO or HNCA) may prove useful in resolving spin systems which overlap (Markley *et al.*, 2003). The HNCO experiment can be used to predict secondary structure, and does so by correlating the amide ^{1}H and ^{15}N chemical shifts of one residue with the ^{13}CO chemical shift of the preceding residue (Grzesiek & Bax, 1993; Kay *et al.*, 1990; Muhandiram & Kay, 1994). Here, magnetization is passed from ^{1}H to ^{15}N and to the ^{13}C by way of the ^{15}N-^{13}CO J coupling and then passed back via ^{15}N to ^{1}H for detection (refer to Fig. 4a); the chemical shift is evolved on all 3 nuclei which results in a three-dimensional spectrum (Grzesiek & Bax, 1993; Kay *et al.*, 1990; Muhandiram & Kay, 1994). HNCA, on the other hand, correlates the intraresidue $^{13}C_\alpha$ chemical shift with the amide ^{1}H and ^{15}N shifts (Farmer *et al.*, 1992; Grzesiek & Bax, 1993; Kay *et al.*, 1990). For this particular experiment, magnetization is passed from ^{1}H to ^{15}N and to $^{13}C_\alpha$ via the ^{15}N-$^{13}C_\alpha$ J coupling and then passed back to the ^{15}N and ^{1}H hydrogen for detection as demonstrated in Fig. 4b (Farmer *et al.*, 1992; Grzesiek & Bax, 1993; Kay *et al.*, 1990). In addition, this experiment provides sequential connectivities by transferring the magnetic coherence from ^{15}N to $^{13}C_\alpha$ of the previous amino acid (Cavanagh *et al.*, 1995). Because the amide Nitrogen is coupled to both the C_α of its own residue and that of the preceding residue, peaks for both C_α's will be visible in the spectrum; peaks with a greater intensity, usually correspond to the C_α's that are directly bonded to the amide Nitrogen (Farmer *et al.*, 1992; Grzesiek & Bax, 1993; Kay *et al.*, 1990). The chemical shift is then evolved for ^{1}H, ^{15}N and $^{13}C_\alpha$, resulting in a 3-dimensional spectrum (Farmer *et al.*, 1992; Grzesiek & Bax, 1993; Kay *et al.*, 1990).

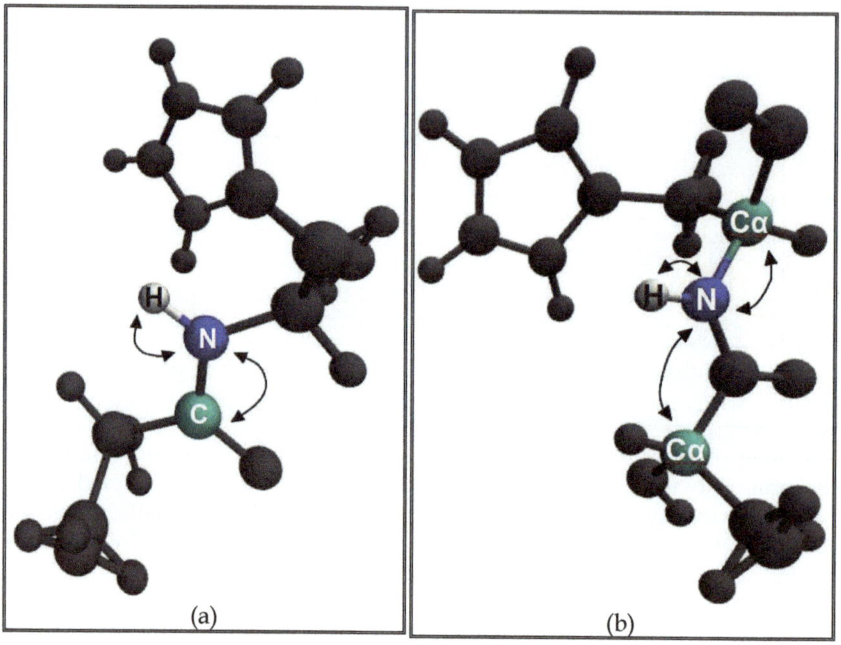

Fig. 4. Visualization of magnetization transfer in the Triple-resonance HNCO (a) and HNCA (b) experiments.

4. A brief discussion of Nuclear Overhauser Effect (NOE) and its implication in modern NMR spectroscopy

Even more structural information can be acquired by examining how the spins of different protons affect neighboring protons (Berg *et al.*, 2002). This is possible through the Nuclear Overhauser Effect (NOE) observed by NMR spectroscopy, which states that an interaction between nuclei is proportional to the inverse sixth power of the distance between them (Berg *et al.*, 2002). Therefore, the distance between nuclei is determined according to the intensity of the peak. By inducing a transient magnetization in a sample through radio-frequency pulse, it is possible to both alter the spin of one nucleus and examine the effect on the spin of a neighboring nucleus (Berg *et al.*, 2002). NOE differs from J coupling in that it identifies pairs of protons that are within close proximity relative to the protein's three-dimensional structure, even if they are not close together with regard to the primary sequence (Berg *et al.*, 2002). J coupling, on the other hand, is only observed when atoms are connected by 2 to 3 covalent bonds, as mentioned in section three. The two-dimensional spectrum acquired by Nuclear Overhauser Enhancement SpectroscopY (NOESY) graphically displays pairs of protons that are close in proximity within the three-dimensional structure of the protein (Berg *et al.*, 2002). As long as nuclei are within ~5Å, the magnetization from an excited nucleus is transferred to that of an unexcited nucleus (Berg *et al.*, 2002). Fig. 5 provides an example of a one-dimensional NOESY spectrum. The diagonal corresponds to a one-dimensional spectrum, whereas the off-diagonal peaks identify the pairs of protons that are within 5Å of each other (Fig. 5). Similarly, in a two-dimensional NOESY spectrum, off-diagonal peaks reveal short proton-proton distances (refer to Fig. 6).

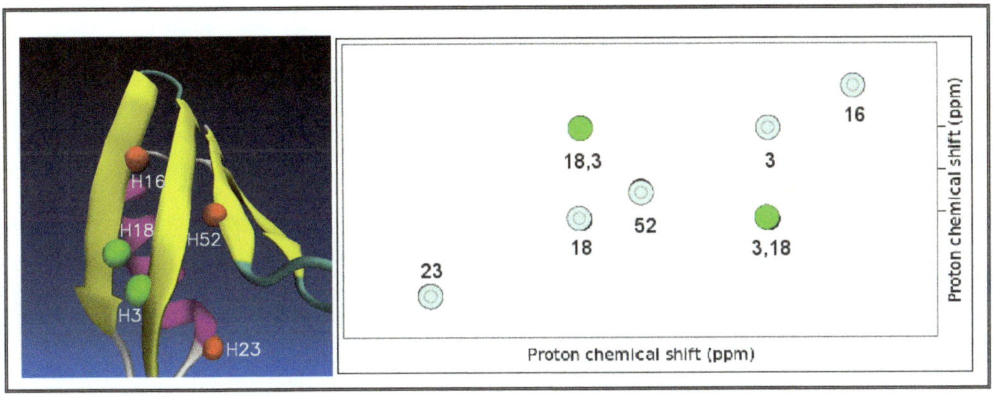

Fig. 5. Example of a one-dimensional NOESY spectrum. The five diagonal peaks correspond to the five protons in the image to the left.. The peaks above the diagonal and the symmetrically related one below reveal that proton H18 is close to proton H3.

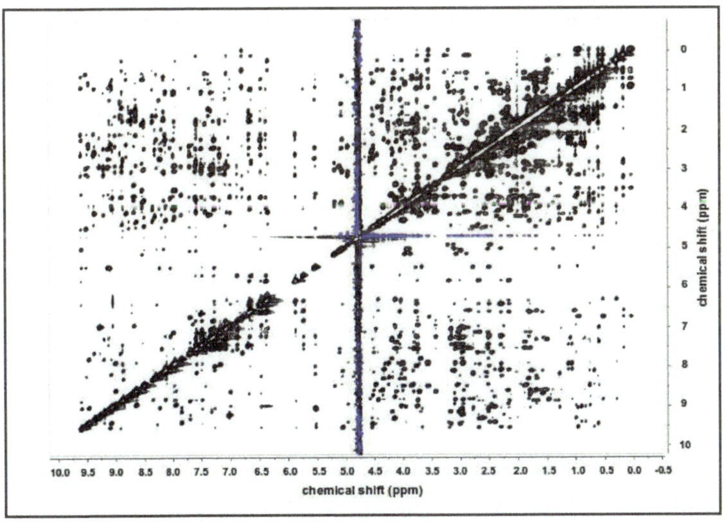

Fig. 6. Example of a two-dimensional NOESY spectrum of the Rubredoxin protein. Each off-diagonal peak corresponds to a short proton-proton separation. This spectrum reveals hundreds of such short proton-proton distances, which can be used to determine the three-dimensional structure of this protein. [Figure courtesy of John Glushka, Complex Carbohydrate Research Center, The University of Georgia.]

Three-dimensional protein structures may be calculated nearly uniquely if a sufficient number of distance constraints are applied, and are reconstructed such that proton pairs identified from NOESY spectra are close to one another in the three-dimensional structure (Berg *et al.*, 2002). Families of related structures may also be generated in cases where: not enough constraints are experimentally accessible to fully describe the structure; distances obtained from NOESY analysis are only approximate; as opposed to utilizing a single molecule, experimental observations are made on a number of molecules in solution which may have slightly different structures at any given moment (Berg *et al.*, 2002). It is important to note that the efficient and accurate assignment of NOEs for structure determination is highly dependent upon the completeness and precision of the chemical shift assignments (Markley *et al.*, 2003).

5. NMR approaches for the study of internal dynamics

The investigation of protein dynamics relies mostly upon the use of NMR techniques. This is due to the fact that biological functions span a range of timescales for which various NMR experiments are sensitive. Here we briefly introduce a number of NMR methods, and summarize for each the timescales for which they are capable of acquiring experimental data (refer to Fig. 7).

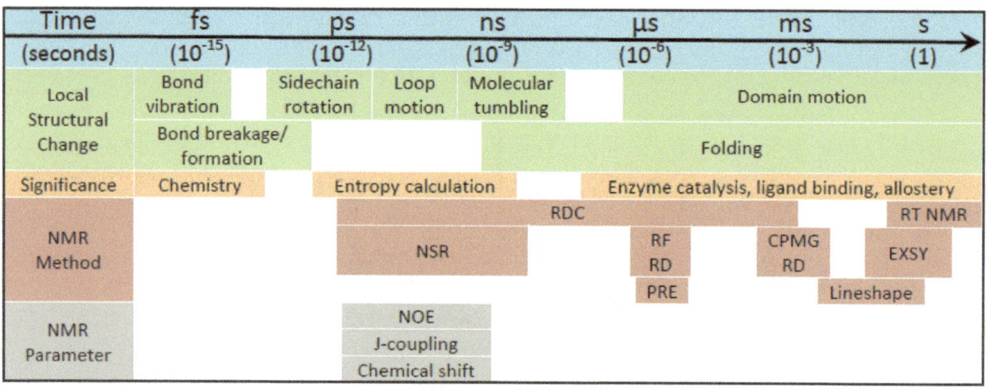

Fig. 7. Timescale of protein dynamic events, and the appropriate experimental methods that can be used to detect them. Acronyms for NMR methods: RDC – Residual Dipolar Coupling; RT NMR - Real Time NMR; NSR – Nuclear Spin Relaxation; RF RD – Rotating Frame Relaxation Dispersion; CPMG RD – Carr-Purcell Meiboom-Gill Relaxation Dispersion; EXSY – Exchange SpectroscopY; and PRE – Paramagnetic Relaxation Enhancement.

Real-Time NMR (RT NMR) – encompasses slower dynamic processes in the seconds range (Kleckner & Foster, 2011) and was originally developed as a method to follow protein folding, combining the availability of high resolution data with kinetics experiments to allow for detailed examination of protein structure during different steps of the folding process (Zeeb & Balbach, 2004). The experiment consists of physically initiating a process of interest and subsequently acquiring a sequence of NMR spectra, which typically demonstrate a progressive weakening of an initial set of signals along with a gradual strengthening of a new set of signals - the result of time-dependent changes from differing conformations and or local structures (Kleckner & Foster, 2011).

EXchange SpectroscopY (EXSY) - is used to quantify exchange kinetics for dynamic processes in the 10 millisecond to 5 second timescale (Kleckner & Foster, 2011) and encompasses slow conformational changes such as domain movements (Key *et al.*, 2009) and ligand binding (Demers & Mittermaier, 2009). In this experiment, kinetic information is obtained from a quantitative analysis of the magnetization transfer and spectral broadening that results from the exchange between bound and free states in a partially ligand-saturated sample (Demers & Mittermaier, 2009).

Lineshape Analysis - is another approach reporting exchange events that take place roughly between 10 milliseconds and 0.1 seconds (Kleckner & Foster, 2011). In this experiment, chemical exchange processes are identified by characteristic changes of the NMR line shape (Jeener *et al.*, 1979), which are typically the result of spectra acquired along a titration coordinate (in the form of ligand concentration, temperature or pH for example) allowing for the observation of their incremental effect upon the NMR observables (Kleckner & Foster, 2011).

Carr-Purcell Meiboom-Gill Relaxation Dispersion (CPMG RD) - is a technique used to obtain kinetic, thermodynamic, and structural information and applies to exchange processes occurring in the 0.3 to 10 millisecond time frame (Kleckner & Foster, 2011; Loria *et al.*, 2008). The purpose of this particular experiment is to use a series of spin echo pulses to transverse magnetization during a relaxation delay in order to refocus exchange broadening (Kleckner & Foster, 2011).

Rotating Frame Relaxation Dispersion (RF RD) - may be used to study exchange processes that occur within the 20 to 100 microsecond range (Kleckner & Foster, 2011; Loria *et al.*, 2008). This particular experiment is very similar to that of CPMG, the only difference is in the range of the spin echo pulses used (25-1200 Hz for CPMG and 1-50kHz for RFRD), this allows RF RD to study exchange events via the same principles as CPMG, but on a faster timescale (Kleckner & Foster, 2011).

Paramagnetic Relaxation Enhancement (PRE) - is a method which allows for the study of protein dynamics within the microsecond timescale and is most appropriate for the examination of non-specific interactions and complexes between binding partners (Clore & Iwahara, 2009; Kleckner & Foster, 2011). This approach results from the identification of the magnetic dipole interaction between a nucleus and an unpaired electron (Kleckner & Foster, 2011).

Nuclear Spin Relaxation (NSR) - may be used to study the details of protein dynamics on fast timescales (picoseconds to nanoseconds) and is made possible by the presence of NMR probes throughout the molecule (Case, 2001). This experiment is based on the weak coupling between spin variables and molecular motion which are then manifested into much slower relaxation of the spins which can be readily studied (Case, 2001).

Full characterization of inter-molecular dynamics has been limited to NMR spectroscopic study of the protein (or complex of proteins) of interest and are typically performed in separate steps - with the protein's structure determined first under the assumption of rigidity, and its motion characterized later. Although structure determination protocols based on the assumption of molecular rigidity will produce a single structure, the degree of similarity between the static model of a protein structure and the many conformations of the dynamic model is not always clear and is poorly investigated. Another source of data capable of studying both structure and dynamics on timescales ranging from picoseconds to microseconds is that of Residual Dipolar Couplings or RDCs, which we discuss in further detail in the section that follows.

6. Emerging methods in simultaneous study of structure and dynamics of proteins

Recent advances in NMR spectroscopy have enabled the acquisition and analysis of data other than the traditional distance constraints. These new sources of data include Residual Dipolar Couplings (RDC), Pseudo Contact Shifts (PCS), and Paramagnetic Relaxation Enhancement (PRE), which provide orientational restraints as well as distance restraints. The introduction of orientational restraints has produced a shift in paradigm of structure determination that has necessitated alternative approaches to the analysis of NMR data. In this section the utility of orientational restraints has been discussed with references to the software development track that has been the subject of additional investigations. Finally, the focus of this section will be aimed at RDC data.

Historically, the use of RDCs has been limited by two factors: data acquisition, and data analysis. The introduction of a variety of alignment media, combined with advances in instrumentation and data acquisition have mitigated the experimental limitations in obtaining RDCs. The major bottleneck in utilization of RDC data in recent years has been attributed to a lack of powerful and yet user-friendly RDC analysis tools capable of extracting the pertinent information embedded within this complex source of data.

6.1 Residual Dipolar Couplings (RDCs)

Residual dipolar couplings (RDCs) have been observed as early as 1963 (Saupe & Englert, 1963) in nematic environments. A number of recent applications (Al-Hashimi & Patel, 2002; Bax *et al.*, 2001; Blackledge, 2005; de Alba & Tjandra, 2002; Prestegard *et al.*, 2000; Tolman, 2001; Zhou *et al.*, 1999) have reignited their wide use in application to a broad spectrum of biomolecules. RDCs arise from the interaction of two magnetically active nuclei in the presence of the external magnetic field of an NMR instrument (Bax *et al.*, 2001; Prestegard *et al.*, 2000; Tjandra *et al.*, 1996; Tolman *et al.*, 1995). This interaction is normally reduced to zero, due to the isotropic tumbling of molecules in their aqueous environment. The introduction of partial order to the molecular alignment, by minutely limiting their isotropic tumbling, will resurrect the RDC observable. This partial order can be introduced by either magnetic anisotropy of the molecule (Prestegard *et al.*, 2000), a crystalline aqueous solution (Prestegard & Kishore, 2001), or incorporation of artificial tags with magnetic anisotropy susceptibility such as Lanthanide (Nitz *et al.*, 2004). RDCs are measured relatively easily and represent an abundant source of highly precise information, such as the relative orientations of different inter-nuclear vectors within a molecule. Equation 1 describes the time average observable of the RDC interaction between a pair of spin ½ nuclei.

$$D_{ij} = \frac{-\mu_0 \gamma_i \gamma_j h}{(2\pi r)^3} \left\langle \frac{3\cos^2(\theta_{ij}(t)) - 1}{2} \right\rangle \tag{1}$$

Here, D_{ij} denotes the residual dipolar coupling in units of Hz between nuclei i and j, γ_i and γ_j are nuclear gyromagnetic ratios of the two interacting nuclei, r is the internuclear distance (assumed fixed for directly bonded atoms) and $\theta_{ij}(t)$ is the time dependent angle of the internuclear vector with respect to the external magnetic field. The angle brackets signify the time average of the quantity.

Residual dipolar couplings serve as an abundant source of orientational information for the inter-nuclear vectors within a molecule (Bax *et al.*, 2001; Prestegard *et al.*, 2000; Tolman *et al.*, 1995). They may also be acquired very rapidly and accurately by a number of techniques including, direct measurement of splittings in coupled heteronuclear single quantum coherence spectra (HSQC) (Bax *et al.*, 1994; Bodenhausen & Ruben, 1980; Cavanagh *et al*, 1995; Tolman & Prestegard, 1996). RDCs provide structural (Bax *et al.*, 2001; Cornilescu *et al.*, 1999; Delaglio *et al.*, 2000) and motional (Al-Hashimi *et al.*, 2002b; Bax, 2003;Bernardo & Blackledge, 2004a, 2004b; Blackledge, 2005; Clore & Schwieters, 2003; O'Neil-Cabello *et al.*, 2004; Tolman, 2001; Yi *et al.*, 2004) information in a biologically relevant timescale and have been used in studies of carbohydrates (Adeyeye *et al.*, 2003; Azurmendi & Bush, 2002; Azurmendi *et al.*, 2002; Tian *et al.*, 2001a), nucleic acids (Al-Hashimi *et al.*, 2000a, 2002a, 2002b; Tjandra *et al.*, 2000; Vermeulen et *al.*, 2000) and proteins (Andrec *et al.*, 2001; Assfalg *et al.*, 2003; Bertini *et al.*, 2003; Clore & Bewley, 2002; Cornilescu *et al.*, 1999; Fowler *et al.*, 2000; Tian *et al.*, 2001b). The use of RDCs as the main source of structural information has led to a significant reduction in data collection and analysis, while providing the possibility of simultaneous resonance assignment, structure determination, and identification of dynamical regions (Bernado & Blackledge, 2004b; Prestegard *et al.*, 2005; Shealy *et al.*, 2011; Tian *et al.*, 2001b; Valafar *et al.*, 2005). As a result, the impact of these developments has enabled direct investigation of protein backbone structures (Bernado & Blackledge, 2004b; Clore *et al.*, 1999; Fowler *et al.*, 2000; Tian *et al.*, 2001b; Valafar *et al.*, 2005). Applications of RDCs have also extended into the structural elucidation of traditionally complex proteins such as membrane proteins (Opella & Marassi, 2004; Park *et al.*, 2009) and homo-multimeric proteins (Wang *et al.*, 2008). In fact, the utility of RDCs has extended well beyond the community of NMR spectroscopists. In several instances, RDCs have been used to validate and refine X-ray protein structures (Bansal *et al.*, 2008; Langmead & Donald, 2003; Valafar & Prestegard, 2003, 2004) as well as modeled protein structures (Bansal *et al.*, 2008; Raman *et al.*, 2010a; Rohl & Baker, 2002; Valafar & Prestegard, 2003, 2004).

6.2 Molecular frame, alignment frame and order tensor

Proper understanding and interpretation of the Order Tensor Matrix (OTM) is central to the study of structure and dynamics of biological macromolecules from orientational restraints, and therefore requires a brief discussion. Upon successful determination of a structure, its atomic coordinates are described within some arbitrarily selected coordinate system. Since this structure is independent of any rotation or displacement within a given frame, the selection of a coordinate system is inconsequential. This arbitrary coordinate system is denoted as the "molecular frame" (*MF*). On the other hand, since RDC data are capable of describing the preferred alignment of a molecule, a more descriptive frame can be selected in which the atomic coordinates of the molecule of interest are described in the appropriate orientation. This more descriptive frame is defined as the "principal alignment frame" (*PAF*). Rotation of the molecule within this frame is consequential in the representation of its order tensor while any translation in space is not. Alignment properties of a molecule can be described in the form of a Saupe order tensor matrix (Saupe & Englert, 1963; Valafar & Prestegard, 2004). Reformulation of Equation 1 in the matrix form collects and defines the Saupe order tensor matrix (or *OTM* for short) as represented by S in Equation 2. The entity v in this equation, represents the Cartesian coordinates of the normalized vector, and describes the relationship between a pair of interacting nuclei. Jacobi transformation (Press *et al.*, 2002) of this symmetric and traceless matrix can separate two important information contents of the molecular alignment as shown in Equation 3. In this equation, a 3×3 order tensor represented by the elements s_{ij} can be decomposed to produce the diagonal form of the order tensor matrix and a corresponding rotation matrix denoted by R. The three elements of the resulting diagonal matrix (S_{xx}, S_{yy} and S_{zz} also referred to as the order parameters) represent the strength of alignment along each of the principal axes x, y and z within the *PAF*. Comparison of the order parameters obtained from different regions of a macromolecular complex can provide information regarding their rigidity with respect to each other. Analysis of the R matrix in turn can provide the preferred direction of orientation with reference to the starting molecular frame. The preferred alignment can be identified through the decomposition of the rotation matrix R (shown in Equation 3) into three distinct rotational operators along z, y and z axes of the *PAF*. These three rotations denoted by α, β, and γ fully define the orientational relationship between the arbitrary *MF* and the *PAF*, and can be used to assemble molecular complexes. In summary, an order tensor encapsulates five independent pieces of information (α, β, γ, S_{yy}, S_{zz} and S_{xx}=- S_{yy}- S_{zz}). Careful study of the order tensor matrix can provide the preferred alignment of the molecule with respect to the molecular frame (α, β, γ) and strength of alignment (S_{yy}, S_{zz}) along each of the principal axes of alignment within the *PAF*. When RDC data are assigned to specific locations in a given structure, the elements of the order tensor can be obtained (Blackledge, 2005; Clore *et al.*, 1998; Dosset *et al.*, 2001; Losonczi *et al.*, 1999; Valafar & Prestegard, 2004). Equally as well, given a structure and the elements of the order tensor, theoretical RDC data can be calculated easily.

$$D = -\left(\frac{\mu_0 \gamma_i \gamma_j h}{(2\pi r)^3}\right) v' \cdot S \cdot v \tag{2}$$

$$S = \begin{bmatrix} s_{xx} & s_{xy} & s_{xz} \\ s_{xy} & s_{yy} & s_{yz} \\ s_{xz} & s_{yz} & s_{zz} \end{bmatrix} = R(\alpha,\beta,\gamma) \cdot \begin{bmatrix} S_{xx} & 0 & 0 \\ 0 & S_{yy} & 0 \\ 0 & 0 & S_{zz} \end{bmatrix} \cdot R(\alpha,\beta,\gamma)^T \tag{3}$$

The attainment of an order tensor is an important requisite step in extracting structural information from RDC data.

6.3 Structure determination by RDCs

Structure determination approaches from RDC data utilize the rotational component of order tensor matrices in order to assemble a protein from rigid structural elements. The rigid structural elements can consist of units as small as peptide planes (Bernado & Blackledge, 2004a; Bryson *et al.*, 2008) or as large as individual structural domains (Delaglio *et al.*, 2000; Fowler *et al.*, 2000). Nearly all of the existing NMR data analysis packages such as Xplor-NIH (Schwieters *et al.*, 2003), and CNS (Brunger, 2007) have been modified to incorporate orientational restraints as part of their analysis. However, this shift in paradigm from distance-based to orientation-based structure determination, has necessitated the development of appropriate analyses. A number of such software packages have been introduced in recent years such as REDCAT (Valafar & Prestegard, 2004), REDCRAFT (Bryson *et al.*, 2008) and others (Delaglio *et al.*, 2000; Fowler *et al.*, 2000).

The prospect of structure determination of macromolecules from orientational restraints has many advantages. First, a carefully selected set of RDC data originating from the backbone of a protein can be used to directly investigate structural parameters such as the backbone torsion angles. For example, backbone N-H and C_α-H_α RDCs can be used to directly restrain the backbone structure of a protein (Bryson *et al.*, 2008; Marassi & Opella, 2002, 2003; Prestegard *et al.*, 2005; Tian *et al.*, 2001b; Tjandra *et al.*, 1997; Valafar *et al.*, 2005; Wang & Donald, 2004). However, in order to address degeneracies (Al-Hashimi *et al.*, 2000b) and variable sensitivity of RDC data (Bryson *et al.*, 2008), it is necessary to acquire orientational restraints from two or more independent alignment media. Therefore it can be argued that in theory, the structure of any protein can be determined with as little as two RDC data per residue from two alignment media. The main reason for this reduction in the data requirement is based on independent investigation of backbone structure from investigation of the side chains. This significant reduction in the amount of required data is of paramount importance in reducing the cost (temporal and financial) of structure determination, and extending the applicability of NMR spectroscopy to challenging proteins such as membrane proteins. Nearly 30% of the human proteome is predicted to consist of membrane proteins. Despite their functional importance and frequency of occurrence, only ~100 unique membrane proteins have been structurally characterized and included in the PDB database (Berman *et al.*, 2000). Their low level of inclusion is because they neither crystallize for X-ray crystallography nor produce the conventional NOE data (12-15 NOEs per residue) through NMR spectroscopy that has been required for successful structure determination. On the other hand, acquisition of two or three RDC data per residue for membrane proteins or large deuterated proteins is feasible with today's technology. Furthermore, because of the direct relationship between the RDC data and backbone conformation of proteins, it is easy to theoretically support the sufficiency of two or three RDC data points per residue for meaningful structure determination.

6.4 Simultaneous study of structure and dynamics

Study of internal dynamics of macromolecules has been one of the long standing challenges in structural and molecular biology. While the importance of elucidating the structure of pharmaceutical targeted proteins is widely accepted, the importance of understanding their associated internal dynamics is less widely recognized. This neglect is due, in part, to a lack of experimental methods capable of probing dynamics on biologically relevant timescales. Although, various techniques for the study of internal dynamics exist (Henzler-Wildman &

Kern, 2007), they usually apply to faster timescales or provide little information regarding conformational changes for slower dynamics.

Traditionally, full characterization of inter-molecular dynamics by NMR spectroscopy is separated from structure elucidation, increasing the cost of these studies. Furthermore, conceptually, it is difficult to separate structure from dynamics since the two are intimately related and any attempt in structure elucidation that disregards the dynamics (or vice versa) can produce faulty results. A structure that has been determined from data perturbed by internal dynamics, is likely to produce a compromised structure. Relying on a false structure to study internal dynamics is likely to produce an inaccurate model of the internal motion. Disregarding internal dynamics during the course of structure determination may have catastrophic effects. Fig. 8 demonstrates the effects of structure determination of a mobile terminal helix while disregarding internal dynamics (Bryson *et al.*, 2008). Although the entire helix maintains its secondary structure, the recovered structure from traditional methods does not bear any resemblance to the actual structure (Fig. 8 green).

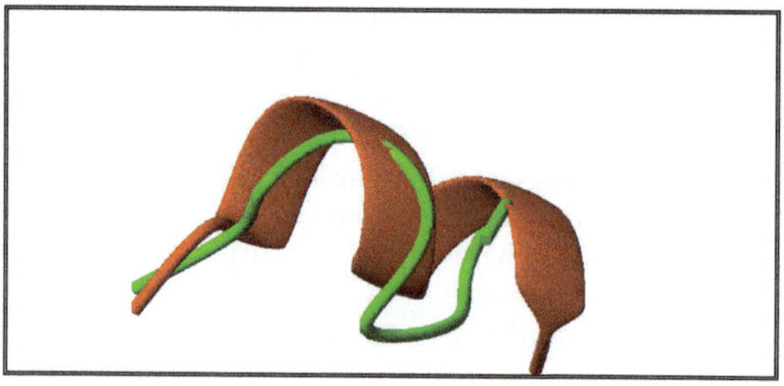

Fig. 8. The structure of a mobile terminal helix (red) that has been determined by conventional approaches (green).

As discussed in section 6.3, proper analysis of RDC data can reveal information regarding the relative stability of various internal regions of a protein. This information can be obtained by comparing the order parameters that are obtained independently from each suspect region. Similar order parameters reported from different regions of a protein imply internal rigidity, while dissimilar order parameters can be interpreted as presence of internal dynamics. In this regard, RDCs prove to be exceptional probes for the inspection of dynamics in biomolecules on timescales that are biologically relevant (Bouvignies *et al.*, 2005). Therefore proper investigation of structure and dynamics of proteins from RDC data should theoretically be possible. However, proper treatment of RDCs requires analysis software packages that are specifically designed for this purpose. Study of structure and dynamics of RDCs with traditional software packages such as Xplor-NIH (Schwieters *et al.*, 2003) or CNS (Brunger, 2007) can be very daunting. Here we utilize the membrane bound form of the bacteriophage Pf1 protein (mbPf1) to illustrate the point. The structure of the mbPf1 protein consists of two helices, a longer transmembrane helix and a short amphipathic helix. A two-state jump model of motion has been applied to the amphipathic helix of this protein. Appropriate averaged RDC data have been generated to reflect the effect of the modeled internal dynamics. Fig. 9 provides an illustration of the two states of the amphipathic helix in red and green. An attempt to recover the structure of this protein from RDC data after insisting on helical secondary structural elements and using conventional analysis methods, produces the ensemble of structures shown in cyan in Fig. 9a. A recently introduced RDC analysis method named REDCRAFT, has demonstrated the possibility of successful study of structure and dynamics of proteins. Application of REDCRAFT analysis to the two-state jump problem successfully recovers the structure and orientation of the two states as shown in Fig. 9b in cyan. Note that the reconstructed states in Fig. 9b exhibit less than 1.3Å bb-rmsd.

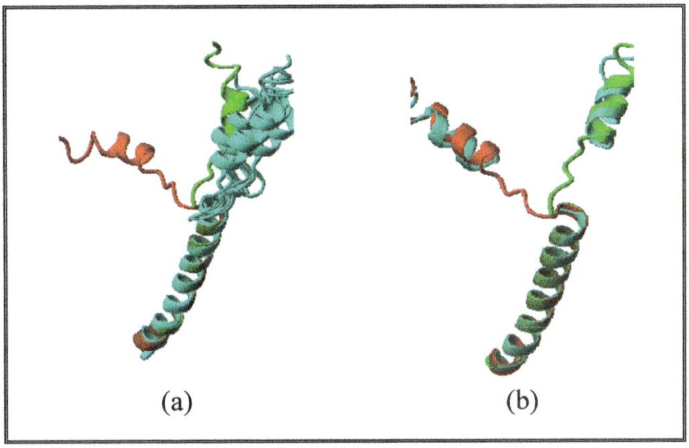

Fig. 9. Structure of the membrane bound form of the bacteriophage coat protein pf1 with a hypothetical and simulated model of dynamics. The two states of the amphipathic helix are illustrated in red and green. Attempts to recover the structure of this protein from RDC data using conventional analysis methods produces an ensemble of cyan structures (a) while REDCRAFT successfully recovers the structure and orientation of the two states in cyan (b).

7. A comparison of NOE versus RDC based approaches to study structure and dynamics of proteins

Over the past few years, the utility of residual dipolar couplings (RDCs) for structure determination has increased precipitously. This explosion can be attributed to its distinct advantages over the traditional distance constraints (Bryson *et al.*, 2008; Prestegard *et al.*, 2005, 2001; Valafar *et al.*, 2005). Generally, RDC data are more precise, easier to measure, and are capable of providing informative structural and dynamic information. The direct relationship between a carefully selected set of RDC data and structural parameters, such as backbone torsion angles, is another notable advantage of RDC data over NOE data. Given the alignment of an unknown protein, a single RDC datum can limit the orientation of the corresponding vector to within two symmetrical cones as illustrated by Tjandra and Bax (Tjandra & Bax, 1997; Tjandra *et al.*, 1996, 1997). In addition, the number of NOE requirements for an unambiguous recovery of a structure is heavily related to structural complexity of the protein, which is unknown *a priori* to structure determination. Fig. 10 illustrates the tertiary structure of two proteins (3LAY and 1A1Z) of similar sizes, which clearly require disparate amounts of NOE data for successful description of their structures. Even the assembly of their secondary structural elements will require a different number of distance constraints. The lack of an understanding, as to the amount of data that is required, will have a direct impact on the cost and success of protein structure characterization. RDC data, on the other hand, are more suitable for theoretically understanding data-requirements, independent of structural complexity. As mentioned previously, strategically collected data can directly constrain a related torsion angle. Therefore, it is of no surprise that 2-3 RDC data points per residue should suffice for successful determination of a protein's backbone structure, regardless of the structural complexity. *A priori* knowledge of data requirements is helpful for many reasons. Proper understanding of data requirements allows establishing the completeness of the acquired data prior to analysis.

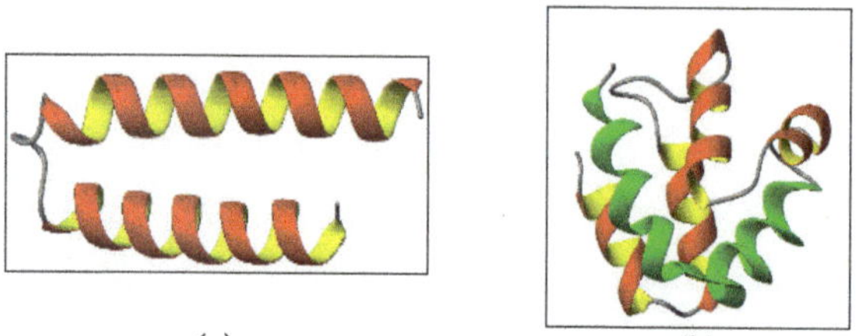

(a) (b)

Fig. 10. Two helical proteins 3LAY (a) and 1A1Z (b) of nearly the same size and different structural complexity.

Another advantage in using RDCs is that the collected data from one portion of a structure, can act as a constraint on any other part of the structure, since all measurements are made with respect to a global point of reference (the common order tensor of the molecule). RDC data collected from the N-terminus of a protein must report the same order tensor as that described by the C-terminus of that protein. This underlying, global relationship between all RDCs significantly enhances their efficacy as global structural constraints. Furthermore, any discrepancy in the assumption regarding the rigidity of a protein can at least be evaluated. This is not the case with NOE data. For example, referring to Fig. 10, once the helical fragments are folded with the measured short-range NOEs, they do not infer any structural restraints for other parts of the molecule.

Piecewise structure determination from NOE data is not always possible, and is often very unlikely, since NOE interactions are normally observed between two atoms that may be anywhere along the backbone of a protein. Therefore, the structure of the entire protein, including that of the side chains, needs to be addressed simultaneously. Simultaneous investigation of the entire protein leads to an exponentially expanding search space that is riddled with many local minima. Although simulated annealing approaches can in theory resolve entrapment in local minima, in practice this requires a large number of redundant structure determination sessions in the hopes of discovering a more suitable structure. The use of RDCs enables the construction of a protein's backbone structure incrementally, through the addition of one amino acid at a time. This progressive strategy is computationally more convenient and allows the direct investigation of the backbone with reduced risk of entrapment in local minima. Addition of the side chains can take place after structure determination of the backbone, thereby benefiting from significant reduction in complexity of the solution-space.

Fig. 11 provides direct evidence of the functional importance of RDC data in high-resolution structure determination. Here we have utilized the solution state structure of Ubiquitin/UIM fusion protein (2KDI) to generate precise NOE and RDC (backbone N-H and C_α-H_α) data using typical order tensors. The computed data were then corrupted through the addition of uniformly distributed noise. The original 2KDI structure was also used to generate 5000 random derivative structures with 0-23Å of deviation, as measured over the backbone atoms with respect to 2KDI. Fig. 11 illustrates the fitness to the simulated data of the 5000 randomly generated structures, versus their backbone deviation (bb-rmsd) from the 2KDI structure. Fitness to the experimental data (RDC in blue and NOE in magenta) is plotted on the vertical axes while bb-rmsd to the high-resolution structure is plotted on the horizontal axis. This figure can be used to ascertain the information content of NOEs versus RDCs in guiding any protein folding strategy. Several conclusions can be made from Fig. 11. First, this figure suggests that backbone N-H and C_α-H_α RDCs are sufficient enough to obtain a protein structure. Second, Fig. 11 suggests that NOEs tend to plateau (lose sensitivity) as the calculated structure approaches the actual structure, while RDCs become more sensitive. Therefore, NOEs may be indiscriminate probes when operating in the range of 0-3.5Å from the actual structure. In contrast, the use of RDC data may very well provide structures within less than 1.0Å from the actual structure. This observation is in agreement with the community wide consensus that X-ray structures fit RDC data better than the NOE based NMR structure of proteins. The final conclusion is that RDCs may be an indispensable source of data in high-resolution structure determination by NMR spectroscopy.

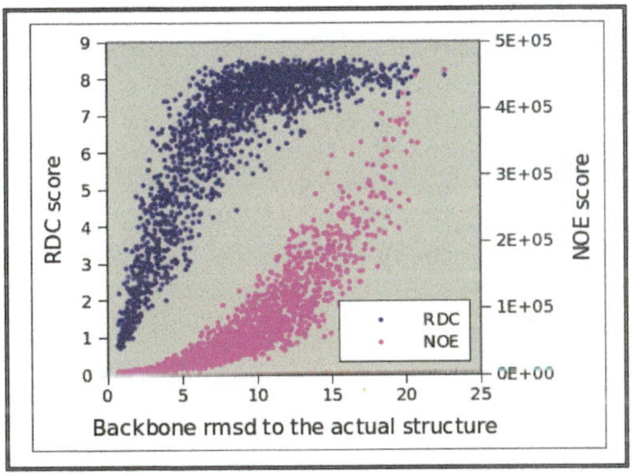

Fig. 11. NOE and RDC fitness of 5000 structures generated randomly from a known structure versus their backbone rmsd to the actual structure.

8. The role of NMR spectroscopy in the era of Computational Biology

Structure determination utilizing traditional NMR techniques, relies on the measurement of Nuclear Overhauser Effects (NOEs) and scalar couplings in order to derive distance and torsion angle constraints, respectively (Montelione *et al.*, 2000). Although NOE constraints will continue to be important for high-throughput structure determination, the measurement of residual dipolar couplings (RDCs) will prove valuable in structure genomics efforts (Montelione *et al.*, 2000). By providing new structural information in qualitative form (Montelione *et al.*, 2000), RDC experiments result in orientational constraints that are complementary to the distance-based constraints available through

NOEs (Prestegard, 1998). Exceptionally challenging proteins, such as membrane bound/associated or glycosylated proteins, are refractory for structure determination by traditional approaches (NMR and X-ray crystallography). This can be attributed to either an insufficient number of the required conventional (distance or NOE) constraints in the case of NMR spectroscopy, or the failure to produce a diffraction-quality crystal for X-ray crystallography. Residual Dipolar Couplings (RDC), are a new type of data that have been anticipated to be instrumental in structural characterization of large proteins, membrane proteins or homo-multimeric protein complexes, to name a few. This is in part due to their rich information content. RDCs also possess the potential for integrating structure determination by NMR spectroscopy (Bryson *et al.*, 2008; Park *et al.*, 2009; Prestegard *et al.*, 2005; Tian *et al.*, 2001b; Valafar *et al.*, 2005), X-ray crystallography (Bansal *et al.*, 2008; Ulmer *et al.*, 2003; Valafar & Prestegard, 2003), and computational modeling (Bansal *et al.*, 2008; Raman *et al.*, 2010b; Valafar & Prestegard, 2003) methods into one unified approach for structural elucidation of biological macromolecules. Because RDCs may be used to characterize the structure and dynamics of challenging proteins, it presents a viable, cost-effective method with the benefit of producing rapid, comprehensive and automated results (Al-Hashimi *et al.*, 2002b; Bailor *et al.*, 2007; Liu *et al.*, 2010; Park *et al.*, 2009; Prestegard *et al.*, 2000; Tian *et al.*, 2001a; Wang *et al.*, 2007).

Another important development to consider, is the automated analysis of NMR data. Many of the interactive tasks related to spectral analysis which are currently performed by experts could, in principle, be performed more efficiently using computational systems (Montelione *et al.*, 2000). This has in fact been demonstrated with proteins ranging from 50 to 200 residues in length (Moseley & Montelione, 1999).

9. Concluding remarks

Here we conclude by summarizing some of the limitations related to modern NMR spectroscopy and briefly describe a method which may help to mitigate the limitations of NMR spectroscopy with respect to large molecules and macromolecular assemblies.

9.1 Limitations of modern NMR spectroscopy

Similar to other methods of structure determination, the accuracy of protein structures determined by NMR are dependent upon the extent and quality of the data that can be obtained (Liu & Hsu, 2005; Montelione *et al.*, 2000). NMR spectroscopy is considered relatively insensitive, typically requiring samples of about 1mM protein concentration; preventing studies of proteins with very low solubilities, thereby limiting certain experimental designs (Montelione *et al.*, 2000). Limitations such as these effect constraints on pulse sequence design and sample stability (Montelione *et al.*, 2000). Although multiple samples may be utilized for the process of structure determination, each sample must be stable (with regard to precipitation, aggregation, and other types of degradation) for any amount of time ranging from days to weeks (Montelione *et al.*, 2000). Furthermore, manual analyses of these multiple NMR datasets are not only laborious and time-consuming, but require expertise (Montelione *et al.*, 2000). Although recent developments hold great promise in reducing the amount of time in structure determination using automated analysis of NMR assignments and 3-dimensional structure (Moseley & Montelione, 1999), general methods for automated analysis of side chain resonance assignment are not yet well developed (Liu & Hsu, 2005; Montelione *et al.*, 2000).

Yet another limitation of NMR analysis is that the density of constraints is occasionally insufficient for accurate structural analysis (Montelione *et al.*, 2000). More specifically, general methods for cross-validation similar to the free R-factor (a statistical measurement used in crystallographic studies for evaluating how well a structure model fits the diffraction data) are not currently available (Montelione *et al.*, 2000). With regard to Residual Dipolar Couplings (RDCs), current limitations include: the efficient identification of alignment media (Montelione *et al.*, 2000), available methods for data extraction and analysis (Jung & Lee, 2004),

The major challenge of NMR spectroscopy, however, is in reducing the amount of time in data collection for structure determination (Liu & Hsu, 2005). The construction of new high-field magnets for enhanced sensitivity, exemplify technological advancements that are of particular interest, yet it is the performance of the NMR probe (used to detect NMR signals) for which the sensitivity of the acquired NMR data depends (Montelione *et al.*, 2000). This can be improved through the introduction of new probes, but is also dependent upon advancements in partial deuteration, which can improve the signal to noise ratios that result from sharper linewidths and longer transverse relaxation times (Gardner & Kay, 1998; Montelione *et al.*, 2000). Transverse Relaxation-Optimized SpectroscopY (TROSY) is another novel technique that may provide significant sensitivity for large proteins by slowly relaxing NMR transitions (Montelione *et al.*, 2000; Pervushin *et al.*, 1997; Wider & Wüthrich, 1999; Wüthrich, 1998). Finally, the proper combination of various sources of data can be very beneficial in overcoming some of the fundamental challenges in NMR spectroscopy. Based on results shown in Fig. 11, it can be concluded that NOE data are much more effective in guiding the protein structure from an extended state to near its native conformation. However, NOE data alone seem to lose structural sensitivity around 3Å from the native structure. RDCs on the other hand seem to provide the needed sensitivity as the structure determination converges toward the native conformation. It is therefore reasonable to speculate that the most effective approach to protein folding from NMR data consists of initial rounds of structure determination by NOE data, followed by structure refinement guided by RDC data in the absence of NOE constraints.

9.2 Contribution of TROSY in mitigating limitations of NMR spectroscopy

Many biologically relevant macromolecules and macromolecular complexes are simply too large for traditional NMR spectroscopy studies, with molecular masses beyond the practical range (Fernandez & Wider, 2003). In fact, conventional NMR based techniques often identify two main problems associated with the solution state study of large molecules and macromolecular assemblies: 1. the large number of acquired resonances causes signals to overlap, making spectral analysis very difficult; and 2. because NMR signals of larger molecules relax faster, it often results in line broadening, poor spectral sensitivity, and eventually no NMR signals (Fernandez & Wider, 2003). TROSY is a technique that has profoundly extended the size limit of macromolecules able to be investigated by NMR, making the analyses of molecular systems of up to 1 kDa possible (Fiaux *et al.*, 2002; Riek *et al.*, 2002). The application of this particular technique has made a wide range of novel applications possible (Fernandez & Wider, 2003) by providing much better sensitivity and line width for large proteins by reducing transverse nuclear spin relaxation during chemical shift evolution (Pervushin *et al.*, 1997). This makes the critical step of resonance assignment possible and allows backbone assignment and secondary structure to be obtained, as first demonstrated with a homo-octameric protein of 110 kDa (Salzmann *et al.*, 2000). In addition, applications of TROSY for side chain resonance assignments have also been demonstrated (Hilty *et al.*, 2002). TROSY has even been incorporated into experimental studies which focus on the dynamics of macromolecules (Zhu *et al.*, 2000). Moreover, TROSY has proven successful in determining structures for some of the most difficult of proteins - membrane proteins (Fernandez *et al.*, 2001a, 2001b). Because TROSY also provides a wide range of NMR measurements with regard to the functional properties of larger macromolecular complexes, it demonstrates great potential for providing clues to the physiological roles of novel proteins and may prove beneficial for drug discovery (Fernandez & Wider, 2003). Other practical applications of TROSY include: the discovery of scalar spin-spin couplings across hydrogen bonds (Cordier & Grzesiek, 1999; Dingley & Grzesiek, 1998; Pervushin *et al.*, 1998), which can be utilized for structure refinement; the measurement of dipolar couplings in large molecules, to determine much larger 3D structures by NMR (Evenäs *et al.*, 2001; Lerche *et al.*, 1999; Yang *et al.*, 1999); and increasing the sensitivity of some triple-resonance experiments for ^{13}C and ^{15}N labeled nucleic acids, so as to increase the range of their functionality to even bigger oligonucleotides (Brutscher & Simorre, 2001; Fiala *et al.*, 2000; Riek *et al.*, 2001; Simon *et al.*, 2001).

In short, the many applications of TROSY and the data made available by its experiments, will contribute significantly by providing important information aimed at solving both future and present biological problems related to the structure and function of large and complex biological molecules.

10. References

Abola E, Kuhn P, Earnest T et al. (2000). Automation of x-ray crystallography. *Nat Struct Biol*, 7 Suppl: 973-977.

Adams PD, Grosse-Kunstleve RW, & Brunger AT (2003). Computational Aspects of High-throughput Crystallographic Macromolecular Structure Determination, In: *Structural bioinformatics*, Philip E. Bourne, Helge Weissig, editors, pp. 75-87, New Jersey: Wiley-Liss, Inc.

Adeyeye J, Azurmendi HF, Stroop CJM et al. (2003). Conformation of the hexasaccharide repeating subunit from the vibrio cholerae o139 capsular polysaccharide. *Biochemistry*, 42: 3979-3988.

Al-Hashimi HM, Bolon PJ, & Prestegard JH (2000a). Molecular symmetry as an aid to geometry determination in ligand protein complexes. *J Magn Reson*, 142: 153-158.

Al-Hashimi HM, Gorin A, Majumdar A et al. (2002a). Towards structural genomics of rna: rapid nmr resonance assignment and simultaneous rna tertiary structure determination using residual dipolar couplings. *J Mol Biol*, 318: 637-649.

Al-Hashimi HM, Gosser Y, Gorin A et al. (2002b). Concerted motions in hiv-1 tar rna may allow access to bound state conformations: rna dynamics from nmr residual dipolar couplings. *J. Mol. Biol.*, 315: pp. 95-102.

Al-Hashimi HM & Patel DJ (2002). Residual dipolar couplings: synergy between nmr and structural genomics. *J Biomol NMR*, 22: 1-8.

Al-Hashimi HM, Valafar H, Terrell M et al. (2000b). Variation of molecular alignment as a means of resolving orientational ambiguities in protein structures from dipolar couplings. *J Magn Reson*, 143: 402-406.

Andrec M, Du PC, & Levy RM (2001). Protein backbone structure determination using only residual dipolar couplings from one ordering medium. *J Biomol NMR*, 21: 335-347.

Assfalg M, Bertini I, Turano P, et al. (2003). N-15-h-1 residual dipolar coupling analysis of native and alkaline-k79a saccharomyces cerevisiae cytochrome c. *Biophys J*, 84: 3917-3923.

Azurmendi HF & Bush CA (2002). Conformational studies of blood group a and blood group b oligosaccharides using nmr residual dipolar couplings. *Carbohydr Res*, 337: 905-915.

Azurmendi HF, Martin-Pastor M, & Bush CA (2002). Conformational studies of lewis x and lewis a trisaccharides using nmr residual dipolar couplings. *Biopolymers*, 63: 89-98.

Bailor MH, Musselman C, Hansen AL et al. (2007). Characterizing the relative orientation and dynamics of rna a-form helices using nmr residual dipolar couplings. *Nature Protocols*, 2: pp. 1536-1546.

Bansal S, Miao X, Adams MWW et al. (2008). Rapid classification of protein structure models using unassigned backbone rdcs and probability density profile analysis (pdpa). *J Magn Reson*, 192: 60-68.

Bax A (2003). Weak alignment offers new nmr opportunities to study protein structure and dynamics. *Protein Science*, 12: 1-16.

Bax A, Kontaxis G, & Tjandra N (2001). Dipolar couplings in macromolecular structure determination. *Methods Enzymol*, 339: 127-174.

Bax A, Vuister GW, Grzesiek S, et al. (1994) Measurement of homonuclear and heteronuclear j-couplings from quantitative j-correlation. In: *Nuclear magnetic resonance*, pt c, pp. 79-105.

Berg J, Tymoczko J, & Stryer L (2002). In: *Biochemistry*, New York: W H Freeman.

Berman HM, Westbrook J, Feng Z et al. (2000). The protein data bank. *Nucleic Acids Res*, 28: 235-242.

Bernado P & Blackledge M (2004a). Anisotropic small amplitude peptide plane dynamics in proteins from residual dipolar couplings. *J Am Chem Soc*, 126: 4907-4920.

Bernado P & Blackledge M (2004b). Local dynamic amplitudes on the protein backbone from dipolar couplings: toward the elucidation of slower motions in biomolecules. *J Am Chem Soc*, 126: 7760-7761.

Bertini I, Luchinat C, Turano P et al. (2003). The magnetic properties of myoglobin as studied by nmr spectroscopy. *Chemistry-a European Journal*, 9: 2316-2322.

Blackledge M (2005). Recent progress in the study of biomolecular structure and dynamics in solution from residual dipolar couplings. *Progress in Nuclear Magnetic Resonance Spectroscopy*, 46: 23-61.

Bodenhausen G & Ruben DJ (1980). Chemical physics letters;natural abundance n-15 nmr by enhanced heteronuclear spectroscopy. *Chemical Physics Letters*, 69: 185-189.

Bouvignies G, Bernado P, & Blackledge M (2005). Protein backbone dynamics from n-hn dipolar couplings in partially aligned systems: a comparison of motional models in the presence of structural noise. *J Magn Reson*, 173: 328-338.

Brunger AT (2007). Version 1.2 of the crystallography and nmr system. *Nat Protoc*, 2: 2728-2733.

Brutscher B & Simorre JP (2001). Transverse relaxation optimized hcn experiment for nucleic acids: combining the advantages of trosy and mq spin evolution. *J Biomol NMR*, 21: 367-372.

Bryson M, Tian F, Prestegard JH et al. (2008). Redcraft: a tool for simultaneous characterization of protein backbone structure and motion from rdc data. *J Magn Reson*, 191: 322-334.

Case, DA (2001). Molecular dynamics and nmr spin relaxation in proteins. *Acc. Chem. Res*, 35: 325-331.

Cavanagh J, Fairbrother WJ, Palmer III AG et al. (1995). In: *Protein nmr spectroscopy principles and practice*, p. 587.

Christendat D, Yee A, Dharamsi A et al. (2000). Structural proteomics of an archaeon. *Nat Struct Biol*, 7: 903-909.

Clore GM & Bewley CA (2002). Using conjoined rigid body/torsion angle simulated annealing to determine the relative orientation of covalently linked protein domains from dipolar couplings. *J. Magn. Reson.*, 154: 329-335.

Clore GM, Gronenborn AM, & Bax A (1998). A robust method for determining the magnitude of the fully asymmetric alignment tensor of oriented macromolecules in the absence of structural information. *J Magn Reson*, 133: 216-221.

Clore GM & Iwahara J (2009). Theory, practice, and applications of paramagnetic relaxation enhancement for the characterization of transient low-population states of biological macromolecules and their complexes. *Chem. Rev*, 109: 4108-4139.

Clore GM, Starich MR, Bewley CA et al. (1999). Impact of residual dipolar couplings on the accuracy of nmr structures determined from a minimal number of noe restraints. *J Am Chem Soc*, 121: 6513-6514.

Clore GM & Schwieters CD (2003). Journal of the american chemical society;docking of protein-protein complexes on the basis of highly ambiguous intermolecular distance restraints derived from h- 1(n)/n-15 chemical shift mapping and backbone n-15-h-1 residual dipolar couplings using conjo. *J Am Chem Soc*, 125: 2902-2912.

Cordier F & Grzesiek S (1999). Direct observation of hydrogen bonds in proteins by interresidue 3hjnc scalar couplings. *J. Am. Chem. Soc*, 121: 1601-1602.

Cornilescu G, Delaglio F, & Bax A (1999). Protein backbone angle restraints from searching a database for chemical shift and sequence homology. *J Biomol NMR*, 13: 289-302.

de Alba E & Tjandra N (2002). Nmr dipolar couplings for the structure determination of biopolymers in solution. *Progress in Nuclear Magnetic Resonance Spectroscopy*, 40: 175-197.

Delaglio F, Kontaxis G, & Bax A (2000). Protein structure determination using molecular fragment replacement and nmr dipolar couplings. *J Am Chem Soc*, 122: 2142-2143.

Demers J & Mittermaier A (2009). Binding mechanism of an sh3 domain studied by nmr and itc. *J Am Chem Soc*, 131: 4355-4367.

Dingley AJ & Grzesiek S (1998). Direct observation of hydrogen bonds in nucleic acid base pairs by internucleotide 2jnn couplings. *J. Am. Chem. Soc*, 120: 8293-8297.

Dosset P, Hus JC, Marion D et al (2001). Journal of biomolecular nmr;a novel interactive tool for rigid-body modeling of multi-domain macromolecules using residual dipolar couplings. *J Biomol NMR*, 20: 223-231.

Evenäs J, Mittermaier A, Yang D et al. (2001). Measurement of (13)c(alpha)-(13)c(beta) dipolar couplings in (15)n,(13)c,(2)h-labeled proteins: application to domain orientation in maltose binding protein. *J Am Chem Soc*, 123: 2858-2864.

Farmer BT, Venters RA, Spicer LD et al. (1992). A refocused and optimized hnca: increased sensitivity and resolution in large macromolecules, pp. 195-202.

Fernandez C, Adeishvili K, & Wüthrich K (2001a). Transverse relaxation-optimized nmr spectroscopy with the outer membrane protein ompx in dihexanoyl phosphatidylcholine micelles. *Proc Natl Acad Sci U S A*, 98: 2358-2363.

Fernandez C, Hilty C, Bonjour S et al. (2001b). Solution nmr studies of the integral membrane proteins ompx and ompa from escherichia coli. *FEBS Lett*, 504: 173-178.

Fernandez C & Wider G (2003). Trosy in nmr studies of the structure and function of large biological macromolecules. *Current Opinion in Structural Biology*, 13: 570-580.

Fiala R, Czernek J, & Sklenár V (2000). Transverse relaxation optimized triple-resonance nmr experiments for nucleic acids. *J Biomol NMR*, 16: 291-302.

Fiaux J, Bertelsen EB, Horwich AL et al. (2002). Nmr analysis of a 900k groel-groes complex, 418: 207-211.

Fowler CA, Tian F, Al-Hashimi HM et al. (2000). Rapid determination of protein folds using residual dipolar couplings. *J Mol Biol*, 304: 447-460.

Gardner KH & Kay LE (1998). The use of 2h, 13c, 15n multidimensional nmr to study the structure and dynamics of proteins. *Annu Rev Biophys Biomol Struct*, 27: 357-406.

Grzesiek S & Bax A (1993). Amino acid type determination in the sequential assignment procedure of uniformly 13c/15n-enriched proteins. *J Biomol NMR*, 3: 185-204.

Henzler-Wildman K & Kern D (2007). Dynamic personalities of proteins. *Nature*, 450: 964-972.

Hilty C, Fernández C, Wider G et al. (2002). Side chain nmr assignments in the membrane protein ompx reconstituted in dhpc micelles. *J Biomol NMR*, 23: 289-301.

Hornak J (1997). In: *The basics of nmr*, Retrieved from <www.cis.rit.edu/htbooks/nmr>

Jeener J, Meier BH, Bachmann P et al. (1979). Investigation of exchange processes by two-dimensional nmr spectroscopy. *J Chem Phys*, 71: 4546-4553.

Jung J & Lee W (2004). Structure-based functional discovery of proteins: structural proteomics. *J Biochem Mol Biol*, 37: 28-34.

Kay LE, Ikura M, Tschudin R et al. (1990). Three-dimensional triple-resonance nmr spectroscopy of isotopically enriched proteins. *Journal of Magnetic Resonance (1969)*, 89: 496-514.

Key J, Scheuermann TH, Anderson PC et al. (2009). Principles of ligand binding within a completely buried cavity in hif2alpha pas-b. *J Am Chem Soc*, 131: 17647-17654.

Kleckner IR & Foster MP (2011). An introduction to nmr-based approaches for measuring protein dynamics. *Biochimica et Biophysica Acta (BBA) - Proteins & Proteomics*, 1814: 942-968.

Langmead CJ & Donald BR (2003). 3d structural homology detection via unassigned residual dipolar couplings. *Proc IEEE Comput Soc Bioinform Conf*, 2: p. pp. 209-217.

Lerche MH, Meissner A, Poulsen FM et al. (1999). Pulse sequences for measurement of one-bond (15)n-(1)h coupling constants in the protein backbone. *J Magn Reson*, 140: 259-263.

Liu H & Hsu J (2005). Recent developments in structural proteomics for protein structure determination. *Proteomics*, 5: 2056-2068.

Liu Y, Kahn RA, & Prestegard JH (2010). Dynamic structure of membrane-anchored arf*gtp. *Nat Struct Mol Biol*, 17: 876-881.

Loria JP, Berlow RB, & Watt ED (2008). Characterization of enzyme motions by solution nmr relaxation dispersion. *Acc. Chem. Res*, 41: 214-221.

Losonczi JA, Andrec M, Fischer MWF et al. (1999). Order matrix analysis of residual dipolar couplings using singular value decomposition. *J Magn Reson*, 138: 334-342.

Marassi FM & Opella SJ (2002). Simultaneous resonance assignment and structure determination in the solid-state nmr spectrum of a membrane protein in lipid bilayers. *Biophys J*, 82: 467A-467A.

Marassi FM & Opella SJ (2003). Simultaneous assignment and structure determination of a membrane protein from nmr orientational restraints. *Protein Science*, 12: 403-411.

Markley JL, Ulrich EL, Westler WM, et al. (2003). Macromolecular Structure Determination by NMR Spectroscopy, In: *Structural bioinformatics*, Bourne PE, Weissig H, editors, pp. 89-113, New Jersey: Wiley-Liss, Inc.

Montelione GT, Zheng DY, Huang YPJ et al. (2000). Protein nmr spectroscopy in structural genomics. *Nature Structural Biology*, 7: 982-985.

Moseley HN & Montelione GT (1999). Automated analysis of nmr assignments and structures for proteins. *Curr Opin Struct Biol*, 9: 635-642.

Muhandiram D & Kay L (1994). Gradient-enhanced triple-resonance three-dimensional nmr experiments with improved sensitivity. *Journal of Magnetic Resonance, Series B*, 103: 203-216.

Nitz M, Sherawat M, Franz KJ, et al. (2004). Structural origin of the high affinity of a chemically evolved lanthanide-binding peptide. *Angew. Chem. Int. Ed. Engl.*, 43: pp. 3682-3685.

O'Neil-Cabello E, Bryce DL, Nikonowicz EP et al. (2004). Measurement of five dipolar couplings from a single 3d nmr multiplet applied to the study of rna dynamics, 126: 66-67.

Ooi L (2010). In: *Principles of x-ray crystallography*, New York: Oxford University Press.

Opella SJ & Marassi FM (2004). Structure determination of membrane proteins by nmr spectroscopy. *Chemical Reviews*, 104: 3587-3606.

Park SH, Son WS, Mukhopadhyay R et al. (2009). Phage-induced alignment of membrane proteins enables the measurement and structural analysis of residual dipolar couplings with dipolar waves and lambda-maps. *J Am Chem Soc*, 131: 14140-14141.

Pervushin K, Ono A, Fernandez C et al. (1998). Nmr scalar couplings across watson-crick base pair hydrogen bonds in dna observed by transverse relaxation-optimized spectroscopy. *Proc Natl Acad Sci U S A*, 95: 14147-14151.

Pervushin K, Riek R, Wider G et al. (1997). Attenuated t-2 relaxation by mutual cancellation of dipole-dipole coupling and chemical shift anisotropy indicates an avenue to nmr structures of very large biological macromolecules in solution, 94: 12366-12371.

Press W, Teukolsky SA, Vetterling WT et al. (2002). In: *Numerical recipes in c, the art of scientific computing*.

Prestegard JH (1998). New techniques in structural nmr--anisotropic interactions. *Nat Struct Biol*, 5 Suppl: 517-522.

Prestegard JH, al-Hashimi HM, & Tolman JR (2000). Nmr structures of biomolecules using field oriented media and residual dipolar couplings. *Q Rev Biophys*, 33: 371-424.

Prestegard JH, Mayer KL, Valafar H et al. (2005). Determination of protein backbone structures from residual dipolar couplings. *Methods Enzymol*, 394: 175-209.

Prestegard JH, Valafar H, Glushka J et al. (2001). Nuclear magnetic resonance in the era of structural genomics. *Biochemistry*, 40: 8677-8685.

Prestegard JH & Kishore A (2001). Current opinion in structural biology;partial alignment of biomolecules: an aid to nmr characterization. *Current Opinion in Structural Biology*, 5: 584-590.

Raman S, Huang YJ, Mao B et al. (2010a). Accurate automated protein nmr structure determination using unassigned noesy data. *J. Am. Chem. Soc.*, 132: pp. 202-207.

Raman S, Lange OF, Rossi P et al. (2010b). Nmr structure determination for larger proteins using backbone-only data. *Science*, 327: 1014-1018.

Riek R, Fiaux J, Bertelsen EB et al. (2002). Solution nmr techniques for large molecular and supramolecular structures. *J Am Chem Soc*, 124: 12144-12153.

Riek R, Pervushin K, Fernandez C et al. (2001). [(13)c,(13)c]- and [(13)c,(1)h]-trosy in a triple resonance experiment for ribose-base and intrabase correlations in nucleic acids. *J Am Chem Soc*, 123: 658-664.

Rohl CA & Baker D (2002). De novo determination of protein backbone structure from residual dipolar couplings using rosetta. *J Am Chem Soc*, 124: 2723-2729.

Sali A, Glaeser R, Earnest T et al. (2003). From words to literature in structural proteomics. *Nature*, 422: 216-225.

Salzmann M, Pervushin K, Wider G et al. (2000). Nmr assignment and secondary structure determination of an octameric 110 kda protein using trosy in triple resonance experiments. *J. Am. Chem. Soc*, 122: 7543-7548.

Saupe A & Englert G (1963). High-resolution nuclear magnetic resonance spectra of orientated molecules. *Phys Rev Lett*, 11: 462-464.

Schwieters CD, Kuszewski JJ, Tjandra N et al. (2003). The xplor-nih nmr molecular structure determination package. *J Magn Reson*, 160: 65-73.

Shealy P, Liu Y, Simin M et al. (2011). Backbone resonance assignment and order tensor estimation using residual dipolar couplings. *J Biomol NMR*, 50: 357-369.

Simon B, Zanier K, & Sattler M (2001). A trosy relayed hcch-cosy experiment for correlating adenine h2/h8 resonances in uniformly 13c-labeled rna molecules. *J Biomol NMR*, 20: 173-176.

Tian F, Al-Hashimi HM, Craighead JL et al. (2001a). Conformational analysis of a flexible oligosaccharide using residual dipolar couplings. *J Am Chem Soc*, 123: 485-492.

Tian F, Valafar H, & Prestegard JH (2001b). A dipolar coupling based strategy for simultaneous resonance assignment and structure determination of protein backbones. *J Am Chem Soc*, 123: 11791-11796.

Tjandra N & Bax A (1997). Direct measurement of distances and angles in biomolecules by nmr in a dilute liquid crystalline medium. *Science*, 278: 1111-1114.

Tjandra N, Grzesiek S, & Bax A (1996). Magnetic field dependence of nitrogen-proton j splittings in n-15-enriched human ubiquitin resulting from relaxation interference and residual dipolar coupling. *J Am Chem Soc*, 118: 6264-6272.

Tjandra N, Omichinski JG, Gronenborn AM et al. (1997). Use of dipolar h-1-n-15 and h-1-c-13 couplings in the structure determination of magnetically oriented macromolecules in solution. *Nature Structural Biology*, 4: 732-738.

Tjandra N, Tate S, Ono A et al. (2000). The nmr structure of a dna dodecamer in an aqueous dilute liquid crystalline phase. *J. Am. Chem. Soc.*, 122: 6190-6200.

Tolman JR (2001). Dipolar couplings as a probe of molecular dynamics and structure in solution. *Curr Opin Struct Biol*, 11: 532-539.

Tolman JR, Flanagan JM, Kennedy MA et al. (1995). Nuclear magnetic dipole interactions in field-oriented proteins - information for structure determination in solution. *Proc Natl Acad Sci U S A*, 92: 9279-9283.

Tolman JR & Prestegard JH (1996). Measurement of amide n-15-h-1 one-bond couplings in proteins using accordion heteronuclear-shift-correlation experiments. *Journal of Magnetic Resonance Series B*, 112: 269-274.

Ulmer TS, Ramirez BE, Delaglio F et al. (2003). Evaluation of backbone proton positions and dynamics in a small protein by liquid crystal nmr spectroscopy. *J Am Chem Soc*, 125: 9179-9191.

Valafar H, Mayer K, Bougault C et al. (2005). Backbone solution structures of proteins using residual dipolar couplings: application to a novel structural genomics target. *J Struct Funct Genomics*, 5: 241-254.

Valafar H & Prestegard JH (2003). Rapid classification of a protein fold family using a statistical analysis of dipolar couplings. *Bioinformatics*, 19: 1549-1555.

Valafar H & Prestegard J (2004). Redcat: a residual dipolar coupling analysis tool. *J Magn Reson*, 167: 228-241.

Vermeulen A, Zhou HJ & Pardi A (2000). Determining dna global structure and dna bending by application of nmr residual dipolar couplings. *J Am Chem Soc*, 122: 9638-9647.

Wang J, Walsh JD, Kuszewski J et al. (2007). Periodicity, planarity, and pixel (3p): a program using the intrinsic residual dipolar coupling periodicity-to-peptide plane correlation and phi/psi angles to derive protein backbone structures. *J Magn Reson*, 189: 90-103.

Wang LC & Donald BR (2004). Exact solutions for internuclear vectors and backbone dihedral angles from nh residual dipolar couplings in two media, and their application in a systematic search algorithm for determining protein backbone structure. *J Biomol NMR*, 29: 223-242.

Wang X, Bansal S, Jiang M et al. (2008). Rdc-assisted modeling of symmetric protein homo-oligomers. *Protein Sci.*, 17: p. pp. 899-907.

Wider G & Wüthrich K (1999). Nmr spectroscopy of large molecules and multimolecular assemblies in solution. *Curr Opin Struct Biol*, 9: 594-601.

Wilson MA & Brunger AT (2000). The 1.0 a crystal structure of ca(2+)-bound calmodulin: an analysis of disorder and implications for functionally relevant plasticity. *J Mol Biol*, 301: 1237-1256.

Wishart DS & Nip AM (1998). Protein chemical shift analysis: a practical guide. *Biochem Cell Biol*, 76: 153-163.

Wishart DS & Sykes BD (1994). The 13c chemical-shift index: a simple method for the identification of protein secondary structure using 13c chemical-shift data. *J Biomol NMR*, 4: 171-180.

Wishart DS, Sykes BD, & Richards FM (1991). Relationship between nuclear magnetic resonance chemical shift and protein secondary structure. *J Mol Biol*, 222: 311-333.

Wishart DS, Sykes BD, & Richards FM (1992). The chemical shift index: a fast and simple method for the assignment of protein secondary structure through nmr spectroscopy. *Biochemistry*, 31: 1647-1651.

Wüthrich K (1998). The second decade--into the third millenium. *Nat Struct Biol*, 5 Suppl: 492-495.

Yang DW, Venters RA, Mueller G et al. (1999). Trosy-based hnco pulse sequences for the measurement of 1hn-15n, 15n-13co, 1hn-13co, 13co-13ca and 1hn-13cî± dipolar couplings in 15n, 13c, 2h-labeled proteins. *J. Biomol. NMR*, 14: 333-343.

Yi X, Venot A, Glushka J et al. (2004). Glycosidic torsional motions in a bicelle-associated disaccharide from residual dipolar couplings. *J Am Chem Soc*, 126: 13636-13638.

Zeeb M & Balbach J (2004). Protein folding studied by real-time nmr spectroscopy. *Methods*, 34: 65-74.

Zhou HJ, Vermeulen A, Jucker FM et al. (1999). Incorporating residual dipolar couplings into the nmr solution structure determination of nucleic acids. *Biopolymers*, 52: 168-180.

Zhu G, Xia Y, Nicholson LK et al. (2000). Protein dynamics measurements by trosy-based nmr experiments. *J Magn Reson*, 143: 423-426.

Section 2

Structure Prediction

4

An Exhaustive Shape-Based Approach for Proteins' Secondary, Tertiary and Quaternary Structures Indexing, Retrieval and Docking

Eric Paquet and Herna L. Viktor
National Research Council & University of Ottawa, Canada

1. Introduction

Over the past ten years, the number of three-dimensional protein structures has grown exponentially (Holm, 2008). This is due, mainly, to the advent of high throughput systems. Consequently, molecular biologists need systems to enable them to effectively store, manage and explore these vast repositories of three-dimensional structures. They want to determine if an unknown structure is in fact a new one, if it has been subjected to a mutation, and/or to which family it possibly belongs. Furthermore, they require the ability to find similar proteins in terms of functionalities. Importantly, they aim to find docking sites. That is, they aim to determine the possible sites for the binding of two proteins, namely the ligand and the receptor, in order to form a stable complex. This similarity in functionality, and specifically the task to find docking sites, are related to outer the shape of the protein (Binkowski & Joachimiak., 2008). The outer shape (or envelope), in part, determines whether two proteins may have similar functionalities and may thus aid us to determine the location of such protein binding sites. The previously introduced docking problem may be better understood from the perspective of drug design. Most diseases and drugs work on the same basic principle. When we become ill, a foreign protein docks itself on a healthy protein and modified its functionality. Such a docking is possible if the two proteins have two sub-regions that are compatible in terms of three-dimensional shape, a bit like two pieces of a puzzle. Drugs are designed to act in a similar way. Namely, a drug docks on the same active site and prevents the docking of foreign proteins which can potentially cause illness (Paquet and Viktor, 2010).

Consequently, one of the main objectives of macromolecular docking, also known as protein-protein docking, is to find compatible regions in between two proteins from a geometrical point of view; the better the fit, the better the efficiency of the docking. There are currently many approaches that have been developed in order to simulate in silico, i.e. with a computer and with algorithms, this docking of two macromolecules. We shall briefly review them in order to enhance their most salient features, their potential and their main weaknesses. We shall limit ourselves to the most recent works, which at any rate, capture most previous advances.

This chapter is organized as follow. In Section 2, approaches based on correlation techniques are presented, while in Section 3 various methods based on spherical harmonics and orthogonal polynomial are reviewed. Section 4 briefly describes shape signatures based on curvature invariants and Section 5 provides an overview of distance-based techniques. In Section 6, techniques based on alignment and Monte Carlo simulations are explored while the specific contribution of computer graphics is outlined in Section 7. Section 8 completes this chapter by presenting some high throughput methods. The main conclusions are presented in Section 9.

2. Correlation-based techniques

One of the earliest approaches for macromolecule docking is based on finding correlations. Such methods aim to determine the best alignment, in terms of translations, in between two macromolecules. In order to evaluate the correlation, a representation or shape must be associated with each macromolecule. Such a representation may take various forms, amongst which the binary and the volumetric representations are the most common. Then, given a representation, the product of their shapes is evaluated for each relative translation in order to determine their relative overlapping. Similar regions, that is, regions that present a substantial overlap, are usually characterized by a high value of their correlation. The position of the correlation peak determines the relative translation in between the two macromolecules. Because the relative translation is unknown, the translation space must be searched exhaustively. In order to be more efficient from a computational point of view, the correlation is implemented with the Fourier transform. To further reduce the complexity of the calculations, the Fourier transform is evaluated with the fast Fourier transform algorithm.

Many methods distinguish themselves by the representation they use in order to describe the proteins. For instance, (Katchalskip-Katzir et al., 1992) use a binary representation for the macromolecule. Here, the points inside the boundary and on the boundary have a value of one, while external points have a value of zero. Such a representation tends to be oversensitive and only allows for exact marching. In an attempt to further speed up the calculations, (Nukuda et al., 2007) implement the correlation on an IBM Blue Gene supercomputer. By also taking the rotations into account, they sample the rotation space and compute a separate correlation for each sampled rotation. Such an exhaustive approach does not scale well, even with a supercomputer. A similar approach is used by (Kasakov et al., 2006), with the difference being that the binary representation is replaced by a continuous one, which is obtained from the molecular potentials associated with the constituent atoms. Their method provides a representation both in terms of shape and physicochemical properties and offers more robustness, since the potentials are real valued. Nevertheless, the scalability problem remains unsolved. Sukhwani et al. (Sukhwani et al., 2008) also use more sophisticated molecular potentials and parallelize the calculation of the correlation. However, the problems associated with an exhaustive search in the rotation space are still untargeted. In order to address this problem, (Vadja & Kozakov, 2009) and (Gray, 2006) propose to replace the exhaustive sampling of the rotation space by a Monte Carlo optimization of a cost function associated with the three-dimensional rotation.

Consequently, a transition probability is associated with the process. Rotations are assimilated to a random walk and are accepted, or rejected, according to a Metropolis criterion based on the transition probability. Gray, in particular, uses the simulated annealing algorithm in order to optimize the cost function. Such Monte Carlo approaches allow one to explore efficiently, and parsimoniously, the rotation space and are consequently much more scalable. The rotation may also be obtained directly in a non iterative way. For instance, (Katebi et al., 2009) determine the rotation in between two macromolecules, with the singular value decomposition (SVD) based on the relative position of the constituent carbon atoms. This approach leads to an oversimplification of the shape, which is not necessarily compatible with docking. Here, SVD has been selected against principal component analysis (PCA) because of its statistical robustness.

Finally, (Bonvin et al., 2006) present general considerations, applicable to all these methods, about the flexibility of proteins. It is well known that most macromolecules are flexible. This flexibility must be taken into account when simulating docking. To this end, Bonvin et al. propose to use various conformations of the same protein, as obtained from nuclear magnetic resonance (NMR), and to repeat the calculations for each one of them. This approach requires a large amount of experimentally determined structures in order to provide realistic results. This, in turn, involves an exhaustive search which eventually does not scale well. Furthermore, NMR is currently limited to relatively small macromolecules. Bonvin et al., also propose to parameterize the deformations of the macromolecules. However, their parameterization appears to be relatively arbitrary and their proposal thus would need further evaluation to determine the feasibility and suitability thereof.

3. Angular basis, spherical harmonics and orthogonal polynomials

In order to address the above-mentioned problems associated with the correlation-based techniques, methods based on spherical harmonics and orthogonal polynomials have been developed. The aim, here, is to provide a representation more adapted to rotation. One of the seminal works was presented by Canterakis (Canterakis, 1994).

Sael & Kihara (Sael & Kihara, 2010a) use a coordinate system centred on the barycentre of the macromolecules. They then project each one of them on an orthogonal basis, formed by the product of a spherical harmonic and an orthogonal polynomial which is, in this particular case, the Zernike polynomial. The resulting basis is known as the 3D Zernike polynomial. Then, rotation invariants are calculated from this representation. The invariants associated with each macromolecule are compared to each other using the Euclidian distance. Although relatively efficient, this approach is limited to global comparison and is not suitable for docking in its present form. Based on their similar work, (Sael & Kihara, 2010b) have extended their approach to docking. Pockets are extracted from the macromolecules and are individually describes in terms of the 3D Zernike polynomials, from which rotation invariants are calculated. In order to determine a docking region, each pair of invariants, that is, each "pair of pockets", must be independently compared. This limits the scalability of the method. Furthermore, it has been pointed out that pockets are far from being the only candidates for docking. This means that many potentially interesting docking regions are potentially ignored by the algorithm. Consequently, (Sael & Kihara, 2010d) expanded their work by randomly select an ensemble of small regions on the macromolecules. Each region is represented in terms of 3D Zernike polynomials, from which a set of rotation invariants are calculated. In order to alleviate the scalability problem which occurs when each pair of invariants is compared, they cluster the invariants with a self organizing map, also known as the Kohonen map. Here, the number of clusters are not predetermined but determined by the algorithm itself. In order to avoid the problems associated with the border effect, they use a boundless Kohonen map with either a spherical or a toroidal topology. Being unsupervised, the Kohonen map may lead to nonsensical results. Furthermore, the 3D Zernike polynomials are more adapted to closed shapes like macromolecules than to an open regions or patches. Finally, (Sael et al., 2008) study the impact of the metric when comparing the rotation invariants. They compare the Euclidian distance, the Manhattan distance and the correlation coefficient, which all seem to provide comparable results in this particular case.

Venkatraman et al. (Venkatraman et al., 2009) also use a global approach, based on 3D Zernike polynomials. The shape is represented by a mixture of Gaussians centred on the constituent carbon atoms. This analytical representation seems relatively adapted to represent the shape of a macromolecule. Nevertheless, the limitations associated with the global approach of Sael et al. remain. Mak et al. (Mak et al., 2008) also employ a global approach similar to Sael et al. However, in contrast with Venkatraman et al., they use a binary voxelised representation for the shape of the macromolecule, which presents a lower expressive power.

Sovic et al. (Sovic et al., 2010) propose a different approach based on the 3D Zernike polynomials. Each macromolecule is represented with a subset of 3D Zernike polynomials. As pointed out earlier, these polynomials are particularly suited for closed shapes presenting a certain degree of spherical symmetry, as many macromolecules do. Then, each pair of macromolecules is compared in that particular representation. Translation invariance is obtained from a standard correlation while rotation invariance is obtained by exhaustively searching the rotation space. Sovic et al. claim that it is computationally less expensive to perform the alignment in the 3D Zernike polynomials representation than it is in the standard Euclidian representation. However, they do not provide evidence that their approach is scalable.

Scalability is a particularly acute issue for docking. There are currently more than 76,000 (28 October 2011) macromolecular structures that have been experimentally determined and which may be found in the Protein Data Bank (PDB) (Dutta et al., 2009). Let us assume, for instance, that we partition each structure in a thousand of structural regions or patches. Then, evaluating any potential match in between two patches would involve 2,812,499,962,500,000 comparisons, if one is to consider all the combinations of two patches selected from a set of 75,000,000 patches. Of course, this is an oversimplification of the problem, but it shows that the importance of the scalability should not be underestimated.

Furthermore, (Ritchie, 2005; Mavridis & Ritchie, 2010) propose a new approach which follows Canterakis' scheme. Here, the shape is represented in terms of spherical harmonics for the angular part and in terms of orthogonal polynomials for the radial part. However, they replace the Zernike polynomials by either the spherical Bessel Gaussian-type orbitals (GTO) which is formed by the product of a Gaussian and a Laguerre polynomial, the exponential type orbitals (ETO) which is formed by the product of an exponential and of a Jacobi polynomial or the Bessel type orbitals. They use various representations for the shape of the macromolecule, including the electronic density. These orthogonal polynomials seem to be more adapted to the geometry of macromolecules than the Zernike polynomial, although no definitive conclusions may be drawn at this stage.

4. Curvature invariants

Geppert et al. (Geppert et al., 2010) extract local curvature invariants, in terms of rigid transformations, from the envelope. They use geometric hashing in order to structure the invariants and in order to compare them efficiently. Nevertheless, geometric hashing is notoriously known for its lack of scalability and robustness. Furthermore, the expressive power of local curvature invariants is relatively low. Geppert et al. claim that the local curvature invariants are oblivious against a change of conformation (robustness against the macromolecule flexibility). This is certainly true for the regions where there is a limited bending or deformation, but it does not hold in general. Ranganath et al. (Ranganath et al., 2007) also extract local curvature invariants, in terms of rigid transformations, from the envelopes of the macromolecules. The calculation of the curvature invariant is based on the wavelet transform. Although the calculation of the curvature invariant is more robust when performed with the wavelet transform, the limitations inherent to these invariants remain.

5. Distance-based and graph-based methods

Liu et al. (Liu et al., 2009) propose to sample the envelope of each macromolecule with a certain number of landmarks. The landmarks are first obtained by randomly sampling points on the surface of the macromolecule and then by clustering them in order to obtain an informative sample. Then, the distances in between each pair of landmarks are calculated and their probability distribution is determined. Here, the distance is defined has the shortest path in between two landmark points within the macromolecular shape. This distance is also known as the geodesic distance which can be obtained, for instance, with the Dijkstra algorithm (Dijkstra, 1959). This probability constitutes a unique geometrical signature associated with the global shape of the macromolecule. The distance, as previously defined, is invariant under rigid transformations, invariant to isometries and relatively oblivious to a small set of general transformations.

Chi (Chi, 2004) follows an approach inspired by image indexing. He computes the distance in between the constituent carbon atoms associated with each macromolecule and creates a distance matrix. Such a matrix is invariant under rigid transformations. Then, the texture associated with the distance matrix is analyzed in terms of uniformity, entropy, homogeneity, contrast, correlation and cluster tendency. These measures constitute the geometrical signature of the macromolecule. It should be noted that the carbon atoms provide a very low resolution representation of the shape of the macromolecule. This approach has been extended by (Chen et al., 2010). They propose to analyze the texture associated with the distance matrix in terms of principal component analysis (PCA). Although more efficient from an information theory point of view, this approach still provide a very low resolution representation of the global shape of the macromolecule.

Furthermore, (Zhang et al. 2009), (Peng & Tsay, 2010), (Novosad et al., 2010) and (Reddy et al., 2011) also use the relative position of the constituent carbon atoms of the macromolecules. The distance matrix is replaced by a graph associated with the topology of the carbon atoms. It should be noted that the graph is highly sensitive to the original topology of the carbon atoms. Such a graph is invariant under rigid transformation. By using partial matching of graphs, they evaluate possible docking configurations. Nevertheless, the expressive power of a graph based on the topology of the carbon atoms is relatively low. Furthermore, extensive partial graph matching may lead to an excessive computational complexity.

Borgwardt et al. (Borgwardt et al., 2005) follow the same approach then in (Zhang et al., 2009; Peng & Tsay, 2010; Novosad et al., 2010; Reddy et al., 2005) for the construction of the graphs associated with the macromolecules. The graphs are compared with a graph kernel which probes the graphs with a random walk. Although the limitations previously outlined remain, it is interesting to note that there is a strong fundamental connection in between the shape and the random walk, which should be further explored. Such a connexion might lead to better metrics for shapes comparison.

6. Alignment and Monte Carlo based methods

In their respective papers, (Wang et al., 2007) and (Hoffmann et al., 2010) propose to find the best global alignment, in terms of rigid transformations, in between two macromolecules. Wang et al. define a cost function which evaluates the local discrepancy in between the shape of two macromolecules. Then, they minimize the cost function in terms of translation and rotation in order to find the best alignment. The rotations are either represented in terms of matrices or in terms of quaternions. Algorithms based on the later are often computationally more efficient. In order to avoid local minima, a Monte Carlo approach is used for the optimization. A transition probability function is defined which allows to explore efficiently and parsimoniously the transformation space and to escape local minima as required during the exploration stage. Wang et al.'s optimization approach is based on the simulated annealing technique, which mimics the physical optimization of the crystalline structure associated with a solid during his cooling phase. Hoffmann et al. propose to represent the shape of the macromolecule with a mixture of Gaussians centred on the constituent carbon atoms and to perform a pre-alignment of each pair of macromolecules with principal component analysis (PCA). The pre-alignment facilitates the convergence of the simulated annealing algorithm toward the global minimum and reduces the computational complexity, since PCA is a linear algorithm. Both methods are relevant for precise global comparison and have some degree of relevance for docking.

7. Computer graphics based methods

In addition to the previous methods based on invariant patterns recognition, graph theory and computer vision, methods originating from computer graphics have also been suggested. For instance, (Zauhar, 2003) proposes to randomly sample a set of points on the envelope of a macromolecule. Next, for each one of these points, he propagates a virtual optical ray inside the envelope up to a predetermined number of reflexions. The probability distribution associated with the length of the rays constitutes a geometrical signature for the global shape of the macromolecule. Such a representation is invariant under rigid transformations. It is unlikely that such a method may be applied for docking, because the patches involve in docking are not closed shapes and consequently are not suitable for virtual light ray propagation: most light rays would escape the patch without any reflexion.

Shapira et al. (Shapira et al., 2008) propose a robust definition of the diameter for a closed object called the shape diameter function (SDF). Given a point on the surface of a closed object, a cone is constructed. The apex of the cone is the given point and the axis of the cone is the inner normal associated with the given point. The opening angle for the cone is typically 120 degrees. Then, the SDF is defined as the weighted average of the lengths of the optical rays that propagate from the apex to the closest point on the surface of the shape under the constraint that they remain inside the cone. The weight is defined as the inverse of the angle in between the ray and the axis of the cone.

The SDF may be defined for each point of the object or for a given subset. In order to describe the shape, a histogram of its associated SDFs is constructed which constitutes its geometric signature. With the aim of having an analytical form for the histogram, a mixture of Gaussians is constructed from the later. Such a signature is invariant under rigid transformations and oblivious under more general transformations. The degree of obliviousness is far from clear, though. Furthermore, the signature seems to be deficient from an expressiveness point of view. With the intention of addressing the expressiveness problem, (Gal et al., 2007) define two bidimensional histograms which encapsulate additional information about the shape. This is done so as to obtain a better discrimination in between shapes. Given a set of points on the surface of the object, the first histogram is constructed from the SDF at each point and from the distance in between each point and the barycentre. The second histogram is constructed from the SDF and from the centricity function calculated at each point. Given a point on the surface of the object, the centricity function (CF) is defined as the average geodesic distance from that point to all the other points in a given neighbourhood. It should be noted that the CF is invariant under isometry. Gal et al.'s approach seems to provide better discrimination. In addition, the SDF appears to provide a natural partition or segmentation of simple shapes, in the sense that points belonging to the same part tend to have a comparable value for their SDF.

Fang et al. (Fang et al., 2009) have applied the SDF to the description of macromolecular shapes. The points forming the envelope of the protein are sampled by clustering them with the K-mean algorithm, with the purpose of obtaining a more informative sample as opposed to random sampling. For each point belonging to the sample, they compute the SDF. Finally, they construct a histogram of the later. This histogram constitutes the global geometry signature for the envelope. Signatures and consequently shapes are compared with the Euclidian distance.

Lo et al. (Lo et al., 2010) propose to describe the envelope of macromolecules with solid angles. The solid angles are computed for each point of the envelope. Then, the surface is partitioned, or segmented, by clustering the solid angles based on the observation that points belonging to a given pocket tend to have a comparable solid angle. In order to speed up the calculation of the solid angles, Lo et al. take advantage of the graphical processing unit (GPU) and of the Compute Unified Device Architecture (CUDA) parallel technology. Docking candidates are found by exhaustively matching the various patches. The expressiveness power of the solid angle is very low. Also, the algorithm, in its present form, may only detect certain types of pockets, while ignoring a large proportion of potentially interesting docking sites.

8. High throughput methods

Shibberu & Holder (Shibberu & Holder, 2011) present an approach for proteins alignment. They compute the distance in between each pair of constituent carbon atoms and construct a distance matrix. A continuous cut-off function is applied to the distance matrix, in order to obtain a contact matrix. The quadratic form constructed from this contact matrix defines a generalized inner product. With the help of the Eigen decomposition of the contact matrix, it is possible to associate a unique Eigen value to each residue or carbon atom. A scoring matrix, which measures the distance in between each pair of Eigen values associated with two distinct proteins, is constructed in order to evaluate the quality of the alignment. The carbon atoms provide only a low resolution description of the surface of the protein. The assignment of an Eigen value to a specific residue might be subject to a certain level of ambiguity and might be computationally unstable, for instance, when some Eigen values are very similar.

Hue et al. (Hue et al., 2010) propose a method based on support vector machine (SVM) which aims to address the scalability problem. Instead of predicting the docking regions, they try to predict the outcome of the docking in between two macromolecules. The SVM algorithm determines the optimal partition, in terms of a hyperplane, in between two classes: the interacting macromolecules and the non-interacting ones. The SVM is trained with a training set of macromolecules for which the presence or the absence of interaction is known experimentally. For each pair of macromolecules in the training set, an alignment is performed and a feature vector, here a similarity matrix, is associated with the quality of the alignment. These feature vectors span a vector space which is partitioned by the SVD with a hyperplane into two classes, namely interacting and non-interacting macromolecules. Once the training is completed, the presence of the absence of interaction in between two proteins may be determined by aligning them. This is accomplished by computing the feature vector associated with the quality of the alignment and by determining the region, or class, to which the feature vector belongs. In order to improve the efficiency of the method, the metric used to quantify the alignment, the Euclidian inner product, is replaced by a nonlinear kernel. This allows one to improve the classification accuracy by introducing a non linearity in what is, otherwise, a linear partition of the classification space. Further, a singular value decomposition (SVD) of the similarity matrix is performed, in order to reduce the complexity associated with the evaluation of the nonlinear kernel. The method may contribute to reduce the computational complexity. However, the precision-recall curves tend to indicate that there is a very high proportion of misclassification involved in the process. In addition, a global alignment is far from being the most efficient method to characterize the degree of interaction in between two macromolecules. Nevertheless, a similar approach could be followed if the alignment is replaced by a more efficient geometrical signature. The limitations, associated with a hyperplane, may be overcome by replacing the SVP with a Bayesian technique (Rodriguez & Schmidler, 2009).

Finally, (Paquet & Viktor, 2010; Paquet & Viktor, 2011) perform a virtual fragmentation of the macromolecules. For each fragment, they estimate a translation and rotation invariant intrinsic reference frame which is statistically optimal in terms of shape. Then, they evaluate various probability distributions associated with the spatial distribution of the elementary surface elements forming the fragments. A geometrical signature is constructed from these probability distributions. This signature is invariant under rigid transformation and oblivious under more general transformations. The signatures are either exhaustively compared pair-wise, with a Euclidian metric, or clustered with a K-means algorithm in order to reduce the computational complexity. The method may be applied to both local and global analysis. It is compatible with various representations including the envelope, the constituent carbon atoms and the van der Waal representation, among others. The method is scalable and a search may be performed in a database of a few hundred thousand of fragments or macromolecules, in approximately a second on a portable computer. A typical query against the 1tyv protein is shown in Fig. 1.

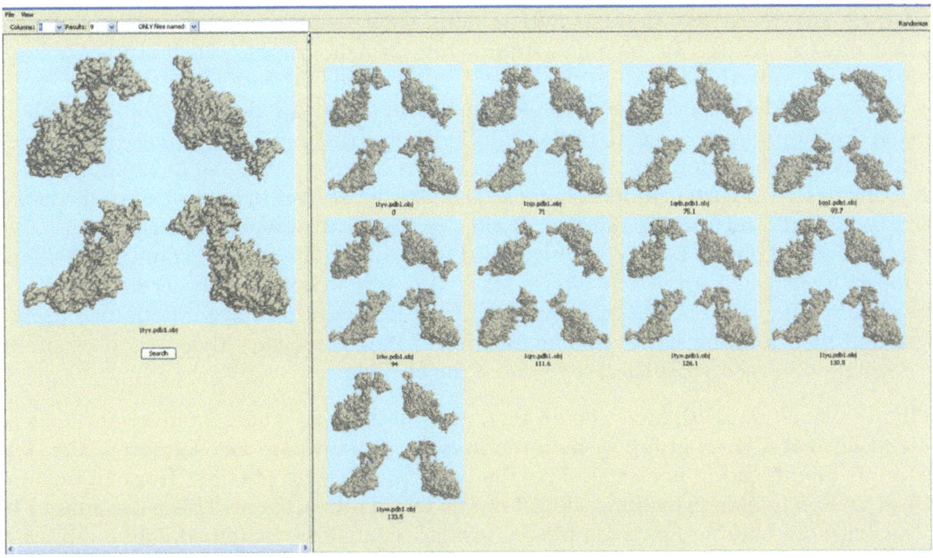

Fig. 1. Retrieval of all members from the P22 tailspike protein from Salmonella phage P22 conformation, using the 1tyv structure as a query, with a precision 100% and recall 100% from a database of more than 35,000 structures.

9. Conclusions

In this section, we summarize our main conclusions about the various algorithms. Further, we identify some potentially important research directions and challenges. Docking is by no means a trivial problem. That is, one has to identify two regions on two complex shapes which may present intricate deformations, while possessing very little a priori knowledge about the location, shape and deformations associated with these regions. The situation is further complicated by the combinatorics of the search. In order to find candidates for docking, one first has to search through large databases of macromolecular structures. Subsequently, for each potentially interesting pair of proteins, a large combination of regions, or docking sites, has to be considered.

We may only conclude that, unless one has an important a priori knowledge about the interaction, it is important to develop approaches that may rapidly determine if two proteins might, or not, interact independently of the details of their interaction. Such a binary classification can be obtained, for instance, with machine learning methods based on support vector machine (SVM) as well as their Bayesian generalization, relevance vector machine (RVM), which provides more robust probabilistic classification (Damoulas et al., 2008).

Because of the large number of regions that are compared, it is important to develop efficient methods to do so. Large scale clustering and constrained clustering (Basu et al., 2008) methods might be able to target such a problem. Constrained clustering, for instance, is particularly adapted to enforce physicochemical constraints.

Macromolecules are not rigid, but deformable entities. As we saw earlier, some methods, up to a certain extent, take this fact into account. It may come as a surprise, that so few methods do, indeed, take these deformations into account. Such a situation might originate from the fact that most structures are obtained from X-ray crystallography which implicitly implies rigidity: this is indeed the case for the crystal, but it is not the case for the protein, in vivo. The present models are not entirely adequate, in order to describe the deformations associated with proteins. Instead, one should be able to distinguish in between two types of transformations: the deformations on the surface of a protein and the flexibility associated with its backbone or carbon atoms chain(s). The later is far from arbitrary, since the bending occurs at a small and specific number of sites or hinges. Consequently, a protein behaves like an articulated object. Following this line of thought, (Rodriguez & Schmidler, 2009) have developed a Bayesian approach which takes this phenomenon into account and which allows to align the backbone of two proteins. They determine a prior for both the number of hinges and their location while, for the likelihood, they associate either a rigid or an affine transformation to each articulated segment of the backbone.

Finally, (Raviv et al., 2010) have shown that, even in the case where an invariance has been obtained against a large group of deformations, as in their case for isometries, the lack of constraints might lead to nonsensical results. Indeed, although possible, from an invariance point of view, some deformations should not be taken into account. This is because of their nonexistence in nature. For example, a protein cannot be flattened, although such a deformation may be allowed under certain groups of transformations. The authors of this chapter foreseen that a volumetric approach might resolve the problem. For macromolecules, a volumetric representation might be obtained, for instance, from the electrostatic, or from the van der Waal potential, associated with each constituent atom.

10. References

Binkowski, T. A. & Joachimiak, A. (2008). Protein Functional Surfaces: Global Shape Matching and Local Spatial Alignments of Ligand Binding Sites. *BMC Structural Biology*, Vol. 8, 23 pp.

Holm, L. et al. (2008). Searching protein structure databases with DaliLite v.3. Bioinformatics, Vol. 24, pp. 2780–2781

Vajda, S. & Kozakov, D. (2009). Convergence and combination of methods in protein-protein docking. *Current Opinion in Structural Biology*, Vol. 19, pp. 164-170

Katchalskip-Katzir, E. et al. (1992). Molecular surface recognition: Determination of geometric fit between proteins and their ligands by correlation techniques. *Proc. Natl. Acad. Sci.*, Vol. 89, pp. 2195-2199

Kazakov D. et al. (2006). PIPER: An FFT-Based Protein Docking Program with Pairwise Potential. *Proteins: Structure, Function, and Bioinformatics*, Vol. 65, 392-406

Nukuda, A. et al. (2007). High Performance 3D Convolution for Protein Docking on IBM Blue Gene. *The Fifth International Symposium on Parallel and Distributed Processing and Applications (ISPA)*, Springer LNCS 4742, pp. 958-969, Niagara Falls, ON, Canada, August 29-31, 2007

Katebi A. R. et al. (2009). Computational Testing of Proteins-Protein Interactions. *Bioinformatics and Biomedicine Workshop, IEEE Int. Conf. on Bioinformatics and Biomedicine*, pp. 144-151, Washington, D. C., USA, November 1-4, 2009

Sukhwani, B. & Herbordt, M. C. (2008). Acceleration of a Production Rigid Molecule Docking Code. *Int. Conf. on Field Programmable Logic and Applications*, pp. 341-346, Heidelberg, Germany, September 8-10, 2008

Gray, J. J. (2006). High-resolution protein-protein docking. *Current Opinion in Structural Biology*, Vol. 16, pp. 183-193

Bonvin, A. M. J. J (2006). Flexible protein-protein docking, *Current Opinion in Structural Biology*, Vol. 16, pp. 194-200

Sael, L. & Kihara, D. (2010a). Improved protein surface comparison and application to low-resolution protein structure data. *Bioinformatics*, Vol. 11 (Suppl. 11): S2, 12 pp.

Sael, L. & Kihara, D. (2010b). Binding Ligand Prediction for Proteins Using Partial Matching of Local Surface Patches. *Int. J. Mol. Sci.*, Vol. 11, pp. 5009-5026

Sael, L. & Kihara, D. (2010c). Characterization and Classification of Local Protein Surfaces Using Self-organizing Map. *International Journal of Knowledge Discovery in Bioinformatics*, Vol. 1, pp. 32-47

Venkatraman, V. et al. (2009). Protein-protein docking using region-based 3D Zernike descriptors. *BMC Bioinformatics*, Vol. 10, 21 pp.

Sael et al., L. (2008). Fast protein tertiary structure retrieval based on global surface shape similarity. *Proteins*, pp. 1259-1273

Canterakis, N. (1994). 3D Zernike Moments and Zernike Affine Invariants for 3D Image Analysis and Recognition, *ESPRIT Basic Research Workshop on Visual Invariances*, 8 pp.

Mak, L. et al. (2008). An extension of spherical harmonics to region-based rotationally invariant descriptors for molecular shape description and comparison. *Journal of Molecular Graphics and Modelling*, Vol. 26, pp. 1035-1045

Sovic, I et al. (2010). Parallel Protein Docking Tool. *33rd International Convention on Information and Communication Technology, Electronics and Microelectronics (MIPRO)*, pp. 1333-1338, May 24-28, Opatija, Croatia, 2010

Ritchie, D. W. (2005). High-order analytic translation matrix elements for real-space six-dimensional polar Fourier correlations. *J. Appl. Cryst.*, Vol. 34, pp. 808-818

Mavridis L. & Ritchie, D. W. (2010). 3D-BLAST: 3D Protein Structure Alignment, Comparison, and Classification using Spherical Polar Fourier Correlations. *Pacific Symposium on Biocomputing*, Vol. 15, pp. 281-292, January 4-8, Kamuela, Hawaii, USA, 2010

Geppert, T. et al. (2010). Protein-Protein Docking by Shape-Complementarity and Property Matching. *Journal of Computational Chemistry*, Vol. 31, pp. 1919-1928

Liu, Y.-S. et al. (2009). IDSS: deformation invariant signatures for molecular shape comparison. *BMC Bioinformatics*, Vol. 10, 14 pp.

Chi, P.-H. (2004). A Fast Protein Structure Retrieval System Using Image-Based Distance Matrices and Multidimensional Index. *Fourth IEEE Symposium on Bioinformatics and Bioengineering (BIBE)*, pp. 522-529, May 19-21, Taichung, Taiwan, 2004

Chen, Y. et al. (2010). 2nd PCA on 3D Protein Structure Similarity. *IEEE Fifth International Conference on Bio-Inspired Computing: Theories and Applications (BIC-TA)*, pp. 253-257, September 8-10, Liverpool, United Kingdom, 2010

Zhang, T. et al. (2009). A Graph-Based Approach for Protein-Protein Docking, *2nd International Conference on Biomedical Engineering and Informatics (BMEI)*, pp. 1-8, October 17-19, Tianjin, China, 2009

Peng, S.-L. & Tsay, Y.-W. (2010). Measuring Protein Structural Similarity by Maximum Common Edge Subgraphs. *Sixth International Conference on Intelligent Computing (ICIC)*, Springer LNAI 6216, pp. 100-107, August 18-21, Changsha, China, 2010

Novosad, T. et al. (2010). Searching Protein 3-D Structures for Optimal Structure Alignment Using Intelligent Algorithms and Data Structures. *IEEE Transactions on Information Technology in Biomedicine*, Vol. 14, pp. 1378-1386

Borgwardt, K. M. et al. (2005). Protein function prediction via graph kernels. *Bioinformatics*, Vol. 21, pp. i47-i56

Reddy, A. S. et al. (2011). Analysis of HIV Protease Binding Pockets Based on 3D Shape and Electrostatic Potential Descriptors. *Chem. Biol. Drug Des.*, Vol. 77, pp. 137-151

Wang, C. et al. (2007). Protein-Protein Docking with Backbone Flexibility. *J. Mol. Biol.*, Vol. 373, pp. 503-519

Hoffmann, B. et al. (2010). A new protein binding pocket similarity measure based on comparison of clouds of atoms in 3D: application to ligand prediction. *BMC Bioinformatics*, Vol. 11, 16 pp.

Zauhar, R. J. (2003). Shape Signatures: A New Approach to Computer-Aided Ligand- and Receptor-Based Drug Design. *J. Med. Chem.*, Vol. 46, pp. 5674-5690

Shapira, L. et al. (2008). Consistent mesh partitioning and skeletonisation using the shape diameter function. *Visual Comput.*, Vol. 24, pp. 249-259

Gal, R. et al. (2007), Pose-Oblivious Shape Signature. *IEEE Transactions on Visualization and Computer Graphics*, Vol. 13, pp. 261-271

Fang, Y. et al. (2009). Three dimensional shape comparison of flexible proteins using the local-diameter descriptor. *BMC Structural Biology*, Vol. 9, 15 pp.

Ranganath, A. et al. (2007). Efficient Shape Descriptors for Feature Extraction in 3D Protein Structures. *In Silico Biology*, Vol. 7, 169-174 (2007).

Lo, Y.-T. et al. (2010). Using Solid Angles to Detect Protein Docking Regions by CUDA Parallel Algorithms. *International Symposium on Parallel and Distributed Processing with Applications*, pp. 536-541, April 19-23, Atlanta, GA, USA, 2010

Shibberu, Y. & Holder, A. (2011). A Spectral Approach to Protein Structure Alignment. *IEEE/ACM Transactions on Computational Biology and Bioinformatics*, Vol. 8, pp. 867-875

Hue, M. et al. (2010). Large-scale prediction of protein-protein interactions from structures. *BMC Bioinformatics*, Vol. 11, 9 pp.

Paquet, E. & Viktor, H. L. (2010). Addressing the Docking Problem: Finding Similar 3-D Protein Envelopes for Computer-aided Drug Design, *Advances in Computational Biology*, Advances in Experimental Medicine and Biology 680, Springer, ISBN 978-1-4419-5912-6, pp. 447-454

Paquet, E. & Viktor, H. L. (2011). Multimodal Representations, Indexing, Unexpectedness and Proteins. *Twenty-fourth International Conference on Industrial, Engineering and Other Applications of Applied Intelligent Systems (IEA/AIE 2011)*, Springer Lecture Notes in Artificial Intelligence (LNAI), pp. 85-94, June 28 - July 1, Syracuse, NY, USA, 2011

Dijkstra, E. W. (1959). A note on two problems in connexion with graphs. *Numerische Mathematik*, Vol. 1, pp. 269–271

Dutta, S. et al. (2009). Data Deposition and Annotation at the Worldwide Protein Data Bank. Molecular Biotechnology, Vol. 42, pp. 1-13

Rodriguez, A. & Schmidler, S. C. (2009). Bayesian Protein Structure Alignment. Submitted to *Annals of Applied Statistics*

Basu, S. (Ed.) et al. (2008). *Constrained Clustering: Advances in Algorithms, Theory, and Applications*. Chapman and Hall/CRC, ISBB: 1584889969, Boca Raton, USA

Raviv, D. et al. (2010). Volumetric Heat Kernel Signatures. *ACM Workshop on 3D Object Retrieval (3DOR)*, pp. 39-44, October 25, Firenze, Italy, 2010

Damoulas, T. et al. (2008). Inferring Sparse Kernel Combinations and Relevance Vectors: an Application to Subcellular Localization of Proteins. *Seventh International Conference on Machine Learning and Applications(ICMLA)*, pp. 577-582, December 11-13, San Diego, USA, 2008

5

Refinement of Protein Tertiary Structure by Using Spin-Spin Coupling Constants from Nuclear Magnetic Resonance Measurements

Jürgen M. Schmidt[1] and Frank Löhr[2]
[1]*School of Biosciences, University of Kent, Canterbury, Kent*
[2]*Institute of Biophysical Chemistry, Center for Biomolecular Magnetic Resonance, Goethe-University, Frankfurt am Main*
[1]*United Kingdom*
[2]*Germany*

1. Introduction

Modelling protein structure seems a challenging enterprise because the number of structure parameters required ordinarily exceeds the amount of independent data points available from experimental observations. Expressing the *predominant conformation* of a protein in terms of a *geometry model*, a polypeptide chain consisting of N atoms would command $3N - 6$ Cartesian coordinates be fixed. Even for small proteins, this becomes a daunting number. Fortunately, so-called holonomic constraints limit the number of variables, leaving substantially fewer, truly relevant parameters for folding the polypeptide chain into its native tertiary structure. For example, adjusting bond lengths and the many angles between the covalent bonds connecting the atoms is of little concern and appropriate standard values can be inserted from tableworks (Pople & Gordon, 1967; Engh & Huber, 1991, 2006). Table 1 exemplifies for the 147-residue protein *Desulfovibrio vulgaris* flavodoxin how the number of truly independent *internal rotational degrees of freedom* amounts to less than one-tenth of the Cartesian coordinate set size.

IUPAC-IUB conventions (1970) define three mainchain torsions for the polypeptide backbone. Protein structure determination primarily seeks to adjust the values of those two mainchain torsion angles that chiefly determine the fold of the chain, that is, ϕ and ψ in each amino-acid residue. Peptide-bond geometry, as described by mainchain torsion angle ω, is normally assumed to be *trans*- or *cis*-planar, fixed at 180° or 0° angles, respectively, and thus does not give rise to rotational variability. In a next instance, the sidechain torsion angle χ_1 is of interest.

Nuclear magnetic resonance (NMR) spectroscopy is uniquely positioned to help determine internal orientational constraints in a molecule, be these atom-atom distances (Wüthrich, 1986; Neuhaus & Williamson, 1989), relative bond orientations (Reif et al., 1997, 2000;

Schwalbe et al., 2001), relative protein domain orientations (Tjandra & Bax, 1997; Fischer et al., 1999) or rotational states of torsion angles (Pachler, 1963, 1964; Hansen et al., 1975; Bystrov, 1976; Ejchart, 1999). Especially powerful at measuring short-range interaction, NMR thus complements X-ray crystallography that is stronger at determining correlations over longer distances.

Approach	Parameters	Items	Coord's
top-down	**Cartesian coordinates** ($3N - 6$)		**6411**
	less bond length constraints ($N - 1$)	- 2138	4273
	less bond angle constraints		
	- at **tetrahedral centers** (5 bond angles to fix)	- 417×5	2188
	– N-terminus	*(1)*	
	– mainchain C^{α} (all residues)	*(147)*	
	– sidechain C^{β} (all residues excl Gly)	*(129)*	
	– sidechain C^{γ} (select residues)	*(85)*	
	– sidechain C^{δ} (select residues)	*(47)*	
	– sidechain C^{ε} (Met, Lys) and N^{ζ} (Lys)	*(8)*	
	- at **planar centers** (3 bond angles to fix)	- 450×3	838
	– mainchain N′ and C′ (excl N- but incl C-terminus)	*(293)*	
	– sidechain nitrogen and carbon	*(157)*	
	- at **angled centers** (1 bond angle to fix)	- 24×1	814
	– sidechain hydroxyl / thiol groups / sulfide bonds	*(24)*	
	less fixed or irrelevant torsion angles		
	– mainchain ω of peptide bonds	- 146	668
	– sidechain methyl	- 84	584
	– sidechain amide / guanidinium groups	- 26	558
	– sidechain hydroxyl / thiol groups	- 24	534
	– N-/C-termini	- 2	**532**
bottom-up	**Variable torsion angles of relevance**		
	– mainchain ϕ and ψ ($2R - 2$)	292	292
	– sidechain χ_1 (all residues excl Gly, Ala)	+ 112	404
	– sidechain χ_2 (select residues)	+ 84	488
	– sidechain χ_3 (select residues)	+ 30	518
	– sidechain χ_4 (Arg, Lys, Pro)	+ 14	**532**

Table 1. Coordinate statistics for *D. vulgaris* flavodoxin (R = 147 residues, N = 2139 atoms)

We here focus on *high-field NMR in aqueous solution* which yields best resolution of the protein signals. Concentrated at around 1 mM, the dissolved protein is exposed to strong fields of 11.7-23.4 T magnetic flux density, giving rise to nominal resonance frequencies in the proton band (^{1}H) of 500-1000 MHz. Other biologically relevant stable isotopes ^{13}C and ^{15}N, respectively, resonate at roughly one-quarter and one-tenth of ^{1}H at a given field.

NMR experiments then consist of radio-frequency pulse trains that generate controlled transitions of spin states (Ernst et al., 1987). These transitions manifest in resonance lines whose spectrum location and detailed shape reveal information about the molecular environment in which the probed atoms are embedded.

Rotatory states of torsion angles are accessible through measurement of so-called *J*-coupling constants which lend NMR signals a splitting fine structure due to bond-electron mediated pair interactions between NMR active nuclei. Put simply, the presence of each neighbouring (coupled) spin splits the resonance line of the observed spin into two. This splitting is field-*in*dependent (hence coupling 'constant') but does depend on local molecular geometry.

Of particular interest to structure determination are three-bond $^{3}J_{XY}$ coupling constants as their magnitudes relate to dihedral angles subtended by the three covalent bonds connecting the coupled pair of nuclei X and Y, which can be any combination of ^{1}H, ^{15}N, or ^{13}C, prevalent in the topology of the torsion considered (Fig. 1).

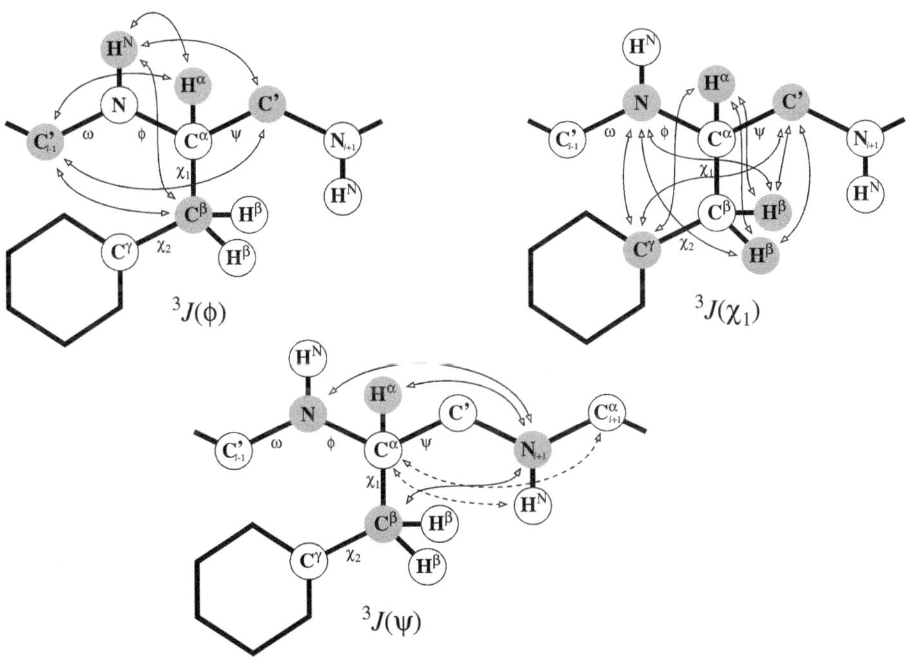

Fig. 1. ^{3}J spin-spin coupling topologies in ^{13}C,^{15}N enriched polypeptides. Arrows denote three-bond connectivities, grouped by conformational relevance to ϕ, ψ, and χ_1 torsions.

Multiple *J* data collected for a given torsion angle *overdetermine* that angle and thus allow its value to be retrieved. Six coupling constants, $^{3}J_{HNH\alpha}$, $^{3}J_{HNC'}$, $^{3}J_{HNC\beta}$, $^{3}J_{C'H\alpha}$, $^{3}J_{C'C'}$, and $^{3}J_{C'C\beta}$, determine torsion angle $\phi(C'_{i-1}\text{-}N'_i\text{-}C^\alpha_i\text{-}C'_i)$ of residue i (Wang & Bax, 1996; Blümel et al., 1998). Another three couplings, $^3J_{H\alpha N'}$, $^3J_{C\beta N'}$, and $^3J_{N'N'}$, would yield $\psi(N'_i\text{-}C^\alpha_i\text{-}C'_i\text{-}N'_{i+1})$. Of qualitative concern only, the planar *trans*- or *cis*-configured peptide bonds at torsion angle $\omega(C^\alpha_i\text{-}C'_i\text{-}N'_{i+1}\text{-}C^\alpha_{i+1})$ can be verified by two couplings, $^3J_{C\alpha HN}$ and $^3J_{C\alpha C\alpha}$, which appear to reflect ψ geometry also (Hennig et al., 2000). Of numerous sidechain torsions encountered in amino acids, $\chi_1(N'_i\text{-}C^\alpha_i\text{-}C^\beta_i\text{-}C^\gamma_i)$ impacts most on the spatial orientation of the sidechain. Up to nine distinct coupling constants are accessible in amino acids (Pérez et al., 2001).

Karplus (1963) suggested the dependence of ^3J on dihedral angle θ subtended by the three bonds that connect the coupled nuclei follow the empirical relation

$$^3J(\theta) = C_0 + C_1 \cos\theta + C_2 \cos 2\theta \tag{1}$$

where C_m are Karplus coefficients in Hz empirically calibrated for the ^3J types encountered in amino acids (Bystrov, 1976; Wang & Bax, 1995, 1996; Hu & Bax, 1996, 1997; Schmidt et al., 1999, Pérez et al., 2001). Multiples of 60° increments to θ establish the phase relation between the actual internuclear dihedral angle θ and the IUPAC-defined torsion ϕ, ψ, or χ_1 (Fig. 2).

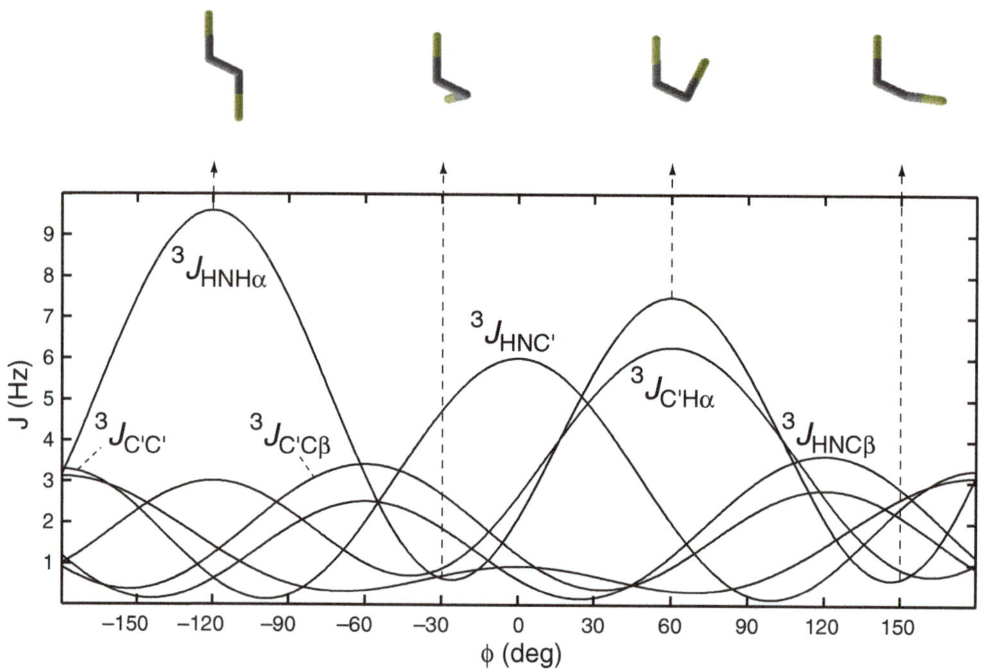

Fig. 2. Angular dependence of the protein ϕ-torsion related ^3J coupling constants. ^3J are at maximum when the bonds between the coupled nuclei are *trans*-oriented, and at minimum for perpendicular orientations. For $^3J_{HNH\alpha}$, the panels show from left to right the internuclear angle $\theta_{HNH\alpha} = \phi - 60°$ in the situations ±180°, -90°, ±0°, and +90°.

The coefficients C_0 signify mean J values obtained for a complete torsion-angle revolution, also referred to as *conformation averaged J* coupling constants. The differences ($C_2 - C_1$) are the largest deflections in J from the mean, where *primary* and *secondary* maxima of the curves at ±180° and ±0°, respectively, differ by $2C_1$.

Given sets of up to six ^3J parameters per torsion, the challenge is now to find that angle value in best agreement with the experimental data. Numerical methods exploiting the *redundance* present in large data pools permit the *self-consistent calibration* of Karplus coefficients during the course of the angle refinement (Schmidt et al., 1999). This obviates the need for traditional referencing of conformations derived from X-ray data.

We previously collected for flavodoxin values for the six possible coupling constants $^3J(\phi)$ and determined the protein's mainchain torsion angles ϕ (Schmidt et al., 1999). Discrepancy in ϕ between our NMR solution structure and comparison coordinates from X-ray crystallography (Walsh et al., 1998; Artali et al. 2002) is only 5° on average, which is smaller than the molecular dynamical angular libration due to thermal effects, indicating that both NMR solution and X-ray crystal structures of flavodoxin are very similar indeed.

Here, we record ^3J data for the enzyme Ribonuclease T1, a 104-residue protein (11 kDa) from *Aspergillus oryzae* (RNase T1, EC 3.1.27.3) that cleaves single-stranded RNA 3′-side of guanine nucleotides, and determine the majority of the ϕ torsion angles in the enzyme.

2. Materials and methods

One stringent requirement for efficient protein NMR analysis is that the protein sample be artifically enriched in the stable non-radioactive isotopes ^{15}N and ^{13}C, a process nowadays commonly applied in protein expression by recombinant technologies (Kainosho, 1997). Whilst oxygen does not play any role in protein NMR practice, the ^1H isotope offering greatest sensitivity is ubiquitous and abundant. Sometimes, it is being depleted by ^2H replacement in order to alleviate adverse signal relaxation effects occuring in large protein samples (above approximately 250 amino acids) exhibiting slow rotational tumbling rates (longer than approximately 10 ns rad^{-1}).

2.1 Protein sample

Uniformly ^{13}C,^{15}N-labeled RNase T1 (Lys25 isoenzyme) was obtained following established protocols (Quaas et al. 1988a,b; Spitzner et al., 2001) and used at 2-mM concentration in aqueous solution of pH 5.5 (containing 10% D_2O). All NMR spectra were recorded at 308 K. Prerequisite to any protein NMR analysis is the assignment of resonance signals to individual nuclei, not unlike a fingerprint of the molecule. Values quoted are chemical shifts in parts per million (ppm) from the respective ^1H, ^{13}C, or ^{15}N band base frequency. Resonance assignments for RNase T1 (Fig. 3) are available from the BioMagRes database for ^1H (BMRB-133; Hoffmann & Rüterjans, 1988) and for ^{15}N (BMRB-1658; Schmidt et al., 1991), and ^{13}C chemical shifts were given by Pfeiffer et al. (1996b). Comparison ϕ torsion angles were calculated from crystal coordinates, resolved at 0.15 nm, of RNase T1 complexed with Ca^{2+} (PDB-9RNT; Martinez-Oyanedel et al., 1991).

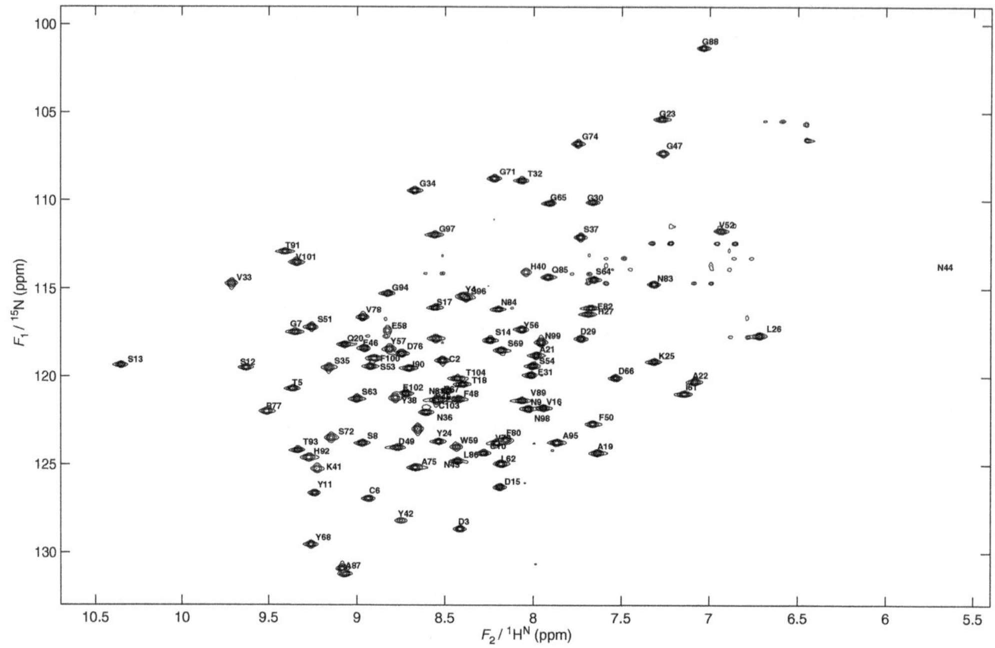

Fig. 3. ^{1}H,^{15}N-chemical shift correlation spectrum of RNase T1 at 900 MHz and 308 K.

2.2 Data collection

The focus of this chapter lies on protein structure and interpretation of NMR data. The reader interested in the details of NMR pulse-sequence design and the setup of multi-dimensional NMR experiments is referred to specialist textbooks (Roberts, 1993; Evans, 1995; Keeler, 2005; Hoch & Stern, 1996; Cavanagh et al., 2007) and to the literature including contained bibliography on the measurement of each of the six protein ϕ related ^3J coupling constants, as given in the separate NMR Experiments section at the end of the chapter.

Measurements of J coupling constants are subject to two principal concepts. One family of NMR experiments exploits *frequency modulation*, such as COSY, DQF-COSY, and E.COSY (Griesinger et al., 1987), giving rise to characteristic multiplet line splittings (Fig. 4). The other technique uses *amplitude modulation*, subdivided into J-modulation (Billeter et al., 1992) and J-correlation (Bax et al., 1994), encoding J in time-dependent signal intensity. Both approaches naturally have their pros and cons. While frequency-modulation schemes produce NMR signals of weaker intensity and larger space requirement, at times leading to signal overlap, amplitude-modulation schemes chiefly fail to measure small coupling constants.

To minimize signal overlap, protein NMR spectra are often recorded in 3D mode by exploiting the resonance of an auxiliary spin and selecting 2D projections at that frequency. Example graphs in Fig. 4 show such an experimental multiplet pattern, together with its best-fit least-squares lineshape reconstruction and the residual error signal.

2.3 Data evaluation

Whichever the chosen approach, the particular method applied to extract *J* values as structure parameters from NMR spectra must be considered an integral part of the analysis also. The authors gathered experience with both methods, albeit with an undeniable preference for E.COSY-type spectra which perform particularly robust in connection with computer-assisted lineshape analysis (Schmidt, 1997a; Löhr et al., 2000).

Contour plots of NMR signals for *J* analysis recorded in E.COSY mode show characteristic tilts (Griesinger et al., 1987), where a prominent large one-bond coupling spreads out the multiplet along the vertical spectrum dimension, so as to permit reading the sought smaller ^3J coupling off the frequency difference between the two multiplet halves in the horizontal dimension.

In the example of Fig. 4, the $^1J_{C\alpha H\alpha}$ of typically 143 Hz is exploited to split the signal into two halves along the F_1 dimension, given here by the $^{13}C^{\alpha}$ resonance frequency, and the small $^3J_{C'H\alpha}$ coupling results from the frequency difference between both halves along the perpendicular F_2 dimension, given here by the $^{13}C'$ carbonyl resonance.

The parameter record lists optimized values for coupling constants $^1J_{C\alpha H\alpha}$ and $^3J_{C'H\alpha}$ as the primary E.COSY components responsible for the tilted appearance of the signal shape in vertical and horizontal directions, respectively. Other parameters include, apart from amplitude scaling, line widths and line asymmetries in both dimensions, a second unresolved splitting pair.

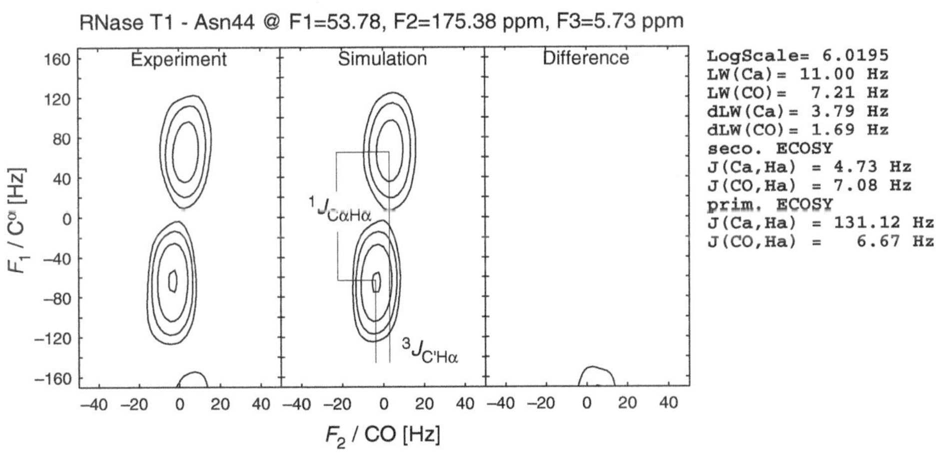

Fig. 4. Contour plot of a 500-MHz H(N)CA,CO[HA]-E.COSY multiplet recorded for RNase T1. Example analysis for extracting the $^3J_{C'H\alpha}$ coupling constant from the 2D $^{13}C^{\alpha}$,$^{13}C'$ correlation signal (F_1,F_2) taken at the resonance of the amide $^1H^N$ proton in the third dimension (F_3). The three nuclei eliciting the three spectrum dimensions all couple with the $^1H^{\alpha}$ spin (Fig. 1.), whose presence, following the E.COSY principle, is seen only as splittings, not as another frequency dimension.

3. Results

3.1 RNase T1

Using a uniformly ^{15}N and ^{13}C stable-isotope labeled sample of the enzyme RNase T1, a total 512 ^3J values were collected, related to the polypeptide mainchain torsion angles ϕ in 82 out of the total 104 amino-acid residues.

Even without fitting quantitative torsion angles to the ^3J data, qualitative inspection of the J values permits insights into some details of the protein's secondary structure already. For example, residue Asn44 in RNase T1 exhibits a very large $^3J_{C'H\alpha}$ coupling of 6.67 Hz (Fig. 4), second only to that seen in Asn84. This is irreconcilable with a ϕ torsion in the negative value range (Fig. 2). In addition, the $^1J_{C\alpha H\alpha}$ coupling of only 131 Hz (Fig. 4) falls well short of the expected average and supports a positive ϕ torsion, too. Consequently, Asn44 must exhibit a positive value for its ϕ torsion.

Signals for consecutive residues Asn43 and Asn44 in RNase T1 could hardly be more different. A value of 2.94 Hz for $^3J_{HNC\beta}$ in Asn43 (Fig. 5A) contrasts the lowly 0.47 Hz in Asn44 (Fig. 5B). Very different $^1J_{C\alpha C\beta}$ couplings also suggest differing backbone geometries. While the near-average $^1J_{C\alpha C\beta}$ coupling of 35.8 Hz in Asn43 is common with negative ϕ torsions, the unusually large 43.7-Hz coupling in Asn44 agrees better with a positive ϕ value (Schmidt et al., 2009). Both being asparagines, this cannot be a residue-type specific effect on the J couplings (Schmidt, 2007a).

The above-average value of 2.66 Hz seen for $^3J_{C'C\beta}$ in Tyr45 (Fig. 5C) is consistent with a type-I β turn spanning both Tyr45 and Glu46. This tyrosine's aromatic ring system is a critical component in nucleotide recognition and binding.

The consecutive residues Ala87 and Gly88 in RNase T1 form a β bulge (Chan et al., 1993). An unusually small $^3J_{HNC'}$ coupling near zero in Ala87 (Fig. 5D) and an unusually large 2.9-Hz $^3J_{C'C'}$ coupling in Gly88 (Fig. 5E) hint at ϕ torsion angles near +90°/−90° and 180° (Fig. 2), respectively, corroborating the distorted geometry in the central portion of a β strand.

Coupling type	$\Delta\phi$ (deg)	C_0 (Hz)	C_1 (Hz)	C_2 (Hz)	*trans* J(180°) (Hz)	*gauche* J(±60°) (Hz)
$^3J_{HNH\alpha}$	-60°	5.67	-0.71	3.37	9.74	3.63
$^3J_{HNC'}$	180°	1.79	-0.75	1.56	4.09	0.64
$^3J_{HNC\beta}$	60°	2.32	-1.64	1.91	5.86	0.54
$^3J_{C'H\alpha}$	120°	3.21	-2.17	2.05	7.43	1.10
$^3J_{C'C'}$	0°	1.43	-0.96	0.76	3.15	0.58
$^3J_{C'C\beta}$	-120°	1.33	-0.51	0.91	2.75	0.62

Table 2. Karplus coefficients for Eq. 1, optimized against experimental ^3J data for RNase T1

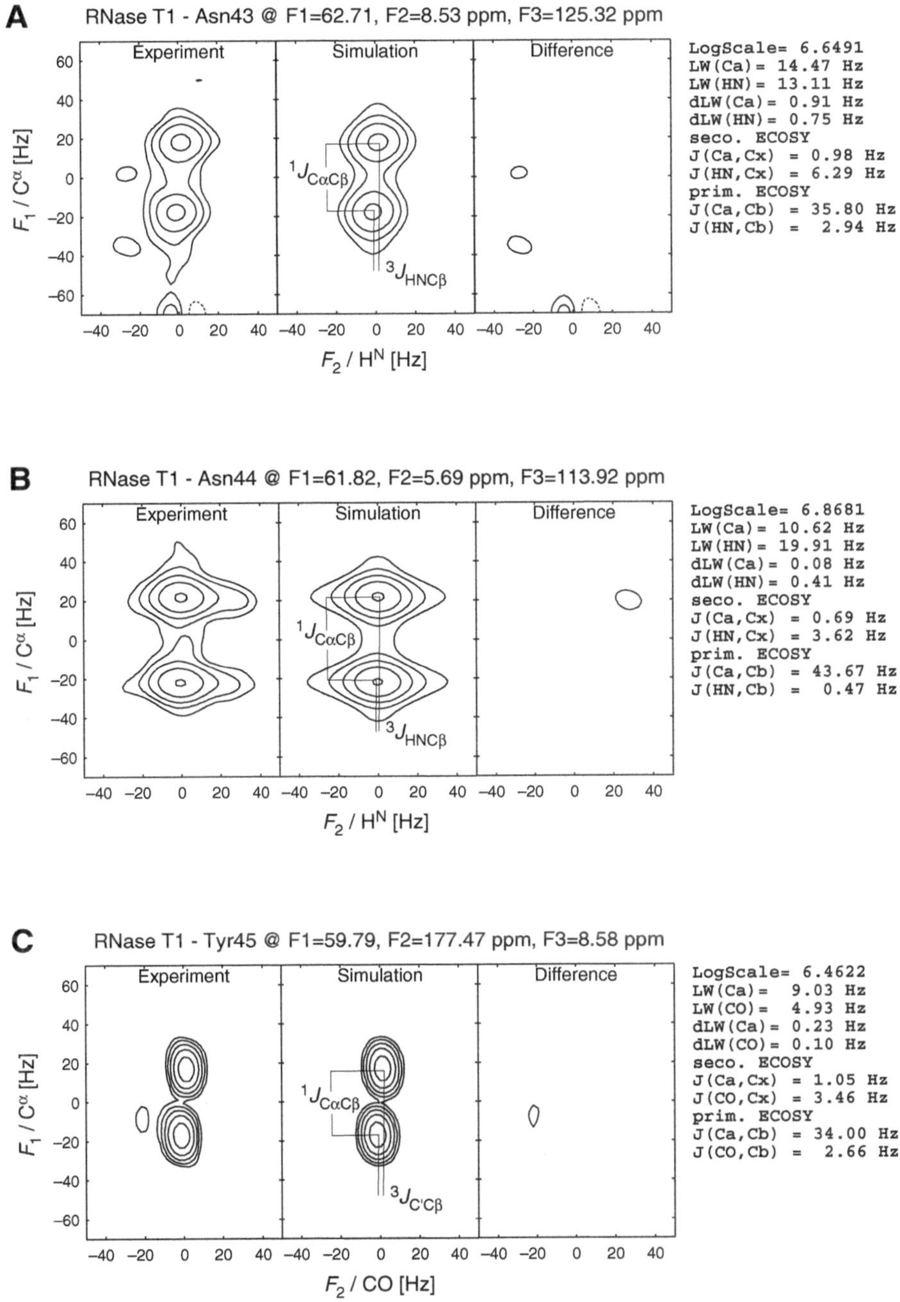
A
RNase T1 - Asn43 @ F1=62.71, F2=8.53 ppm, F3=125.32 ppm
Experiment
Simulation
Difference
$^1J_{C\alpha C\beta}$
$^3J_{HNC\beta}$
F_1 / C^α [Hz]
F_2 / H^N [Hz]
LogScale= 6.6491
LW(Ca)= 14.47 Hz
LW(HN)= 13.11 Hz
dLW(Ca)= 0.91 Hz
dLW(HN)= 0.75 Hz
seco. ECOSY
J(Ca,Cx) = 0.98 Hz
J(HN,Cx) = 6.29 Hz
prim. ECOSY
J(Ca,Cb) = 35.80 Hz
J(HN,Cb) = 2.94 Hz
B
RNase T1 - Asn44 @ F1=61.82, F2=5.69 ppm, F3=113.92 ppm
Experiment
Simulation
Difference
$^1J_{C\alpha C\beta}$
$^3J_{HNC\beta}$
F_1 / C^α [Hz]
F_2 / H^N [Hz]
LogScale= 6.8681
LW(Ca)= 10.62 Hz
LW(HN)= 19.91 Hz
dLW(Ca)= 0.08 Hz
dLW(HN)= 0.41 Hz
seco. ECOSY
J(Ca,Cx) = 0.69 Hz
J(HN,Cx) = 3.62 Hz
prim. ECOSY
J(Ca,Cb) = 43.67 Hz
J(HN,Cb) = 0.47 Hz
C
RNase T1 - Tyr45 @ F1=59.79, F2=177.47 ppm, F3=8.58 ppm
Experiment
Simulation
Difference
$^1J_{C\alpha C\beta}$
$^3J_{C'C\beta}$
F_1 / C^α [Hz]
F_2 / CO [Hz]
LogScale= 6.4622
LW(Ca)= 9.03 Hz
LW(CO)= 4.93 Hz
dLW(Ca)= 0.23 Hz
dLW(CO)= 0.10 Hz
seco. ECOSY
J(Ca,Cx) = 1.05 Hz
J(CO,Cx) = 3.46 Hz
prim. ECOSY
J(Ca,Cb) = 34.00 Hz
J(CO,Cb) = 2.66 Hz

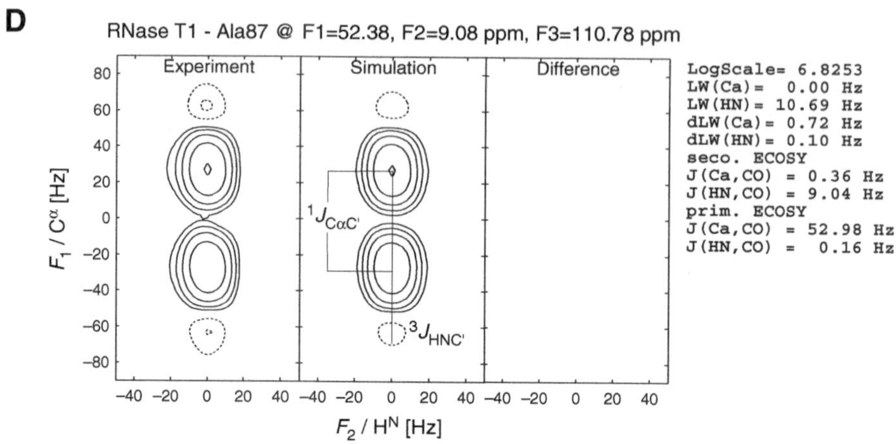

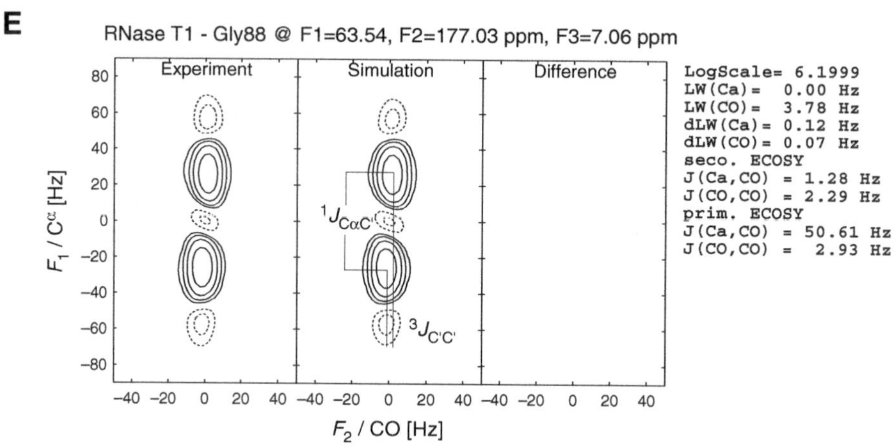

Fig. 5. Example 2D E.COSY multiplet sections from 3D NMR spectra recorded for RNase T1. *Panels A-C* (previous page): $^3J_{HNC\beta}$ and $^3J_{C'C\beta}$ evaluations exploiting the $^1J_{C\alpha C\beta}$ coupling of typically 35 Hz; *Panels D-E*: $^3J_{HNC'}$ and $^3J_{C'C'}$ evaluations exploiting the $^1J_{C\alpha C'}$ coupling of typically 53 Hz. Solid and dotted lines are positive and negative contours, respectively.

Eventually, the 82 ϕ torsion angles were fitted simultaneously to the pool of all 512 ^3J data. Self-consistently optimized in conjunction with the torsion-angle values, the Karplus curves shown in Fig. 6 and respective coefficients summarized in Table 2 represent the best fit to the ^3J data available for RNase T1 exclusively, yet, would be similar with other proteins also.

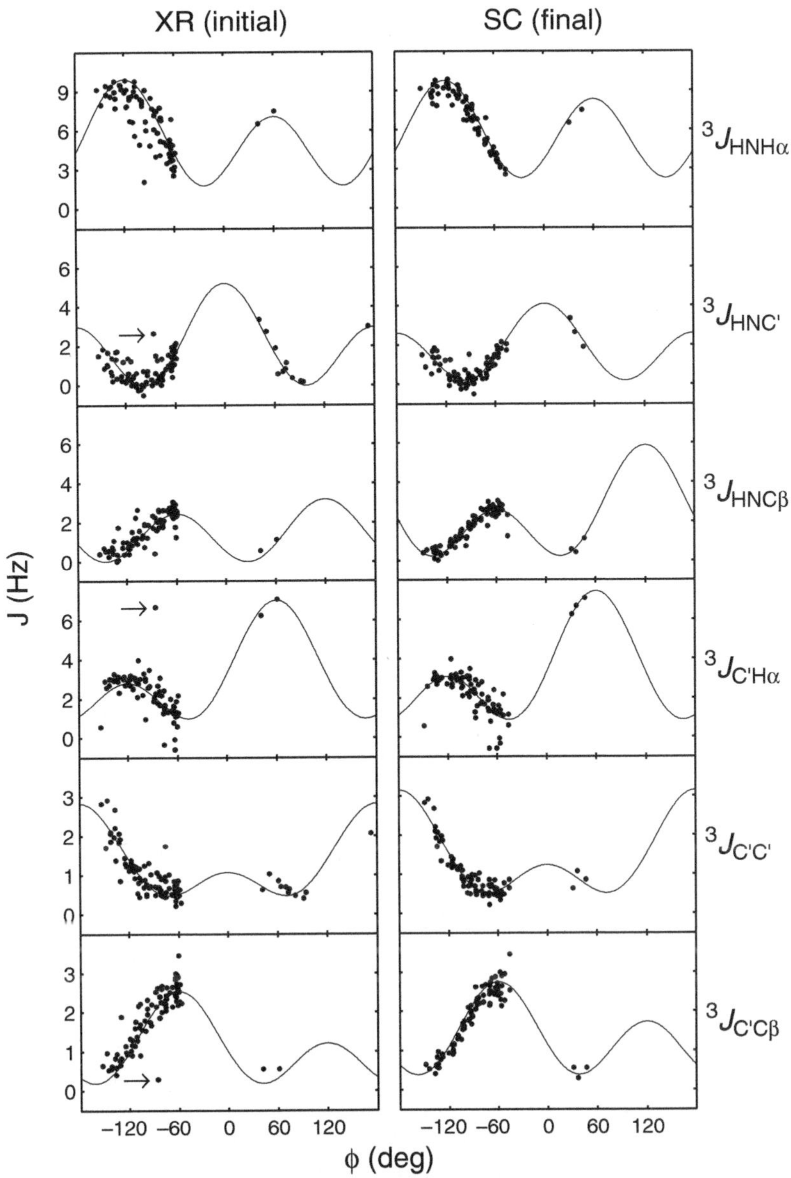

Fig. 6. Optimization of protein ϕ torsions on the basis of ^3J coupling constants. *XR panels*: Experimental data for RNase T1 (dots) plotted against initial torsion values calculated from crystal structure coordinates (PDB-9RNT). *SC panels*: The same data plotted against torsion values iteratively optimized by referring exclusively and simultaneously to all ^3J data, adjusting both torsion angles and Karplus coefficients in a self-consistent manner (Schmidt et al., 1999). Arrows point at Asn44 data that appear to be outliers initially.

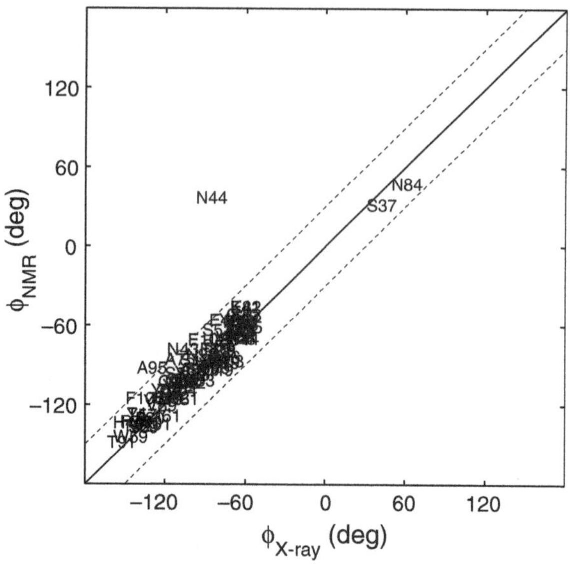

Fig. 7. Comparison of mainchain torsions ϕ in RNase T1 as inferred from self-consistent *J* coupling analysis and from crystallographic data (PDB-9RNT) by Martinez-Oyanedel et al. (1991). The majority of items agree within a tolerance of ±30° (dashed bounds). In the final optimized set, the Asn44 ϕ torsion angle has flipped from -86° to +37°.

Two facts rationalize the markedly reduced scatter seen in the right-hand-side $^3J_{\mathrm{HNH}\alpha}$ panel in Fig. 6 compared with the other five coupling types: Firstly, being larger than $^3J_{\mathrm{HC}}$ or $^3J_{\mathrm{CC}}$ on an absolute scale, $^3J_{\mathrm{HH}}$ values dominate the fit and, secondly, ϕ torsion values in regular protein structure typically fall on the steep slope of the $^3J_{\mathrm{HH}}$ curve, making regression sensitive to small changes in $^3J_{\mathrm{HH}}$. The residual scatter is due to unaccounted substituent effects in the different amino-acid types (Schmidt, 2007a), unaccounted torsion-angle dynamics (Brüschweiler & Case, 1994; Pérez et al., 2001), and random experimental error.

Assigned to Asn44 in RNase T1, the solitary data point in the left-hand-side $^3J_{\mathrm{C'H}\alpha}(\phi)$ panel in Fig. 6, near the upper border, is elevated by around 3 Hz above the crowd. Apparently, other large $^3J_{\mathrm{C'H}\alpha}$ coupling constants around 6-7 Hz connect with positive ϕ torsion angles. Contrasting the negative ϕ value found for Asn44 in crystal structure PDB-9RNT (Martinez-Oyanedel et al., 1991), the conformation in solution that emerged from our *J* coupling analysis differs from that in the solid state (Fig. 7). Similarity between relative *J* coupling values for Asn44 and those residues that exhibit positive ϕ angles clearly suggest that ϕ_{44} be rotated to a positive value also. Indeed, self-consistent optimization of ϕ torsions referencing the set of six ϕ-related coupling types converges at +37° for Asn44, while improving dramatically the error between observed coupling constants and those predicted from the crystal-structure angles.

The region around residues 42-48 of the enzyme engages in the recognition and binding of the substrate nucleotide (Fig. 8). Inspection of eight crystal structures of RNase T1 in a variety of complexes reveals that Asn44 adopts a negative ϕ value in only those structures presenting a free binding site or, rather, a calcium, zinc, or vanadate metal ion bound to the *apo*-enzyme (9RNT, 8RNT, 3RNT). Similarly, the inadequate adenosine nucleotide complex (6RNT) also resembles the *apo*-enzyme. Even though the distant ribose phosphate interacts with catalytic residues His40 and Glu58 in this latter structure, the nucleobase is directed away from the enzyme, so as to leave the recognition loop unoccupied, for RNase T1 binding is G not A specific.

However, four other crystal structures of RNase T1 in various complexes, notably those with a bound guanine nucleotide (1RLS, 1RNT, 2RNT, 5RNT), do exhibit positive ϕ angles for Asn44 (Fig. 7). Also ϕ angles in the adjacent positions Asn43 and Tyr45 differ between the two groups, by approximately 40° and 10°, respectively.

The observations suggest that binding of the correct substrate elicits a concerted conformation change involving torsion ϕ_{44}, and likely ψ_{44}, too, as well as torsions ψ_{43} and ϕ_{45} in the preceding and subsequent residue, respectively. Yet, the NMR evidence tells a different story: Torsion angle ϕ_{44} is positive already in our sample of the free enzyme in aqueous solution that was subjected to NMR measurement! Conformational variability of the Asn44 backbone, such as continual flips between positive and negative values, can be ruled out on the basis of the large ${}^3J_{C'H\alpha}$ coupling value observed, as such conformational averaging would reduce the J value towards its mean of about half the size (Table 2).

A likely explanation for the negative angle value in some of the crystal structures would be packing effects through molecules in adjacent grid cells distorting the conformation of the rather exposed 42-48 loop region at the protein surface. Another possible explanation could be that the metal agents added to aid the crystallization process obstruct the binding site differently than the natural substrate would do.

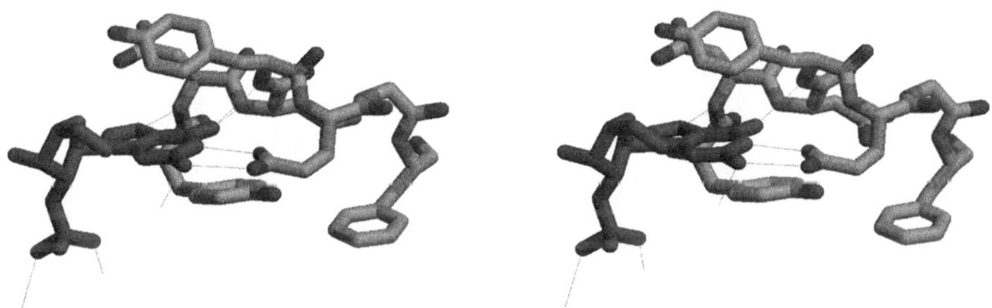

Fig. 8. Stereo view of residues 42-48 in the crystal structure of the RNase T1:2′GMP inhibitor complex (PDB-1RNT, Arni et al., 1988). Involved in nucleotide recognition and binding, Asn43 and Asn44 backbone amide groups interact with the guanine five- and six-ring, respectively, while the Glu46 sidechain carboxylate probes the presence of the correct hydrogen bonding capacity at the Watson-Crick edge of the guanine. Tyr42 and Tyr45 sandwich the guanine ring plane.

3.2 Flavodoxin

${}^3J$ coupling data from a similar study targeting backbone torsion angles ψ in flavodoxin suggested subtle differences between the structures of this protein in aqueous solution

(NMR) and in the crystalline solid state (X-ray). For example, on the basis of *J* coupling constants the orientation of peptide plane Ala39/Ser40 shown in the X-ray structure needed to be adjusted by about +10° in order to fulfil the experimental data (Löhr et al., 2001).

A conformational hot spot is found in flavodoxin's so-called "60-loop", the region protruding around the isoalloxazine ring system, made up of Gly61-Asp62-Asp63-Ser64. Results from our analysis on the exclusive basis of *J*-coupling constants agree with what has been known for some time already from X-ray studies on flavodoxin (Watenpaugh et al., 1976). These crystal structures of flavodoxin can be grouped according to 2FX2, 3FX2, 1J8Q, and 'WFX2'[1], on one side, and 4FX2, 5FX2, 1BU5A, and 1BU5B, on the other. The former group represents oxidized species of the FMN cofactor, the latter includes reduced (5FX2) and semi-reduced (4FX2) quinone forms. 1BU5A and 1BU5B (Walsh et al., 1998) are two chains in an asymmetric unit of *apo*-flavodoxin-riboflavin complex lacking the ribose phosphate.

Conformations of torsions ψ_{61} and ϕ_{62} in flavodoxin differ by almost 180° between both structure sets, implying a flip of the intervening peptide plane with connected changes in the hydrogen bonding network. Our NMR data on the oxidized species support the torsion angles derived from the first group. Most notably, Asp62 adopts a rare positive ϕ torsion angle responsible for an elevated $^3J_{C'H\alpha}$ coupling constant of 5.1 Hz. As it is also connected with an unusual negative ψ torsion, this data point stands out in a graph of $^3J_{C'H\alpha}(\psi)$ (not shown). Albeit a non-glycine residue, Asp62 exhibits a secondary-structure feature normally characteristic of glycine as frequently found in a type-II' β turn motif (Creighton, 1993). But then, the surrounding torsion angles are too distorted to form a proper reverse turn. Variability in this loop region is also corroborated by an analysis of coupling constants related to the sidechain χ_1 torsion angle (Schmidt, 2007b).

4. Discussion, conclusion and scope

Complete determination of a protein fold from scratch is inherently impossible on the sole basis of *J* coupling restraints alone, owing to their short-range nature of interaction. Even though ϕ torsion angles can be determined fairly accurately, and values of ψ can at least be narrowed to plausible ranges, while fixing ω torsions at the planar 180° value, the chaining of the distinct amino-acid fragments is likely to cause errors to accumulate in the process. A few long-range restraints, such as those provided by the measurement of NOE effects, will normally be required to ensure the correct fold over the whole polypeptide chain. In fact, the present results derived from our *J*-coupling data are consistent with an independent previous investigation into the RNase T1 solution structure by means of the traditional measurement of NOE effects that were subsequently converted into proton-proton distance constraints (Pfeiffer et al., 1996a).

4.1 Data correlation and redundance

The ϕ torsion angle in each amino-acid residue is supported by up to six ^3J coupling constants, which, in aggregate, determine just one internal rotational degree of freedom of the molecular model, the value of ϕ. Even though the *J* data are all independent

[1] 'WFX2' signifies a preliminary X-ray coordinate set of *D. vulgaris* flavodoxin not available from the PDB and kindly provided by Martin Walsh, University College Galway, Ireland.

observations, they do not represent six independent structure parameters. Owing to their simultaneous dependence on the same torsion, their values are highly correlated, or anti-correlated, for that matter (Table 3). Somewhat limited by pairs of couplings exhibiting correlated angle dependence (Fig 2), the amount of independent structure information available reduces to effectively 3 or 4, rather than 6, yet still exceeds the single rotational degree sought to be fixed. Precisely this redundance, however, is key to determining torsion angles accurately.

	$^3J_{HNH\alpha}$	$^3J_{HNC'}$	$^3J_{HNC\beta}$	$^3J_{C'H\alpha}$	$^3J_{C'C'}$	$^3J_{C'C\beta}$
$^3J_{HNH\alpha}$	6.77 ± 2.12	-0.82	-1.20	1.18	0.81	-1.13
$^3J_{HNC'}$	-44%	0.89 ± 0.74	-0.11	0.27	0.27	-0.19
$^3J_{HNC\beta}$	-79%	-2%	1.62 ± 0.85	-0.82	-0.62	0.77
$^3J_{C'H\alpha}$	54%	8%	-61%	2.32 ± 1.29	0.47	-0.84
$^3J_{C'C'}$	58%	16%	-76%	30%	0.97 ± 0.58	-0.59
$^3J_{C'C\beta}$	-79%	-6%	89%	-69%	-77%	1.83 ± 0.78

Table 3.Variance-covariance matrix for 512 ϕ-related ^3J coupling constants in RNase T1[2]

4.2 Molecular dynamics effects

At times, the excess information contained in a set of observables can give insights into dynamic effects that may prevail in the molecular structure (Schmidt, 1997b). Conformational dynamics frequently complicate the analysis of amino-acid sidechain torsion angles χ_1 and need be taken into account for satisfactory interpretation of experimental J data related to that angle type. A variety of angular-mobility models can be applied to cases in which a single fixed torsion does not explain the observed data satisfactorily. Analyses commonly assume the χ_1 torsion either to dwell preferentially in energetically favourable staggered states (Pachler, 1963, 1964; Hansen et al., 1975) or to librate about a mean value according to a Gaussian probability profile (Jardetzky, 1980; Karimi-Nejad et al., 1994; Brüschweiler & Case, 1994).

In the first approach, amino-acid sidechain torsions are analyzed assuming the presence of interconverting staggered-rotamer conformations of χ_1 = -60°, ±180° and +60° to solve for the respective populations, p_1, p_2, and p_3, by linear combinations of so-called *trans* and *gauche* coupling values (Table 2). The Gaussian model, however, typically limits the torsion to one predominant conformation, yet, allowing for larger angular variability, at times. To this end, angular standard deviations of approximately 60° can be considered to represent a fully revolving torsion. Both models are somewhat complementary, the staggered-rotamer one being easier to apply, whereas the Gaussian one usually fits the data better (Schmidt, 1997b).

[2] Diagonal: mean and standard deviation (square-root of variance, in Hz). Upper triangle: standard deviation attributed to joint variation in both J types (square-root of covariance, in Hz), negative signs indicating antivariance. Lower triangle: pair-correlation coefficients.

4.3 ^1J and ^2J couplings in protein structure analysis

Interest is arising in short-range ^1J and ^2J coupling constants (Wienk et al., 2003; Löhr et al., 2011) as these are comparatively easier to measure than ^3J, albeit less well understood regarding their conformational dependence (Schmidt et al., 2009, 2010). However, by simply recognizing qualitative classes of small, medium, and large magnitudes for the respective parameters, it was feasible to pinpoint the boundaries of secondary-structure elements in a protein (Schmidt et al., 2011). The suggested procedure of *J*-indexing also exploits data redundance as each torsion angle is surrounded by large numbers of ^1J and ^2J coupling constants (Fig. 9).

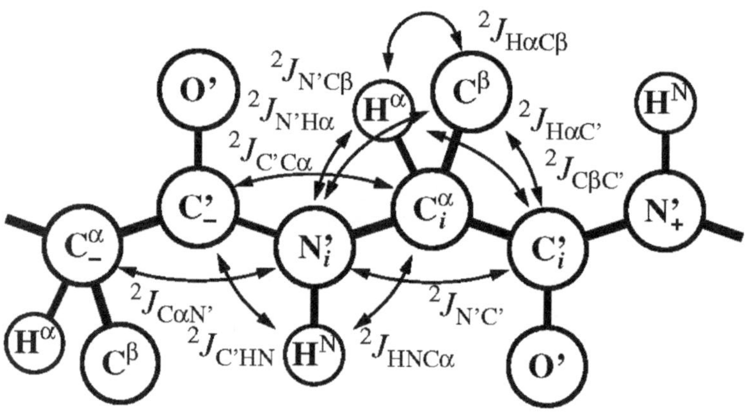

Fig. 9. The ten types of ^2J coupling constants encountered in the protein backbone.

Protein ^1J-, ^2J-, and ^3J-coupling constant data related to one-, two-, and three-bond interaction, respectively, continue being deposited by NMR spectroscopists with the Biological Magnetic Resonance Data Bank (BMRB, Ulrich et al., 2008) (Fig. 10).

4.4 Accuracy issues

Quantitative ^3J-coupling analysis in proteins appears to have reached a level of detail and accuracy at which a change of a few degrees in a torsion angle, comparable to thermal librational amplitudes, makes a noticeable difference, allowing, for example, genuine differences between NMR-based solution and X-ray based crystal structures to be detected.

The process of defining useful conformational constraints on the basis of *J* coupling constants usually requires, first, extracting accurate values for the *spin-system* related property *J* from NMR spectra and, second, translating these values into *molecular-geometry* related dihedral-angle values within the framework of a specified model of molecular structure and possibly dynamics. Naturally, both these stages come with their inherent inaccuracies which will impact on the final result (Fig. 11).

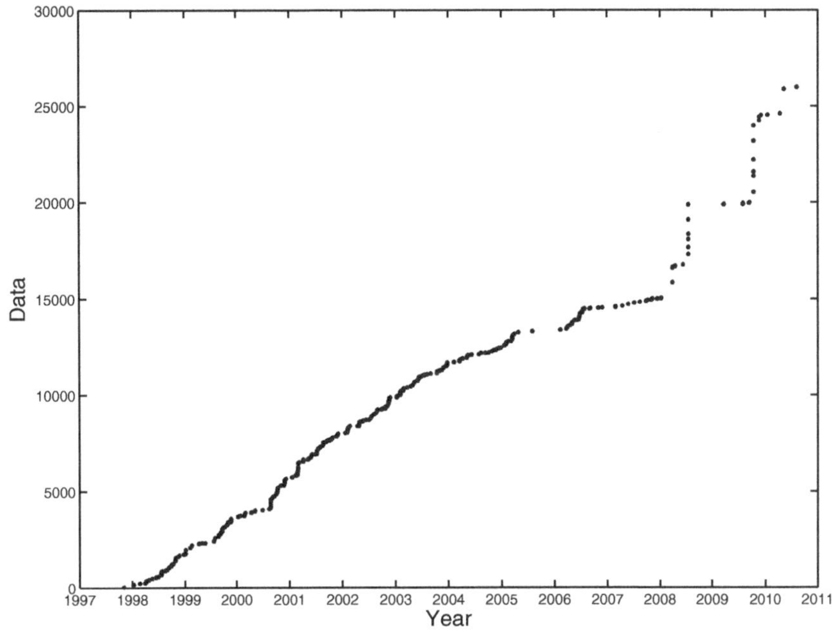

Fig. 10. Cumulative data volume submitted to the BMRB in category protein *J* coupling constants. The current number of deposited sets is around 350, totaling more than 25,000 values, the majority of which are $^3J_{HNH\alpha}$ coupling constants. Recent leaps reflect large-scale depositions of hitherto less popular ^1J and ^2J parameters which the authors of this chapter have set out to explore regarding utility in protein structure determination (Schmidt et al., 2009, 2010).

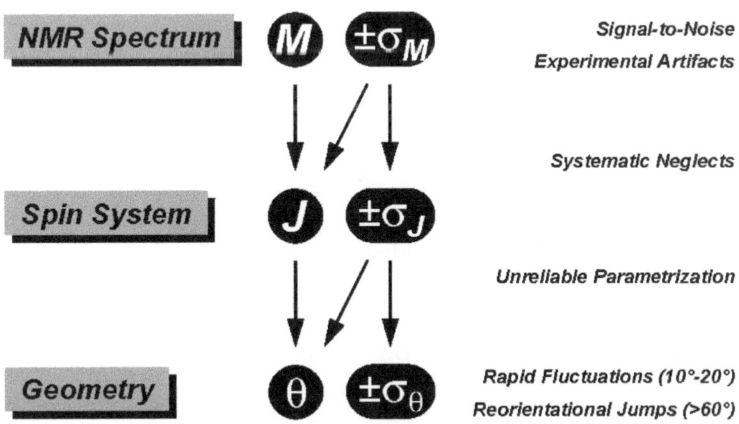

Fig. 11. Progression of procedural errors and random uncertainties into molecular structure.

A word of caution is advised though. Even if both above mentioned evaluation stages as well as the recording of NMR spectra were void of systematic error, i.e., not limited by their technical approach, molecules don't always do us the favour of revealing an unambiguous structure. If conformational dynamics are at play, such ambiguity is not to be confused with lack of accuracy or precision, and the geometry of the molecule must not be squeezed into a 'strait-jacket', as it were. Rather we need to develop our understanding of molecular structure, if not re-define the term, in a way that integrates both dynamic as well as static aspects of conformation. Research into partially or intrinsically disordered protein structures is witness to such development (Dyson & Wright, 1998).

5. NMR Experiments

J coupling measurements were carried out on NMR spectrometers manufactured by Bruker, Rheinstetten, Germany, with nominal fields ranging from 500 to 900 MHz proton frequency. All instruments were fitted with cryogenically cooled triple-resonance z-axis pulsed-field-gradient probes, except the 500-MHz instruments were equipped with room-temperature triple-resonance three-axis pulsed-field-gradient probes.

Experiment names of the exclusive correlation spectroscopy (E.COSY) follow a convention that identifies by uppercase lettering those protein nuclei that are active in generating the spectrum dimensions, in round parentheses those used for magnetization relay only, and in square brackets the presence of those passive spins that give rise to the sought *J* coupling interaction (Wang & Bax, 1995).

Each coupling constant was measured at least twice, and mean values quoted carry two standard deviations, the former denoting the *intra*-spectrum, i.e., variation across all amino-acid residues, and the latter denoting the standard error or *inter*-spectrum rms variation between the repeat experiments, i.e., a measure of reproducibility.

5.1 $^3J(H^N,H^\alpha)$ coupling constants

$^3J_{HNH\alpha}$ coupling constants originated from 3D-HA[HB,HN](CACO)NH quantitative *J* correlation spectra recorded at 500 MHz (Löhr et al., 1999). The coupling constant for residue *i* was evaluated in spectrum planes at the resonance N_{i+1} by summing the data points comprising the $H^N{}_i,H^N{}_{i+1}$ cross peak and referring its intensity to the $H^\alpha{}_i,H^N{}_{i+1}$ auto peak according to $I_{cross}/I_{auto} = -\tan^2(\pi J\Delta)$, with the magnetization-transfer delay, Δ, set at 17 ms. Raw coupling constants were corrected for H/D exchange by an average factor of 1.049, due to scaling all I_{cross}/I_{auto} ratios for the 10-% D_2O solvent component needed for field-frequency locking. Both datasets yielded 77 values of corrected $^3J_{HNH\alpha}$ averaging 6.77 ± 2.17 ± 0.24 Hz.

5.2 $^3J(H^N,C')$ coupling constants

$^3J_{HNC'}$ coupling constants were determined by least-squares fitting 2D projections of E.COSY-type C^α,H^N multiplets (Schmidt et al., 1997a) recorded at 900 MHz using a [$^{15}N,^1H$]-TROSY variant (Pervushin et al., 2000) of 3D-(H)CANNH[CO] spectra (Löhr & Rüterjans, 1995). To minimize transverse relaxation effects during the constant-time period, C^α

chemical shifts and $^{1}J_{C\alpha C'}$ couplings were simultaneously evolved in a $^{1}H^{\alpha}$,$^{13}C^{\alpha}$ multiple-quantum constant-time t_1 period while spin-locking $^{1}H^{\alpha}$ magnetization using a 3-kHz spin-lock field. Insufficient to cover the complete H^{α} shift range uniformly, poorly decoupled C^{α} resonances near the excitation band edges were omitted from evaluation. Glycine patterns were disregarded entirely as the NMR experiment is not optimized for their twin-H^{α} spin topology. The highly reproducible results comprised 70 values of $^{3}J_{HNC'}$ averaging 0.75 ± 0.71 ± 0.06 Hz. For many of those residues that needed to be ignored in (H)CANNH evaluations for reasons of lineshape distortion or overlap, $^{3}J_{HNC'}$ coupling constants were obtained using a simpler 3D-heteronuclear relayed-E.COSY approach at 500 MHz in combination with more sophisticated antiphase 2D-multiplet line-shape analysis (Schmidt et al., 1996). The 76 values averaged 0.86 ± 0.72 ± 0.17 Hz, slightly larger than those from (H)CANNH. Additional values were extracted from a [^{15}N,^{1}H]-TROSY version of ct-HNCA[CO]-E.COSY spectra (Wang & Bax, 1995) recorded at 600 MHz, averaging 0.81 ± 0.79 ± 0.21 Hz and bringing the total number of items to 92, including data for glycine. Rmsd between all six spectra was 0.22 Hz.

5.3 $^{3}J(H^{N},C^{\beta})$ coupling constants

$^{3}J_{HNC\beta}$ coupling constants were determined by simulating 2D projections of C^{α},H^{N} multiplets resulting from a [^{15}N,^{1}H]-TROSY version of the 3D-HNCA[CB]-E.COSY experiment (Wang & Bax, 1996) carried out at 800 MHz. Glycines generally lacking the C^{β} nucleus were ignored. Overlapping C^{α}_{i} and C^{α}_{i-1} resonances prevented a few coupling constants from being obtained. Coincident C^{α} and C^{β} chemical-shift ranges in serine and threonine residues precluded selective excitation of either nucleus and respective data were disregarded. Thus, the two recordings yielded only 58 values, averaging 1.50 ± 0.94 ± 0.09 Hz. Additional $^{3}J_{HNC\beta}$ coupling constants resulted however from 3D [^{15}N,^{1}H]-TROSY-HNCB quantitative J correlation experiments (F. Löhr, unpublished) performed at 600 and 800 MHz. Dephasing of $^{1}H^{N}$ coherence due to passive $^{3}J_{HNH\alpha}$ couplings during J evolution delay Δ, which was set at either 40 or 45 ms, was avoided by employing BIRD refocusing elements (Garbow et al., 1982) in the centre of the Δ period. Three spectra yielded 68 values of $^{3}J_{HNC\beta}$ averaging 1.71 ± 0.79 ± 0.18 Hz between quantitative J correlation data. Between all five sets, 81 values were determined at 1.62 ± 0.85 ± 0.14 Hz.

5.4 $^{3}J(C'_{i-1},H^{\alpha}_{i})$ coupling constants

$^{3}J_{C'H\alpha}$ coupling constants resulted from fitting 2D projections of C^{α}_{i},C'_{i-1} multiplets in 3D-H(N)CA,CO[HA]-E.COSY spectra (Löhr & Rüterjans, 1997; Löhr et al., 1997) recorded at 500 MHz. The total 81 values collected averaged 2.33 ± 1.29 ± 0.16 Hz.

5.5 $^{3}J(C'_{i-1},C'_{i})$ coupling constants

$^{3}J_{C'C'}$ coupling constants resulted from fitting 2D projections of C^{α}_{i},C'_{i-1} multiplets in 3D-H(N)CA,CO[CO]-E.COSY spectra (Löhr et al., 1997) recorded at 500 MHz. Glycine couplings included, the two datasets comprised 90 $^{3}J_{C'C'}$ values averaging 0.84 ± 0.65 ± 0.16 Hz. Additional $^{3}J_{C'C'}$ coupling constants resulted from quantitative J-correlation experiments using a [^{15}N,^{1}H]-TROSY version of the 3D HN(CO)CO pulse sequence (Hu & Bax, 1996). Two spectra were acquired at 500 MHz with the *J*-evolution period Δ set at either 50 or 60 ms. The spectra allowed both sequential coupling constants $^{3}J(C'_{i-2},C'_{i-1})$ and $^{3}J(C'_{i-1},C'_{i})$ to be evaluated, yielding 78 values averaging 1.03 ± 0.61 ± 0.23 Hz. The final set comprised 98 coupling constants of 0.95 ± 0.63 ± 0.26 Hz.

5.6 $^{3}J(C'_{i-1},C^{\beta}_{i})$ coupling constants

$^{3}J_{C'C\beta}$ coupling constants were measured at 500 MHz by H(N)CA,CO[CB]-E.COSY (Löhr et al., 1997), yielding only 59 values averaging 1.96 ± 0.86 ± 0.16 Hz. As with $^{3}J_{HNC\beta}$, determining $^{3}J_{C'C\beta}$ coupling constants with these E.COSY-type experiments fails for serine, threonine, and some leucine residues, all exhibiting downfield-shifted $^{13}C^{\beta}$ resonances, so as to overlap with C^{α} chemical-shift ranges, preventing selective excitation. In contrast, HN(CO)CB quantitative *J* correlation (Hu & Bax, 1997) provides values for all residue types as long as the active coupling magnitude exceeds a certain threshold determined primarily by the signal-to-noise ratio. Spectra were acquired at 500 MHz with delays Δ set at either 37.5 or 38.5 ms to match $2/^{1}J(C',C^{\alpha})$. The ^{13}C carrier frequency was positioned at either 26 or 60 ppm. Quantitative *J* correlation yielded 74 data, averaging 1.83 ± 0.75 ± 0.10 Hz. Additional $^{3}J_{C'C\beta}$ values including those in Ser and Thr residues were obtained from C^{α}_{i},C'_{i-1} multiplet projections in H(N)CO,CA[CA]-E.COSY recorded at 800 MHz (F. Löhr, unpublished). The 76 values averaged 1.66 ± 0.80 ± 0.20 Hz, and the grand average over all ten sets was 1.82 ± 0.78 ± 0.20 Hz, totaling 82 values.

6. Supplementary material

^{3}J coupling constants related to the ϕ-torsion angles in RNase T1 as determined in the present work are deposited with the BioMagRes Database (accession number BMRB-16469), available at http://www.bmrb.wisc.edu/cgi-bin/explore.cgi?format=raw&bmrbId=16469

7. Acknowledgements

Norman Spitzner (Frankfurt university) is thanked for preparing a doubly-labelled RNase T1 sample. Robert Pritchard (graduate from Kent university) is thanked for helping evaluate RNase T1 spectra in the course of his undergraduate final-year project. Financial support by the Access to Research Infrastructures activity in the 7th Framework Programme of the EC (Project number: 261863, Bio-NMR) is gratefully acknowledged.

8. References

Arni, R., Heinemann, U., Maslowska, M., Tokuoda, R. & Saenger, W. (1988) Restrained least-squares refinement of the crystal structure of the ribonuclease T1*2'-guanylic acid complex at 1.9 Å resolution. *Acta Crystallographica Section B*, Vol. 43, pp. 534-554, ISSN 0108-7681 (p) 1600-5740 (e)

Artali, R., Bombieri, G., Meneghetti, F., Gilardi, G., Sadeghi, S.J., Cavazzini, D. & Rossi, G.L. (2002) Comparison of the refined crystal structure of wild-type (1.34 Å) flavodoxin from *Desulfovibrio vulgaris* and the S35C mutant (1.44 Å) at 100 K. *Acta Crystallographica Section D*, Vol. 58, pp. 1787-1792, ISSN 0907-4449 (p) 1399-0047 (e)

Bax, A., Vuister, G.W., Grzesiek, S., Delaglio, F., Wang, A.C., Tschudin, R. & Zhu, G. (1994) Measurement of homo- and heteronuclear J couplings from quantitative J correlation. Methods in Enzymology, Vol. 239, pp. 79-105, ISBN 0-12-182140-4

Billeter, M., Neri, D., Otting, G., Qian, Y.Q. & Wüthrich, K. (1992) Precise vicinal coupling constants $^3J_{HN\alpha}$ from nonlinear fits of *J*-modulated $[^{15}N,^{1}H]$-COSY-experiments. *Journal of Biomolecular NMR*, Vol. 2, pp. 257-274, ISSN 0925-2738 (p) 1573-5001 (e)

Blümel, M., Schmidt, J.M., Löhr, F. & Rüterjans, H. (1998) Quantitative ϕ torsion angle analysis in *Desulfovibrio vulgaris* flavodoxin based on six ϕ related ^3J couplings. *European Biophysics Journal*, Vol. 27, pp. 321-334, ISSN 0175-7571 (p) 1432-1017 (e)

Brüschweiler, R. & Case, D.A. (1994) Adding harmonic motion to the Karplus relation for spin-spin coupling. *Journal of the American Chemical Society*, Vol. 116, pp. 11199-11200, ISSN 0002-7863 (p) 1520-5126 (e)

Bystrov, V.F. (1976) Spin-spin coupling and the conformational states of peptide systems. *Progress in NMR Spectroscopy*, Vol. 10, pp. 41-81, ISSN 0079-6565

Cavanagh, J., Fairbrother, W.J., Palmer III, A.G., Rance, M. & Skelton, N.J. (2007) *Protein NMR spectroscopy: principles and practice*, Academic Press, San Diego, California, ISBN 978-0-12-164491-8 / 0-12-164491-X

Chan, A.W.E., Hutchinson, E.G. & Harris, D. (1993) Identification, classification, and analysis of beta-bulges in proteins. *Protein Science*, Vol. 2, 1574-1590, ISSN 0961-8368 (p) 1469-896x (e)

Creighton, T.E. (1993) *Proteins: structures and molecular properties*, 2nd ed., W. H. Freeman and Co., New York, ISBN 0-7167-7030-X

Dyson, H.J. & Wright, P.E. (1998) Equilibrium NMR studies of unfolded and partially folded proteins. *Nature Structural Biology, Supplement*, Vol. 5, pp. 499-503, ISSN 1072-8368

Ejchart, A. (1999) Scalar couplings in structure determination of proteins. *Bulletin of the Polish Academy of Sciences-Chemistry*, Vol. 47, pp. 1-19, ISSN 0239-7285

Engh, R.A. & Huber, R. (1991) Accurate bond and angle parameters for X-ray protein structure refinement. *Acta Crystallographica Section A*, Vol. 47, pp. 392-400, ISSN 0108-7673 (p) 1600-5724 (e)

Engh, R.A. & Huber, R. (2006) Structure quality and target parameters in *International Tables for Crystallography* Vol. F, Chapter 18.3, pp. 382–392, ISBN 978-1-4020-4969-9

Ernst, R.R., Bodenhausen, G. & Wokaun, A. (1987) *Principles of nuclear magnetic resonance in one and two dimensions*, Oxford University Press, Oxford, ISBN 0-19-855629-2

Evans, J.N.S. (1995) *Biomolecular NMR spectroscopy*, Oxford University Press, Oxford, ISBN 0-19-854766-8

Fischer, M.W.F., Losonczi, J.A., Weaver, J.L. & Prestegard, J.H. (1999) Domain orientation and dynamics in multidomain proteins from residual dipolar couplings. *Biochemistry*, Vol. 38, pp. 9013-9022, ISSN 0006-2960 (p) 1520-4995 (e)

Garbow, J.R., Weitekamp, D.P. & Pines, A. (1982) Bilinear rotation decoupling of homonuclear scalar interactions. *Chemical Physics Letters*, Vol. 93, pp. 504-509, ISSN 0009-2614

Griesinger, C., Sørensen, O.W. & Ernst, R.R. (1987) Practical aspects of the E. COSY technique. Measurement of scalar spin-spin coupling constants in peptides. *Journal of Magnetic Resonance*, Vol. 75, pp. 474-492, ISSN 0022-2364

Hansen, P.E., Feeney, J. & Roberts, G.C.K. (1975) Long range ^{13}C-^{1}H spin-spin coupling constants in amino acids. Conformational applications. *Journal of Magnetic Resonance*, Vol. 17, pp. 249-261, ISSN 0022-2364

Hennig, M., Bermel, W., Schwalbe, H. & Griesinger, C. (2000) Determination of ψ torsion angle restraints from $^{3}J(C_{\alpha},C_{\alpha})$ and $^{3}J(C_{\alpha},H_{N})$ coupling constants in proteins. *Journal of the American Chemical Society*, Vol. 122, pp. 6268–6277, ISSN 0002-7863 (p) 1520-5126 (e)

Hoch, J.C. & Stern, A.S. (1996) *NMR data processing*, John Wiley and Sons, New York, ISBN 978-0-471-03900-6

Hoffmann, E. & Rüterjans, H. (1988) Two-dimensional ^{1}H-NMR investigation of ribonuclease T1. *European Journal of Biochemistry*, Vol. 177, pp. 539-560, ISSN 0014-2956 (p) 1432-1033 (e)

Hu, J.-S. & Bax, A. (1996) Measurement of three-bond ^{13}C-^{13}C J couplings between carbonyl and carbonyl/ carboxyl carbons in isotopically enriched proteins. *Journal of the American Chemical Society*, Vol. 118, pp. 8170-8171, ISSN 0002-7863 (p) 1520-5126 (e)

Hu, J.-S. & Bax, A. (1997) Determination of ϕ and χ^{1} angles in proteins from ^{13}C-^{13}C three-bond couplings measured by three-dimensional heteronuclear NMR. How planar is the peptide bond? *Journal of the American Chemical Society*, Vol. 119, pp. 6360-6368, ISSN 0002-7863 (p) 1520-5126 (e)

IUPAC-IUB Commision on Biochemical Nomenclature (1970) Abbreviations and symbols for the description of the conformation of poplypeptide chains. Tentative rules (1969). *Biochemistry*, Vol. 9, pp. 3471-3479, ISSN 0006-2960

Jardetzky, O. (1980) On the nature of molecular conformations inferred from high-resolution NMR. *Biochimica et Biophysica Acta*, Vol. 621, pp. 227-232, ISSN 0006-3002

Kainosho, M. (1997) Isotope labelling of macromolecules for structural determinations. *Nature Structural Biology*, Vol. 4, Suppl. S, pp. 858-861, ISSN 1072-8368

Karimi-Nejad, Y., Schmidt, J.M., Rüterjans, H., Schwalbe, H. & Griesinger, C. (1994) Conformation of valine side chains in ribonuclease T1 determined by NMR studies of homonuclear and heteronuclear ^{3}J coupling constants. *Biochemistry*, Vol. 33, pp. 5481-5492, ISSN 0006-2960 (p) 1520-4995 (e)

Karplus, M. (1963) Vicinal proton coupling in nuclear magnetic resonance. *Journal of the American Chemical Society*, Vol. 85, pp. 2870-2871, ISSN 0002-7863 (p) 1520-5126 (e)

Keeler, J. (2005) *Understanding NMR spectroscopy*, John Wiley and Sons, Chichester, West Sussex, England, ISBN 0-470-01787-2

Löhr, F. & Rüterjans, H. (1995) (H)NCAHA and (H)CANNH experiments for the determination of vicinal coupling constants related to the ϕ-torsion angle. *Journal of Biomolecular NMR*, Vol. 5, pp. 25-36, ISSN 0925-2738 (p) 1573-5001 (e)

Löhr, F. & Rüterjans, H. (1997) A sensitive method for the measurement of three-bond C',H^{α} J coupling in uniformly ^{13}C- and ^{15}N-enriched proteins. *Journal of the American Chemical Society*, Vol. 119, pp. 1468-1469, ISSN 0002-7863 (p) 1520-5126 (e)

Löhr, F., Blümel, M., Schmidt, J.M. & Rüterjans, H. (1997) Application of H(N)CA,CO-E.COSY experiments for calibrating the ϕ-angular dependences of vicinal $J(C'_{i-1},H^{\alpha}_{i})$, $J(C'_{i-1},C'_{i})$, and $J(C'_{i-1},C^{\beta}_{i})$ in proteins. *Journal of Biomolecular NMR*, Vol. 10, pp. 107-118, ISSN 0925-2738 (p) 1573-5001 (e)

Löhr, F., Schmidt J.M. & Rüterjans, H. (1999) Simultaneous measurement of $^{3}J_{HN,H\alpha}$ and $^{3}J_{H\alpha,H\beta}$ coupling constants in ^{13}C,^{15}N-labeled proteins. *Journal of the American Chemical Society*, Vol. 121, pp. 11821-11826, ISSN 0002-7863 (p) 1520-5126 (e)

Löhr, F., Pérez, C., Köhler, R., Rüterjans, H. & Schmidt, J.M. (2000) Heteronuclear relayed E.COSY revisited: Determination of $^{3}J(H^{\alpha},C^{\gamma})$ couplings in Asx and aromatic residues in proteins. *Journal of Biomolecular NMR*, Vol. 18, pp. 13-22, ISSN 0925-2738 (p) 1573-5001 (e)

Löhr, F., Schmidt, J.M., Maurer, S. & Rüterjans, H. (2001) Improved measurement of $^{3}J(H^{\alpha}{}_{i}, N_{i+1})$ coupling constants in H_2O dissolved proteins. *Journal of Magnetic Resonance*, Vol. 153, pp. 75-81, ISSN 1090-7807

Löhr, F., Reckel, S., Stefer, S., Dötsch, V. & Schmidt, J.M. (2011) Improved accuracy in measuring one-bond and two-bond ^{15}N,$^{13}C^{\alpha}$ coupling constants in proteins by double-inphase/antiphase (DIPAP) spectroscopy. *Journal of Biomolecular NMR*, Vol. 50, pp. 167-190, ISSN 0925-2738 (p) 1573-5001 (e)

Martinez-Oyanedel, J., Choe, H.W., Heinemann, U. & Saenger, W. (1991) Ribonuclease T1 with free recognition and catalytic site: Crystal structure analysis at 1.5 Å resolution. *Journal of Molecular Biology*, Vol. 222, pp. 335-352, ISSN 0022-2836

Neuhaus, D. & Williamson, M.P. (1989) *The nuclear Overhauser effect in structural and conformational analysis*, VCH Publishers, New York, ISBN 0-89573-343-9

Pachler, K.G.R. (1963) Nuclear magnetic resonance study of some α-amino acids—I. Coupling constants in alkaline and acidic medium. *Spectrochimica Acta*, Vol. 19, pp. 2085-2092, ISSN 0371-1951

Pachler, K.G.R. (1964) Nuclear magnetic resonance study of some α-amino acids—II. Rotational isomerism. *Spectrochimica Acta*, Vol. 20, pp. 581-587, ISSN 0371-1951

Pérez, C., Löhr, F., Rüterjans, H. & Schmidt, J.M. (2001) Self-consistent Karplus parametrization of ^{3}J couplings depending on the polypeptide sidechain torsion χ_1. *Journal of the American Chemical Society*, Vol. 123, pp. 7081-7093, ISSN 0002-7863 (p) 1520-5126 (e)

Pervushin, K. (2000) Impact of transverse relaxation optimized spectroscopy (TROSY) on NMR as a technique in structural biology. *Quarterly Reviews of Biophysics*, Vol. 33, pp. 161–197, ISSN 0033-5835

Pfeiffer, S., Karimi-Nejad, Y. & Rüterjans, H. (1996a) Limits of NMR structure determination using variable target function calculations: ribonuclease T1, a case study. *Journal of Molecular Biology*, Vol. 266, pp. 400-423, ISSN 0022-2836

Pfeiffer, S., Engelke, J. & Rüterjans, H. (1996b) Complete ^{1}H, ^{15}N and ^{13}C resonance assignment of ribonuclease T1: Secondary structure and backbone dynamics as derived from the chemical shifts. *Quarterly Magnetic Resonance in Biological Medicine*, Vol. 3, pp. 69-87

Pople, J.A. & Gordon, M. (1967) Molecular orbital theory of the electronic structure of organic compounds. I. Substituent effects and dipole moments. *Journal of the American Chemical Society*, Vol. 89, pp. 4253-4261, ISSN 0002-7863 (p) 1520-5126 (e)

Quaas, R., McKeown, Y., Stanssens, P., Frank, R., Blöcker, H. & Hahn, U. (1988) Expression of the chemically synthesized gene for ribonuclease T1 in Escherichia coli using a secretion cloning vector. *European Journal of Biochemistry*, Vol. 173, pp. 617-622, ISSN 0014-2956 (p), 1432-1033 (e)

Quaas, R., Grunert, H.-P., Kimura, M. & Hahn, U. (1988) Expression of ribonuclease T1 in Escherichia coli and rapid purification of the enzyme. *Nucleosides & Nucleotides*, Vol. 7, pp. 619-623, ISSN 0732-8311

Reif, B., Hennig, M. & Griesinger, C. (1997) Direct measurement of angles between bond vectors in high resolution NMR. *Science*, Vol. 276, pp. 1230-1233, ISSN 0036-8075 (p) 1095-9203 (e)

Reif, B., Diener, A., Hennig, M., Maurer, M. & Griesinger, C. (2000) Cross correlated relaxation for the measurement of angles between tensorial interactions, *Journal of Magnetic Resonance*, Vol. 143, pp. 45-68, ISSN 1090-7807

Roberts, G.C.K. (ed.) (1993) *NMR of macromolecules: A practical approach*, Oxford University Press, Oxford, ISBN 978-0199632244

Schmidt, J.M., Thüring, H., Werner, A., Rüterjans, H., Quaas, R. & Hahn, U. (1991) Two-dimensional ^{1}H,^{15}N-NMR-investigation of uniformly ^{15}N-labeled ribonuclease T1 – Complete assignment of ^{15}N resonances. *European Journal of Biochemistry*, Vol. 197, pp. 643-653, ISSN 0014-2956 (p) 1432-1033 (e)

Schmidt, J.M., Löhr, F. & Rüterjans, H. (1996) Heteronuclear relayed E.COSY applied to the determination of accurate $^{3}J(H^{N},C')$ and $^{3}J(H^{\beta},C')$ coupling constants in *Desulfovibrio vulgaris* flavodoxin. *Journal of Biomolecular NMR*, Vol. 7, pp. 142-152, ISSN 0925-2738 (p) 1573-5001 (e)

Schmidt, J.M. (1997a) Conformational equilibria in polypeptides. I. Determination of accurate $^{3}J_{HC}$ coupling constants in antamanide by 2D NMR multiplet simulation. *Journal of Magnetic Resonance*, Vol. 124, pp. 298-309, ISSN 1090-7807

Schmidt, J.M. (1997b) Conformational equilibria in polypeptides. II. Dihedral-angle distribution in antamanide based on three-bond coupling information. *Journal of Magnetic Resonance*, Vol. 124, pp. 310-322, ISSN 1090-7807

Schmidt, J.M., Blümel, M., Löhr, F. & Rüterjans, H. (1999) Self-consistent ^{3}J coupling analysis for the joint calibration of Karplus coefficients and ϕ-torsion angles. *Journal of Biomolecular NMR*, Vol. 14, pp. 1-12, ISSN 0925-2738 (p) 1573-5001 (e)

Schmidt, J.M. (2007a) A versatile component-coupling model to account for substituent effects. Application to polypeptide ϕ and χ_1 torsion related ^3J data. *Journal of Magnetic Resonance*, Vol. 186, pp. 34-50, ISSN 1090-7807

Schmidt, J.M. (2007b) Asymmetric Karplus curves for the protein side-chain ^3J couplings. *Journal of Biomolecular NMR*, Vol. 37, pp. 287-301, ISSN 0925-2738 (p) 1573-5001 (e)

Schmidt, J.M., Howard, M.J., Maestre-Martínez, M., Pérez, C.S. & Löhr, F. (2009) Variation in protein C^{α}-related one-bond *J* couplings. *Magnetic Resonance in Chemistry*, Vol. 47, pp. 16-30, ISSN 0749-1581

Schmidt, J.M., Hua, Y. & Löhr, F. (2010) Correlation of ^2J couplings with protein secondary structure. *Proteins*, Vol. 78, pp. 1544-1562, ISSN 0887-3585

Schmidt, J.M., Zhou, S., Rowe, M.L., Howard, M.J., Williamson, R.A. & Löhr, F. (2011) One-bond and two-bond *J* couplings help annotate protein secondary-structure motifs: *J*-coupling indexing applied to human endoplasmic reticulum protein ERp18. *Proteins*, Vol. 79, pp. 428-443, ISSN 0887-3585

Schwalbe, H., Carlomagno, T., Hennig, M., Junker, J., Reif, B., Richter, C. & Griesinger, C. (2001) Cross-correlated relaxtion for measurement of angles between tensorial interactions. *Methods in Enzymology*, Vol. 338, pp. 35-81, ISBN 0-12-182239-7

Spitzner, N., Löhr, F., Pfeiffer, S., Koumanov, A., Karshikov, A. & Rüterjans, H. (2001) Ionization properties of titratable groups in ribonuclease T1 – pKa values in the native state determined by two-dimensional heteronuclear NMR spectroscopy. *European Journal of Biochemistry*, Vol. 30, pp. 186-195, ISSN 0014-2956 (p) 1432-1033 (e)

Tjandra, N. & Bax, A. (1997) Direct measurement of distances and angles in biomolecules by NMR in a dilute liquid crystalline medium. *Science*, Vol. 278, pp. 1111-1114, ISSN 0036-8075 (p) 1095-9203 (e)

Ulrich, E.L., Akutsu, H., Doreleijers, J.F., Harano, Y., Ioannidis, Y.E., Lin, J., Livny, M., Mading, S., Maziuk, D., Miller, Z., Nakatani, E., Schulte, C.F., Tolmie, D.E., Wenger, R.K., Yao, H. & Markley, J.L. (2008) BioMagResBank. *Nucleic Acids Research*, Vol. 36, pp. D402-D408, ISSN 0305-1048 (p) 1362-4962 (e)

Walsh, M.A., McCarthy, A., O'Farrell, P.A., McArdle, P., Cunningham, P.D., Mayhew, S.G. & Higgins, T.M. (1998) X-ray crystal structure of the *Desulfovibrio vulgaris* (Hildenborough) apoflavodoxin-riboflavin complex. *European Journal of Biochemistry*, Vol. 258, pp. 362-371, ISSN 0014-2956 (p) 1432-1033 (e)

Wang, A.C. & Bax, A. (1995) Reparametrization of the Karplus relation for $^3J(H^{\alpha}$-N) and $^3J(H^N$-C') in peptides from uniformly $^{13}C/^{15}N$-enriched human ubiquitin. *Journal of the American Chemical Society*, Vol. 118, pp. 1810-1813, ISSN 0002-7863 (p) 1520-5126 (e)

Wang, A.C. & Bax, A. (1996) Determination of the backbone dihedral angle ϕ in human ubiquitin from reparametrized empirical Karplus equations. *Journal of the American Chemical Society*, Vol. 118, pp. 2483-2494, ISSN 0002-7863 (p) 1520-5126 (e)

Watenpaugh, K.D., Sieker, L.C. & Jensen, L.H. (1976) A crystallographic structural study of the oxidation states of Desulfovibrio vulgaris flavodoxin in *Flavins and Flavoproteins* (T.P. Singer ed.), pp. 405-410, Elsevier, Amsterdam, ISBN 0-444-41458-8

Watt, W., Tulinsky, A., Swenson, R.P. & Watenpaugh, K.D. (1991) Comparison of the crystal structures of a flavodoxin in its three oxidation states at cryogenic temperatures. *Journal of Molecular Biology*, Vol. 218, pp. 195-208, ISSN 0022-2836

Wienk, H.L.J., Martínez, M.M., Yalloway, G.N., Schmidt, J.M., Pérez, C., Rüterjans, H. & Löhr, F. (2003) Simultaneous measurement of protein one-bond and two-bond nitrogen-carbon coupling constants using an internally referenced quantitative *J*-correlated [^{15}N,^{1}H]-TROSY-HNC experiment. *Journal of Biomolecular NMR*, Vol. 25, pp. 133-145, ISSN 0925-2738 (p) 1573-5001 (e)

Wüthrich, K. (1986) *NMR of proteins and nucleic acids*, John Wiley and Sons, New York, ISBN 0-471-82893-9

6

Protein Structure Alphabetic Alignment

Jiaan Yang[1,*] and Wei-Hua Lee[2]
[1]*MicrotechNano LLC*
[2]*Department of Pediatrics and Anatomy, Indiana University School of Medicine, Indianapolis, USA*

1. Introduction

This study presents a fast approach to compare protein 3D structures with protein structure alphabetic alignment method. First, the folding shape of 5 consecutive residues is represented by protein folding shape code (PFSC) (Yang, 2008) and thus protein folding conformation can be completely described by PFSC. With complete description for folding shape along the backbone, any protein with given 3D structure can be converted into an alphabetic string and aligned for comparison. Consequently, this approach is able to provide a unique score to assess the global similarity in structure while it supplies an alignment table for analysis of local structure. Several sets of proteins with diverse homology or different degrees in complexity are compared. The results demonstrate that this approach provides an efficient method for protein structure alignment which is significant for protein structure search with high throughput screening of protein database.

Comparison of protein structures is a challenging task because of complication of 3D structure which involves ambiguous procedure in analysis. First, protein structure obviously is not a simple geometric subject. It is not easily to superimpose two proteins together because the specific emphasis of one portion of structures may cause other parts with similar structures to orient toward different directions in geometric space. In practice, an individual turning point in protein may overshadow entire similarity between two structures. Second, it is hard to develop a uniform process to compare the proteins with different homologies. For protein structures with identical amino acid sequence or with mutation in sequence, the comparison often requires sensitivity to distinguish the conformers with higher similarity in structure. However, for proteins with drastic difference in structural conformation, the good comparison expects a consistent procedure to evaluate the similarity in variant cases. Significant variation of protein conformation is primarily determinate by sequence difference, which affects the formation of hydrogen bond, van der Waals force interaction and disulfide bridge. Also, the protein conformation may be changed by other factors, such as solvent effect, protein-protein interaction, ligand docking and so on. From view of topological order of secondary structure, if two structures belong to different categories in protein classification, such as under different families, superfamilies,

* Corresponding Author

folds and class, the structural comparison becomes more difficult. An ideal method should have a consistent process to assess the similarity for proteins with various homologies in structures.

Many established methods for protein structure comparison were developed and evaluated (Kolodny et al., 2005). DALI method (Holm & Sander, 1993; Holm & Park, 2000) is frequently used in protein structure comparison based on the alignment of distance matrices. LGA method (Zemla, 2003) generates the different local superposition to detect the regions where the structures are more similar. CE method (Shindyalov & Bourne, 1998) is for calculating pairwise structure alignments. Two proteins are aligned by using characteristics of local geometry between C-alpha positions. Heuristics are used in defining a set of optimal paths joining aligned fragment pairs with gaps. The path with the best RMSD is subject to dynamic programming to achieve an optimal alignment. 3D-BLAST method (Mavridis & Ritchie, 2010) is developed to align the protein structures using 3D spherical polar Fourier for protein shape. There are many of well known methods, including DAL (Kryshtafovych et al., 2005; Hvidsten et al., 2003), MAMMOTH (Ortiz et al., 2002), ProSup (Lackner et al., 2000), VAST (Madej et al., 1995; Gibrat et al., 1996), SSAP (Taylor & Orengo, 1989), STRUCTAL (Subbiah et al., 1993), LSQMAN (Kleywegt & Jones, 1994), SSM (Krissinel & Henrick, 2004), FlexProt (Shatsky et al., 2002), FATCAT(Yuzhen & Adam, (2003) and TM-align/score (Zhang Y & Skolnick J. 2005).

For optimistic solution, most of methods attempt to find out higher number of equivalent residues while obtain lower value of root-mean-square deviation (RMSD) through superimposition of protein 3D structures or alignment of structural fragments. Unfortunately, it is tough to optimize these two parameters simultaneously because the intention of higher number of equivalent residues leans higher RMSD, or the favor of lower RMSD leads less number of equivalent residues. In protein structural superimposition two factors, the cutoff distance for RMSD and the initiative focusing location, may be artificially adjusted. These artificial factors are not unique for various methods and they may be changed on case-by-case basis with using same method. Apparently it directly affects the outcome of protein structural comparison. So, it is not surprised that with different methods or even same method, it may produce different values of RMSD and different numbers of equivalent residues. Consequently, different methods may generate unlike rank of similarity in assessment of proteins structures.

The structural alignment is a popular approach for protein comparison which has been developed by different strategies. First strategy is the rigid body alignment, which directly superimposes two proteins with possible best fitting to obtain the lowest RMSD and higher number of equivalent residues. Second strategy is the non-rigid body alignment, which allows smaller structural fragments of proteins with certain flexibility to orient or shift for better fittings, and then adopts various algorithms of measurement for similarity. However, no matter how the protein structure is partitioned, the acquisition of optimum result still involves obtaining the lowest RMSD and highest number of equivalent residues, which are two of contradictory adjustments. The attempt of direct alignment of geometric objects is difficult because no unique resolution is able to handle a geometric object of more than three points with no double superposition. In order to avoid direct alignment of geometric objects, the structural alphabetic alignment is a solution.

The earliest application of structural alphabets was the reorganization of the secondary structure in protein, and then adopted letter "A" for α-helix, "E" for β-strand and "C" for coil. Furthermore, the structural alphabetic methods (Brevern et al., 2000; Kolodny et al., 2002; Micheletti et al., 2000; Rooman et al., 1990; Schuchhardt et al., 1996; Unger et al., 1989; Sander et al., 2006; Tung et al., 2007; Ku & Hu, 2008; Karplus et al., 2003; Murphy et al., 2000) have been developed for more detail assignment for representative folding shapes. Different approaches in structure alphabets defined different length of peptide and adopted different number of prototypes for folding shapes. With pentapeptide motif, Protein Blocks (PBs) method determined 16 of folding shapes and use alphabets represent these primary prototypes (Kolodny et al., 2002). Thus, it was applied to protein structural alignment (Brevern, 2005; Joseph et al., 2011). Based on different designs in structural alphabets, a variety of methods of structural alphabetic alignment have been developed (Ku & Hu, 2008; Karplus et al., 2003; Tyagi et al., 2006; Melo & Marti-Renom, 2006; Friedberg et al., 2007; Tyagi et al., 2006; Guyon et al., 2004; Sacan et al., 2008; Wang & Zheng, 2008). The performance of structural alphabetic alignment approaches are significantly faster than the methods based on 3D structural comparison, and the unambiguousness is avoided during structural superimposition. However, to date the prototypes of folding shapes in structural alphabetic methods are obtained by observations from training database, and then the primary motifs for folding patterns are determined by statistics judgment. With training database, the experimental observations may collect most of folding patterns with higher frequency of appearance in protein, but may leave out certain folding shapes as leak because of its rare appearance in proteins. Also, each prototype of folding pattern or alphabet is isolated without association meaning. A recently developed structural alphabets approach, protein folding shape code (PFSC) (Yang, 2008), overcomes the shortcomings, which is comprised by complete folding patterns for motif of five residues, and all folding patterns have the meaningful interrelated relationship.

In this study, a set of 27 PFSC vectors is used to describe the folding shapes of protein structure, and to apply to structural alignment. The 27 PFSC vectors are rigorously obtained by mathematical derivation to cover an enclosed space, and represent all possible folding shapes for any five of successive Cα atoms (Yang, 2008). The 27 PFSC vectors are symbolized by 26 alphabetic letters plus $ symbol, which are capable completely to describe the change of protein folding shapes along protein backbone from N-terminus to C-terminus without gap. With complete description of folding shape for any given protein 3D structure, a consistent method for alignment of protein structures is developed, which is able to assess the structural similarity with various homologies.

2. Method

2.1 Conversion of alphabet description

The protein 3D structure is first converted into alphabetic description with protein folding shape code (PFSC) (Yang, 2008). With PFSC approach, a set of 27 PFSC vectors represent all possible folding shapes for each five successive Cα atoms. The 27 PFSC vectors, prototypes of folding shapes and alphabets are shown on top of Fig.1. The 27 PFSC vectors are able to map all possible folding shapes, including the regular secondary structure and irregular coil and loop. The 27 PFSC alphabetic codes are able to describe the change of protein folding shapes along based on five successive Cα atoms. It provides a complete alphabetic description of protein structural conformation from N-terminus to C-terminus without gap. To take protein structure of 8DFR (PDB ID) as sample, the folding shape of each of each five successive Cα atoms is converted into one of 27 PFSC alphabetic letters along protein backbone. Consequently, the structural folding conformation is expressed by the PFSC alphabetic description and is demonstrated on bottom of Fig. 1.

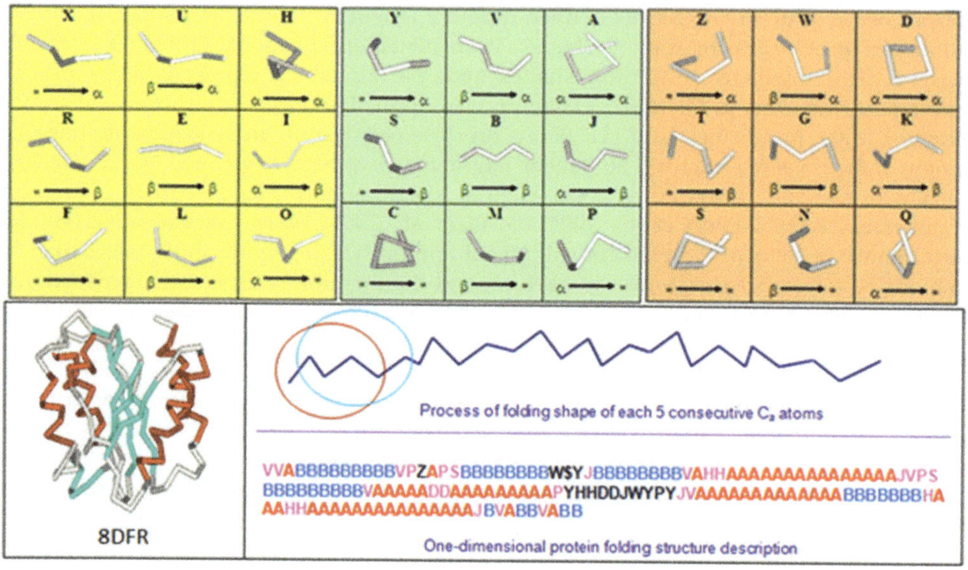

Fig. 1. The 27 protein folding shape code and the conversion of protein alphabetic description. Top: Three blocks represent three regions of pitch distance of motif for five residues; the nine vectors in each block represent the nine folding shape patterns determined by two torsion angles; each vector is simultaneously represented by a letter, a folding shape pattern and an arrow. The vector characteristic is represented by an arrow line. The "α", "β" or "*"at each end of vector indicates the folding features similar to α-helix, β-strand or random coil respectively. Bottom: 8DFR (PDB ID) is a sample to illustrate how protein backbone conformation is converted into PFSC alphabetic description. The folding shape of each five successive C-alpha atoms in a protein backbone from N-terminal to C-terminal is converted into alphabetic description. "A" represents a typical alpha helix with red color and "B" beta strand with blue. The folding shape is derived from secondary structure in pink color, and shape for loop or coil in black.

2.2 Protein Folding Shape Alignment (PFSA)

With one-dimensional PFSC alphabetic description, the protein conformation structures are able to be compared by protein folding shape alignment (PFSA) approach (Yang, 2011). Similarly as sequence alignment, the PFSC alphabetic strings for proteins are aligned to match the similarity. The Needleman-Wunsch algorithm of dynamic programming technique (Needleman SB & Wunsch, 1970) is used in the PFSA for structural alignment. Therefore, the structural similarity of two proteins is able to be discovered by structural alphabetic alignment with PFSA approach.

In PFSA approach, a substitution matrix for 27 PFSC vectors is defined according relationship of vector similarity. Within substitution matrix S, each element of similarity matrix S[i, j] is determined by the similarity between PFSC[i] and PFSC[j], which is determined by the integrated relationship of 27 PFSC vectors (Yang, 2008). For identical folding shape, the value S[i, i] = 2; for analogous folding shape, the value S[i, j] = 1 and for different folding shape, the value S[i, j] = 0. The substitution matrix S is displayed in Table 1. In next step, a similarity matrix for two proteins is constructed. According substitution matrix S, all elements of similarity matrix M are able to be determined. It assumes that m and n are the lengths of amino acid sequence for protein A and B respectively. Thus the lengths PFSC strings for protein A and B are m-4 and n-4. With the protein folding shape strings of protein A[3…m-2] and protein B[3...n-2], a similarity matrix M with (m-4) x (n-4) dimension is constructed for a pair proteins of A and B in structural alignment. The third step is to obtain a sum matrix by computing the elements of the similarity matrix according the Needleman-Wunsch algorithm. With the sum matrix, an optimized structural alignment is obtained based on tracing elements from the largest value to smaller value. When the track shifts from diagonal in the sum matrix, it actually tries to reduce the mismatch by insertion of gap for match of identical or analogous folding shape.

	A	B	C	D	E	F	G	H	I	J	K	L	M	N	O	P	Q	R	S	T	U	V	W	X	Y	Z	$
A	2			1				1		1						1						1			1		
B		2			1		1			1			1						1			1					
C			2			1							1			1			1						1		1
D	1			2							1						1						1			1	
E		1			2				1			1						1			1						
F			1			2						1			1			1						1			
G		1					2				1			1						1			1				
H	1							2	1						1						1			1			
I					1			1	2	1					1			1									
J	1	1							1	2	1					1			1								
K				1			1			1	2						1			1							
L					1	1						2	1		1						1						
M		1	1									1	2	1		1						1					
N							1						1	2			1						1				1
O						1		1	1			1			2	1											
P	1		1							1			1		1	2	1										
Q				1							1			1		1	2										1
R					1	1			1									2	1					1			
S		1	1							1								1	2	1					1		
T							1				1								1	2						1	1
U					1			1				1									2	1		1			
V	1	1											1								1	2	1		1		
W				1			1							1								1	2			1	
X						1		1										1			1			2	1		
Y	1		1																1			1		1	2	1	
Z				1																1			1		1	2	1
$			1											1			1			1						1	2

Table 1. The substitution matrix of 27 PFSC vectors. The top row and the left column list the 27 PFSC letter. The value of element in substitution matrix is 2 for identical folding shape code; 1 for analogues folding shape code; empty means zero for different folding shape code.

2.3 Similarity score

With optimized alignment, the protein structural similarity score is calculated. Each match of identical folding shape is assigned by 2; analogous folding shape 1; different folding shape 0; penalty of open a gap -2 and penalty of extended a gap -0.25. The value of protein folding structure alignment score (PFSA-S) is determined by the total contribution of identical folding shapes, analogous folding shapes and gaps. The score is normalized with below function.

$$PFSA\text{-}S = \frac{2 \times ID_{FS} + 1 \times AN_{FS} - (2 \times GPO + 0.25 \times GPE)}{2 \times TSQ}$$

Here ID_{FS} is the number of identical folding shapes, AN_{FS} the number of analogous folding shape, *GPO* the number of open gaps, *GPE* the number of extended gaps and *TSQ* is the length of PFSC of protein. The denominator in formula, 2 x *TSQ*, assures the value of PFSA-S to equal numeral one for comparison of two identical structures. When similarity between two protein structures decreases, the value of PFSA-S will decrease. When two proteins have less similarity, the structural alignment produces larger number of gaps, which may give negative value for PFSA-S and signify no noteworthy similarity existing. For normalization, the value of PFSA-S is limited to larger or equal to zero, so any negative value of PFSA-S is converted as zero. Therefore, the PFSA approach provides a normalized score between one and zero to evaluate the protein structural similarity.

2.4 Alignment table

With comparison of one-dimensional alphabetic strings for protein folding conformation, the PFSA alignment table is generated. There are two types of alignment tables, i.e. sequence-dependence mode and sequence-independence mode. For same protein or proteins with mutation, the structural alignment for conformation analysis may prefer the sequence-dependent mode because gap insertion is not necessary. For proteins with different sequence and size, the structural alignment takes the advantage of the sequence-independent mode, which allows inserting gaps to obtain the best match in local structural similarity.

The PFSA alignment table possesses several features. First, the alignment table is able explicitly to reveal the similarity and dissimilarity for local structure. Second, the alignment table exhibits how all similar fragments are matched or shifted with insertion of gaps. Third, it intuitively display how the structural folding shape associates with the corresponding residue of five consecutive amino acids, which is able to assist the analysis of relationship between amino acid sequence-structure-function in protein.

3. Results

3.1 Conformation analysis

Protein structure 1M2F (PDB ID) has 25 conformers obtained by NMR spectroscopy and show in Fig. 2(A). 1M2E (PDB ID) in Fig. 2(B) is the average structural models of 25 conformers of 1M2F (Williams et al., 2002). All of these structures apparently have identical sequence and similar 3D structural conformations. To differentiate the structures with higher similarity requests a tool with higher sensitivity to distinguish each conformer in global and local structure. With PFSA approach, each conformer of protein 1M2F and the structure of 1M2E are converted into one-dimensional PFSC alphabetic description, and then are aligned for comparison. The PFSA alignment table is displayed in Table 2.

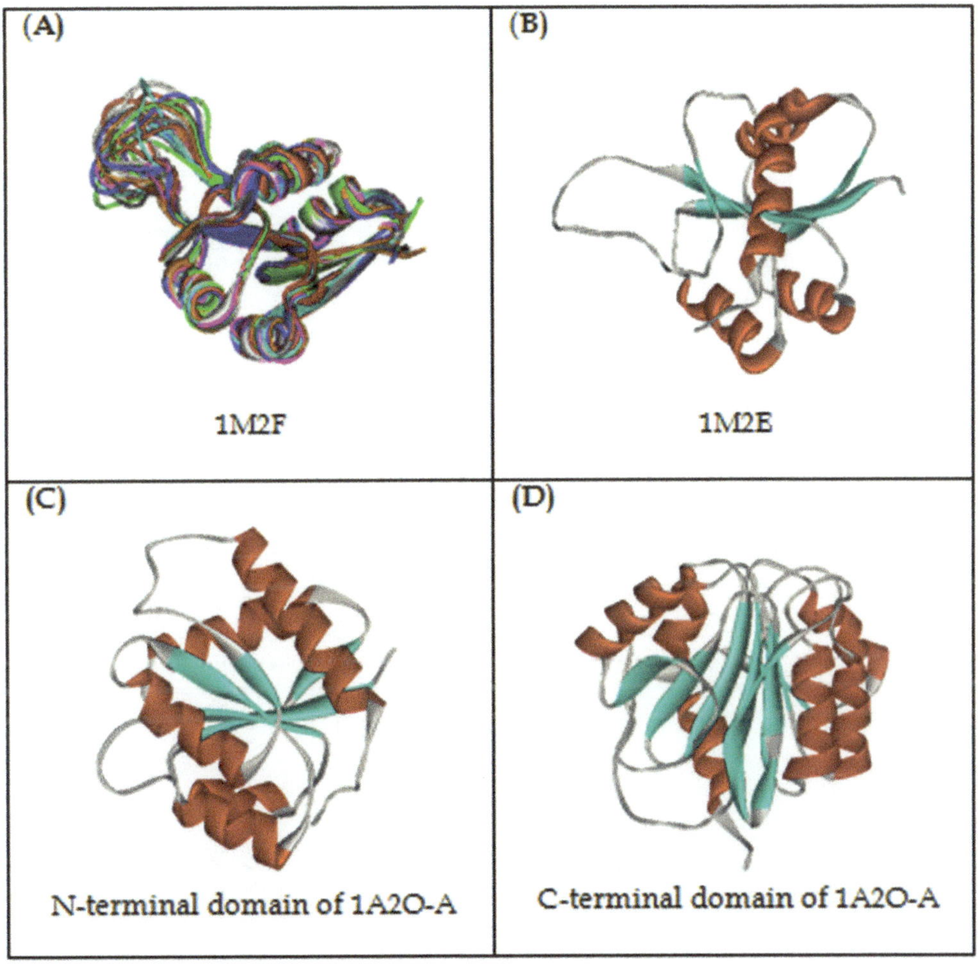

Fig. 2. Images of protein for structural comparisons. (A): 25 conformers of protein 1M2F. (B): protein structure of 1M2E. (C) and (D): N-terminal domain and C- terminal domain in chain A of 1A2O.

	0 100
	1 2 3 4 5 6 7 8 9 0 1 2 3
Rule	12345678901234567890123456789012345678901234567890123456789012345678901234567890123456789012345678901234567890123456789012345678901234 5
Seq.	MLSQIAICIWVESTAILQDCQRALSADRYQLQVCESGEMLLEYAQTHRDQIDCLILVAANPSFRAVVQQLCFEGVVVPAIVVGDRDSEDPDEPAKEQLYHSAELHLGIHQLEQLPYQVDAALAEFLRLAPVETMA
1M2E-A	--WYBBBBBBBVJVAAAAADAAAABVAAJBBBBBVAAAAAAAAAADJVAAJVJBBBBVHJVAAAAAAAAAAAADJBBWSBBBVJBBBBBJVAJBBWYAPCYAABBBBHAAJVAAADDAAAAAAAAAAAJVJBBB--
1M2F-1	--WYIBBBBBBVJVAAAAADAAAABVAAJBBBBBVAAAAAAAAAADJVAAJVJBBBBVHJVAAAAAAAAAAAADJBBWSBBBVJBBVJBBVAJVJBVHPCYAABBBBHAAJVAAADAAAAAAAAAAAADJVJVJB--
1M2F-7	--WYJBBBBBBVJVAAAAADAAAABVAABBBBBBVAAAAAAADAADJVAAJVJBBBBVHJVAAAAAAAAADAADJBBWSBBBVJBBBBBBVAJBVPYAPCYAAJBBBBVAJVAAADDAADAAAAAAADJVJVJB--
1M2F-2	--WYBBBBBBBVJVAAAAADAAAABVAAJBBBBBVAAAAAAADAADJVAAJVJBBBBVHJVAAAAAAAAAAAADJBBWSBBBVJBBHPSBVAJBVPYAQCYAABBBBHAAJVAAADDAAAAAAAAAAADJVJWSB--
1M2F-18	--PYBBBBBBBVJVAAAAADAAAABVAAJBBBBBVAAAAAAAAAADJVAAJVJBBBBVHJVAAAAAAAAAAAADJBBWSBBBVJBBBBHBVAJBVAAAPGYAAJBBBHAAJVAAADDAAAAAAAAAAADJVJVJB--
1M2F-25	--WYIBBBBBBVJVAAAAADAADABVAAJBBBBBVAAAAAAAAAADJVAAJWSBBBBVHJVAAAAAAAAAAAADJBBWSBBBVJVJVJBBVAJBVPYAPCYAAHBBBHAAJVAAADDAADAAAAAAAJVJVJB--
1M2F-8	--WYIBBBBBBVJVADAAAAAADABVAAJBBBBBVAAAAAAAAAADJVAAJWSBBBBVHJVAAAAAAAAAAAADJBBWSBBBVJBBHJBBVAJBVPYAQCYAAJBBBHAAJVAAADDAAAAAAAAAAADJVJVJB--
1M2F-20	--PYBBBBBBBVJVAAAAAAAADABVAAJBBBBBVAAAAAAAAAADJVAAJWSBBBBVHJVAAAAAAAAAAAADJBBWSBBBVAJBBHBBVAJVJWZHPCYAABBBBHAAAAAAAADAAAAAAAAAAADJVJVJB--
1M2F-13	--PSBBBBBBBVJVAAAAADAAAABVAAJBBBBBVAAAAAAADAADJVAAJVJBBBBVHJVAAAAAAAAAAAADJBBWSBBBVJBBHJBBAAJBBVHHPCYAAHBBBHAAJVAAADDAAAAAAAAAAADJVJVJB--
1M2F-15	--PYBBBBBBBVJVAAAAAAAADAJVAABBBBBBVAAAAAAADAADAAAAJVJBBBBVHHAAAAAAAAAAAADJBBWSBBBVJBBBBBBVAJBVAAAPCYAAJBBVHAAJVAAADDAAAAAAAAAAAJVJVJB--
1M2F-14	--WYBBBBBBBVJVAAAAAAAAAABVAAJBBBBBVAAAAAAAAAADJVAAJWSBBBBVHJVAAAAAAAAAAAADJBBWSBBBVABBHJBVQYBBVPYAQCYAABBBVHAAAAAAAAAAAAAAAAAAAJVJBBB--
1M2F-10	--AJBBBBBBBVJVAAAAAAAAAABVAABBBBBBVAAAAAAADAADJVAAJVJBBBBVHJVAAAAAAAAADAADJBBWSBBBVAJWSBVJVAAJBBVHPGYAABBBBHAAJVAAADDAAAAAAAAAAAJVJBBB--
1M2F-12	--PYBBBBBBBVJVAAAAAAAADABVAAJBBBBBVAAAAAAADAADJVAAJWSBBBBVHJVAAAAAAAAAAAADJBBWSBBBVJBHAJBVAAJBVAAAPCYAABBBBHAAJVAAADDAADAAAAAAAJVJVJB--
1M2F-3	--WYIBBBBBBVJVAAAAADAAAABVAABBBBBBVAAAAAAADAADJVAAJWSBBBBVHJVAAAAAAAAAAAADJBBWSBBBVJBBBWYJVHJBVAAAPCYAAHBBBHAAJVAAADAAAAAAAAAAAAJVJWSB--
1M2F-16	--PSBBBBBBBVJVAAAAAAAADABVAAJBBBBBVAAAAAAADAADJVAAJWSBBBBVHHAAAAAAAAAAAADJBBWSBBBVJBBVAJVADJBBVHHPCYAABBBBHAAJVAAADDAAAAAAAAAAAJVJBBB--
1M2F-24	--PSBBBBBBBVJVAAAAAAAADAJVAABBBBBBVAAAAAAADAADJVAAJVJBBBBVAHAJVAAAAAAAAAADJBBWSBBBVJBBBBBBVDJBVPYHPGYAABBBBHAAJVAAADDAAAAAAAAAAADJVJVJB--
1M2F-9	--WYBBBBBBBVJVAAAAAAAADABVAAJBBBBBVAAAAAAAAAADJVAAJWSBBBBVHJVAAAAAAAAAAAADJBBWSBBBVJBVHHJBVAJVHPYHPGYAABBBVJVAJVAAADDAAAAAAADAAAAJVJBBB--
1M2F-21	--PYBBBBBBBVJVAAAAAAAADABVAAJBBBBBVAAAAAAADAADJVAAJWSBBBBVHJVAAAAAAAAAAAADJBBWSBBBVJBBBBBBVAHHHPYAQGYAABBBBHAAAAAAADDAADAAAAAAAJVJVJB--
1M2F-11	--PYBBBBBBBVJVAAAAADAAAABVAABBBBBBVAAAAAAADAADAAAAJWSBBBBVHHAAAAAAAAAAAADJBBWSBBBWYJWSBVJVAJBBBVHPCYAAJBBBHAAJVAAADDAADAAAAAAAJVJVJB--
1M2F-17	--WYIBBBBBBVJVAAAAADAAAABVAAJBBBBBVAAAAAAADAADJVAAJVJBBBBVHJVAAAAAAAAAAAADJBBWSBBBVABWSBVJVAAHHPYAPGYAAHBBBHAAAAAAADDAADAAAAAAAJVJVJB--
1M2F-23	--PSBBBBBBBVJVAAAAADAAAABVAAJBBBBBVAAAAAAADAADJVAAJVJBBBBVAHAJVAAAAAAAAAADJBBWSBBBVABBHPYBVAABVAAAPCYAABBBUHAAJVAAADDAAAAAAAAAAAJVJBBB--
1M2F-19	--WSBBBBBBBVJVAAAAADAAAABVAAJBBBBBVAAAAAAADAADJVAAJVJBBBBVHHAAAAAAAAAAAADJBBWSBBBVJBBHBWYPZJBBPYHPGYAAHBBBHAAJVAAADDAAAAAAAAAAAJVJVJB--
1M2F-5	--WYJBBBBBBVJVAAAAAAAADABVAAJBBBBBVAAAAAAADAADJVAAJVJBBBBVHJVAAAAAAAAAAAADJBBWSBBBVAHHAJBBVAAHHPYHPGYAABBBBHAAJVAAADDAAAAAAAAAAADJVJWSB--
1M2F-22	--WYIBBBBBBVJVAAAAAAAADABVAABBBBBBVAAAAAAADAADJVAAJVJBBBBVHHAAAAAAAAAAAADJBBWSBBBVAHHHJBBVAAHBBHHPCYAABBBBHAAJVAAADDAADAAAAAAAJVJVJB--
1M2F-6	--WYHBBBBBBVJVAAAAAAAADABVAAJBBBBBVAAAAAAADAADJVAAJWSBBBBVHHAAAAAAAAAAAADJBBWSBBBVJBBBVHBVJWYJVHAPGYAAHBBVHAAAAAAAADAAAAAAAAAAAJVJVJB--
1M2F-4	--WCSBBBBBBVJVAAAAAAAADAJWZABBBBBBVAAAAAAADAADJVAAJWSBBBBVHHAAAAAAAAAAAADJBBWSBBBVJBBBBHJVJVAJBAAPCYAABBBBHAAJVAAAADAADAAAAAAADJVJVJB--

Table 2. PFSA Alignment Table: Conformation alignment of 1M2E-A and 25 conformers of 1M2F. The names of conformers of 1M2F and 1M2E are listed on left column. The amino acid sequence and rule for number of residue are listed on top rows. The protein folding shape code (PFSC) for each conformer is listed following the structure name. The typical α-helices are remarked with red color, the β-strands with blue color and the tertian fragments with black. Also, the analogue structures with secondary structure are remarked with pink color.The PFSA alignment table has the capability for analysis of local structure. First, the alignment table does not only align the secondary structure (font with red and blue colors), but it also aligns the attributes of tertian structure (font with black color). Second, the alignment table exhibits the detail element alignment according PFSC within each of fragment of secondary structure. The font with pink color indicates the alteration inside the secondary structure or the flexible terminal of secondary structure. Third, the alignment table is able intuitively to reveal the location of motifs with structural stability or flexibility. For example, in regions of fragment of residues (50-54) and fragment of residues (82-106) the conformations show the fluctuation in 25 conformers, which indicates these two regions with more flexible than remainder of protein segments.

The PFSA approach has capability for evaluation of global similarity. It provides PFSA-S as score to assess the global structural similarity. The 1M2E in Fig. 2(B), as average structure, is compared with each of 25 conformers of 1M2F in Fig. 2(A). The similarity scores are listed with descending order of PFSA-S in Table 3, including the number of identical and number of analogous folding shapes. Also, the results are compared with LGA method (Zemla, 2003). Both of PFSA-S and PFSA alignment table explicitly display the structural difference in protein conformation analysis. Apparently, the PFSA approach has ability to differentiate each conformer with its appropriate sensitivity.

Name	PFSA				LGA		
	PFSA-S	ID_FS	AN_FS	Gaps	N	RMSD	GDT_TS
1M2F-1	0.964	118	11	0	133	0.93	96.111
1M2F-7	0.962	117	12	0	131	0.85	95.741
1M2F-2	0.956	120	7	0	135	0.91	96.296
1M2F-18	0.956	120	7	0	134	0.89	95.370
1M2F-25	0.954	116	12	0	134	0.94	95.000
1M2F-8	0.952	115	13	0	135	0.79	97.037
1M2F-20	0.952	112	17	0	133	0.94	95.926
1M2F-13	0.950	117	10	0	131	0.95	95.370
1M2F-15	0.948	113	15	0	130	0.80	95.185
1M2F-14	0.943	113	14	0	133	0.91	95.962
1M2F-10	0.939	114	12	0	135	1.09	95.556
1M2F-12	0.939	114	12	0	133	0.92	95.370
1M2F-3	0.937	116	9	0	129	0.74	94.074
1M2F-16	0.937	113	13	0	133	0.70	96.296
1M2F-24	0.937	110	17	0	133	0.89	95.185
1M2F-9	0.931	113	12	0	132	1.14	95.000
1M2F-21	0.931	113	12	0	130	0.80	95.556
1M2F-11	0.929	109	17	0	134	0.84	95.926
1M2F-17	0.927	114	10	0	133	0.80	96.296
1M2F-23	0.926	113	11	0	132	1.00	93.704
1M2F-19	0.920	116	6	0	134	0.95	96.111
1M2F-5	0.918	112	11	0	134	1.04	95.556
1M2F-22	0.914	110	13	0	131	0.85	95.000
1M2F-6	0.906	106	17	0	130	1.05	92.963
1M2F-4	0.893	106	20	0	130	1.01	94.444

Table 3. Comparison of 25 conformers of 1M2F to average structural model of 1M2E with PFSA approach and LGA method. All data are sorted by values of PFSA-S. Left column lists the names of 25 conformers of 1M2F. PFSA: approach of protein folding structure alignment; PFSA-S: PFSA score for structural similarity; ID_FS: number of identical folding shapes; AN_FS: number of analogous folding shapes and Gaps: number of insertion gaps. LGA: LGA method (Zemla, 2003); GDT_TS: an estimation of the percent of residues (largest set) that can fit under the distance cutoff of 1, 2, 4 and 8 Å. N: number of superimposed residues under a cutoff distance and RMSD: root mean square deviation of all corresponding C-alpha atoms.

3.2 Domain-domain comparison

The proteins belong to different categories in the structural classification of protein (SCOP) (Murzin et al., 1995) are compared. The structures 1M2E in Fig. 2(B) is compared with N-terminal domain of chain A of 1A2O (1A2O-A) in Fig. 2(C) and then its C-terminal domain in Fig. 2(D) respectively. Although, all of three structures are classified as the class of alpha and beta proteins (α/β), they belong to two of different folds in SCOP. Both structures of 1M2E and N-terminal domain of 1A2O-A belong to Flavodoxin-like fold, but the C-terminal domain of 1A2O-A belongs to Methylesterase CheB fold. The summary of structural classification of 1M2E, N-terminal domain and C-terminal domain of 1A2O-A is listed in Table 4.

First, the alignment table provides the detail information of alignment for local structural fragments. Table 5 shows the comparison of 1M2E and N-terminal domain 1A2O-A while Table 6 shows the comparison of 1M2E and C-terminal domain of 1A2O-A. The alignment tables in Table 5 and Table 6 display how the fragments with similar local folding shapes are matched up with insertion of gaps. In alignment table, the aligned identical protein folding shape code is marketed with " | ", the analogue with "*", the different with "^" and the insertion with "+". Actually, the alignment table shows the optimized structural alignment with matching all local structural fragments between two proteins. Second, the PFSA-S provides the quantitative assessment of similarity for global structural comparisons. The PFSC-S values are listed in Table 4, including the numbers of identity and analog of folding shapes, and the number of insertion gaps. In contrast to C-terminal domain, the comparison of N-terminal domain of 1A2O-A and 1M2E have higher PFSA-S similarity score (0.7214 vs. 0.2109), larger number of identical and analogous folding shapes and less number of gaps. The results reflect the homologous difference of these two pairs of proteins in structure classification.

Name	1M2E	1A2O-A	
		N- terminal domain	C- terminal domain
Class	Alpha and beta proteins (α/β)	Alpha and beta proteins (α/β)	Alpha and beta proteins (α/β)
Fold	Flavodoxin-like	Flavodoxin-like	Methylesterase CheB
Superfamily	CheY-like	CheY-like	Methylesterase CheB
Family	Circadian clock protein KaiA	CheY-related	Methylesterase CheB
Protein	Circadian clock protein KaiA	Methylesterase CheB,	Methylesterase CheB
PFSA-S	1.000	0.7214	0.2109
Number of Identity	131	69	59
Number of Analog	0	38	43
Number of Gaps	0	3	25

Table 4. Comparison of structure of 1M2E with N-terminal domain and C-terminal domain of Chain A of 1A2O. Top rows show the structural classification of 1M2E, N-terminal domain and N-terminal domain of 1A2O-A. The bottom four rows list the values of structural similarity of comparison of 1M2E with N-terminal domain and N-terminal domain of 1A2O-A respectively with using PFSA approach

```
Seq          1:    -MLSQIAICIWVESTAILQDCCRALSADRYQLQVCESGEMLLEYAQTHRDQIDCLILVAANPSFRAVVQQLCFEGVVVPAIVVGDRDSEDPDEPAKEQLY
1M2E-A             ---WYBBBBBBBVJVAAAAADAAAABVAAJBBBBBVAAAAAAAAAAADJVAAJVJBBBBVHJVAAAAAAAAAAADJBBWSBBBVJBBBBBJVAJBBWYAPC
                      +^^|||||**|**|||||*||||*||**|^^|||||||||||||||*^*|**||*^*^^^^^*|**||||||**|^^*|||**|^^^^**|*^^^*|**
1A2O-A_N_DoM          VJBBBBBBVJVAAAAAAAAAAAAJVAJBBWYBBVAAAAAAAAAAADJBBVJBBBBVHBVQSWYADDAAAAAAQSBWSBBBBBBBPZAAAAAAAAAAAQS
Seq          1:    MSKIRVLSVDDSALMRQIMTEIINSHSDMEMVATAPDPLVARDLIKKFNPDVLTLDVEMPRMDGLDFLEKLMRLRPMPVVMVSSLTGKGSEVTLRALELG
```

```
Seq        100:    HSAELHLGIHQLEQ-LPYQVDAALAEFLRLAPVETMA---
1M2E-A             YAABBBBHAAJVAA-ADDAAAAAAAAAAAJVJBBB-----
                   ^^*||||^^**|||+***|||||||||||***^|*+++++
1A2O-A_N_DoM       BWYBBBBPZPSVAAIVAAAAAAAAAAAAAAAAHBVAAJ--
Seq        100:    AIDFVTKPQLGIREGMLAYSEMIAEKVRTAARARIAAHKP
```

Table 5. PFSA Alignment Table: Protein Folding Structure Alignment of 1M2E and N-Terminus Domain of 1A2O-A. The names of 1M2E and N-terminus domain of 1A2O are listed on left column. Amino acid sequence is listed on top and bottom in each section. The protein folding shape code (PFSC) for each structure is listed following the structural name. The typical α-helices are remarked with red color, the β-strands with blue color and the tertiary fragments with black. Also, the analogue structures with secondary structure are remarked with pink color. The "|" indicates the alignment with identical folding shapes; "*" analogous folding shapes; "^" different folding shapes. The "+" represents the insertion of gaps.

```
Seq            1:   ------------MLSQIAICIWV-ESTAILQDCQRAL--SADRYQLQVCESGEMLLEYAQTHRDQI--DCLIL-VA----ANPSFR---AVVQQ--LCFE
1M2E-A              --------------WYBBBBBBB-VJVAAAAADAAAA--BVAAJBBBBBVAAAAAAAAAAADJVAAJ--VJBBB-BV----HJVAAA---AAAAA--AAAD
                      +^^^^^^^^^^^^^**||||||+^**|||||**|||++^****|||||****|||||||***||*+++*||+*+|*++++^^*|+*+++**||*++*++^
1A2O-A_C_DoM          HJBAJVAPCRWYAAJBBBBBBVDAAAAAAAADAAAHHAAHPSBBBBBBVPYAAAAAAAAAAAAAJVAJB-VHBWYJVJBWYA-JBBBHHAAHBBV--I
Seq          141:   MAAPTTLKAGPLLSSEKLIAIGASTGGTEAIRHVLQPLPLSSPAVIITQHMPPGFTRSFAERLNKLCQISV-KEAEDGERVLPG-HAYIAPGDKHME--L

Seq           74:   GVVV-P-AI-VVGDRDSEDPDEP-A---KEQ------LY-HSAELHL---GIHQLEQLPYQVDAALAEFLRLAPVETMA--------------------
1M2E-A              JBBW-S-BB-BVJBBBBBJVAJB-B---WYA------PC-YAABBBB---HAAJVAAADDAAAAAAAAAAAAJVJBBB----------------------
                    *|+|+*+|*+|**|++++^^^^^+|+++**|++++++^^+*||*|||+++||**||||**|||||+++++^***||++++++++++++++++++++++++
1A2O-A_C_DoM        BB-WGYHBVIBBBB----WCSVQYBVJBVAAAAAAAADDJVAAJBBBBWYHAPSVAAAAAAAAAA-----QSBBBBBHHAAAJVJVAAAAAAAAAAHHJWY
Seq          237:   SG-ANYQIKIHDGP----PVNRHRPSVDVLFHSVAKHAGRNAVGVILTGMGNDGAAGMLAMYQAG     AWTIAQNEASCVVFGMPREAINMGGVSEVV
                    AR-SGANYQIKIHD----GPPVNRHRPSVDVLFHSVAKHAGRNAVGVILTGMGNDGAAGMLAMYQ-----AGAWTIAQNEASCVVFGMPREAINMGGVSE

Seq                 ---------------------
1M2E-A              ---------------------
                    ++++++++++++++++++++
1A2O-A_C_DoM        BVHAAAAAAAAAAAAAPYA--
Seq          327:   VVDLSQVSQQMLAKISAGQAI
```

Table 6. PFSA Alignment Table: Protein Folding Structure Alignment of 1M2E and C-Terminus Domain of 1A2O-A. The names of 1M2E and N-terminus domain of 1A2O are listed on left column. Amino acid sequence is listed on top and bottom in each section. The protein folding shape code (PFSC) for each structure is listed following the structural name. The typical α-helices are remarked with red color, the β-strands with blue color and the tertiary fragments with black. Also, these analogue structures with secondary structure are remarked with pink color. The "|" indicates the alignment with identical folding shapes; "*" analogous folding shapes; "^" different folding shapes. The "+" represents the insertion of gaps.

3.3 Protein comparison

Proteins may be comprised by single domain or multiple domains in the chain structure. To take protein chain-chain in alignment will related to multiple domain comparison. For example, insulin-like growth factor 1 receptor (IGF1R) and insulin receptor (INSR), transmembrane proteins belonging to the tyrosine kinase super-family, have multiple domains in structure. Over the past two decades, rich structural data of IGF1R/INSR has been accumulated, and the sequence alignment was applied in comparison (Werner et al., 2008; McKern, 2006; Garrett, 1998; Pautsch, 1997; Hubbard, 1997; Lou, 2006; Garza-Garcia, 2007). In this study, instead, the folding conformations of IGF1R and INSR are directly aligned for structural comparison. The crystal structures of first three domains of L1-CR-L2 structures of IGF1R (PDB ID: 1IGR) (Hubbard, 1997) and INSR (PDB ID: 2HR7) (Murzin, 1995) are available in PDB. The images of first three domains for IGF1R (1IGR) and INSR (chain A of 2HR7) are displayed in Fig.3. Both L1 and L2 domains consist of a right-handed α-helix conformation. The CR domain is composed of seven modules with eight disulphide-bond connectivity.

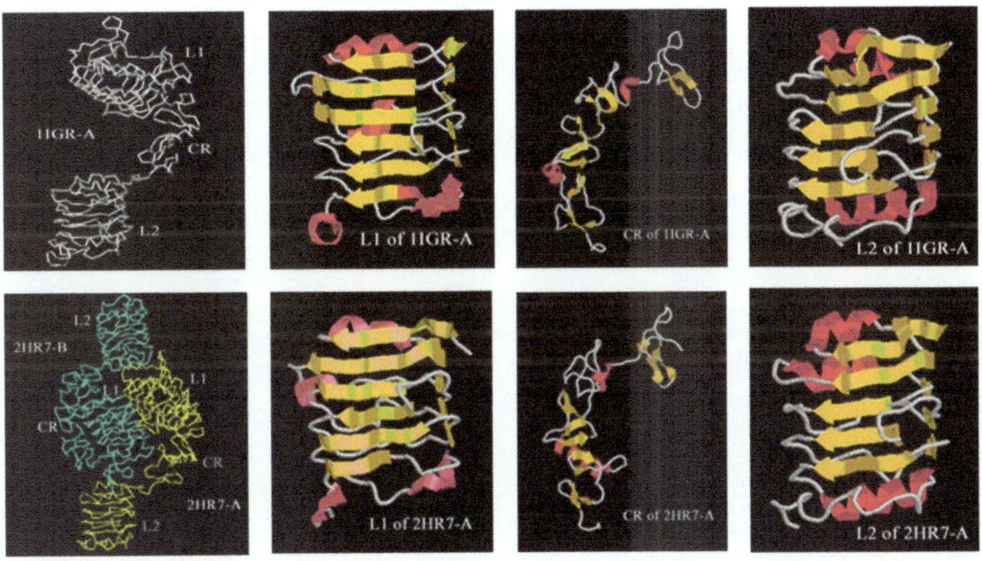

Fig. 3. Structural images of IGF1R (1IGR) and INSR (2HR7). The IGF1R and its L1, CR and L2 domains are shown in the top row; INSR and its L1, CR and L2 domains are shown in the bottom row. The atomic geometric coordinates of 1IGR and 2HR7 are obtained from the protein databank (PDB). The image was drawn with RasWin Molecular Graph

ics V. 2.6. The structural images are displayed by α-carbon backbone or cartoon views.

```
        0000000000000000000000000000000000000000000000000000000000000000000000000000000000000000000000000000
                    1         2         3          4         5         6         7         8         9
1IGR-A  -1234567890123456789012345678901234567 890 1234567890123456789012345678901234567890123456789012345678901234567
Geom.   -LLSTHTLSLLSSLLLLLLLLSSSSLLTLLLLLLLLTLT LSL LTLLLHLLSSLSLLLLLLLSTTLLTLTSLLHSLSSLLTLLLLLSLTLSLLLLSSLLS
PhyChe  -APSHPHPAPBNAONNPBBPANSOPPAHOPBPPPPOBP AAO BOOBPPBPOPPOAOPPPPBPPHPAOPHAPPPNPOPPBHPBPPONOPPPPPAPONPBA
Seq.    -EICGPGIDIRNDYQQLKRLENCTVIEGYLHILLISKA EDY-RSYRFPKLTVITEYLLLFRVAGLESLGDLFPNLTVIRGWKLFYNYALVIFEMTNLKD
PFSC    ---BVPSBBBVAJVAAAAAAAABVJBVPSBBBVJBBBB-VAA-OXBBBVAJVJBVPSBBBWSBVABVJVAADJVJBVJBBBVJBP$YBVJBBWSBVJBWY
        +++|*|||||**|||||||||||||||||||||*^|+*||+^^*||||||||||||||||||||||||||||||||||||||||||||*|||||||||*|*^
PFSC    --YBBPSBBBWYJVAAAAAAAABVJBVPSBBBVJBVABVAAAAAAJBBVAJVJBVPSBBBWSBVABVJVAADJVJBVJBBBVJBP$YJVJBBWSBVABVJ
Seq.    PGEVCPGMDIRNNLTRLHELENCSVIEGHLQILLMFKTRPEDFRDLSFPKLIMITDYLLLFRVYGLESLKDLFPNLTVIRGSRLFFNYALVIFEMVHLKE
PhyChe  PHAPSPHPAPBNNPOBPBAPANSOPPAHBPNPPPPPBOBPAAPBAPOPPBPPPPOAOPPPPBPOHPAOPBAPPPNPOPPBHOBPPPNOPPPPPAPPBPBA
Geom.   HTLSSHTLSLLSSLSLLLLLLSSTSLLTLLLLLLLLLSLHLSLLSLTLHLLLLLSSLLLLLLSLTLLTLLSLLHSLSSLLTTLLLLSLTLSLLLLSLLLL
2HR7-A  4567890123456789012345678901234567890123456789012345678901234567890123456789012345678901234567890123
              1         2         3         4         5         6         7         8         9         0
        0000000000000000000000000000000000000000000000000000000000000000000000000000000000000000000001111

        0011111111111111111111111111111111111111111111111111111111111111111111111111111111111111111111111111
          0         1         2         3         4          5         6         7         8         9
1IGR-A  89012345678901234567890123456789012345678901 23456789012345678901234567890123456789012345678901234567890123456
Geom.   LTLLSLLSLSLTTLLLLLSTSLSLLTSSSLTLLLSTSTSSLLST SLHHLLSTSLSHTSLLLLHLSLLSSLSSLLSLLSLSSSLSLLLSHTSSTLLTSSL
PhyChe  PHPONPBNPOBHPPBPABNPAPSOPOOPAPOPPPAPPONNOPPH NBPPBASHAPSPHOPAABPPSABOOPNNAONOBSPOONBSNBPSPOOSHBBPSOA
Seq.    IGLYNLRNITRGAIRIEKNADLCYLSTVDWSLILDAVSNNYIVG-NKPPKECGDLCPGTMEEKPMCEKTTINNEYNYRCWTTNRCQKMCPSTCGKRACTE
PFSC    BBBVABVJBVAAHIBBWSBVABVAAAAJBVAAAJVJVAABBBAA-PYJVAAJBBBWZAAAAHJBVHHBBBW$YBBWRBBWCZAJVHBBPYAAJWYHBUHD
        |||||||||||**|||**||*||||||||||||||||||||**+****||*||||*|||*|*|||^||||^|*|^^^||||||||||^|||||*^||||
PFSC    BBBVABVJBVAPXIBBVJBVJBVAAAAJBVAAAJVJVAABBBVJVAAAAAABBBBWYAAADHBBVHJBBBWGYJBBBHBWCZAJVHBBWYAAJWZJBUHD
Seq.    LGLYNLMNITRGSVRIEKNNELCYLATIDWSRILDSVEDNHIVLNKDDNEECGDICPGTAKGKTNCPATVINGQFVERCWTHSHCQKVCPTICKSHGCTA
PhyChe  PHPONPPNPOBHOPBPABNNAPSOPPOPAPOBPPAOPAANBPPPNBAANAASHAPSPHOPBHBONSPPOPPNHNPPABSPOBOBSNBPSPOPSBOBHSOP
Geom.   LTLLSLLSLSLTTSLLLLSSLLSLLTSLSLTLLLSTSLSSLLSLSLSSSLLSTSLSHTSTLTLSSSHTSSLSTLLSLLSLSLTLSLLSSHSLSLTLTSST
2HR7-A  4567890123456789012345678901234567890123456789012345678901234567890123456789012345678901234567890123
              1         2         3         4         5         6         7         8         9         0
        1111111111111111111111111111111111111111111111111111111111111111111111111111111111111111111112222

        1112222222222222222222222222222222222222222222222222222222222222222222222222222222222222222222222222
           0         1         2         3         4         5           6          7         8         9
1IGR-A  789012345678901234567890123456789012345678901234567890123456789  01  23456789012345678901234567890123456789012
Geom.   SSLSSLHLSLTTSTTHSSSSTSSTSLLLLLTTSSSHTSHHSSLLLLTLLSSSLSLSTSLLTTL  TT  STLTLSLLSTLSLLLSHTTLLLSTTLTLLSL
PhyChe  NNASSBPASPHOSOPPANAOPSPPSBBOOOPHPSPPPSPPNOOBPAHPBSPABAPSPNPPOPA  OO  AOAHPPPBAHASPNASPOHPPBNHONOPOSP
Seq.    NNECCHPECLGSCSAPDNDTACVACRHYYYAGVCVPACPPNTYRFEGWRCVDRDFCANILSAE--SS--DSEGFVIHDGECMQECPSGFIRNGSQSMYCI
PFSC    DJBBWYAAJVPYBVAAABVAABWYJW$SBW$YJBBVPYPZAJVIWCZAABBHAAAAAAAAAAJ--BB--VHHPSBBW$YBBBAAPCZAJVJVPYJBVJBV
        ||**|||||*^^|*||||||||||**^|||||^|||*^^||||*||||||||||||||||||*++*|++**|^*||||||||*+*++**^*^^^****|*
PFSC    DJVJWYAAJWYJBAAAABVAABWYBVPSBW$YHBBVAIWZAJVJWCZAABBHAAAAAAAAAAAAADJBVHAAHBBBBW$YBBBP-C--YPZAJVJBVAABB
Seq.    EGLCCHSECLGNCSQPDDPTKCVACRNFYLDGRCVETCPPPYYHFQDWRCVNFSFCQDLHHKCKNSRRQGCHQYVIHNNKCIP-E--CPSGYTMNSSNLL
PhyChe  AHPSSBOASPHNSONPAAPOBSPPSBNPOPAHBSPAOSPPPOOBPNAPBSPNPOPSNAPBBBSBNOBBNHSBNOPPBNNBSPP A  SPOHOOPNOONPP
Geom.   LTLSSLTLSLTSSTLHSSHSLSSTSLSLLLSTLSSLSSHHHLLLLLSLLSSSLTLSLSLLLLSLSTLLLTSLLLSLLSSLSLH L  SHTTLSLSTTSLL
2HR7-A  45678901234567890123456789012345678901234567890123456789012345678901234567890123456 7  8901234567890
              1         2         3         4         5         6         7         8          9        0
        2222222222222222222222222222222222222222222222222222222222222222222222222222222222222222222222223
```

<table>
<tr><td></td><td>2222222222333333333333333333333333333333333333333333333333333333333333333333333333333333333333</td></tr>
<tr><td></td><td>0 1 2 3 4 5 6 7 8</td></tr>
<tr><td>1IGR-A</td><td>345 6789012345678901234567890123456789012345678901234567890123456789012345678901234567890123456789</td></tr>
<tr><td>Geom.</td><td>HSL THSHLSSLLLLLSLSLSTSSTTLLLLLTSSLLLTSLLLSLLLTSSLTTLLLSLLTLLLSSSTLSLLLLTLTLSTLTLLLSLLLLLTLLLLLTSLT</td></tr>
<tr><td>PhyChe</td><td>PSA HPSPBPSAAABBOBOPAOPOOPNPPNHSOPPBHNPPPNPBBHNNPPOAPANPPHPPAPPOHOPBPBBOBPPPOPOPPBNPBPPPHAANPAHNOO</td></tr>
<tr><td>Seq.</td><td>PCE---GPCPKVCEEEKKTKTIDSVTSAQMLQGCTIFKGNLLINIRRGNNIASELENFMGLIEVVTGYVKIRHSHALVSLSFLKNLRLILGEEQLEGNYS</td></tr>
<tr><td>PFSC</td><td>JBV---QYHJBBBBVAJVJBBBVAAAAAAAAAJBVJBVPSBBBBBVJBWSVAAAAAADAAABVJBVPYBBBWSBVAJVJBAAJVABVJBBBVJVPZQSBV</td></tr>
<tr><td></td><td>*|++++*|^*|||||*^*||||||||||||**|||||||*|||||*^^^||||||||||||||||||*|||||||||||**|||||||||||||||||||</td></tr>
<tr><td>PFSC</td><td>BB-HHP$YPSBBBBVP$YJBBBVAAAAAAAJVJBVJBVPRBBBBBWYPCSVAAAAAADAAABVJBVPSBBBWSBVAJVJVHAJVABVJBBBVJVPZQSBV</td></tr>
<tr><td>Seq.</td><td>CT-PCLGPCPKVCHLLEGEKTIDSVTSAQELRGCTVINGSLIINIRGGNNLAAELEANLGLIEEISGYLKIRRSYALVSLSFFRKLRLIRGETLEIGNYS</td></tr>
<tr><td>PhyChe</td><td>SO PSPHPSPBPSBPPAHABOPAOPOOPNAPBHSOPPNHOPPPNPBHHNNPPPAPAPNPHPPAAPOHOPBPBBOOPPPOPOPPBBPBPPBHAOPAPHNOO</td></tr>
<tr><td>Geom.</td><td>SS HSLTHSHLSSLLLLTLLSLSTSSTTLLLLTSSSLSTTLLLSLLTTSSLTTLLLTSLTLLLLLTTLLLLLLTLTLSTLTLLLLLLLLLTLSLLLTSLT</td></tr>
<tr><td>2HR7-A</td><td>12 3456789012345678901234567890123456789012345678901234567890123456789012345678901234567890123456789</td></tr>
<tr><td></td><td>1 2 3 4 5 6 7 8 9</td></tr>
<tr><td></td><td>3333333333333333333333333333333333333333333333333333333333333333333333333333333333333333333333333333</td></tr>
<tr><td></td><td>444444444444444444444444444444444444444444444444444444444444444444444444444444444</td></tr>
<tr><td></td><td>9 0 1 2 3 4 5 6 7</td></tr>
<tr><td>1IGR-A</td><td>012345678901234567890123456789012345678901234567890123456789012345678901234567890</td></tr>
<tr><td>Geom.</td><td>LLSLSSLSLLLLLSLSLLSLSLLTTLLLLTLSHLLSSTLLLLLLLSSTSLTLLTLTSLSSLSSTLLTTSL</td></tr>
<tr><td>PhyChe</td><td>POPPANNNPNNPPAPABBNPOPBPHBPOPPPNPBPSPOAPOBPAAPOHOBHBNOBHAPNOBNNHABPOSA</td></tr>
<tr><td>Seq.</td><td>FYVLDNQNLQQLWDWDHRNLTIKAGKMYFAFNPKLCVSEIYRMEEVTGTKGRQSKGDINTRNNGERASCE</td></tr>
<tr><td>PFSC</td><td>JBBVJBVJBVJWCSVAAJBBBBVAAHIBBVJBVAJBAAAAAAAAAAAHHAAHJBAAAJBVAAPSVAHJ</td></tr>
<tr><td></td><td>|||**|||||||**||||||||||**|||||||||||||||||||||||||||^^|||||||||||*+</td></tr>
<tr><td>PFSC</td><td>JBBWSBVJBVJWYJVAAJBBBBVAPXIBBVJBVAJBAAAAAAAAAAAHHAAHJWZAAJBVAAPSVAA-</td></tr>
<tr><td>Seq.</td><td>FYALDNQNLRQLWDWSKHNLTITQGKLFFHYNPKLCLSEIHKMEEVSGTKGRQERNDIALKTNGDKASCE</td></tr>
<tr><td>PhyChe</td><td>POPPANNNPBNPPAPOBBNPOPONHBPPPBONPBPSPOAPBBPAAPOHOBHBNABNAPPPBONHABPOSA</td></tr>
<tr><td>Geom.</td><td>LLTLSSLSLLLLLSLTLLSLSLSLTLLLLLSHLLSLTLLLLLLLSTTSLTLLLLSSLTLLSSTSLTTSL</td></tr>
<tr><td>2HR7-A</td><td>01234567890123456789012345678901234567890123456789012345678901234567890</td></tr>
<tr><td></td><td>0 1 2 3 4 5 6 7</td></tr>
<tr><td></td><td>44444444444444444444444444444444444444444444444444444444444444444444444</td></tr>
</table>

Table 7. PFSA Alignment Table: Insulin-like growth factor 1 receptor (1IGR-A) and insulin receptor (2HR7-A). The 1IGR-A represents the chain A of IGF1R; 2HR7-A for chain A of INSR. Seq: sequence of amino acid. PFSC letter font with various colors: red α-helix; blue β-strand; pink analogous α-helix and β-strand. Numeric rules are displayed at top and bottom in each section. The rules indicate the hundredth, tenth and digit numbers. PhyChe: The physicochemical properties of amino acids. The physicochemical properties are labeled with nonpolar side chain for hydrophobicity: P, acidic side chain: A, basic side chain: B, glycine with unchanged hydrogen as side chain: H. The Polar group of amino acids is divided into subgroup according properties. Amino acid with hydroxyl group –OH: O, with carbonyl and amide group O=CNH2: N, with sulfhydryl group: S. Geom: geometric size and characteristic of amino acid. Tiny: T, small: S, large: L and hardiness: H.

The structural is assessed. The sequence similarity is evaluated by the percentage of identical residues. The structural similarity is quantitatively assessed by PFSA score. The similarity of three domains of L1, CR and L2 for IGF1R and INSR are summarized in Table 8. Overall, two protein structures have 60% of sequence in identity with structural similarity score at 0.860. Furthermore, each pair of domains is compared. The L1 domain has 67% of sequence in identity with structural similarity score at 0.909, the L2 domain 64% of sequence in identity with structural similarity score at 0.929 and CR domain 49% of sequence in identity with structural similarity score at 0.749. The PFSA scores specified that the L1 and L2 domains have higher structural similarity than the CR domains. Also, L2 domains have a higher degree of structural homology than L1, even though L1 has a higher degree of identity of sequence. With PFSA approach, the quantitative assessment of similarity between IGF1R and INSR agrees with previous quality specifications by sequence alignment. However, detail structural features are exposed for comparison.

Domain	Residues	Sequence Identity	Structural Homology				
			PFSA Score	Identical	Analog	GapO	GapE
L1 + CR + L2	1IGR-A (1-459) 2HR7-A (4-469)	60%	0.860	353	68	10	6
L1	1IGR-A (1-147) 2HR7-A (4-154)	67%	0.909	119	20	4	0
CR	1IGR-A (148-299) 2HR7-A (155-309)	49%	0.749	96	30	5	6
L2	1IGR-A (300-459) 2HR7-A (310-469)	64%	0.929	133	16	1	0

Table 8. Quantitative assessment of similarity for domain structures between IGF1R and INSR. PFSA Score: protein folding structural alignment score - a value for structural similarity; Identical: number of identical folding shapes; Analog: number of analogous folding shapes; GapO: number of opened gaps for structural alignment and GapE: number of extended gaps for structural alignment. Sequence identity is obtained with running J Aligner at http://jaligner.sourceforge.net/

4. Discussion

4.1 Feature of PFSA approach

4.1.1 Consistent procedure

The PFSA approach provides an unambiguous procedure for protein comparison based on structural alphabetic alignment. First, the PFSA approach relies on complete assignment of protein conformation. The PFSC provides a complete assignment of protein conformation for any protein with given 3D structure. Without usage of training database, all 27 PFSC are obtained by restrict mathematical derivation. Each PFSC vector or alphabetic letter represents a special folding shape of five successive C_α atoms in protein backbone. The folding shape of each of five successive C_α atoms in protein backbone is assuredly assigned by one among 27 PFSC vectors. Therefore, the protein backbone from N-terminal to C-terminal gets complete alphabetic assignment for folding conformation without gap. Second, the PFSA alignment of alphabetic strings is a consistent process. The PFSA approach is able to avoid the artificial choice of geometric parameters in structural comparison, such as the adjustment of initiative focusing location, cutoff distance for RMSD and the length of segment. Similarly as sequence alignment, the structural alphabetic alignment provides a fast and steady procedure for protein structure comparison. Third, the PFSA approach is able to handle protein comparison in various homologies, i.e. in wider scope of structure difference. This feature is well demonstrated by results of comparison of conformers in Table 2, comparison of different proteins in Table 5 and Table 6, and comparison of protein with complicated structures in Table 7. Furthermore, the PFSA approach is able to categorize the protein structures according structural classification in homology. With structure classification of protein SCOP (Murzin, 1995; Andreeva, 2008) as gold standard, the PFSA assessed the homologous degree for a set of protein structures, and the distribution of similarity scores, PFSA-S, was overall agreed with the categories in SCOP (Yang, 2011).

4.1.2 Normalized score and unique measurement

With normalization of PFSA-S score, the structural similarity of various proteins is easily assessed. If two structural data are an identical protein structure, the PFSA-S equals one. If the structural similarity decreases, the value of PFSA-S decreases. When the value of PFSA-S is near zero or less than zero, two proteins have large difference in conformation shape. The PFSA-S score is normalized by size of protein. In PFSA approach, the length of protein folding shape string is used as the denominator in formula for normalization when the PFSA-S is calculated. If a pair of two proteins is compared, anyone of proteins may be taken as the referent protein. If a set of proteins are compared with a reference protein, the similarity scores are normalized according the length of referent protein. The PFSA approach provides a unique quantitative measurement to evaluate the similarity in protein structural comparison.

4.1.3 Local structural comparison

The PFSA alignment table is able to compare protein structures in detail. The one-dimensional alphabetic string expresses the change of protein folding conformation along backbone. A letter of PFSC represents the folding shape of fragment for five successive amino acids. In alignment table, the protein folding conformations are aligned with similarity. The PFSA alignment table is comprised with the amino acid residues by adhesive to the associated folding shape code. Furthermore, the PFSA alignment table includes physicochemical properties of amino acid residue which are expresses by alphabetic letters as seen in Table 7. Therefore, the PFSA alignment table may become a good tool to study the relationship between sequence-structure-function. The PFSA alignment table has capability to exam the structural similarity as well as dissimilarity. In alignment table, if the local structures match with identical or analogous folding shapes, it reveals the structural similarity; if local structures align with different folding shapes, it exhibits the dissimilarity. Also, some of unmatched local structures are shifted with insertion of gaps to display the dissimilarity. In general, it is hard to straightforwardly expose both of similarity and dissimilarity with protein 3D structural image or computer modeling animation. Protein modeling provides visualization for view of 3D structure, but PFSA alignment table provides digit description for conformation. The combination of application of protein 3D modeling with PFSA alignment table is helpfully to inspect both of similarity and dissimilarity in protein structures.

4.2 Comparison with other methods

Different methods adopt various strategies to study specific geometric parameters for protein structural comparison. With different parameters and approaches, all methods have a common goal trying to evaluate the similarity of protein structures. As complexity, it not surprise there is no unique outcome for protein comparison. In this study, the results from PFSA approach are compared with other methods.

4.2.1 PFSA vs. LGA

LGA method (Zemla, 2003) is an important approach for protein structure comparison. Specially, it is extensively applied for assessment of similarity for protein prediction in Critical Assessment of Techniques for Protein Structure Prediction (CASP) (Kryshtafovych et al., 2007; Moult et al., 2009). The 25 conformer of 1M2F and its average model of 1M2E are compared by both of PFSA approach and LGA method respectively. The results are listed in Table 3, where all structures are ranked by the order of PFSA-S. LGA method and PFSA approach adopt different strategies to assess the structural similarity. LGA method is designed to evaluate the longest continuous segments (LCS) searching for the largest set of 'equivalent' residues that deviate by no more than a specified distance cutoff. GDT_TS is an estimation of the percent of largest set of residues that can fit under selected cutoff distances. A scoring function (LGA_S) was defined as a combination of these values and can be used to evaluate the level of structure similarity of selected regions. However, PFSA takes the fixed length of segment of five successive C_{α} atoms to determine the folding shape, and then directly makes the alignment with structural alphabets. It is not surprised that PFSA and LGA methods present different ranks in structural comparisons. Due to higher similarity, the comparison of 25 conformer of 1M2F requires a tool with sensitivity to distinguish structural perturbation. The PFSA approach provides finer description for folding conformation. Each PFSC code steadily represents the folding shape of five successive residues and each of PFSC vector can be transformed from one to another. 27 PFSC vectors cover all possible folding shapes. Therefore, each conformer of 1M2F acquires a complete assignment along protein backbone, so the alignment is performed with full length of structure from N-terminal to C-terminal. Furthermore, with structural alphabets, the PFSA adopts an unambiguous process in alignment for protein comparison. Except similarity score PFSA-S, with folding shape for each five residues, the PFSA approach provides explicit comparison in alignment table. Therefore, the PFSA approach offers a complementary tool in analysis of protein conformation.

4.2.2 PFSA vs. CE

The combinatorial extension (CE) method (Shindyalov & Bourne, 1998) breaks each structure in the query set into a series of fragments that it then attempts to reassemble into a complete alignment. A series of pairwise combinations of fragments are used to define a similarity matrix through which an optimal path is generated to identify the final alignment.

The size of each aligned fragment pairs is usually set to empirically determined values of 8 and 30 respectively. One group of 20 structures, the quaternary complex of cAMP dependent protein kinase, has certain structural similarities and is compared with the structure of 1ATP-E by CE method and PFSA approach respectively. The results of comparisons between 1ATP-E and 20 of cAMP dependent protein kinases are listed in Table 9 which is sorted in the order by Z Score of CE. Two conclusions are observed from results. First, the ranks of similarity are overall agreed between CE and PFSA, except the structures with number 5, 8, 9, 10 and 18. With PFSA approach, the assessment of similarity is an aggregate of matched folding shape, structural topological distribution, gap and size of protein. Thus, the rank for structural similarity may be adjusted by relative size of compared proteins. 1ATP-E, as reference protein, has sequence length of 335 and other structures, as target proteins, with the matched size may have higher PFSA-S. Referring the length of 1ATP-E, for example, the similarity of structures 5 and 8 is assigned with lower values of PFSA-S because of having large difference in length of 298 and 438; the similarity of structures 18 with higher value of PFSA-S because of having matched length of 366.

		CE					PFSA			
No.	PDB ID	Size	N^A	N^G	RMSD	Z Score	ID_FS	AN_FS	Gaps	PFSA-S
1	1APM-E	350	336	0	0.3	7.9	282	48	5	0.9145
2	1CDK-A	350	336	0	0.4	7.9	271	57	3	0.8968
3	1YDR-E	350	336	0	0.5	7.9	274	43	0	0.9092
4	1CTP-E	350	303	0	1.5	7.4	249	52	19	0.8074
5	1PHK	298	255	28	2.5	7.2	161	83	59	0.5034
6	1KOA	491	258	20	2.7	7.1	205	88	21	0.6122
7	1KOB-A	387	260	20	2.8	7.1	182	99	24	0.6382
8	1AD5-A	438	237	31	2.5	7.0	117	133	102	0.3825
9	1CKI-A	317	260	47	2.8	6.9	128	113	46	0.4725
10	1CSN	298	249	37	2.4	6.8	117	117	43	0.4439
11	1ERK	364	254	55	2.6	6.8	171	106	43	0.5768
12	1FIN-A	298	253	69	2.2	6.8	168	88	52	0.5557
13	1GOL	364	254	55	2.6	6.8	174	99	43	0.5535
14	1JST-A	298	253	69	2.4	6.7	161	96	50	0.5553
15	1IRK	306	244	69	3.3	6.5	136	103	53	0.4552
16	1FGK-A	310	251	54	3.5	6.2	141	81	66	0.4322
17	1FMK	452	245	19	2.8	6.2	122	129	104	0.4040
18	1WFC	366	240	72	3.1	5.6	157	94	54	0.5294
19	1KNY-A	253	112	79	4.3	3.9	114	89	83	0.3592
20	1TIG	94	54	3	4.2	3.9	71	17	228	0.0565

Table 9. Comparison of 20 of quaternary complex of cAMP with strucrure of 1ATP-E (PDB ID: 1ATP, chain E, sequence length 335). **N^A**: number of aligned position; **N^G**: number of non-aligned position; RMSD: results based on C alpha atoms over the length of the alignment; Z Score: measure of the statistical significance of the result relative to an alignment of random structures. ID_FS: number of identical folding shapes; AN_FS: number of analogous folding shapes and Gaps: number of insertion gaps; PFSA-S: PFSA score for structural similarity.

Second, the CE method indicates that 20 protein structures have similar fold as structure of 1ATP-E. However, the PFSA has capability further to distinguish the dissimilarity between 20 structures of cAMP dependent protein kinases. According to CE method, if Z Score is larger than 3.5, the compared proteins have similar fold in structure. The values of Z Score of 20 structures of cAMP are from 3.9 to 7.9, so they all have similar fold structure as 1ATP-E. The values of PFSA-S for 20 structures are distributed in the wide range of 0.9145 - 0.0565. According to PFSA approach, the value of PFSA-S is near one when two structures have high similarity, and on the contrary, the value of PFSA-S is near zero when two structures with less similarity. The PFSA-S value 0.0565 is for comparison between 1ATP-E and No. 20 of structure. PFSA-S near zero indicates that the pair of structures is dissimilar. It is noted that the sequence length of structures 20 is 94. To compare with 1ATP-E, two structures have big difference in length and the alignment generates 228 gaps which give the lower value of PFSA-S. Therefore, the PFSA has ability to distinguish the structural deference in more detail.

4.2.3 PFSA vs. other methods

A set of 10 pairs of proteins with lower structural similarity was recognized as difficult structures for comparison, and was evaluated by VAST (Madej et al., 1995; Gibrat et al., 1996), DALI (Holm & Sander, 1993; Holm & Park, 2000), CE (Shindyalov & Bourne, 1998), Prosup (Lackner et al., 2000) and LGA (Zemla, 2003) methods respectively. The structural similarity was evaluated by two optimistic parameters, i.e. lower RMSD and larger number of equivalent residues. It is apparently, in Table 10, that various methods gave comparative results for each pair of proteins. The results from various methods provide complementary information for protein structural comparison. Overall, the ProSup and LGA methods provided consistent results with restriction of RMSD less than 3.0. The PFSA, however, offers new observation for assessment of similarity of protein structures. First, the similarity is able to be evaluated by a single value of the PFSA-S. In order to compare with other methods, information of (sum of number of identical and analogous shapes) / (number of gaps) / (PFSA-S) is listed in Table 10. The similarity score of PFSA-S is determined by number of identical shapes, number of analogous shapes and number of gaps. Second, the value of score PFSA-S may judge the similarity crossing isolated comparisons, i.e. the values of score PFSA-S from unrelated comparisons can be used to assess the protein homologous degree. Each pair of proteins in Table 10 is a lonely comparison without common reference structure, but the value of PFSA-S may indicate which pair of structures has higher similarity. In Table 10, the results of each pair of comparison are sorted according the values of PFSA-S descendingly. For example, the pair of comparison of 1CEW-I and 1MOL-A has the PFSA-S = 0.564 and the pair of comparison of 1CID and 2RHE has the PFSA-S = 0.384. A conclusion may be obtained that the pair of 1CEW-I and 1MOL-A has higher structural similarity than the pair of 1CID and 2RHE. Third, relative size of compared proteins makes the contribution to structural similarity in PFSA approach. With various methods, the value of RMSD is often used to make judgment of structural similarity. For example, the pair of comparison of 1CEW-I and 1MOL-A have the RMSD (VAST: 2.0, DALI: 2.3, CE: 2.3, ProSup: 1.9and LGA: 2.0); the pair of comparison of 1TEN and 3HHR-B have the RMSD (VAST: 1.6, DALI: 1.9, CE: 1.9, ProSup: 1.7 and LGA: 1.9). Both pairs have lower RMSD than other remaining pairs and have overall agreement with various methods. However, PFSA approach distinguishes these two pairs by PFSA-S. With PFSA-S = 0.456, the pair of 1TEN and 3HHR-B is ranked below other five pairs, including the pair of 1CEW-I and 1MOL-A with PFSA-A = 0.564. The separation is explained by a factor that the pair of 1CEW-I and 1MOL-A has comparable length of sequence (108 : 94), but the pair of 1TEN and 3HHR-B has larger different in length (99 : 195). The contribution of relative difference of size is counted in PFSA approach. Therefore, with normalization of PFSA-S, for separated comparisons, the similarity degree still can be evaluated without common reference protein.

Proteins	N1	Proteins	N2	VAST	DALI	CE	ProSup	LGA	PFSA
1CEW-I	108	1MOL-A	94	75/2.0	81/2.3	81/2.3	76/1.9	79/2.0	69/24/0.564
1FXI-A	96	1UBQ	76	48/2.1	52/2.5	64/3.8	54/2.6	61/2.6	59/28/0.538
1BGE-B	159	2GMF-A	121	71/2.3	94/3.3	107/3.9	87/2.4	91/2.5	92/47/0.526
1CRL	534	1EDE	310	186/3.7	212/3.6	219/3.8	161/2.6	182/2.6	264/237/0.475
3HLA-B	99	2RHE	114	58/2.3	74/3.0	83/3.3	71/2.7	74/2.5	69/25/0.473
1TEN	99	3HHR-B	195	76/1.5	86/1.9	87/1.9	85/1.7	87/1.9	72/71/0.456
1TIE	166	4FGF	124	76/1.6	114/3.1	116/2.9	101/2.4	104/2.3	88/35/0.456
2AZA-A	129	1PAZ	110	70/2.1	82/3.0	84/2.9	82/2.6	80/2.2	81/36/0.453
2SIM	381	1NSB-A	390	299/4.2	289/3.2	275/3.0	248/2.6	269/2.6	249/129/0.428
1CID	177	2RHE	114	78/2.0	96/3.1	97/2.9	84/2.3	93/2.3	75/61/0.384

Table 10. Comparison of 10 pairs of proteins with lower structural similarity using various methods. N1 and N2 are the lengths of proteins. For methods of VAST, DALL, CE, ProSup and LGA, the results are presented as (Number of identical residues) / (RMSD). For PFSA approach, results are presented as (sum of number of identical and analogous shapes) / (number of gaps) / (PFSA-S). The results of each pair of comparison are sorted according values of PFSA-S descendingly.

5. Conclusion

The PFSA approach adopts the vector of folding shape of five residues as element, and the geometric feature of folding shape is embedded by alphabets as representation. With application of alphabets, the alignment of protein structures is straightforward and steady. This study demonstrates two advantages in PFSA approach. First, 27 PFSC vectors are able to cover all possible folding shapes of five successive Cα atoms in protein. This is fundamental important because it offers a complete description of folding conformation for any protein with given 3D structure. Second, with consistent procedure, the PFSA approach generates unique score for similarity and detail information in alignment table, which provides new observation for the protein structure comparison.

6. Acknowledgments

This work was supported in part by a grant from the Indiana Spinal Cord and Brain Injury Research Fund (2009-2011). The algorithms of PFSC and PFSA have been coded with Java (J2SE v.1.5.0_07) computer language. Requests for additional information will be accepted via e-mail to info@proteinshape.com, jiaan@microtechnano.com, jiaanyang@comcast.net or via Website: http://www.proteinshape.com.

7. References

Andreeva A, Howorth D, Chandonia JM, Brenner SE, Hubbard TJP, Chothia C & Murzin AG. (2008). Data growth and its impact on the SCOP database: new developments. Nucleic Acids Res. 36 (Database issue), 419–425.

Brevern AG, (2005). New assessment of a structural alphabet. *Silico Biol.*, 5, 283-289

Brevern AG, Etchebest C & Hazout S. (2000). Bayesian probabilistic approach for predicting backbone structures in terms of protein blocks. *Proteins*, 41, 271-287

Friedberg I, Harder T, Kolodny R, Sitbon E, Li Z, Godzik A. (2007). Using an alignment of fragment strings for comparing protein structures, *Bioinformatics*, 23, 219-224

Garrett TPJ, McKern NM, Lou M, Frenkel MJ, Bentley JD, Lovrecz GO, Elleman TC, Cosgrove LJ & Ward CW. (1998). Crystal structure of the first three domains of the type-1 insulin-like growth factor receptor. *Nature*, 394, 395-399

Garza-Garcia A, Patel DS, Gems D & Driscoll PC. (2007). RILM: a web-based resource to aid comparative and functional analysis of the insulin and IGF-1 receptor family. *Hum Mutat.*, 28, (7), 660-668

Gibrat JF, Madej T & Bryant SH. (1996). Surprising similarities in structure comparison. *Curr. Opin. Struct. Biol.*, 6, 377–385

Guyon F, Camproux AC, Hochez J & Tuffery P. (2004). SA-Search: a web tool for protein structure mining based on a structural alphabet. *Nucleic Acids Res.*, 32, W545-W548

Holm L & Sander C. (1993). Protein structure comparison by alignment of distance matrices. J. Mol. Biol., 233, 123-138

Holm, L. & Park, J. (2000). DaliLite workbench for protein structure comparison. *Bionformatics*, 16,566-567

Hubbard SR. (1997). Crystal structure of the activated insulin Crysreceptor tyrosine kinase in complex with peptide substrate and ATP analog. *EMBO J*, 16, 5572–5581

Hvidsten TR, Kryshtafovych A, Komorowski J & Fidelis K. (2003) A novel approach to fold recognition using sequence-derived properties from sets of structurally similar local fragments of proteins. *Bioinformatics*,19 (Suppl 2):II81–II91

Joseph AP, Srinivasan N, Brevern AG. (2011). Improvement of protein structure comparison using a structural alphabet, *Biochimie.*, 93,(9),1434-1445

Karplus K, Karchin R, Draper J, Casper J, Mandel-Gutfreund Y, Diekhans M & Hughey R. (2003). Combining local-structure, fold-recognition, and new fold methods for protein structure prediction. *Proteins*, 53, Suppl 6:491-6

Kleywegt GJ & Jones TA. (1994). A super position. *ESF/CCP4 Newsletter*, 31, 9-14

Kolodny R, Koehl P & Michael Levitt M. (2005). Comprehensive Evaluation of Protein Structure Alignment Methods: Scoring by Geometric Measures. *J. Mol. Biol*, 346,1173-1188

Kolodny R, Koehl P, Guibas L., & Levitt M. (2002). Small libraries of protein fragments model native protein structures accurately, *J. Mol. Biol.*, 323, 297-307

Krissinel E & Henrick K. (2004). Secondary-structure matching (SSM), a new tool for fast protein structure alignment in three dimensions. *Acta Crystallogr D Biol Crystallogr.*, 60, 2256-2268

Kryshtafovych A, Milostan M, Szajkowski L, Daniluk P & Fidelis K. (2005). CASP6 data processing and automatic evaluation at the protein structure prediction center. *Proteins*, 61, (Suppl 7):19–23

Kryshtafovych A, Prlic A, Dmytriv Z, Daniluk P, Milostan M, Eyrich V, Hubbard T & Fidelis K. (2007). New tools and expanded data analysis capabilities at the protein structure prediction center. *Proteins*, 69, S8:19- 26

Ku SY & Hu YJ, (2008). Protein structure search and local structure characterization, BMC Bioinformatics, 9, 349

Lackner P, Koppensteiner WA. Sippl MJ & Domingues FS. (2000). ProSup: a refined tool for protein structure alignment. *Protein Eng.*, 13, 745–752

Lou M, Garrett TP, McKern NM, Hoyne PA, Epa VC, Bentley JD, Lovrecz GO, Cosgrove LJ, Frenkel MJ &, Ward CW. (2006). The first three domains of the insulin receptor differ structurally from the insulin-like growth factor 1 receptor in the regions governing ligand specificity. *Proc. Natl. Acad. Sci. USA.*, 103, 12429-12434

Madej T, Gibrat JF & Bryant SH. (1995). Threading a database of protein cores. Proteins, 23, 356–369

Mavridis L & Ritchie DW, (2010). 3D-blast: 3d protein structure alignment, comparison, and classification using spherical polar fourier correlations, *Pac Symp Biocomput*, 15, 281-292

McKern NM, Lawrence MC, Streltsov VA, Lou MZ, Adams TE, Lovrecz GO, Elleman TC, Richards KM, Bentley JD, Pilling PA, Hoyne PA, Cartledge KA, Pham TM, Lewis JL, Sankovich SE, Stoichevska V, Da Silva E, Robinson CP, Frenkel MJ, Sparrow LG, Fernley RT, Epa VC & Ward CW. (2006). Structure of the insulin receptor ectodomain reveals a folded-over conformation. *Nature*, 443, 218-221

Melo F & Marti-Renom MA. (2006). Accuracy of Sequence Alignment and Fold Assessment Using Reduced Amino Acid Alphabets, PROTEINS: *Structure, Function, and Bioinformatics*, 63, 986–995

Micheletti C, Seno F, Maritan A. (2000). Recurrent oligomers in proteins: an optimal scheme reconciling accurate and concise backbone representations in automated folding and design studies. *Proteins*, 40, 662-674.

Moult, J., et al. (2009). Critical assessment of methods of protein structure prediction — Round VIII. *Proteins*, 77, (Suppl 9):1–4

Murphy LR, Wallqvist A & Levy RM. (2000). Simplified amino acid alphabets for protein fold recognition and implications for folding, *Protein Eng.*, 13, (3): 149-152

Murzin AG, Brenner SE, Hubbard T & Chothia C. (1995). SCOP: a structural classification of proteins database for the investigation of sequences and structures. *J. Mol. Biol.* 247, 536-540

Needleman SB & Wunsch CD. (1970). A general method applica¬ble to search for similarities in the amino acid sequenc¬es of two proteins. *J Mol Biol*, 48, 442–453

Ortiz AR, Strauss CE, Olmea O. (2002). MAMMOTH (matching molecular models obtained from theory): an automated method for model comparison. *Protein Sci.*, 11:2606–2621

Pautsch A, Zoephel A, Ahorn H, Spevak W, Hauptmann R & Nar H. (2001). Crystal structure of bisphosphorylated IGF-1 receptor kinase: insight into domain movements upon kinase activation. *Structure*. 9, 955-965

Rooman MJ, Rodriguez J, Wodak SJ. (1990). Automatic definition of recurrent local structure motifs in proteins. *J. Mol. Biol.*, 213, 327-336

Sacan A., Toroslu IH & Ferhatosmanoglu H. (2008). Integrated search and alignment of protein structures. *Bioinformatics*, 24, 2872-2879

Sander O, Sommer I & Lengauer T. (2006). Local protein structure prediction using discriminative models, BMC Bioinformatics, 7, 14

Schuchhardt J, Schneider G, Reichelt J, Schomburg D & Wrede P. (1996). Local structural motifs of protein backbones are classified by self-organizing neural networks. *Protein Eng.*, 9, 833-842

Shatsky M, Nussinov R, & Wolfson HJ. (2002). Flexible protein alignment and hinge detection, *Proteins*, 48, 242-256

Shindyalov IN & Bourne PE. (1998). Protein structure alignment by incremental combinatorial extension (CE) of the optimal path. *Protein Eng.*,11,739–747

Subbiah S, Laurents DV & Levitt, M. (1993). Structural similarity of DNA-binding domains of bacteriophage repressors and the globin core. *Curr. Biol.*, 3,141–148

Taylor WR & Orengo CA. (1989). Protein structure alignment. *J. Mol. Biol.*,208, 1–22

Tung CH, Huang JW & Yang JM. (2007). Kappa-alpha plot derived structural alphabet and BLOSUM-like substitution matrix for rapid search of protein structure database. *Genome Biol.*, 8 R31

Tyagi M, Gowri VS, Srinivasan N, Brevern AG & Offmann B. (2006). A substitution matrix for structural alphabet based on structural alignment of homologous proteins and its applications, *Proteins*, 65, (1):32-9

Tyagi M, Sharma P, Swamy C S, Cadet F, Srinivasan N, Brevern A.G and Offmann B. (2006). Protein Block Expert (PBE): a web-based protein structure analysis server using a structural alphabet. *Nucleic Acids Research*, 34, Web Server issue W119–W123.

Unger R, Harel D, Wherland S & Sussman JL. (1989). A 3D building blocks approach to analyzing and predicting structure of proteins. *Proteins*, 5, 355-373

Wang S & Zheng WM. (2008). CLePAPS: fast pair alignment of protein structures based on conformational letters. *J. Bioinform. Comput. Biol.*, 6, 347-366

Werner H, Weinstein D & Bentov I. (2008) Similarities and differences between insulin and IGF-I: structures, receptors, and signaling pathways. *Arch Physiol Biochem*. 114,(1),17-22

Williams SB, Vakonakis I, Golden SS & LiWang AC. (2002). Structure and function from the circadian clock protein KaiA of Synechococcus elongatus: a potential clock input mechanism. *Proc Natl Acad Sci USA*. 26, 99(24), 15357-62

Yang J. (2008). Comprehensive description of protein structures using protein folding shape code. *Proteins*, 71, 3, 1497-1518

Yang, J, (2011). Complete Description of Protein Folding Shapes for Structural Comparison. *In: Series: Protein Biochemistry, Synthesis, Structure and Cellular Functions: Protein Folding*. Edited by Walters EC. New York, Nova Science Publishers, 421-442 (ISBN: 978-1-61761-259-6),

Yuzhen Y, & Adam G. (2003). Flexible structure alignment by chaining aligned fragment pairs allowing twists. *Bioinformatics*, 19(suppl.2), ii246-ii255

Zemla A. (2003). LGA: a method for finding 3D similarities in protein structures. *Nucleic Acids Res*, 31, 3370–3374

Zhang Y & Skolnick J. (2005).TM-align: a protein structure alignment algorithm based on the TM-score. *Nucleic Acids Research*, 33,7, 2302–2309

Section 3

Energy and Thermodynamics

Theoretical Analyses of Photoinduced Electron Transfer from Aromatic Amino Acids to the Excited Flavins in Some Flavoproteins

Kiattisak Lugsanangarm[1], Nadtanet Nunthaboot[2], Somsak Pianwanit[1,3,*], Sirirat Kokpol[1,3] and Fumio Tanaka[1,4,*]
[1]*Department of Chemistry, Faculty of Science, Chulalongkorn University,*
[2]*Department of Chemistry, Faculty of Science, Mahasarakham University, Mahasarakham,*
[3]*Center of Excellence for Petroleum, Petrochemicals, and Advanced Materials, Chulalongkorn University, Bangkok,*
[4]*Laser Biochemistry Division, Institute for Laser Technology, Osaka,*
[1,2,3]*Thailand*
[4]*Japan*

1. Introduction

Electron transfer phenomena have been an important subjects in the fields of physics (Jortner & Bixon, 1999), chemistry (Mataga et al., 2005a, 2005b; Vogler et al., 2011) and biology (Marcus & Sutin, 1985; Gray & Winkler, 1996; Bendal, 1996). Photoinduced electron transfer (PET) plays an essential role in photosynthetic systems (Blankenship, 2002). In the last decade a number of new flavin photoreceptors have been found. Among six families of the photoreceptors, phototropins (Crosson & Moffat, 2001), cryptochromes (Giovani et al., 2003) and BLUF (blue-light sensing using flavin) contain flavins as the reaction center (Masuda & Bauer, 2002). The PET from Tyr to the excited isoalloxazine (Iso*) is considered as an initial step of the photo-regulation for photosynthesis in AppA (Masuda & Bauer, 2002; Laan et al., 2003) and pili-dependent cell motility in TePixD (Kita et al., 2005) and in Slr1694 (Masuda et al., 2004) photoactive bacteria.

Flavoproteins contain flavin mononucleotide (FMN), flavin adenine dinucleotide, and riboflavin as a cofactor and are ubiquitously distributed in various microorganisms, in leafy vegetables and specific tissues of other multicellular plants, and in the milk, brain, kidney, liver and heart of mammals, where they play an essential role in many redox reactions (Frago et al., 2008).

The fluorescence of flavins was first reported by Weber (1950), along with the fluorescence quenching of flavins by various substances, including aromatic amino acids. Since then many researchers have studied the photochemistry of flavins and flavoproteins (Silva &

* Corresponding Authors

Edward, 2006). The quenching of flavin fluorescence by an indole ring was reported with isoalloxazine-$(CH_2)_n$-indole dyads by McCormick (1977). Time-resolved fluorescence spectroscopy of flavins and flavoproteins has been reviewed by Berg and Visser (2001). However, a number of flavoproteins are practically non-fluorescent, but rather they emit fluorescence with very short lifetimes (sub-picoseconds) upon excitation with an ultra-short laser pulse (Mataga et al., 1998, 2000, 2002; Tanaka et al. 2007; Chosrowjan et al., 2007, 2008, 2010). In these flavoproteins tryptophan (Trp) and/or tyrosine (Tyr) residues always exist near the isoalloxazine ring (Iso). The remarkably fast fluorescence quenching in these flavoproteins was demonstrated to be caused by PET from Trp and/or Tyr to the excited state Iso (Iso*), by means of picosecond (Karen et al., 1983, 1987) and femtosecond (Zhong & Zewail, 2001) transient absorption spectroscopy. The PET phenomena in these flavoproteins are similar to the flavin photo-receptors (Crosson & Moffat, 2001; Masuda & Bauer, 2002), but had been discovered before the flavin photoreceptors.

Since the seminal works on electron transfer theory by Marcus (1956a, 1956b, 1964), several researchers have further developed the electron transfer theory (Hush, 1961; Sumi & Marcus, 1986; Bixon & Jortner, 1991, 1993; Bixon et al., 1994; Kakitani & Mataga, 1985; Kakitani et al., 1991, 1992). However, they have been modeled for PET in bulk solution and it is not clear whether these theories can be applicable to PET in proteins. Therefore, it is required to establish a method to quantitatively analyze PET in proteins.

In any electron transfer theories there are several parameters that are difficult to determine experimentally. The PET rates in flavoproteins have been analyzed experimentally with ultrafast fluorescence dynamics and theoretically by an electron transfer theory using the atomic coordinates obtained by molecular dynamics (MD) simulation. The procedure to determine the unknown PET parameters is as follows (Nunthaboot et al., 2008a, 2009a): (1) the time-dependent atomic coordinates of flavoproteins are obtained by MD simulation, (2) the PET rates are then calculated using a PET theory and the atomic coordinates with a set of trial PET parameters, (3) the parameters are then varied until the best-fit between the calculated and observed fluorescence decays is obtained, according to a non-linear least squares method.

In this review article we describe the results of quantitative analyses of PET in wild type (WT) flavodoxin and FMN binding proteins from *Desulfovibrio vulgaris*, Miyazaki F, and three relevant flavodoxin amino acid substitution mutants (isoforms) and two relevant FMN binding protein amino acid substitution mutants, respectively, and discuss the characteristics of the PET mechanism in flavoproteins.

Note that for brevity, unless stated otherwise, reference to flavodoxin and FMN binding proteins in this article refers to those from *Desulfovibrio vulgaris*, Miyazaki F.

2. Method of analysis of Photoinduced Electron Transfer (PET) in flavoproteins

2.1 Electron transfer theory in flavoproteins

All of the original PET theories were modeled for a system in solution. Here, we describe the PET theories that have been used for the flavoproteins. Electrostatic (ES) energy was first introduced by Nunthaboot et al. (2009a).

2.1.1 Marcus-Hush theory

When the jth flavoprotein contains several PET donors, the PET rate from the kth Trp and/or Tyr near Iso to Iso* by Marcus theory as modified by Hush (1961) (MH theory) is expressed by Eq. (1), and the energy diagram for MH theory is shown in Figure 1.

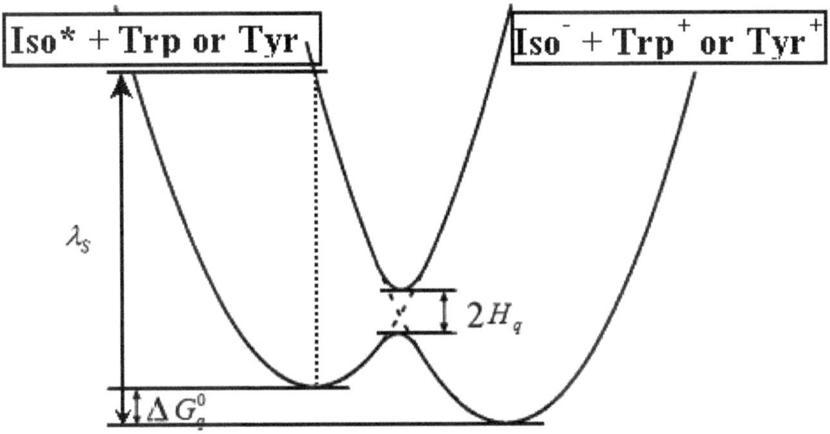

Fig. 1. Energy diagram for Marcus-Hush PET theory

$$k_{MH}^{jk} = \frac{2\pi}{\hbar}\frac{H_q^2}{\sqrt{4\pi\lambda_S^{jk}}}\exp\left[-\frac{\left\{\Delta G_q^0 - e^2/\varepsilon_{DA}R_{jk} + \lambda_S^{jk} + ES_j(k)\right\}^2}{4\lambda_S^{jk}k_BT}\right] \tag{1}$$

In Eq. (1), H_q is the electronic interaction energy between Iso* and Trp (q = Trp) or Tyr (q = Tyr). R_{jk} is the center to center (Rc) distance between Iso and the ET donor k in the jth flavoprotein. $\hbar$, k_B, T and e are the reduced Planck constant, Boltzmann constant, temperature and electron charge, respectively. ε_{DA} is the static dielectric constant of medium between the PET donors and acceptor. $ES_j(k)$ is the net ES energy between the kth aromatic ionic species and all other ionic groups in the jth flavoprotein, as described below. λ_S^{jk} is the solvent reorganization energy of the Iso* and the kth donor in the jth flavoprotein, as shown by Eq. (2);

$$\lambda_S^{jk} = e^2\left(\frac{1}{2a_{Iso}} + \frac{1}{2a_q} - \frac{1}{R_{jk}}\right)\left(\frac{1}{\varepsilon_\infty} - \frac{1}{\varepsilon_{DA}}\right) \tag{2}$$

Here, a_{Iso} and a_q are the radii of Iso and the donor q (Trp or Tyr), assuming these reactants are spherical, and ε_∞ is the optical dielectric constant (a value of 2 being used). ε_{DA} is the static dielectric constant between Iso and a donor. The radii of Iso, Trp and Tyr were determined according to the following procedure. (1) The three dimensional sizes of lumiflavin for Iso, 3-methylindole for Trp, and p-methylphenol for Tyr were obtained by a semi-empirical molecular orbital method (PM3). (2) The volumes of these molecules were determined as asymmetric rotors. (3) The radii of the spheres having the same volumes of the asymmetric rotors are obtained. The obtained radii by this procedure are a_{Iso} = 0.224 nm, a_{Trp} = 0.196 nm and a_{Tyr} = 0.173 nm.

The standard free energy gap (ΔG_q^0) was expressed with the ionization potential (E_{IP}^q) of the PET donor q (Trp or Tyr), as shown in Eq. (3).

$$\Delta G_q^0 = E_{IP}^q - G_{Iso}^0 \tag{3}$$

where G_{Iso}^0 is the standard Gibbs energy related to the electron affinity of Iso*. The values of E_{IP}^q for Trp and Tyr were 7.2 eV and 8.0 eV, respectively (Vorsa et al., 1999).

2.1.2 Kakitani-Mataga (KM) theory

The PET rate by Kakitani & Mataga (KM theory) is expressed as Eq. (4), which describes the PET rate for both adiabatic and non-adiabatic processes, whilst the MH and Bixon-Jortner (BJ) theories (see below) describe only adiabatic processes.

$$k_{KM}^{jk} = \frac{\nu_0^q}{1+\exp\left\{\beta^q(R_{jk} - R_0^q)\right\}}\sqrt{\frac{k_B T}{4\pi\lambda_S^{jk}}}\exp\left[-\frac{\left\{\Delta G_q^0 - e^2/\varepsilon_{DA}R_{jk} + \lambda_S^{jk} + ES_j(k)\right\}^2}{4\lambda_S^{jk}k_B T}\right] \tag{4}$$

Here ν_0^q is an adiabatic frequency, β^q is the PET process coefficient, and R_0^q is a critical distance between the adiabatic and non-adiabatic PET processes. These quantities depend only on q (Trp or Tyr). When $R_{jk} < R_0^q$ the ET process is adiabatic, whereas when $R_{jk} > R_0^q$ it is non-adiabatic. The other quantities are the same as those in the MH theory (section 2.1.1).

2.1.3 Bixon-Jortner (BJ) theory

The BJ theory describes the PET rates from various vibronic states, as shown in Eq. (5), while the MH and KM theories only describe the PET from the lowest vibrational state.

$$k_{BJ}^{jk} = \frac{2\pi}{\hbar}H_q^2\frac{\exp\left\{-\beta(R_{jk} - \sigma_q) - S\right\}}{\sqrt{2\pi\lambda_S^{jk}}}\sum_{i=0}^{n}\frac{S^i}{i!}\exp\left[-\frac{\left\{\Delta G_q^0 - e^2/\varepsilon_{DA}R_{jk} + \lambda_S^{jk} + i\hbar\langle\omega\rangle + ES_j(k)\right\}^2}{4\lambda_S^{jk}k_B T}\right] \tag{5}$$

$S = \lambda_V / \hbar\langle\omega\rangle$ is the vibronic coupling constant, where λ_V is the reorganization energy associated with the average frequency $\hbar\langle\omega\rangle$, n is the number of vibrational modes in the donor and σ_q is the van der Waals contact and is given by Eq. (6).

$$\sigma_q = a_{Iso} + a_q \tag{6}$$

The meanings of all the other notations are the same as that given in the MH theory (section 2.1.1).

2.2 Electrostatic (ES) energy between the photoproducts and ionic groups in a protein

Proteins, including flavoproteins, contain many ionic groups, which may influence the PET rate. The cofactor in the relevant flavoproteins is FMN, which has two negative charges at the phosphate. The ES energy between the Iso anion or donor k^{th} cation, and all the other ionic groups in the jth flavoprotein is expressed by Eq. (7):

$$E_j(k) = \sum_{i=1}^{n_E} \frac{C_k \cdot C_{Glu}}{\varepsilon_0^j R_k(Glu-i)} + \sum_{i=1}^{n_B} \frac{C_k \cdot C_{Asp}}{\varepsilon_0^j R_k(Asp-i)}$$

$$+ \sum_{i=1}^{n_K} \frac{C_k \cdot C_{Lys}}{\varepsilon_0^j R_k(Lys-i)} + \sum_{i=1}^{n_R} \frac{C_k \cdot C_{Arg}}{\varepsilon_0^j R_k(Arg-i)} + \sum_{i=1}^{2} \frac{C_k \cdot C_P}{\varepsilon_0^j R_k(P-i)} \quad (7)$$

where n_E, n_B, n_K and n_R are the numbers of Glu, Asp, Lys and Arg residues, respectively, in the flavoprotein. Here, k = 0 for the Iso anion, and $k > 0$ for the donor cations. ε_0^j is the static dielectric constant inside the entire j^{th} flavoprotein, which should be different from ε_{DA}. C_k is the charge of the aromatic ionic species k, and is $-e$ for k = 0 (Iso anion), $+e$ for $k >$ 1. C_{Glu} (= $-e$), C_{Asp} (= $-e$), C_{Lys} (= $+e$), C_{Arg} (= $+e$) and C_P (= $-e$) are the charges of Glu, Asp, Lys, Arg and phosphate anions, respectively. It was assumed that these groups are all in an ionic state in solution. The pK_a values of the ionic amino acids in water are 4.3 in Glu, 3.9 in Asp, 10.5 in Lys and 12.5 in Arg. However, as residues within proteins these pK_a values may be modified in the range of ± 0.3. His displays a pK_a of 6.0 in water. All fluorescence measurements were performed in 0.1 M phosphate buffer at pH 7.0, where His should be neutral. Distances between the aromatic ionic species k and the ith Glu are denoted as $R_k(Glu-i)$, those between k and the ith Asp are denoted as $R_k(Asp-i)$, and so on. $ES_j(k)$ is expressed in Eq. (8);

$$ES_j(k) = E_j(0) + E_j(k) \quad (8)$$

2.3 Observed ultrafast fluorescence dynamics of flavodoxins and FMN binding proteins

Ultrafast fluorescence dynamics of flavodoxins and FMN binding proteins have been measured by means of a fluorescence up-conversion method (Mataga et al., 2002; Chosrowjan et al., 2007, 2008, 2010). The fluorescence decay functions of the WT flavodoxin, the two single substitution isoforms, Y97F and W59F, and the double substitution, Y97F/W59F (DM), are represented by Eq. (9), whilst the fluorescence decays of the WT FMN binding protein and the four single substitution isoforms, E13T, E13Q, W32Y and W32A, are represented by Eq. (10).

$$F_{FD}^j(t) = \sum_{i=1}^{n} \alpha_{FDi}^j \exp(-t/\tau_{FDi}^j) \quad (j=1, \text{WT}; j=2, \text{Y97F}; j=3, \text{W59F}; j=4, \text{Y97F/W59F}) \quad (9)$$

$$F_{FBP}^j(t) = \sum_{i=1}^{n} \alpha_{FBPi}^j \exp(-t/\tau_{FBPi}^j) \quad (j=1, \text{WT}; j=2, \text{E13T}; j=3, \text{E13Q}; j=4, \text{W32Y}; j=5, \text{W32A}) \quad (10)$$

In Eq. (9), $n = 1$ or 2, and in Eq. (10) $n = 1$ to 3, depending on the protein system, j. The decay parameters are listed in Table 1. The experimental decay of the WT flavodoxin contains an additional lifetime component with 500 ps. However, it was interpreted to be free FMN dissociated from the protein. The average lifetime values, τ_{AV}^{k}, were obtained from $\tau_{AV}^{j} = \sum^{n} \alpha_i^j \tau_i^j$ and are listed in the last line of Table 1. The decays with n greater than 1 display non-exponential function.

Decay parameter	Flavodoxin[b]				FMN binding protein[c]				
	WT	Y97F	W59F	DM[d]	WT	E13T	E13Q	W32Y	W32A
τ_1^j (ps)	0.157	0.254	0.322	18	0.167	0.107	0.134	3.4	30.1
(α_1^j)	(1.0)	(0.85)	(0.83)	(1.0)	(0.96)	(0.86)	(0.85)	(0.23)	(1.0)
τ_2^j (ps)	-	4.0	5.5	-	1.5	1.5	0.746	18.2	-
(α_2^j)	-	(0.15)	(0.17)	-	(0.04)	(0.12)	(0.12)	(0.74)	-
τ_3^j (ps)	-	-	-	-	-	30	30	96	-
(α_3^j)	-	-	-	-	-	(0.02)	(0.03)	(0.03)	-
τ_{AV}^{k} [e] (ps)	0.157	0.816	1.20	18	0.22	0.872	1.10	17.1	30.1

[a]The observed flavodoxin and FMN binding protein decay functions are expressed in Eqs. (9) and (10), respectively.
[b]Data were taken from the work by Mataga et al. (2002).
[c]Data were taken from the works by Chosrowjan et al. (2007, 2008, 2010).
[d]DM denotes the Y97F/W59F double mutant.
[e]Averaged lifetimes were obtained by $\tau_{AV}^{j} = \sum_{i=1}^{n} \alpha_i^j \tau_i^j$.

Table 1. Fluorescence decay parameters of the flavodoxin and FMN binding protein isoforms from *Desulfovibrio vulgaris*, Miyazaki F [a]

2.4 Determination of the PET parameters

The calculated decay function in the jth protein system is expressed by Eq. (11).

$$F_{calc}^{j}(t) = \left\langle \exp\left\{ -\sum_{k=1}^{m} k_{ET}^{jk}(t')t \right\} \right\rangle_{AV} \tag{11}$$

$\langle ... \rangle_{AV}$ means the averaging procedure of the exponential function in Eq. (11) over t'. In Eq. (11) we assumed that the decay function at every instant of time, t', during the MD simulation time range can always be expressed by an exponential function, and thus the MD simulation time range must be much longer than experimental decay time range. In Eq. (11) m is the total number of PET donors in the jth flavoprotein. In MH theory, the unknown PET parameters were H_q (q = Trp and Tyr), G_{Iso}^0, ε_{DA} and ε_0^j, whilst in KM theory they are ν_0^q, β^q, and R_0^q for Trp and Tyr, G_{Iso}^0, ε_{DA} and ε_0^j, and in BJ theory they are H_q (q = Trp or Tyr), β, λ_V, $\hbar\langle\omega\rangle$, G_{Iso}^0, ε_{DA} and ε_0^j. These parameters were determined so as to obtain the minimum value of χ^2, as defined by Eq. (12), by means of a non-linear least squares method, according to the Marquardt algorithm.

$$\chi^2 = \frac{1}{N_j N_F} \sum_{j=1}^{N_j} \sum_{i=1}^{N_F} \frac{\left\{F_{calc}^j(t_i) - F_{obs}^j(t_i)\right\}^2}{F_{calc}^j(t_i)} \tag{12}$$

Here, N_F denotes the number of time intervals in the fluorescence decay, and N_j is the total number of flavoproteins for simultaneous analysis.

3. Flavodoxins from *Desulfovibrio vulgaris*, Miyazaki F

3.1 Homology modeling

Flavodoxins are small flavoproteins with a molecular weight of 15 - 23 kDa that have been isolated from a variety of microorganisms. Flavodoxins are considered to function as electron-transport proteins in various metabolic pathways (Sancho, 2006). They contain one molecule of non covalently-bound FMN (see Chart 1) as a cofactor, and exhibit a highly negative reduction potential for the semiquinone / hydroquinone couple of FMN, and accordingly the semiquinone state is stable. The redox properties of FMN in flavodoxins are considerably different from those of the free FMN.

The biochemical properties of flavodoxin from *Desulfovibrio vulgaris*, strain Miyazaki F were first characterized by Kitamura et al. (1998). The dissociation constant of FMN is 0.38 nM, which is ~1.6-fold higher than that in the related flavodoxin from Desulfovibrio vulgaris Hildenborough (0.24 nM). The redox potential of these two closely related flavodoxins is also slightly different, being E_1 = -434 and -440 mV for the Miyazaki and Hildenborough forms, respectively, for the oxidized-semiquinone reaction of flavodoxin, and E_2 = -151 and -143 mV for the semiquinone-2-electron reduced reaction, respectively (Kitamura et al., 1998). Recently, the three-dimensional structures of numerous flavodoxins have been determined, including Desulfovibrio vulgaris Hildenborough (Watenpauph, 1973) and the flavodoxins from Anacystis nidulans (Drennan et al., 1999), Clostridium beijerinckii (Ludwig et al., 1997), Escherichia coli (Hoover & Ludwig, 1997), Anabaena 7120 (Burkhart et al., 1995) a red algae (Fukuyama et al., 1992) Chondrus crispu (Fukuyama et al., 1990) and H. pylori (Freigang et al., 2002) by X-ray crystallography. The structure of flavodoxin, however, has not yet been determined, although the primary structure is known (Kitamura et al., 1998).

The ultrafast fluorescence dynamics of flavodoxins (Mataga et al., 2002) have been extensively investigated in the WT and the Y97F, W59F and W59F/Y97F (DM) substitution isoforms, as described above.

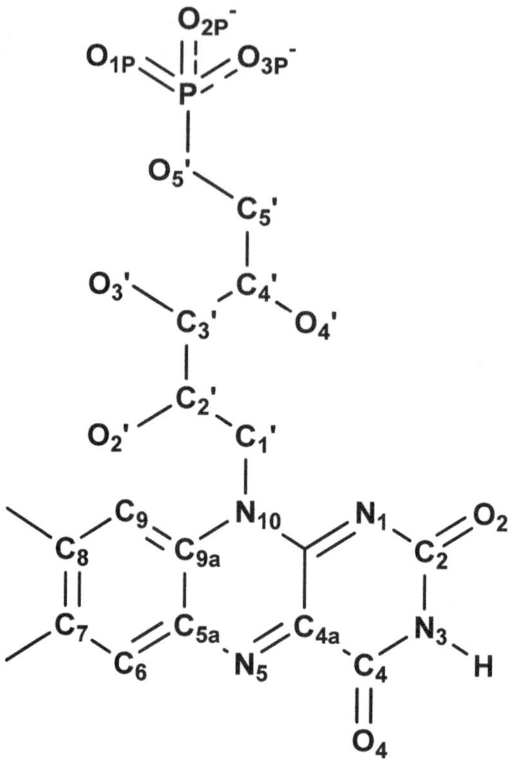

Chart 1. Chemical structure of FMN and its atom notations.

3.2 Three-dimensional structures of four flavodoxin isoforms

The protein structures of the WT, single amino acid substitution (Y97F and W59F) and the double amino acid substitution (W59F/Y97F; DM) isoforms have been determined by homology modeling method with the Modeler Module of the Discovery Studio 2.0 software package (http://www.discoverystudios.com) using the flavodoxin *Desulfovibrio vulgaris*, strain Hildenborough structure (PDB code: 1J8Q) as the template. This protein displays 66% amino acid sequence identity and 79% similarity to the WT flavodoxin of Miyazaki reviewed herein. The validities of the structures were examined with a Verified3D analysis (visit for the method, www.proteinstructures.com by Prof. Salam Al-Karadaghi). Verify3D assigns each residue a structural class based on its location and environment (alpha, beta, loop, polar, apolar etc). Then, a database generated from good structures is used to obtain a score for each of the 20 amino acids in this structural class. Figure 2 shows the Verified3D scores at each amino acid residue, where the quality of the structures is satisfactory.

MD simulations were performed for 10 ns in order to investigate the dynamic properties of the proteins and the important interactions that are involved in the binding of the FMN cofactor to the proteins. Figure 3 shows the three-dimensional structures in water that were obtained by MD simulation.

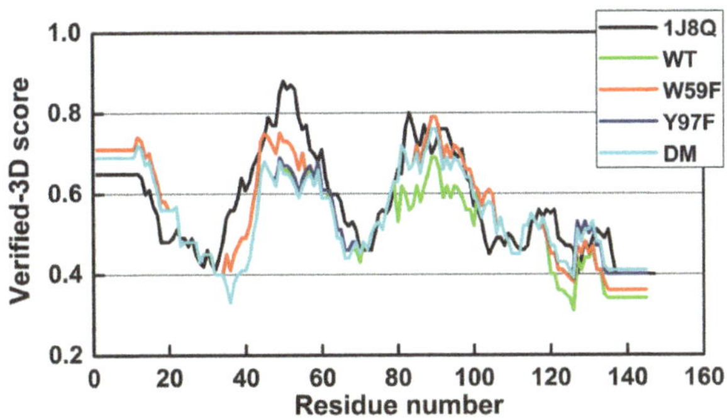

Fig. 2. Verifyed-3D analysis of the template (1J8Q), and the four isoforms of *Desulfobivrio vulgaris*, strain Miyazaki F., that were constructed by the homology modeling. The compatibilities of amino acids in their environments are indicated by the positive scores. Data taken from Lugsanangarm et al. (2011a).

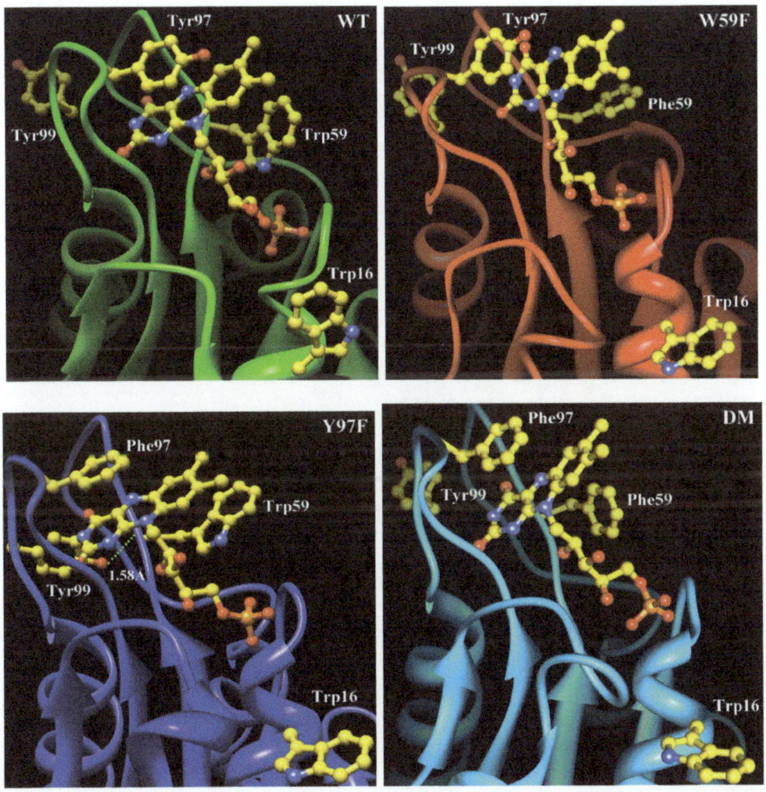

Fig. 3. Structures of four flavodoxin isomers. In the WT isoform, Trp59, Tyr97, Tyr99 and Trp16 are potential PET donors to Iso*, whilst these are Trp59, Tyr99 and Trp16 in the Y97F isomer and Tyr97, Tyr99 and Trp16 in the W59F isomer. In the DM, Tyr99 and Trp16 are the potential PET donors. Data were taken from Lugsanangarm et al. (2011a).

3.3 Decomposition free energy analysis of amino acid residues at the FMN binding site

In order to evaluate the important amino acid residues for FMN binding, the decomposition free energy per amino acid residue has been obtained. Figure 4 shows the decomposition energy of FMN from FMN-apoflavodoxin complexes. The amino acids near FMN are categorized into three groups, the 10-loop, 60-loop and 90-loop regions (see Figure 4). The decomposition energy is highest in the amino acids in the 10-loop regions (Ser9, Thr10, Thr11, Gly12 and Asn13 and Thr14) in all isoforms (Figure 4). All amino acids in the 10-loop region form hydrogen bonds with the FMN side chain viz: Ser9OH with O_{3P}, Thr10NH(peptide) with O_{1P}, Thr11OH with O_{2P}, Thr11NH(peptide) with O_{2P} and O_{1P}, Gly12NH(peptide) with O_{2P}, Asn13NH(peptide) with O_{2P}, Thr14OH with O_{3P} and Thr14NH(peptide) with O_{3P} (see Chart 1 for atom notations). These hydrogen bond interactions are considered to contribute the largest proportion of the decomposition free energy. Among the four flavodoxin isoforms, the decomposition energy is highest in Y97F (-9.30 kcal/mol), followed by W59F (-9.25 kcal/mol), DM (-8.60 kcal/mol) and is lowest in the WT (-8.54 kcal/mol).

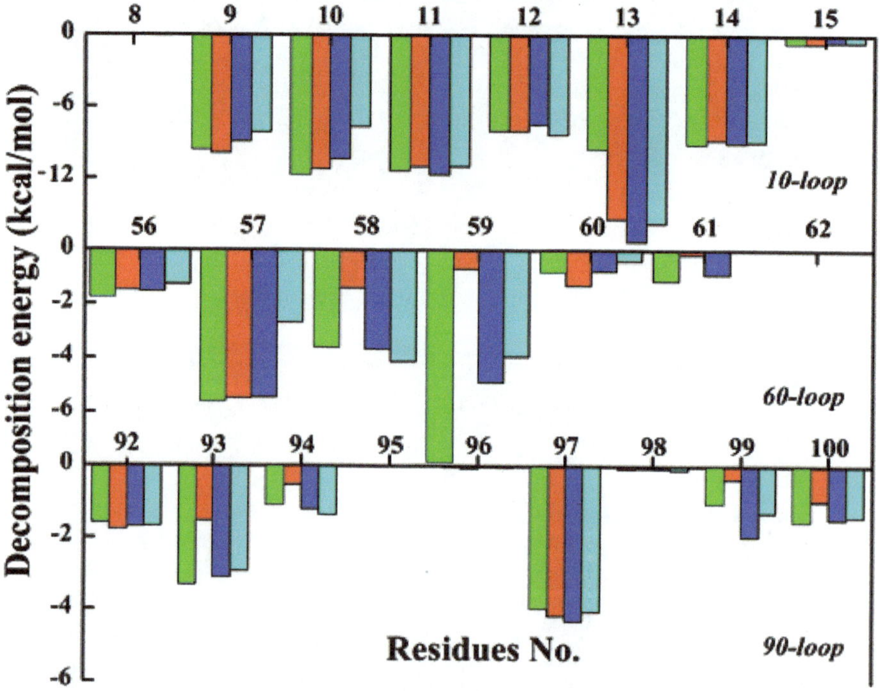

Fig. 4. Decomposition free energy of amino acid residues at the FMN binding site of the four flavodoxin isoforms. The energies are shown with green bars for WT, red bars for W59F, deep blue bars for Y97F and light blue bars for DM. Data were taken from Lugsanangarm et al. (2011a).

3.4 Structural dynamics of flavodoxins

Potential PET donors in the WT flavodoxin are Trp59, Tyr97 and Tyr99 and Trp16. The protein dynamics of these flavodoxin isoforms have been examined by viewing the time-dependent changes in the Rc distances and the inter-planar angles between Iso and these donors. Figure 5 shows the time-evolutions of Rc in the four different flavodoxin isoforms, where the Rc distances clearly fluctuate rapidly but are mostly within ± 10% of the mean values. In the DM the Rc values of Tyr99 and Trp16 vary with long periods in addition to the rapid fluctuation. Since the bulky Tyr97 and Trp59 residues are both replaced by the smaller Phe residue in the DM then the space around Iso may be increased compared to that in the WT, and so may account for the marked fluctuation in the Rc distances of Tyr99 and Trp16. Figure 6 shows the time-evolutions of the inter-planar angles in the WT flavodoxin, where the variation of the inter-planar angles is about ± 30 deg around the mean. The derived mean Rc and edge-to-edge (Re) distances and inter-planar angles over the MD time range are listed in Table 2. The Rc distance was shortest in Tyr97 and then Trp59 in all four flavodoxin isoforms, whilst Tyr99 and Trp16 are quite far from Iso. The inter-planar angle of Trp 59 in the WT is -43 deg, while it is 73 deg in Y97F.

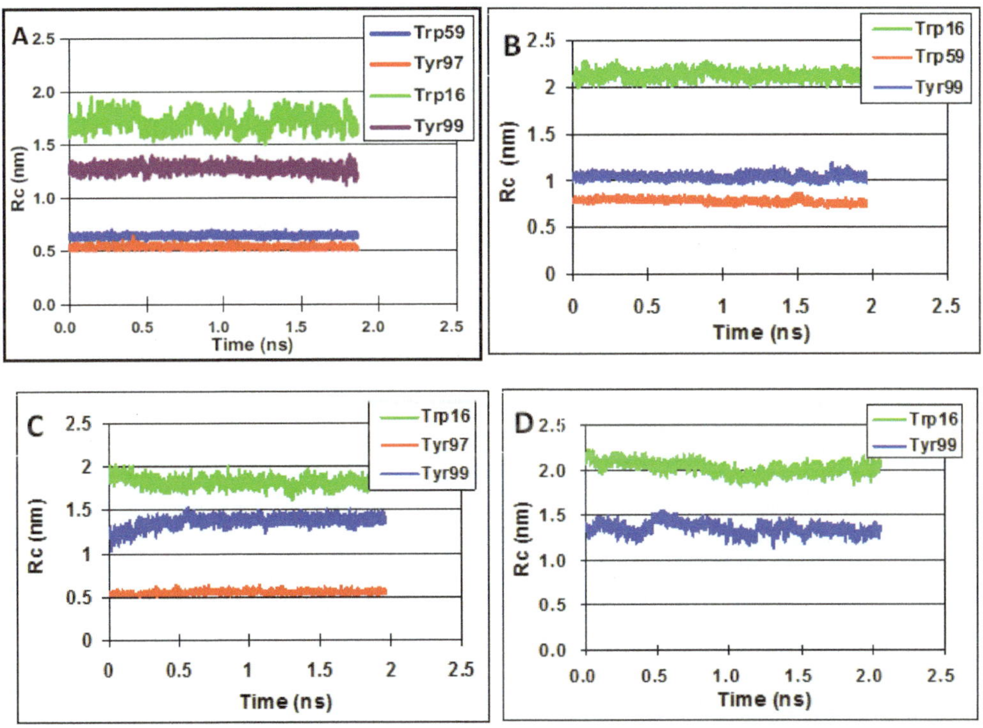

Fig. 5. Time evolution of the Rc distance between Iso and the indicated potential PET donor in the (A) WT, (B) Y97F, (C) W59F and (D) DM (Y97F/W59F) flavodoxin isoforms. Figure 5A was taken from Lugsanangarm et al. 2011b. Figures 5B, 5C and 5D were taken from Lugsanangarm et al. 2011c.

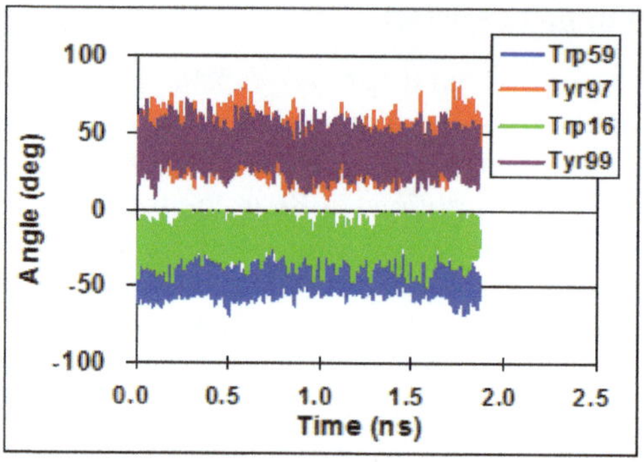

Fig. 6. Time evolution of the inter-planar angle between Iso and the four potential PET donors in the WT flavodoxin. Data were taken from Lugsanangarm et al. (2011b).

Protein	Donor	*Rc*[b] (nm)	*Re*[c] (nm)	Angle[d] (deg)
WT[e]	Trp59	0.642	0.247	-42.8
	Tyr97	0.536	0.301	14
	Tyr99	1.28	0.533	23.5
	Trp16	1.72	1.18	-18.1
Y97F[f]	Trp59	0.858	0.264	73.1
	Tyr99	1.12	0.329	55.7
	Trp16	2.1	1.51	53
W59F[g]	Tyr97	0.577	0.259	14
	Tyr99	1.34	0.513	28.9
	Trp16	1.85	1.42	-24.5
DM[h]	Tyr99	1.35	0.496	-28.6
	Trp16	2.02	1.44	30.9

[a] The means values are listed, which were obtained by taking the average over the MD simulation time (2 ns with 0.1 ps time intervals). Data are taken from Lugsanangarm et al. (2011c).
[b] Center to center and [c] edge to edge distances between Iso and the aromatic amino acids.
[d] Inter-planar angles between Iso and the aromatic amino acids.
[e] The data are taken from Lugsanangarm et al. (2011b).
[f] Tyr97 is replaced by Phe.
[g] Trp59 is replaced by Phe.
[h] Both Tyr97 and Trp59 are replaced by Phe

Table 2. Geometrical factors in the four flavodoxin isoforms[a]. Data were taken from Lugsanangarm et al. (2011c).

3.5 The PET mechanism in flavodoxins

3.5.1 Analysis of PET with crystal structures of flavoproteins

The PET analysis in flavoproteins first starts with their crystal structures (Tanaka et al., 2007, 2008). The logarithms of the averaged PET rate (inverse of the averaged lifetimes) in ten flavoprotein systems are plotted against the Re and Rc distances. The logarithms of the PET rates can be expressed with two straight lines when Rc instead of Re is used (see Figure 7). At longer distances the PET rate rapidly decreases with increasing Rc distances, while at shorter distances it decreases slowly with the same sized increments in the Rc value. When Re is used in place of Rc as the distance measure, no such clear distance-dependence is observed in the flavoprotein systems. According to Moser et al. (1992), the logarithm of the PET rate in photosynthesis systems linearly decreases with increasing Re. However, the time domain of the PET rates in their work is much longer than the one in the flavoprotein systems. It is conceivable then that the logarithm of the PET rate in photosynthesis systems increases more slowly with Rc when the distances become shorter.

The PET in the fast phase with low slope was interpreted to be "Coherent PET", where the PET takes place to the Franck-Condon state of Iso* from Trp or Tyr (Mataga et al., 2002).

Of the ten flavoproteins evaluated, the PET donors with a Rc distance of less than 1 nm were all Trp residues, except for Tyr97 in flavodoxin with an exceptionally low PET rate at an Rc value of 0.57 nm. The low rate in Tyr97 was elucidated by the higher ionization potential of Tyr compared to Trp (Tanaka et al., 2007, 2008). Moreover, the agreement between $\ln k_{ET}^{obs}$ and $\ln k_{ET}^{calc}$ were the highest with KM theory (Figure 7) compared to that MH theory (Sumi & Marcus, 1986) or BJ theories (not shown).

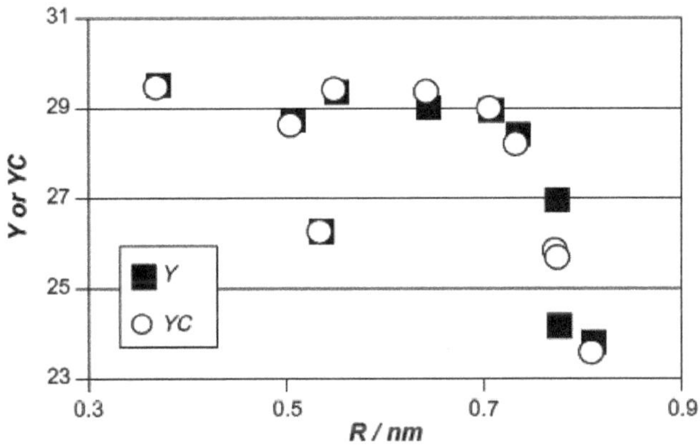

Fig. 7. $\ln k_{ET}$ vs. Rc plot for the observed and KM theory calculated PET rates of 10 flavoprotein systems. Y and YC represent $\ln k_{ET}^{obs}$ and $\ln k_{ET}^{calc}$, respectively, where k_{ET}^{obs} and k_{ET}^{calc} are the observed and KM theory calculated PET rates, respectively. Data are taken from Tanaka et al. (2008).

3.5.2 PET analysis with MD snapshots of four flavodoxin isoforms

The PET analysis from ultrafast fluorescence dynamics was first conducted by Nunthaboot et al. (2008a, 2009a). Time-dependent PET rates in FMN binding proteins were evaluated from the atomic coordinates of the protein as obtained by MD simulation. All PET theories contain several PET parameters that cannot be experimentally determined. Rather these parameters are numerically determined by a non-linear least-square method, as described in Section 2.4.

Fluorescence decay functions of four flavodoxin isoforms (WT, Y97F, W59F and DM) were simultaneously analyzed, with the atomic coordinates of these proteins obtained by MD simulation and KM theory, by Lugsanangarm et al. (2011b, 2011c). The PET parameters common among these flavodoxin systems are listed in Table 3. Ultrafast decay functions of the flavodoxins are expressed by Eq. (9) using the decay parameters listed in Table 1. It is noted that the values of ν_0 and β are quite different between Trp and Tyr, which is related to the electron coupling terms in the KM theory. The quantum basis for the difference is described by Nunthaboot et al. (2008b). In these works it is assumed that the static dielectric constant varies with the protein systems. Table 4 lists the static dielectric constants inside each protein. The dielectric constant of the WT flavodoxin is greatest among the four systems, and that of the DM is the lowest. This is reasonable because Iso in the WT flavodoxin is sandwiched between the polar Trp59 and Tyr97 residues, while both of them are replaced by the non-polar Phe in the DM. Thus, in the DM isoform Iso should be in a relatively non-polar environment, whilst in the Y97F and W59F isoforms the Iso residue may be in a moderately polar environment.

System	ν_0 (ps^{-1})		β (nm^{-1})		R_0 (nm)		G^0_{Iso} (eV)	ε_{DA}
	Trp	Tyr	Trp	Tyr	Trp	Tyr		
Flavodoxin[b]	3090	2460	55.6	9.64	0.772	0.676	7.60	-
FMN binding protein[c]	1016	197	21.0	6.25	0.663	0.499	6.71	2.19

[a]Physical meanings of the PET parameters are described at Section 3.1. The PET parameters in the Table are common among the four isoforms of flavodoxin (WT, W59F, Y97F and DM), and were obtained according to the procedure described at Section 3.4.
[b]For flavodoxins, the four isoforms (WT, Y97F, W59F, DM) were simultaneously analyzed. Data are taken from Lugsanangarm et al. (2011b, 2011c).
[c]For the FMN binding proteins, the five isoforms (WT, E13T, E13Q, W32Y and W32A) were simultaneously analyzed.

Table 3. The best-fit PET parameters[a] . Data are taken from Nunthaboot et al. (2008a, 2009a, 2011).

	Flavodoxin [b]				FMN binding protein [c]				
Variant	WT	Y97F	W59F	DM	WT	E13T	E13Q	W32Y	W32A
ε_0^j	5.85	4.78	4.04	2.28	14.8	5.99	6.69	5.89	6.29

[a] Dielectric constants, ε_0^j, are determined according to the procedure described at Section **3.4**
[b] The WT, Y97F, W59F and DM (Y97F/W59F) flavodoxin isoforms were simultaneously analyzed. Data are taken from Lugsanangarm, et al. (2011b, 2011c)
[c] The WT, E13T, E13Q, W32Y, W32A FMN binding protein isoforms were simultaneously analyzed.

Table 4. Dielectric constant inside the protein [a] Data taken from Lugsanangarm et al. (2011c) for Flavodoxin, and Nunthaboot et al. (2011).

3.5.3 Dynamics of the PET Rate and related physical quantities in flavodoxins

Time-dependent changes in the PET rates of the four flavodoxin isoforms are shown in Figure 8. n the WT and Y97F isoforms, the PET rates from Trp59 are the fastest even though the Rc distance between Iso and Tyr97 in the WT is shorter (see Table 2). The mean PET rates over the MD time range (2 ns with 0.1 ps intervals) are listed in Table 5 along with the other mean physical quantities. The mean PET rate is fastest from Trp59 in WT and then in Y97F, as mentioned above, and is then followed by Tyr97 in the W59F isoform. The PET rates from Trp16 and Tyr99 are always negligibly slow.

The net ES energy, $ES_j(k)$, markedly varied from -0.00159 eV in Trp59 (Y97F) to 3.42 eV in Tyr99 (W59F), while λ_S^{jk} varied from 0.377 eV in Trp16 (DM) to 2.06 eV in Tyr99 (WT), and the ES energy between the donor and acceptor, $-e^2 / \varepsilon_0^j R_{jk}$, varied from -0.652 eV in Tyr97 (W59F) to -0.142 eV in Trp16 (Y97F). The amount of the variation is largest in the net ES energies. The dielectric constant between the Iso anion and the donor cation, ε_{DA}, is not introduced in the PET analysis for flavodoxins.

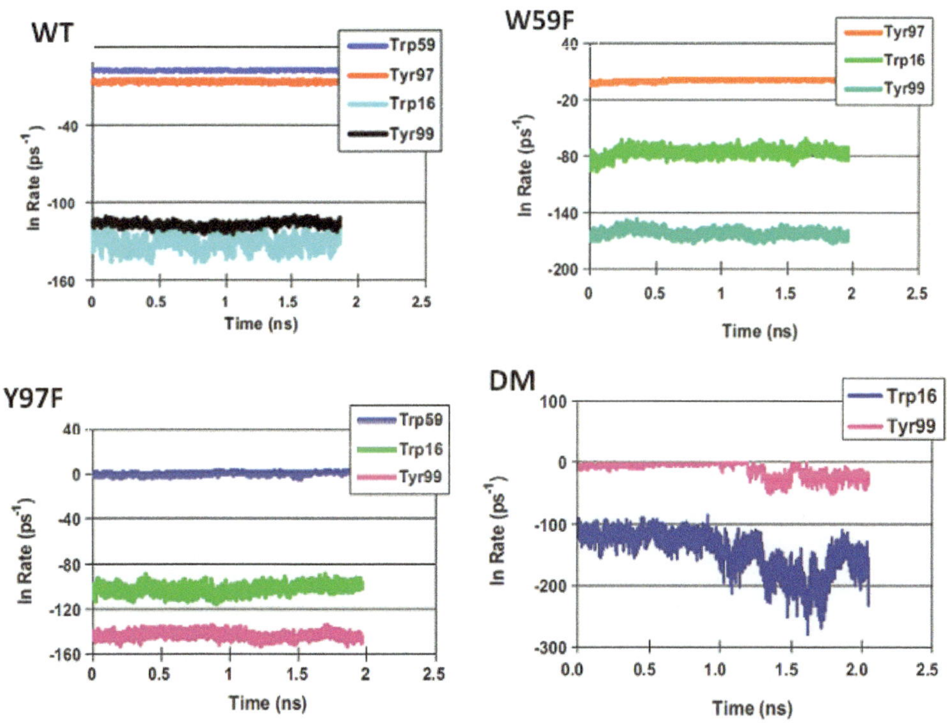

Fig. 8. The PET rates from Trp and/or Tyr to Iso* in four flavodoxin isoforms. Figure for WT was taken from Lugsanangarm (2011b). Figures for W59F, Y97F and DM were taken from Lugsanangarm (2011c).

Physical quantity	WT				W59F		
	Trp59	Tyr97	Trp16	Tyr99	Tyr97	Trp16	Tyr99
k_{KM}^{jk} (ps^{-1})[b]	7.10	1.26×10^{-3}	-	-	3.13	-	-
$\ln k_{KM}^{jk}$	1.96	−6.68	−125	−115	1.14	−69.10	−156
λ_S^{jk} (eV)[c]	1.53	1.54	1.99	2.06	1.20	1.54	1.59
$ES_j(k)$ (eV)[d]	−0.0172	−0.0942	2.73	2.71	−0.219	0.984	3.42
ΔG_q^0 (eV)[e]	−0.467	0.333	−0.467	0.333	0.333	−0.467	0.333
$-e^2 / \varepsilon_0^j R_{jk}$ (eV)[f]	−0.384	−0.460	−0.144	−0.193	−0.652	−0.196	−0.263
$-\Delta G_T^0(jk)$ (eV)[g]	0.868	0.221	−2.12	−2.86	0.583	−0.321	−3.49

Table 5A. Mean physical quantities related to the PET in the WT and W59F[a] flavodoxin isoforms[a]. Data were taken from Lugsanangarm et al. (2011c).

Physical quantity	Y97F			DM	
	Trp59	Trp16	Tyr99	Trp16	Tyr99
k_{KM}^{jk} (ps^{-1})[b]	4.95	-	-	-	7.43×10^{-2}
$\ln k_{KM}^{jk}$	1.60	−96.5	−134	−95.0	−2.60
λ_S^{jk} (eV)[c]	1.47	1.81	1.74	0.377	0.385
$ES_j(k)$ (eV)[d]	−0.00159	1.20	3.26	2.13	0.0131
ΔG_q^0 (eV)[e]	−0.467	−0.467	0.333	−0.467	0.333
$-e^2 / \varepsilon_0^j R_{jk}$ (eV)[f]	−0.386	−0.142	−0.289	−0.313	−0.470
$-\Delta G_T^0(jk)$ (eV)[g]	0.855	−0.594	−3.31	−1.35	0.124

[a]Physical quantities were obtained with the PET parameters listed in Table 3. Mean values are from over the MD time range (2 ns with 0.1 ps intervals).
[b]KM evaluated PET rates are given by Eq. (4).
[c]Solvent reorganization energy, as given by Eq. (2).
[d]Net ES energy, as given by Eq. (8).
[e]Standard free energy gap, as given by Eq. (3).
[f]ES energy between the Iso anion and the donor cation.
[g]Total free energy gap, as given by Eq. (12).

Table 5B. Mean physical quantities related to the PET in the Y97F and DM flavodoxin isoforms[a]. Lugsanangarm et al. (2011c)

4. Protein dynamics of FMN binding proteins

The FMN binding protein from *Desulfovibrio vulgaris* (Miyazaki F) is considered to play an important role in the electron transport process in the bacterium, but the whole picture of the electron flow and coupling of the redox proteins is not yet clear (Kitamura et al., 1998). Three-dimensional structures of the FMN binding protein from *D. Vulgaris* (Miyazaki F) were determined by X-ray crystallography (Suto et al., 2000) and NMR spectroscopy (Liepinsh et al., 1997). According to these structures, Trp32 is the closest residue to Iso followed by Tyr35 and then Trp106. To examine the effect of Trp32 on the PET rate in the FMN binding protein, Trp32 was replaced by Tyr (W32Y) or Ala (W32A), and further the single negative charge at residue 13, glutamate 13 (E13) was replaced by either Thr (E13T) or Gln (E13Q). The crystal structures of E13T and E13Q were determined by X-ray crystallography (Chosrowjan et al. 2010). The dynamic behavior of these FMN binding protein isoforms were studied by MD simulation (Nunthaboot et al., 2008a, 2009a, 2011), and Figure 9 shows snapshots of the WT, E13T, E13Q, W32Y and W32A FMN binding protein isoforms.

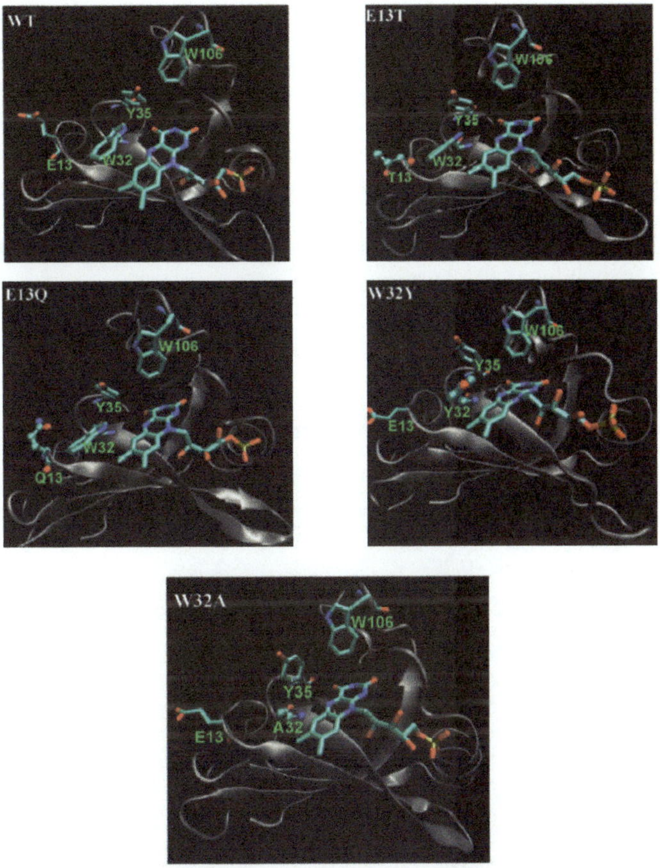

Fig. 9. Protein structures near Iso in the five FMN binding protein isoforms. Trp32, Tyr35 and Trp106 are potential PET donors in the FMN binding protein. Trp32 is replaced by Tyr in W32Y and Ala in W32A. Amino acids at residue position 13 are also shown in the Figures. These structures were obtained by MD simulation.Data were taken from Nunthaboot et al. (2011). (Reproduced by permission of the PCCP Owner Societies).

Mean the donor-acceptor distances over MD time range are summarized in Table 6A, 6B. The WT displays great variations in the Rc distances with long periods, in addition to the instantaneous fluctuations. The mean values of the geometrical factors over the entire MD time range (2 ns with 0.1 ps time intervals) are listed in Table 6. The Rc distance is shortest in Trp32 among the three different aromatic amino acid residue positions (Trp/tyr32, Tyr 35 and Trp106) with mean distances of Trp32 of 0.70, 0.72 and 0.75 nm in the WT, E13T and E13Q isoforms, respectively. The distance between Iso and Tyr32 in W32Y is shorter than that between Iso and Trp32 in the WT. The inter-planar angle between Iso and Trp32 varies from -52 deg in the WT to -38 deg in the E13Q isoform, while that between Iso and Tyr35 varies from 43 deg in W32A to 93 deg in the WT.

4.1 Amino acid at position 13 of the FMN binding proteins

The WT FMN binding protein contains Glu13, with a negative charge at neutral pH, whilst in the E13T and E13Q substitution isoforms the amino acids at this position are Thr13 and Gln13 with neutral charges. The distances between the PET donors or acceptor and amino acid residue 13 of the five FMN binding protein isomers are listed in Table 7. The distances between Iso and side chain of amino acid 13 do not significantly vary between the five FMN binding protein isoforms (range 1.5 - 1.6 nm), nor does that between Trp32 (0.9 - 1.0 nm), Tyr35 (1.0 - 1.2 nm) and Trp106 (1.7 - 1.97 nm) excepting that of Trp106 in the W32Y isoform that was further away (2.13 nm).

Protein system	Rc (nm)[b] Trp32	Rc (nm)[b] Tyr32	Rc (nm)[b] Tyr35	Rc (nm)[b] Trp106	Re (nm)[b] Trp32	Re (nm)[b] Tyr32	Re (nm)[b] Tyr35	Re (nm)[b] Trp106
WT	0.703	--	1.016	1.052	0.261	--	0.425	0.314
(RSD[c])	-0.072	--	-0.097	-0.088	-0.086	--	-0.292	-0.29
E13T	0.724	--	0.872	0.913	0.269	--	0.331	0.269
(RSD[c])	-0.048	--	-0.069	-0.038	-0.079	--	-0.181	-0.111
E13Q	0.748	--	0.854	0.939	0.265	--	0.287	0.294
(RSD[c])	-0.044	--	-0.053	-0.043	-0.095	--	-0.123	-0.131
W32Y	--	0.654	0.826	0.907	--	0.276	0.284	0.251
(RSD[c])	--	-0.05	-0.075	-0.036	--	-0.091	-0.167	-0.11
W32A	--	--	0.769	0.895	--	--	0.29	0.277
(RSD[c])	--	--	-0.082	-0.05	--	--	-0.226	-0.139

[a]Mean values of factors between Iso and the nearby indicated aromatic amino acids are listed. The mean values were obtained by taking an average over the entire MD time range.
[b]Center-to-center distance (Rc) and edge-to-edge (Re) distance.
[c]Relative standard deviation (RSD), obtained from SD/mean value.

Table 6A. Geometrical factor of Iso and the indicated nearby aromatic amino acids of the FMN binding protein isomers[a]. Data were taken from Nunthaboot et al. (2011). (Reproduced by permission of the PCCP Owner Societies).

Protein system	Inter-planar angle (deg) Trp32	Tyr32	Tyr35	Trp106
WT	-52.2	--	93.3	67.4
(RSD[c])	(-0.3)	--	(-0.3)	(-0.1)
E13T	-42.5	--	59.3	85.7
(RSD[c])	(-0.2)	--	(-0.2)	(-0.1)
E13Q	-37.8	--	116.4	79.2
(RSD[c])	(-0.2)	--	(-0.1)	(-0.1)
W32Y	--	28.7	76.7	77.5
(RSD[c])	--	(-0.6)	(-0.1)	(-0.1)
W32A	--	--	42.8	70.9
(RSD[c])	--	--	(-0.6)	(-0.1)

Table 6B. Inter-planar angle factor between Iso and the indicated nearby aromatic amino acids of the FMN binding protein isomers[a] Data were taken from Nunthaboot et al. (2011). (Reproduced by permission of the PCCP Owner Societies).

System	Iso	Trp32	Tyr32	Tyr35	Trp106
WT[b]	1.53 ± 0.10	0.98 ± 0.09	--	0.99 ± 0.16	1.72 ± 0.15
E13T[c]	1.49 ± 0.06	0.92 ± 0.07	--	1.22 ± 0.08	1.97 ± 0.09
E13Q[c]	1.58 ± 0.13	0.98 ± 0.12	--	1.12 ± 0.16	1.84 ± 0.17
W32Y[b]	1.64 ± 0.07	--	1.13 ± 0.08	1.46 ± 0.09	2.13 ± 0.09
W32A[b]	1.60 ± 0.11	--	--	1.24 ± 0.15	1.76 ± 0.16

[a]Mean distances (± 1 standard deviation), averaged over the MDS time range, are shown in units of nm.
[b]Distances were obtained taking the average over all distances between atoms in the aromatic ring and the center of the two oxygen atoms in the side chain of Glu13.
[c]Obtained by taking the average over all distances between the atoms in the aromatic ring and the oxygen atom of the Thr13 (E13T) or Gln13 (E13Q) side chain.

Table 7. Geometry of the amino acid residue at position 13 in the five FMN binding protein isoforms[a]. Data were taken from Nunthaboot et al. (2011). (Reproduced by permission of the PCCP Owner Societies).

4.2 The PET rates and related physical quantities in FMN binding proteins

The common parameters among the five FMN binding protein isoforms are listed in Table 3, where ν_0 = 1016 (ps^{-1}) for Trp and 197 (ps^{-1}) for Tyr, β = 21.0 (nm^{-1}) for Trp and 6.25 (nm^{-1}) for Tyr, R_0 = 0.663 (nm) for Trp and 0.499 (nm) for Tyr. G_{Iso}^0 = 6.71 (eV) and ε_{DA} = 2.19.

These values are quite different from those of the flavodoxins. The time-evolutions of the PET rates in the five different FMN binding protein isoforms over the MD time course are shown in Figure 10. Fluctuations of the PET rate are always marked in Tyr35, but not so much in Trp32. In the WT isoform the PET rates vary with rather long periods in addition to the instantaneous fluctuations, which is in accord with the time-evolution of Rc distances in the WT. The mean PET rate and physical quantities related to the PET rates are listed in Table 8, where the PET rate is observed to always be fastest from Trp32, and then from Trp106 whilst that from Tyr35 is always slow (see also Figure 10). Among the WT, E13T and

Quantity	Donor	WT	E13T	E13Q	W32Y	W32A
k_{KM}^{jk} [b] (ps^{-1})	Trp32	7.10 ± 3.08	17.22 ± 14.76	10.81 ± 10.43	--	--
	Tyr32	--	--	--	1.6 ± 30 x10^{-7}	
	Tyr35	4 ± 95 x 10^{-14}	6.4 ± 400 x10^{-21}	7 ± 200 x10^{-17}	3.7 ± 200 x10^{-14}	5 ± 130 x 10^{-13}
	Trp106	0.082 ± 0.110	0.003 ± 0.003	0.018 ± 0.011	0.192 ± 0.350	0.176 ± 0.599
λ_S^{jk} [c] (eV)	Trp32	0.202 ± 0.005	0.206 ± 0.004	0.208 ± 0.004	--	--
	Tyr32	--	--	--	0.217 ± 0.005	--
	Tyr35	0.249 ± 0.006	0.240 ± 0.005	0.229 ± 0.004	0.236 ± 0.006	0.231 ± 0.006
	Trp106	0.232 ± 0.005	0.223 ± 0.003	0.225 ± 0.003	0.223 ± 0.002	0.222 ± 0.003
$E_j(k)$ [d] (eV)	Iso	0.071 ± 0.013	-0.023 ± 0.024	0.021 ± 0.028	0.074 ± 0.030	0.079 ± 0.026
	Trp32	0.005 ± 0.017	0.335 ± 0.043	0.269 ± 0.032	--	--
	Tyr32	--	--	--	0.123 ± 0.033	
	Tyr35	0.080 ± 0.025	0.472 ± 0.050	0.391 ± 0.041	0.256 ± 0.052	0.242 ± 0.052
	Trp106	-0.140 ± 0.007	-0.326 ± 0.011	-0.297 ± 0.010	0.141 ± 0.039	0.230 ± 0.054
$ES_j(k)$ [e] (eV)	Trp32	0.076 ± 0.010	0.312 ± 0.027	0.290 ± 0.021	--	--
	Tyr32	--	--	--	0.197 ± 0.020	--
	Tyr35	0.150 ± 0.022	0.449 ± 0.041	0.412 ± 0.035	0.330 ± 0.046	0.321 ± 0.043
	Trp106	-0.069 ± 0.017	-0.349 ± 0.025	-0.276 ± 0.029	0.215 ± 0.034	0.309 ± 0.042
$-e^2/\varepsilon_0^{DA}R_{jk}$ [f] (eV)	Trp32	-0.949 ± 0.055	-0.912 ± 0.045	-0.883 ± 0.045	--	--
	Tyr32	--	--	--	-1.009 ± 0.049	--
	Tyr35	-0.660 ± 0.070	-0.759 ± 0.051	-0.883 ± 0.039	-0.803 ± 0.039	-0.863 ± 0.066

Quantity	Donor	WT	E13T	E13Q	W32Y	W32A
	Trp106	-0.627 ± 0.056	-0.722 ± 0.027	-0.703 ± 0.030	-0.728 ± 0.027	-0.728 ± 0.037
$-\Delta G_T^0$ [g] (eV)	Trp32	0.371	0.098	0.090	--	--
	Tyr32	--	--	--	-0.491	--
	Tyr35	-0.792	-0.992	-0.832	-0.829	-0.760
	Trp106	0.194	0.569	0.477	0.011	-0.073

[a] Mean (± SD) values, taken over the MD time range (2 ns with 0.1 ps intervals), are listed. The PET rate is obtained by KM theory.
[b] The PET rate is given by Eq. (4).
[c] Solvent reorganization energy is given by Eq. (2).
[d] ES energy of the Iso anion or the donor cation and other ionic groups, as given by Eq. (7).
[e] Net ES energy, as given by Eq. (8).
[f] ES energy between the Iso anion and a donor cation.
[g] Total standard free energy, as given by Eq. (12).

Table 8. Mean PET rate and its related physical quantities in five FMN binding protein isoforms [a]. Data were taken from Nunthaboot et al. (2011). (Reproduced by permission of the PCCP Owner Societies).

E13Q isoforms, the PET rate from Trp32 was fastest in E13T. The values of λ_S^{jk} do not vary significantly with the donor and protein system (range 0.202 – 0.231 eV). Likewise the ES energies between the Iso anion and the donor cations, $-e^2 / \varepsilon_{DA} R_{jk}$, did not vary much among the donors (range -0.949 eV to -0.627 eV) (Table 8 and Figure 11). In contrast, the net ES energies, $ES_j(k)$, varied from -0.069 eV in Trp106 (WT) to 0.449 eV in Tyr35 (E13T). This remarkable variation in $ES_j(k)$ compared to the other physical quantities is also seen in the flavodoxin isoforms.

4.3 Effect of changing the negative charge of amino acid residue 13 on the PET rate

The PET rate of Trp32 was fastest in all five FMN binding protein isoforms. The ES energies between the Iso anion and ionic groups in the proteins, $E_j(k)$, fell from 0.071 eV in the WT (and similar values in W32Y and W32A) to -0.023 eV and 0.021 eV in E13T and E13Q, respectively (Table 8; Figure 11), suggesting a potential affect of the charge neutralization at residue position 13. In addition, the ES energies between the Trp32 cation and the ionic groups in the proteins increased dramatically from 0.005 eV in the WT, to 0.335 eV and 0.269 eV in the E13T and E13Q isoforms, respectively. In the WT the ES energy between the negative charge of Glu and Trp32 cation should be negative, which contributes to reduce the value of $E_j(k)$. In the neutral charged (at residue 13) E13T and E13Q isoforms the stabilizing energy found in the WT disappears, again supporting the potential importance of the negative charge at residue 13. It is noted that the absolute values of the net ES energies are quite low in the WT, while they are much higher in the other isoforms. Net ES energies of Trp32, from which the PET rate is fastest, are always positive, while those for Trp106 are negative in the WT, E13T and E13Q isoforms.

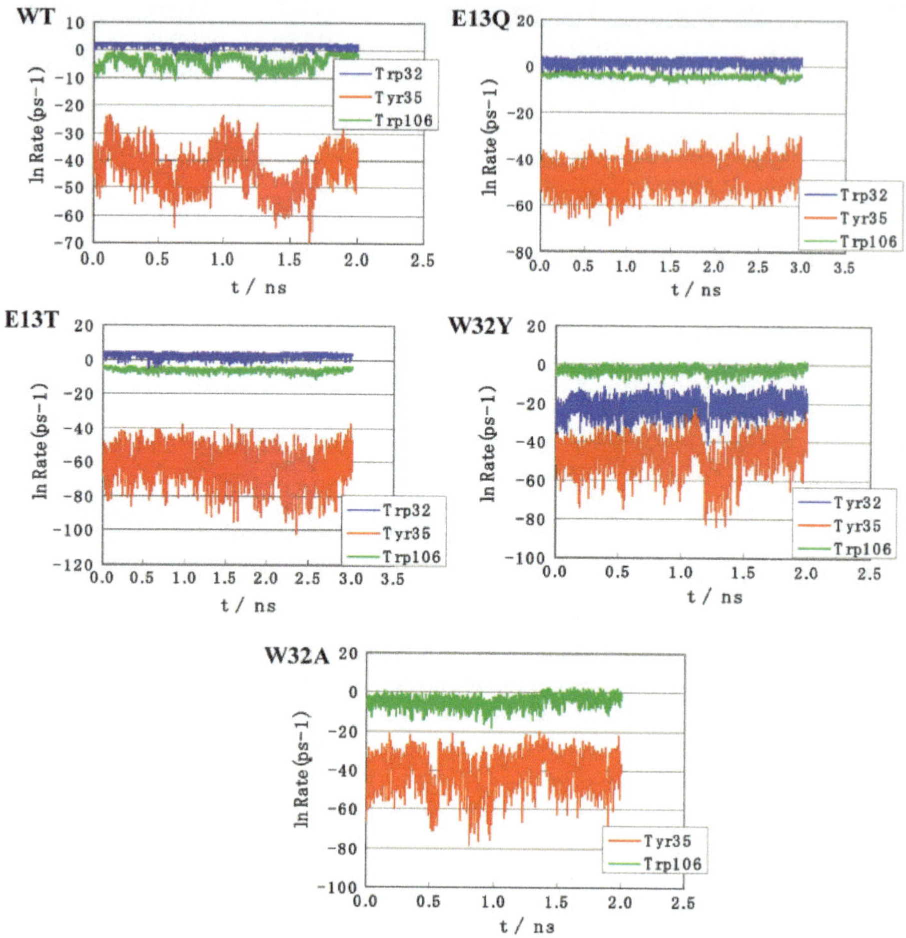

Fig. 10. Time-evolution of the PET rate in the five FMN binding protein isoforms. Data were taken from Nunthaboot et al. (2011). (Reproduced by permission of the PCCP Owner Societies).

5. Energy gap law in flavodoxin and FMN binding protein systems

The total free energy gap of the kth donor in the jth flavoprotein is expressed by Eq. (12);

$$-\Delta G_T^0(jk) \propto -ES_j(k) + e^2 / \varepsilon_{DA} R_{jk} - \Delta G_q^0 \tag{12}$$

When λ_S^{jk} varies with $-\Delta G_T^0(jk)$, the normal energy gap law is modified, as in Eq. (13);

$$\ln k_{KM}^{jk} / \lambda_S^{jk} \propto -\left[\left\{1 + \Delta G_T^0(jk) / \lambda_S^{jk}\right\}^2\right] \tag{13}$$

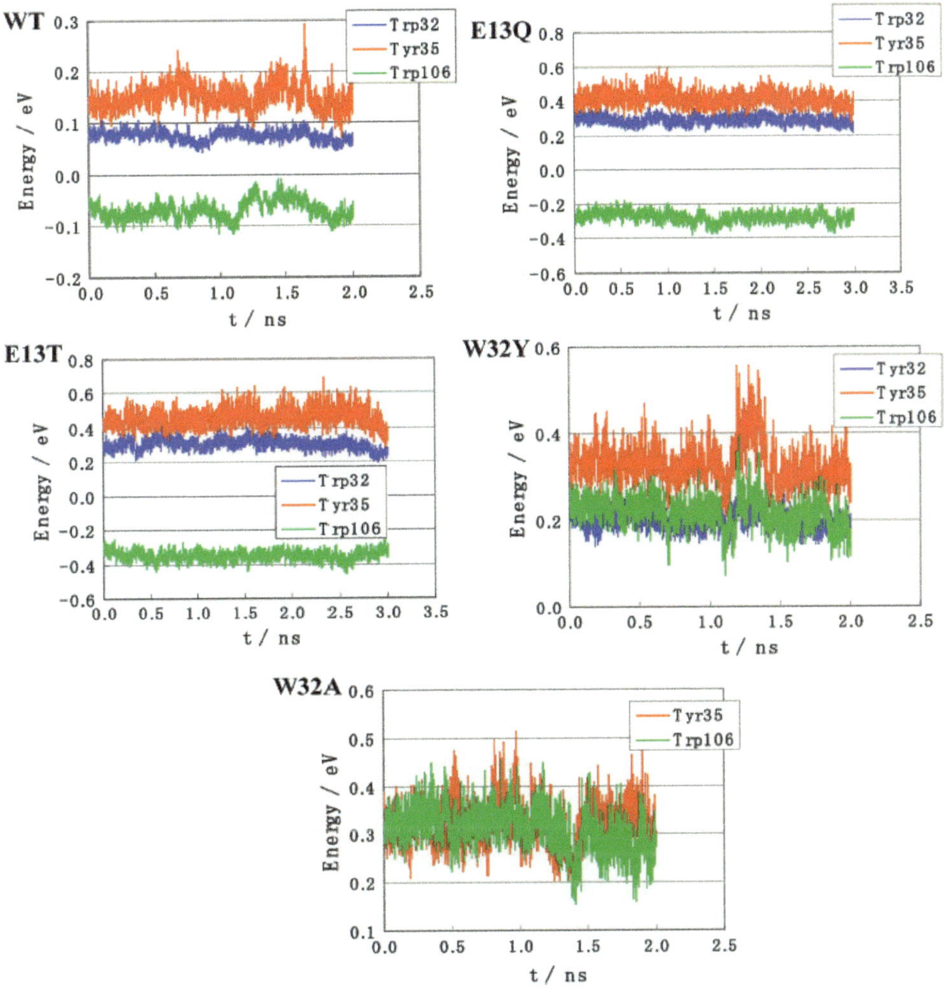

Fig. 11. Net ES energy in the five FMN binding protein isoforms. Data were taken from Nunthaboot et al. (2011). (Reproduced by permission of the PCCP Owner Societies).

Here $ES_j(k)$ is given by Eq. (8), and ΔG_q^0 by Eq. (3). The values of $-\Delta G_T^0(jk)$ are listed in the bottom lines of Table 5 for flavodoxins and Table 8 for FMN binding proteins. Figure 12 shows the modified energy gap law in flavodoxins and FMN binding proteins, as expressed by Eq. (13). The inserts in Figure 12 represent the approximate parabola functions. In the both systems, the PET takes place in the normal region.

6. Concluding remarks on the PET mechanism in flavoproteins

Quantitative analyses of the PET in proteins have been difficult, because all of the current PET theories contain several unknown parameters which cannot be determined experimentally. In the earlier works the PET rate was qualitatively analyzed from the following two aspects.

1. The donor-acceptor distance-dependence of the PET rate (Dutton law).

Hopfield (1974) described biological electron transfer rate in the ground state of a donor in terms of the electron tunneling model. In this model, the rate drops off exponentially with increasing donor-acceptor distance. Hopfield estimated the slope of the logarithm of the rate against the distance to be 14 nm^{-1} for biological electron transfer reactions. Indeed, Moser et al. (1992) have experimentally demonstrated that logarithms of PET rates linearly decrease with the Re distance between PET donors and acceptors in photosynthetic proteins. In accord, the slope of the logarithm of the PET rate against the free energy gap was also around 14 nm^{-1}. Gray & Winkler (1996) have reviewed the experimental works on PET rates in ground state donors from various aspects.

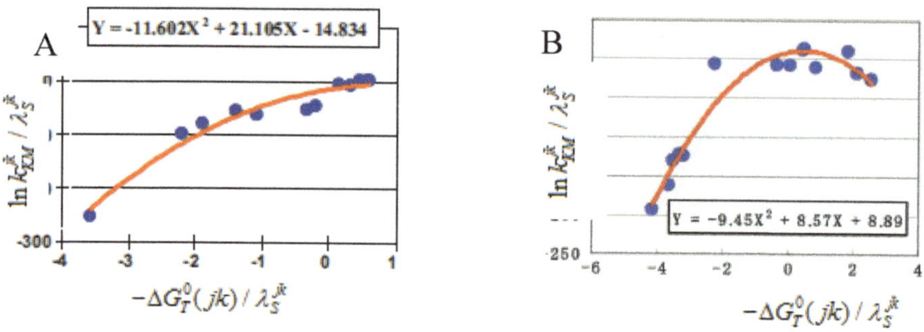

Fig. 12. Modified energy gap law in (A) flavodoxins and (B) FMN binding proteins. Inserts indicate the approximate parabola functions, Y $\ln k_{KM}^{jk} / \lambda_S^{jk}$, and X $-\Delta G_T^0(jk) / \lambda_S^{jk}$. Formally, the value of $\ln k_{KM}^{jk} / \lambda_S^{jk}$ should be maximal when $-\Delta G_T^0(jk) = \lambda_S^{jk}$. Data were taken from Lugsanangarm et al. (2011c) for Figure 12A and Nunthaboot et al. (2011) for Figure 12B. (Reproduced by permission of the PCCP Owner Societies).

2. The free energy gap dependence of PET rate (Energy gap law).

The characteristics of the Marcus theory (1956a, 1956b, 1964) is that the logarithm of the PET rate is a parabolic function of the reorganization energy and the free energy gap (see Eq. (1)), which is common with the other theories (see Eqs. (4) and (5)). As a test for the Marcus theory many researchers have examined the dependence of the logarithmic values of the PET rates on the free energy gap. Rehm & Weller (1969; 1970) first examined the energy gap law with the donor-acceptor systems in organic solvents, but could not find the predicted parabolic dependence. Later Closs et al. (1986) and Mataga et al. (2003) found evidence of the PET processes in the so-called "Inverted region". Interested readers should consult Mataga et al. (2005), who have precisely reviewed the current knowledge of PET in solution.

The energy gap law in proteins was first experimentally demonstrated in the reaction center of the purple bacterium, *Rhodobacter sphaeroides*, by Gunner & Dutton (1989), and in both the plant photosystem I and reaction center of the purple bacterium by Iwaki *et al.* (1996). In these systems, the PET takes place in the normal regions, as in the flavoproteins described above.

We have been trying to quantitatively analyze PET in flavoproteins (Nunthaboot et al., 2008a, 2008b, 2009a, 2009b, 2010, 2011; Lugsanangarm et al., 2011b, 2011c), using the experimental and theoretical approaches of evaluating the ultrafast fluorescence dynamics of Iso in the flavoproteins and using MD simulation based approaches, respectively.The following conclusions have been derived on the mechanisms of PET in the flavoproteins.

1. The donor-acceptor distance-dependent PET rates were analyzed with MH, KM and BJ theories, whereupon the KM theory was found to be the best for describing PET in the flavoproteins.
2. The ultrafast fluorescence decays of flavoproteins are mostly non-exponential. The non-exponential decay of the WT FMN binding protein was first reproduced with MD snapshots and PET theories, taking an average of the single-exponential decay function over the MD time domain (Nunthaboot et al., 2009a). This suggests that the non-exponential behavior in the decays is caused by the fluctuations of the protein structures with short and longer fluctuation periods. Again, KM theory could best reproduce the observed non-exponential decay.
3. The ultrafast experimental decays in several flavoprotein isoforms are satisfactorily reproduced with common PET parameters in the present method (Nunthaboot et al., 2008a, 2010, 2011; Lugsanangarm et al., 2011c).
4. The introduction of ES energy into the PET theories greatly improves the agreement between the observed and (KM theory) calculated decays in the three FMN binding protein isoforms (Nunthaboot et al, 2008a, 2009a).
5. The introduction of the dielectric constant between the donor and acceptor (ε_{DA}) improved the agreement between the observed and (KM theory) calculated decays (Nunthaboot et al., 2011). ε_{DA} is different from the dielectric constant inside the entire protein (ε_0^j), and always much lower than ε_0^j. This is reasonable because normally no amino acid exists between the donor and acceptor.
6. Changes in the single negative charge at residue 13 of the WT FMN binding protein (Glu13) to amino acids with a neutral charge (E13T and E13Q) substantially changed the ultrafast fluorescence decay, which suggests that the ES energy inside the proteins is very important for the PET rate (Chosrowjan et al., 2010; Nunthaboot et al., 2011).

7. Perspective of the quantitative PET analyses

Method of homology modeling has been useful for the determinations of protein structures, which have been experimentally unable (www.proteinstructures.com). The present method for the quantitative analysis of the PET mechanism may be also applicable to photosynthetic systems and flavin photoreceptors, such as AppA (Nunthaboot et al., 2009b; 2010). Most of the flavoproteins function in the electron transport and electron transfer from a substrate to Iso without light. A number of researchers have been working on the mechanisms of the dark electron transfer in proteins (Grey & Winkler, 1996; Beratan et al., 1991; 2008). These works, however, have mostly focused on the electron coupling term, and not discussed much on the nuclear term. ES energy which is in the nuclear term, should also play an important role on the dark electron transfer rates, and redox potentials of Iso in flavoproteins. Determination of all physical quantities contained in both electronic and nuclear terms of an electron transfer theory could explore a new aspect of the mechanisms of PET and dark electron transfer phenomena in proteins.

8. Acknowledgments

The Royal Golden Jubilee Ph.D. Program (3.C.CU/50/S.1), from Chulalongkorn University and The Thailand Research Fund (TRF) and The Ratchadaphiseksomphot Endowment Fund from Chulalongkorn University are acknowledged for financial support. N. N. (Grant No. MRG5380255) acknowledges the funding for New Research from the Thailand Research Fund. We thank Computational Chemistry Unit Cell, Chulalongkorn University and the National Electronics and Computer Technology Center (NECTEC) for computing facilities. The Thai Government Stimulus Package 2 (TKK2555) under the Project for Establishment of Comprehensive Center for Innovative Food, Health Products and Agriculture and The Higher Education Research Promotion is acknowledged.

9. References

Bendall, D. S. (1996). *Protein Electron Transfer*. BIOS Scientific Publishers Ltd., ISBN 1859960405, Oxford, UK

Beratan, D. N., Betts, J. N. & Ounchic, J. N. (1991). Protein electron transfer rates are predicted to be set by the bridging secondary and tertiary structure, Science, Vol. 252, pp. 1285-1288, ISBN 9780199753833

Beratan, D. N. & Balabin, I. A. (2008). Heme-copper oxidases use tunneling pathways, *Proc. Nat. Acad. Sci. USA*, Vol. 105, pp. 403-404, ISBN 0027-8424

Bixon, M. & Jortner, J. (1991). Non-Arrhenius temperature dependence of electron-transfer rates. *J. Phys. Chem.*, Vol. 95, No. 5, pp. 1941-1944, ISBN 0022-3654

Bixon, M. & Jortner, J. (1993). Charge Separation and Recombination in Isolated Supermolecules. *J. Phys. Chem.*, Vol. 97, No. 50, pp. 13061-13066, ISBN 0022-3654

Bixon, M.; Jortner, J.; Cortes, J.; Heitele, H. & Michelbeyerle, M. E. (1994). Energy-Gap Law for Nonradiative and Radiative Charge-Transfer in Isolated and in Solvated Supermolecules. *J. Phys. Chem.*, Vol. 98, No. 30, pp. 7289-7299, ISBN 0022-3654

Blankenship, R. E. (2002). *Molecular Mechanisms of Photosynthesis*. Wiley-Blackwell, ISBN 0632043210, Oxford, UK

Burkhart, B. M.; Ramakrishnan, B.; Yan, H.; Reedstrom, R. J.; Markley, J. L.; Straus, N. A. & Sundaralingam, M. (1995). Structure of the Trigonal Form of Recombinant Oxidized Flavodoxin from Anabaena-7120 at 1.40 Angstrom Resolution. *Acta Crystallogr D-Biol Cryst*, Vol. 51, No. 3, pp. 318-330, ISBN 0907-4449

Chosrowjan, H.; Taniguchi, S.; Mataga, N.; Tanaka, F.; Todoroki, D. & Kitamura, M. (2007). Comparison between ultrafast fluorescence dynamics of FMN binding protein from

Desulfovibrio vulgaris, strain miyazaki, in solution vs crystal phases. *J. Phys. Chem. B*, Vol. 111, No. 30, pp. 8695-8697, ISBN 1520-6106

Chosrowjan, H.; Taniguchi, S.; Mataga, N.; Tanaka, F.; Todoroki, D.& Kitamura, M. (2008). Ultrafast fluorescence dynamics of FMN-binding protein from *Desulfovibrio vulgaris* (Miyazaki F) and its site-directed mutated proteins. *Chem. Phys. Lett.*, Vol. 462, No. 1-3, pp. 121-124, ISBN 0009-2614

Chosrowjan, H.; Taniguchi, S.; Mataga, N.; Nakanishi, T.; Haruyama, Y.; Sato, S.; Kitamura, M. & Tanaka, F. (2010). Effects of the Disappearance of One Charge on Ultrafast Fluorescence Dynamics of the FMN Binding Protein. *J. Phys. Chem. B*, Vol. 114, No. 18, pp. 6175-6182, ISBN 1520-6106

Closs, G. L.; Calcaterra, L. T.; Green, N. J.; Penfield, K. W. & Miller, J. R. (1986). Distance, stereoelectronic effects, and the Marcus inverted region in intramolecular electron transfer in organic radical anions. *J. Phys. Chem.*, Vol. 90, No. 16, pp. 3673-3683, ISBN 0022-3654

Crosson, S. & Moffat, K. (2001). Structure of a flavin-binding plant photoreceptor domain: insights into light-mediated signal transduction. *Proc. Natl. Acad. Sci. U.S.A.*, Vol. 98, No. 6, pp. 2995-3000, ISBN 0027-8424

Drennan, C. L.; Pattridge, K. A.; Weber, C. H.; Metzger, A. L.; Hoover, D. M. & Ludwig, M. L. (1999). Refined structures of oxidized flavodoxin from Anacystis nidulans. *J. Mol. Biol.*, Vol. 294, No. 3, pp. 711-724, ISBN 0022-2836

Frago, S.; Gomez-Moreno, C. & Medina, M. (2008). *Flavins and Flavoproteins 2008*. Prensas Universitarias de Zaragoza, ISBN 8477330174, Spain

Freigang, J.; Diederichs, K.; Schafer, K. P.; Welte, W. & Paul, R. (2002). Crystal structure of oxidized flavodoxin, an essential protein in *Helicobacter pylori*. *Protein Sci.*, Vol. 11, No. 2, pp. 253-261, ISBN 0961-8368

Fukuyama, K.; Matsubara, H. & Rogers, L. J. (1992). Crystal structure of oxidized flavodoxin from a red alga Chondrus crispus refined at 1.8 A resolution. Description of the flavin mononucleotide binding site. *J. Mol. Biol.*, Vol. 225, No. 3, pp. 775-789, ISBN 0022-2836

Fukuyama, K.; Wakabayashi, S.; Matsubara, H. & Rogers, L. J. (1990). Tertiary structure of oxidized flavodoxin from an eukaryotic red alga Chondrus crispus at 2.35-A resolution. Localization of charged residues and implication for interaction with electron transfer partners. *J. Biol. Chem.*, Vol. 265, No. 26, pp. 15804-15812, ISBN 0021-9258

Giovani, B.; Brydin, M.; Ahmad, M. & Brettel, K. (2003). Light-induced electron transfer in a cryptochrome blue-light photoreceptor. *Nat. Struct. Biol. 10*, pp.489–490, ISBN 0878931686

Gray, H. B. & Winkler, J. R. (1996). Electron transfer in proteins. *Annu. Rev. Biochem.*, Vol. 65, No. 1, pp. 537-561, ISBN 0066-4154

Gunner, M. R. & Dutton, P. L. (1989). Temperature and -.DELTA.G.degree. dependence of the electron transfer from BPh.cntdot.- to QA in reaction center protein from Rhodobacter sphaeroides with different quinones as QA. *J. Am. Chem. Soc.*, Vol. 111, No. 9, pp. 3400-3412, ISBN 0002-7863

Hoover, D. M. & Ludwig, M. L. (1997). A flavodoxin that is required for enzyme activation: The structure of oxidized flavodoxin from Escherichia coli at 1.8 angstrom resolution. *Protein Sci.*, Vol. 6, No. 12, pp. 2525-2537, ISBN 0961-8368

Hopfield, J. J. (1974). Electron Transfer Between Biological Molecules by Thermally Activated Tunneling. *Proc. Natl. Acad. Sci. U.S.A.*, Vol. 71, No. 9, pp. 3640-3644, ISBN 0027-8424

Hush, N. S. (1961). Adiabatic theory of outer sphere electron-transfer reactions in solution. *Trans. Faraday Soc.*, Vol. 57, No. 4, pp. 557-580, ISBN 0014-7672

Iwaki, M.; Kumazaki, S.; Yoshihara, K.; Erabi, T. & Itoh, S. (1996). Delta G(0) dependence of the electron transfer rate in the photosynthetic reaction center of plant photosystem I: Natural optimization of reaction between chlorophyll a (A(0)) and quinone. *J. Phys. Chem.*, Vol. 100, No. 25, pp. 10802-10809, ISBN 0022-3654

Jortner, J. & Bixon, M. (1999). *Electron Transfer—from Isolated Molecules to Biomolecules, in Advances in Chemical Physics: Electron Transfer - from Isolated Molecules to Biomolecules. Part 1.* John Wiley & Sons, Inc., ISBN 9780471252924, Hoboken, NJ, U.S.A.

Kakitani, T. & Mataga, N. (1985). New energy gap laws for the charge separation process in the fluorescence quenching reaction and the charge recombination process of ion pairs produced in polar solvents. *J. Phys. Chem.*, Vol. 89, No. 1, pp. 8-10, ISBN 0022-3654

Kakitani, T.; Yoshimori, A. & Mataga, N. (1991). Theoretical Analysis of Energy-Gap Laws of Electron-Transfer Reactions, In: *Electron Transfer in Inorganic, Organic, and Biological Systems*, pp. 45-69, American Chemical Society, ISBN 0-8412-1846-3

Kakitani, T.; Yoshimori, A. & Mataga, N. (1992). Effects of the donor-acceptor distance distribution on the energy gap laws of charge separation and charge recombination reactions in polar solutions. *J. Phys. Chem.*, Vol. 96, No. 13, pp. 5385-5392, ISBN 0022-3654

Karen, A.; Ikeda, N.; Mataga, N. & Tanaka, F. (1983). Picosecond laser photolysis studies of fluorescence quenching mechanisms of flavin: a direct observation of indole-flavin singlet charge transfer state formation in solutions and flavoenzymes. *Photochem. Photobiol.*, Vol. 37, No. 5, pp. 495-502, ISBN 0031-8655

Karen, A.; Sawada, M. T.; Tanaka, F. & Mataga, N. (1987). Dynamics of excited flavoproteins-picosecond laser photolysis studies. *Photochem. Photobiol.*, Vol. 45, No. 1, pp. 49-54, ISBN 1751-1097

Kita, A.; Okajima, K.; Morimoto, Y.; Ikeuchi, M. & Miki, K. (2005). Structure of a Cyanobacterial BLUF Protein, Tll0078, Containing a Novel FAD-binding Blue Light Sensor Domain. *J. Mol. Biol.*, Vol. 349, No. 1, pp. 1-9, ISBN 0022-2836

Kitamura, M.; Kojima, S.; Ogasawara, K.;Nakaya, T.; Sagara, T.; Niki, K.; Miura, K.;Akutsu, H. & Kumagai, I. (1994). Novel FMN-binding protein from *Desulfovibrio vulgaris* (Miyazaki F). Cloning and expression of its gene in Escherichia coli. *J. Biol. Chem.*, Vol. 269, No. 8, pp. 5566-5573, ISBN 0021-9258

Kitamura, M.; Sagara, T.; Taniguchi, M.; Ashida, M.; Ezoe, K.; Kohno, K.; Kojima, S.; Ozawa, K.; Akutsu, H.; Kumagai, I. & Nakaya, T. (1998). Cloning and expression of the gene encoding flavodoxin from *Desulfovibrio vulgaris* (Miyazaki F). *J. Biochem.*, Vol. 123, No. 5, pp. 891-898, ISBN 1756-2651

Laan, W.; van der Horst, M. A.; van Stokkun, I. H. M. & Hellingwerf, K. (2003). Initial Characterization of the Primary Photochemistry of AppA, a Blue-light-using Flavin Adenine Dinucleotide-domain Containing Transcriptional Antirepressor Protein from *Rhodobacter sphaeroides*: A Key Role for Reversible Intramolecular Proton Transfer from the Flavin Adenine Dinucleotide Chromophore to a Conserved Tyrosine? *J. Photochem. Photobiol.*, *78*, 290–297, ISBN 1010-6030.

Liepinsh, E.; Kitamura, M.; Murakami, T.; Nakaya, T. & Otting, G. (1997). Pathway of chymotrypsin evolution suggested by the structure of the FMN-binding protein from *Desulfovibrio vulgaris* (Miyazaki F). *Nat. Struct. Mol. Biol.*, Vol. 4, No. 12, pp. 975-979, ISBN 1072-8368

Ludwig, M. L.; Pattridge, K. A.; Metzger, A. L.; Dixon, M. M.; Eren, M.; Feng, Y. C. & Swenson, R. P. (1997). Control of oxidation-reduction potentials in flavodoxin from Clostridium beijerinckii: The role of conformation changes. *Biochemistry*, Vol. 36, No. 6, pp. 1259-1280, ISBN 0006-2960

Lugsanangarm, K.; Pianwanit, S.; Kokpol, S. & Tanaka, F. (2011a). Homology modelling and molecular dynamics simulations of wild type and mutated flavodoxins from *Desulfovibrio vulgaris* (Miyazaki F): insight into FMN–apoprotein interactions. Mol. Simulat., Vol. 37, No. 14, pp. 1164-1178, ISBN 0892-7022

Lugsanangarm, K.; Pianwanit, S.; Kokpol, S.; Tanaka, F.; Chosrowjan, H.; Taniguchi, S. & Mataga, N. (2011b). Analysis of photoinduced electron transfer in flavodoxin. *J. Photochem. Photobiol. A: Chem.*, Vol. 217, No. 2-3, pp. 333-340, ISBN 1010-6030

Lugsanangarm, K.; Pianwanit, S.; Kokpol, S.; Tanaka, F.; Chosrowjan, H.; Taniguchi, S. & Mataga, N. (2011c). Photoinduced electron transfer in wild type and mutated flavodoxin from *Desulfovibrio vulgaris*, strain Miyazaki F.: Energy gap law. *J. Photochem. Photobiol. A: Chem.*, Vol. 219, No. 1, pp. 32-41, ISBN 1010-6030

Marcus, R. A. (1956a). On the Theory of Oxidation-Reduction Reactions Involving Electron Transfer. I *J. Chem. Phys.*, Vol. 24, No. 5, pp. 966–978, ISBN 0021-9606

Marcus, R. A. (1956b). Electrostatic Free Energy and Other Properties of States Having Nonequilibrium Polarization. I. *J. Chem. Phys.*, Vol. 24, No. 5, pp. 979-989, ISBN 0021-9606

Marcus, R. A. (1964). Chemical and Electrochemical Electron-Transfer Theory. *Annu. Rev. Phys. Chem.*, Vol. 15, No. 1, pp. 155-196, ISBN 0066-426X

Marcus, R. A. & Sutin, N. (1985). Electron transfers in chemistry and biology. *Biochim. Biophys. Acta.*, Vol. 811, No. 3, pp. 265-322, ISBN 0304-4173

Masuda, S. & Bauer, C. E. (2002). AppA is a blue light photoreceptor that antirepresses photosynthesis gene expression in Rhodobacter sphaeroides. *Cell*, Vol. 110, No. 5, pp. 613-623, ISBN 0092-8674

Masuda, S., Hasegawa, K., Ishii, A. & Ono, T. (2004). Light-Induced Structural Changes in a Putative Blue-Light Receptor with a Novel FAD Binding Fold Sensor of Blue-Light Using FAD (BLUF); Slr1694 of Synechocystis sp. PCC6803. *Biochemistry*, Vol. 43, No. 18, pp. 5304-5313, ISBN 0006-2960

Mataga, N.; Chosrowjan, H.; Shibata, Y. & Tanaka, F. (1998). Ultrafast fluorescence quenching dynamics of flavin chromophores in protein nanospace. *J. Phys. Chem. B*, Vol. 102, No. 37, pp. 7081-7084, ISBN 1089-5647

Mataga, N.; Chosrowjan, H.; Shibata, Y.; Tanaka, F.; Nishina, Y. & Shiga, K. (2000). Dynamics and mechanisms of ultrafast fluorescence quenching reactions of flavin chromophores in protein nanospace. *J. Phys. Chem. B*, Vol. 104, No. 45, pp. 10667-10677, ISBN 1089-5647

Mataga, N.; Chosrowjan, H.; Taniguchi, S.; Tanaka, F.; Kido, N. & Kitamura, M. (2002). Femtosecond fluorescence dynamics of flavoproteins: Comparative studies on flavodoxin, its site-directed mutants, and riboflavin binding protein regarding ultrafast electron transfer in protein nanospaces. *J. Phys. Chem. B*, Vol. 106, No. 35, pp. 8917-8920, ISBN 1520-6106

Mataga, N.; Taniguchi, S.; Chosrowjan, H.; Osuka, A. & Yoshida, N. (2003). Ultrafast charge separation and radiationless relaxation processes from higher excited electronic states of directly linked porphyrin-acceptor dyads. *Photochem. Photobiol. Sci.*, Vol. 2, No. 5, pp. 493-500, ISBN 1474-905X

Mataga, N.; Taniguchi, S.; Chosrowjan, H.; Osuka, A. & Kurotori, K. (2005). Observations of the whole bell-shaped energy gap law in the intra-molecular charge separation (CS) from S-2 state of directly linked Zn-porphyrin-imide dyads: Examinations of wider range of energy gap (-ΔGcs) for the CS rates in normal regions. *Chem. Phys. Lett.*, Vol. 403, No. 1-3, pp. 163-168, ISBN 0009-2614

Mataga, N.; Chosrowjan, H. & Taniguchi, S. (2005). Ultrafast charge transfer in excited electronic states and investigations into fundamental problems of exciplex chemistry: Our early studies and recent developments. *J. Photochem. Photobiol. C Photochem. Rev.*, Vol. 6, No. 1, pp. 37-79, ISBN 1389-5567

McCormick, D. B. (1977). Interactions of flavins with amino acid residues: assessments from spectral and photochemical studies. *Photochem. Photobiol.*, Vol. 26, No. 2, pp. 169-182, ISBN 1751-1097

Moser, C. C.; Keske, J. M.; Warncke, K.;Farid, R. S. & Dutton, P. L. (1992). Nature of biological electron transfer. *Nature*, Vol. 355, No. 6363, pp. 796-802, ISBN 0028-0836

Nunthaboot, N.; Tanaka, F.; Kokpol, S.; Chosrowjan, H.; Taniguchi, S. & Mataga, N. (2008a). Simultaneous analysis of ultrafast fluorescence decays of FMN binding protein and its mutated proteins by molecular dynamic simulation and electron transfer theory. *J. Phys. Chem. B*, Vol. 112, No. 41, pp. 13121-13127, ISBN 1520-6106

Nunthaboot, N.; Tanaka, F.; Kokpol, S.; Chosrowjan, H.; Taniguchi, S. & Mataga, N. (2008b). Quantum mechanical study of photoinduced charge transfer in FMN binding protein. *J. Phys. Chem. B*, Vol. 112, No. 49, pp. 15837-15843, ISBN 1520-6106

Nunthaboot, N.; Tanaka, F.; Kokpol, S.; Chosrowjan, H.; Taniguchi, S. & Mataga, N. (2009a). Simulation of ultrafast non-exponential fluorescence decay induced by electron transfer in FMN binding protein. *J. Photochem. Photobiol. A: Chem.*, Vol. 201, No. 2-3, pp. 191-196, ISBN 1010-6030

Nunthaboot, N.; Tanaka, F. & Kokpol, S. (2009b). Analysis of photoinduced electron transfer in AppA. *J. Photochem. Photobiol. A: Chem.*, Vol. 207, No. 2-3, pp. 274-281, ISBN 1010-6030

Nunthaboot, N.; Tanaka, F. & Kokpol, S. (2010). Simultaneous analysis of photoinduced electron transfer in wild type and mutated AppAs. *J. Photochem. Photobiol. A: Chem.*, Vol. 209, No. 1, pp. 79-87, ISBN 1010-6030

Nunthaboot, N.; Pianwanit, S.; Kokpol, S. & Tanaka, F. (2011). Simultaneous analyses of photoinduced electron transfer in the wild type and four single substitution isomers of the FMN binding protein from *Desulfovibrio vulgaris*, Miyazaki F. *Phys. Chem. Chem. Phys.*, Vol. 13, No. 13, pp. 6085-6097, ISBN 1463-9076

Rehm, D.; Weller, A. & Bunsen., B. (1969). Kinetics and mechanism of electron transfer in fluorescence quenching in acetonitrile. *Ber. Bunsenges, Phys. Chem.*, Vol. 73, No. 1, pp. 834 – 839

Rehm, D. & Weller, A. (1970). Kinetics of fluorescence quenching by electron and hydrogen-atom transfer. *Isr. J. Chem.*, Vol. 8, No. 1, pp. 259 - 271

Sancho, J. (2006). Flavodoxins: sequence, folding, binding, function and beyond. *Cell. Mol. Life Sci.*, Vol. 63, No. 7-8, pp. 855-864, 1420-682X

Silva, E. & Edwards, A. M. (2006). *Flavins: Photochemistry and Photobiology*. RSC Publishing, ISBN 0-85404-331-4, Cambridge, London

Sumi, H. & Marcus, R. A. (1986). Dynamical effects in electron transfer reactions *J. Chem. Phys.*, Vol. 84, No. 9, pp. 4894-4914, ISBN 00219606

Suto, K.; Kawagoe, K.; Shibata, N.; Morimoto, Y.; Higuchi, Y.; Kitamura, M.; Nakaya, T. & Yasuoka, N. (2000). How do the X-ray structure and the NMR structure of FMN-binding protein differ? *Acta Crystallogr. D*, Vol. 56, No. 3, pp. 368-371, ISBN 0907-4449

Tanaka, F.; Chosrowjan, H.; Taniguchi, S.; Mataga, N.; Sato, K.; Nishina, Y. & Shiga, K. (2007). Donor-acceptor distance-dependence of photoinduced electron-transfer rate in flavoproteins. *J. Phys. Chem. B*, Vol. 111, No. 20, pp. 5694-5699, ISBN 1520-6106

Tanaka, F.; Rujkorakarn, R.; Chosrowjan, H.; Taniguchi, S. & Mataga, N. (2008). Analyses of donor-acceptor distance-dependent rates of photo-induced electron transfer in flavoproteins with three kinds of electron transfer theories. *Chem. Phys.*, Vol. 348, No. 1-3, pp. 237-241, ISBN 0301-0104

van der Berg, P. A. & Visser, A. J. W. G. (2001). *New Trends in Fluorescence Spectroscopy. Applications to Chemical and Life Sciences*. First ed.; Springer, ISBN 3540677798, Berlin

Vogler, A.; Scandola, F.; Rehorek, D. & Kunkely, H. (2011). *Photoinduced Electron Transfer (Topics in Current Chemistry)*. Springer, ISBN 3540525688, Berlin

Vorsa, V.; Kono, T.; Willey, K. F. & Winograd, N. (1999). Femtosecond Photoionization of Ion Beam Desorbed Aliphatic and Aromatic Amino Acids: Fragmentation via α-Cleavage Reactions. *J. Phys. Chem. B*, Vol. 103, No. 37, pp. 7889-7895, ISBN 1520-6106

Watenpaugh, K. D.; Sieker, L. C. & Jensen, L. H. (1973). The binding of riboflavin-5'-phosphate in a flavoprotein: flavodoxin at 2.0-Angstrom resolution. *Proc. Natl. Acad. Sci. U.S.A.*, Vol. 70, No. 12, pp. 3857-3860, ISBN 0027-8424

Weber, G. (1950). Fluorescence of riboflavin and flavin-adenine dinucleotide. *Biochem. J.*, Vol. 47, No. 1, pp. 114–121

Zhong, D. P. & Zewail, A. H. (2001). Femtosecond dynamics of flavoproteins: Charge separation and recombination in riboflavine (vitamin B-2)-binding protein and in glucose oxidase enzyme. *Proc. Natl. Acad. Sci. U.S.A.*, Vol. 98, No. 21, pp. 11867-11872, ISBN 0027-8424

8

On the Relationship Between Residue Solvent Exposure and Thermal Fluctuations in Proteins

Yu-Tung Chien[1], Jenn-Kang Hwang[2] and Shao-Wei Huang[1]
[1]Department of Medical Informatics, Tzu Chi University, Hualien, Taiwan,
[2]Institute of Bioinformatics and Systems Biology, National Chiao Tung University, Hsinchu, Taiwan, R.O.C.

1. Introduction

The dynamic properties such as temperature factors or B-factors of a crystallographic protein structure result from a complex network of various interactions. To compute temperature factors, the straightforward way is to integrate the long time trajectory of the protein structure using molecular dynamics (Levitt & Warshel, 1975; Warshel, 1976; McCammon et al., 1977) and to compute the root mean square fluctuations of the trajectory, which takes into account all possible interactions in terms of empirical force field. Recently, the elastic network model (Tirion, 1996; Bahar et al., 1997; Ming et al., 2002) has been shown to be quite successful in computing temperature factors. Compared with the usual molecular mechanical force field, the elastic network model is a much simpler model. It is based on a mechanical model that each Cα atom is connected through a single-parameter harmonic potential function to its surrounding atoms that are within a cut-off distance usually in the range of 7 to 10 Å. The mathematical operations in the elastic network model are simply the matrix inversion and diagonalization, no trajectory computation required. It is somewhat surprising that a mechanical model as simple as the elastic network model, which uses only single parameter, i.e., the cut-off distance, can compute dynamics properties such as B-factors with accuracy comparable to the more complex molecular dynamics method.

The study of solvent accessible surface area (ASA) has been one of the most important topics in computation biology due to the fact that the residues interacting with other biological molecules are located on protein surface (Connolly, 1983). ASA has been used in the studies of protein stability (Gromiha et al., 1999), protein folding (Eisenberg & McLachlan, 1986), and fold recognition (Liu et al., 2007). The relationships between thermal fluctuation and some concepts related to ASA have been studied. B-factor has shown to correlate to local packing densities (Halle, 2002), atomic distance to protein center-of-mass (Shih et al., 2007; Lu et al., 2008) and residue depth (Zhang et al., 2009). Residue flexibility was known to be correlated to its ASA in single protein (Sheriff et al., 1985) or small dataset (Carugo & Argos,

1997). In a related study, the impact of the ASA of immediate and further neighbors of investigated residues was also noted (Zhang et al., 2009). In this study, we further discuss the relationship between ASA and theoretical thermal fluctuations which are not reported in previous studies.

Since the elastic network model is mainly based on information of local packing of each Cα atom, one may expect that relative atomic solvent accessibility may qualitatively reflect atomic fluctuation. In this work, we showed that this relation goes beyond just a qualitative one - the profiles of the temperature factors of crystallographic structures are very similar to those of the smoothed amino acid solvent accessibility. Our results show that protein dynamical properties can be inferred directly from the static structural properties without assuming an additional mechanical model. Another interesting corollary from our results is that one may predict temperatures factors from protein sequences with a prediction program of solvent accessibility.

2. Comparison between B-factor and actual RSA value

In this section, we compare B-factor and relative solvent accessibility (RSA) derived from protein structure. First, we discuss B-RSA relationships when different window sizes are used in smoothing RSA. Second, we show in detail how smoothed RSA (sRSA) are better correlated to B-factor than unsmoothed RSA (uRSA) in several example proteins and the phenomenon is generally true in a large dataset. Third, B-RSA correlations are calculated separately for residues located on protein surface, in the core, and in between.

2.1 Methods

2.1.1 The B-factor and the relative solvent accessibility profiles

The amino acid solvent accessible areas of the proteins are obtained from DSSP (Kabsch & Sander, 1983). For each protein, we can compute its relative solvent accessibility (RSA) profile. The RSA of the ith amino acid of type x, a_i^x, is computed from

$$a_i^x = 100A_i / A_0^x \quad (1)$$

where A_i is the solvent accessible area of the ith amino acid and A_0^x is the maximal solvent accessible area of amino acid of type x. The RSA profile of a protein of N residues is denoted as $A_u = (a_1, a_2, \ldots a_N)$.

The *smoothed* RSA profile is computed by averaging the RSA of each amino acid together with those of its n flanking amino acids. In the case of the terminal amino acid, its one-sided flanking amino acids are used twice in averaging. The smoothed RSA profile of a protein of N residues is denoted as $A_u' = (a_1', a_2', \ldots a_N')$, where a_i' is the smoothed relative solvent accessibility. For convenience, we will refer to the original RSA as uRSA and the smoothed RSA as sRSA. The predicted RSA is denoted as pRSA and the smoothed pRSA as spRSA. The spRSA is computed from pRSA by the smoothing method mentioned above. The B-factors are extracted from the PDB files of x-ray protein structures. The B-factor profile is given as $B = (b_1, b_2, \ldots b_N)$, where b_i is the B-factor of the Cα atom of the ith residue.

2.1.2 The z-scores of the B-factor and the relative solvent accessibility

For easy comparison, the B-value, the uRSA, the sRSA, the pRSA and the spRSA profiles are normalized to their corresponding z-scores

$$z_x = \frac{x - \bar{x}}{\sigma_x} \tag{2}$$

where $\bar{x}$ and σ_x are the mean and standard deviation of x, respectively. Here x is b, a or a'. We will use the notations z_B, z_{uRSA}, z_{sRSA}, z_{pRSA} and z_{spRSA} and to denote the z values of B, uRSA, sRSA, pRSA and spRSA, respectively. Their corresponding profiles are denoted by the vectors $\mathbf{z}_B$, $\mathbf{z}_{uRSA}$, $\mathbf{z}_{sRSA}$, $\mathbf{z}_{pRSA}$ and $\mathbf{z}_{spRSA}$. In terms of $\mathbf{z}$, the correlation coefficient c between, for example, $\mathbf{z}_{sRSA}$ and $\mathbf{z}_B$ is computed by $\mathbf{z}_{rRSA} \cdot \mathbf{z}_B$. If $c = 1$, these profiles are perfectly correlated; if $c = 0$, they are completely independent of each other; $c = -1$, they are perfectly anti-correlated.

2.1.3 Data set

The data set (PR972) is selected from PDB-REPRDB (Noguchi & Akiyama, 2003), including 972 protein chains with sequence length larger than 60 residues and pair wise sequence identity smaller than 25%. All structures are solved by x-ray crystallography with resolution better than 2.0 Å and *R*-factors smaller than 0.2. Detail list of the data set can be found in our previous study on protein structure-dynamics relationship (Lu et al., 2008).

2.2 Selection of smoothing window size of sRSA profiles

The thermal fluctuation of a single residue is affected by its flanking amino acids because of various interactions between their atoms, for example, hydrogen bonds, van der Waals interaction, electrostatic force, etc. Here, we show that the correlation between B-factor and RSA increases obviously with proper smoothing window size.

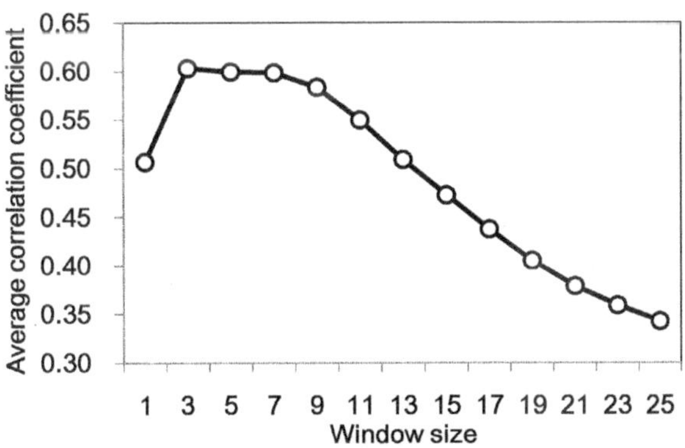

Fig. 1. Average correlation coefficients between normalized B-factor and RSA profiles based on the PR972 dataset. The RSA profiles are smoothed by different window sizes, from 0 (unsmoothed) to 25.

Figure 1 shows the average correlation coefficients between B-factor and RSA profiles smoothed by different window sizes, ranging from 1 (unsmoothed) to 25. The calculations are done based on the PR972 dataset. The average correlation coefficient between B-factor and RSA increases greatly from 0.51 (unsmoothed) to 0.60 (window size = 3). It is interesting that the results are similar when the window sizes are between 3 to 9 and become worse when the window size is larger than 9. It suggests that the thermal fluctuation of a residue is mostly affected by the neighbouring residues in this range (window size from 3 to 9). When the window size is too large, it seems that the information from distant residues has little correlation with the thermal fluctuation.

Based on the calculations shown in Figure 1, the window size is set to 3 (with the highest average correlation coefficient of 0.60) in the following sections.

2.3 The effects of smoothing on the correlation between B-factor and RSA

Figure 2 shows the B-factor, unsmoothed RSA (uRSA) and smoothed RSA (sRSA) profiles of a typical example: 5-carboxymethyl-2-hydroxymuconate isomerase (1OTG:C). The uRSA profile has a more rugged shape when compared with the B-value profile. If the rugged fine structures of uRSA profile are smoothed out, the global shapes of the B-factor and the sRSA profiles are seen to be quite similar (Figure 1B). The ruggedness of the uRSA profile is due to that the solvent accessibility of an amino acid may be quite different from its immediate flanking amino acids, but the B values of an amino acid and its immediate flanking amino acids appear to be more correlated.

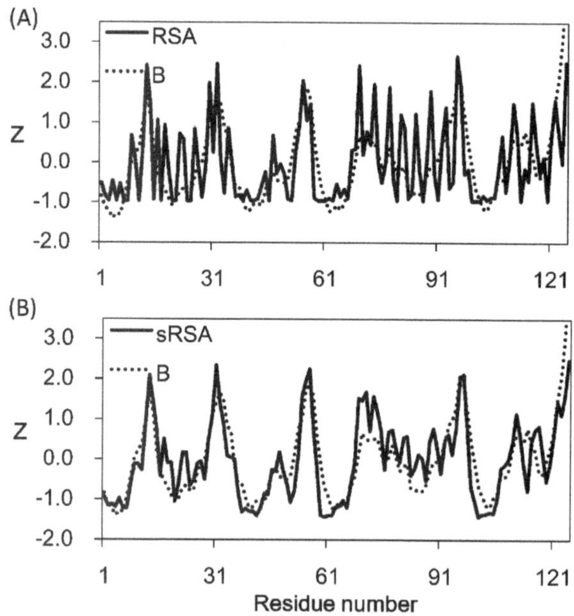

Fig. 2. Comparison of the B-value, uRSA and sRSA profiles of 5-carboxymethyl-2-hydroxymuconate isomerase (1OTG:C): (A) the uRSA (solid line) and the B-factors (dotted line) profiles; (B) the sRSA (solid line) and the B-factors (dotted line) profiles.

We computed the correlation coefficient between the B-factor, uRSA and sRSA profiles for the PR972 data set. Figure 3A shows the distribution of correlation coefficient between the B- factor and the uRSA profiles. The median of the correlation coefficients is 0.52 and 62% of the proteins have a correlation coefficient ≥ 0.5. Figure 3B shows the distribution of correlation coefficient between the B-factor and the sRSA profiles. The median of the correlation coefficients is now improved to 0.65 and 86% of the proteins have a correlation coefficient ≥ 0.5.

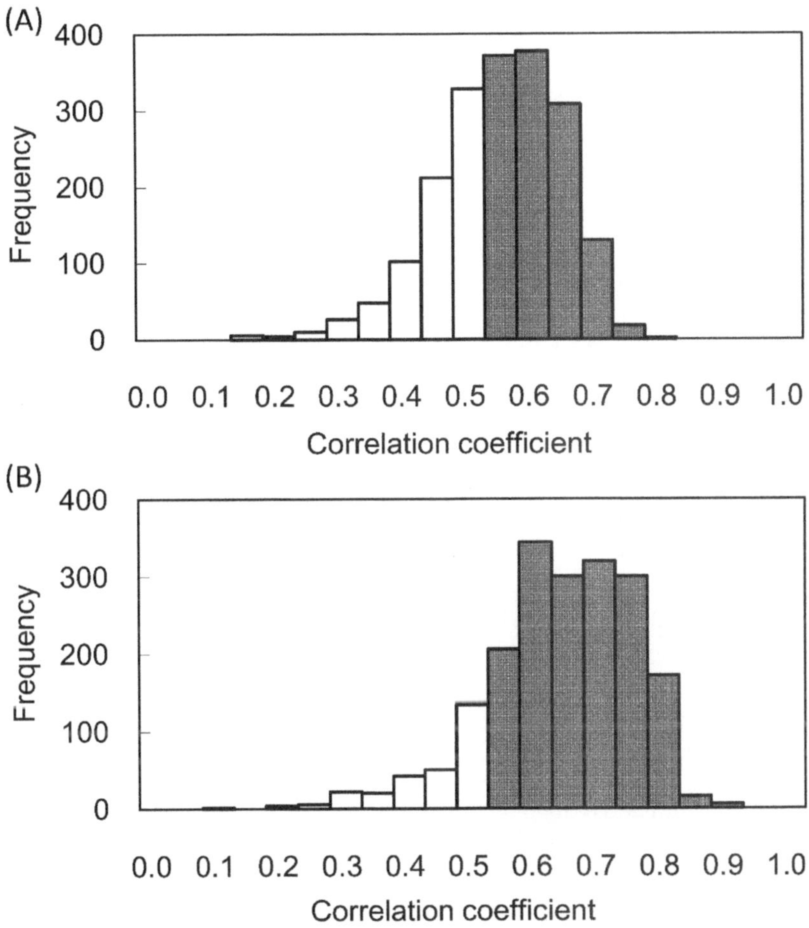

Fig. 3. The distribution of correlation coefficients of (A) the uRSA and (B) the sRSA profiles for the nonhomologous data set of 972 chains (PR972). The shaded areas indicate the sequences that have correlation coefficients ≥ 0.5.

2.4 Examples

Figure 4 shows some examples of the uRSA and sRSA profiles together with the B-value profiles: iron (II) superoxide dismutase (1ISA:B), dimethylsulfoxide reductase (1EU1:A) and the α/β domain of 6-phosphogluconate dehydrogenase (2PGD), Serratia endonuclease (1SMN:B). The B-factor profiles are very similar to sRSA profiles. Figure 4 shows the correlation between z_{sRSA} and z_B of the corresponding proteins.

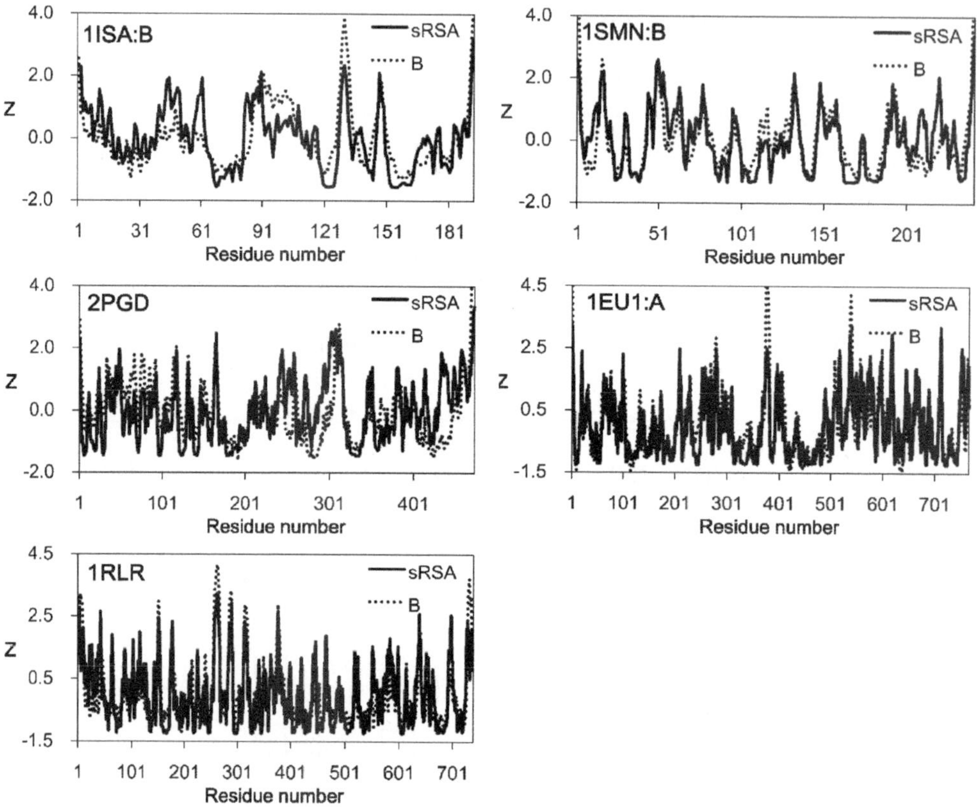

Fig. 4. Comparison of the crystallographic B-factors (dotted lines) of Cα atoms and the sRSA profiles (solid lines) of the following proteins: 1ISA:B, iron (II) superoxide dismutase; 1EU1:A, a dimethylsulfoxide reductase; 2PGD, the α/β domain of 6-phosphogluconate dehydrogenase; 1SMN:B, a Serratia endonuclease; 1RLR, the the α/β domain of ribonucleotide reductase protein R1. The sRSA profiles are normalized to the scale of the crystallographic B-factors.

2.5 Comparison of B-RSA relationship for residues in different structural contexts

Proteins are strongly interacting with their environment and protein dynamics have been shown to be affected or "slaved" to the solvent around them (Fenimore et al., 2004). Since we have observed close relationship between B-factor and RSA, it is interesting to examine the relationships separately for the surface residues, the partially buried residues and the residues deeply buried in protein core.

The residues in the PR972 dataset are grouped based on the local rigidity of each residue. The rigidity score is computed using the WCN model (Lin et al., 2008) which is a cutoff-free contact number model. The rigidity scores range from 0 to 1 and we use two bin sizes, bin size=0.1 and bin size=0.2, to separate the residues into 10 and 5 groups respectively. Figure 5 shows the correlation coefficients between B values and sRSA for each group (bin size=0.1 shown in filled circle, bin size=0.2 shown in rhomb). The results clearly show that the correlations between B values and sRSA decrease as the rigidity scores increase (correlation coefficient=-0.77). In other words, B values are better correlated to sRSA for residues located in less crowded environment or on the surface of the protein.

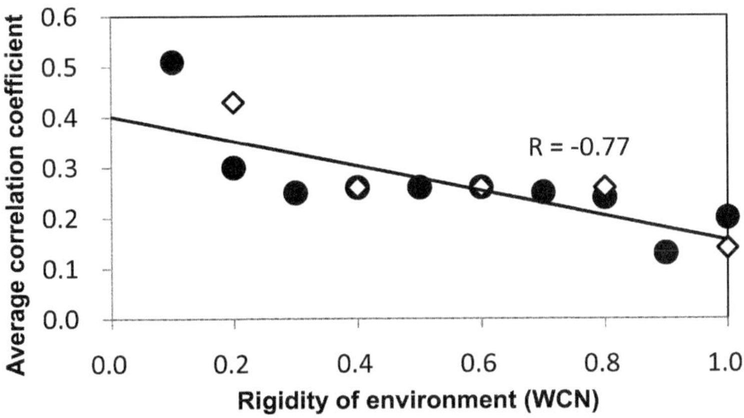

Fig. 5. The correlation coefficients between B and sRSA profiles for residue groups with increasing rigidity scores. The bin size=0.1 (filled circle) and bin size = 0.2 (rhomb) are used to separate the residues into groups. The linear regression is based on the groups of bin size=0.1.The statistics are done using the PR972 dataset.

3. Comparison between theoretical thermal fluctuation and actual RSA value

Thermal fluctuations in protein can be computed theoretically from its structure. Here, we further compare RSA profiles with theoretical thermal fluctuations computed by two methods, the Protein-fixed-point (PFP) model (Shih et al., 2007; Lu et al., 2008) and Weighted-contact number (WCN) model (Lin et al., 2008). These two models are shown to be able to reproduce B-factor from protein structure. The WCN model was further applied to the prediction of NMR order parameters (Huang et al., 2008) and the identification of enzyme active sites (Huang et al., 2011). The basic idea of the PFP model is assume that thermal fluctuation of a residue is related to the squared distance between the residue and the center-of-mass of the protein:

$$B_i \propto r_i^2 \tag{3}$$

where B_i is the theoretical thermal fluctuation of residue i and r_i is the distance between residue i and protein center-of-mass. The WCN model is a cutoff-free contact number model. Unlike traditional contact number calculation, the WCN model assumes that all residues interact with each other in a protein. The contact number of a residue is contributed by all other residues but each contact term is weighted inversely by the squared distance between them:

$$w_i = \sum_{j \neq i}^{N} 1/r_{ij}^2 \tag{4}$$

where w_i is the weighted contact number score of residue *i*, *N* is the total residue number of the protein, and r_{ij} is the distance between residue *i* and *j*.

Experimental thermal fluctuations are affected by different experimental conditions. For example, two structures of rubredoxin, 1CAA and 1IRO, have almost identical structure (RMSD = 0.44Å) but their B-factor profiles are dissimilar. Figure 6 compares their B-factor and WCN profiles, clearly showing that similar X-ray structures may have very different B-factor profiles.

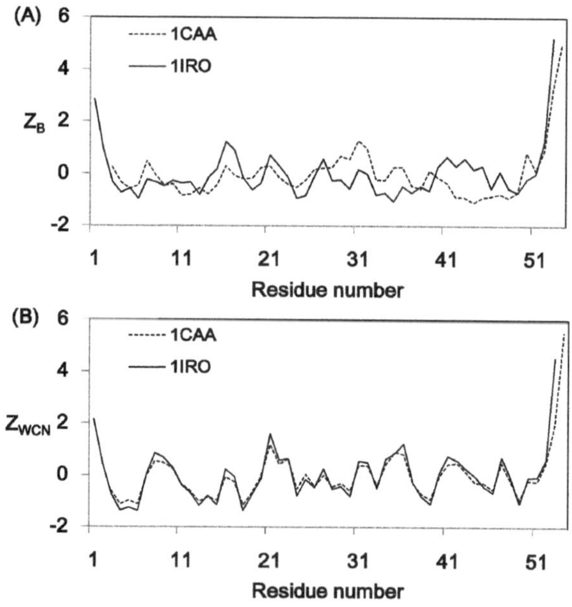

Fig. 6. Comparison of Z-scores of (A) B-factor profiles and (B) WCN profiles of two structures of rubredoxin (PDB id: 1CAA and 1IRO)

Figure 7 shows the distributions of correlation coefficients between sRSA and thermal fluctuations derived from different methods, including experimental B-factor, PFP model, and WCN model. The average correlation coefficients based on the PR972 dataset are 0.72 (WCN) and 0.55 (PFP) which are higher than that of B-factor (0.52). The correlations between sRSA and theoretical thermal fluctuations are more obvious than that of sRSA and B-factor. One reason may be that B-factor is affected by experimental conditions as shown in the example illustrated in figure 6.

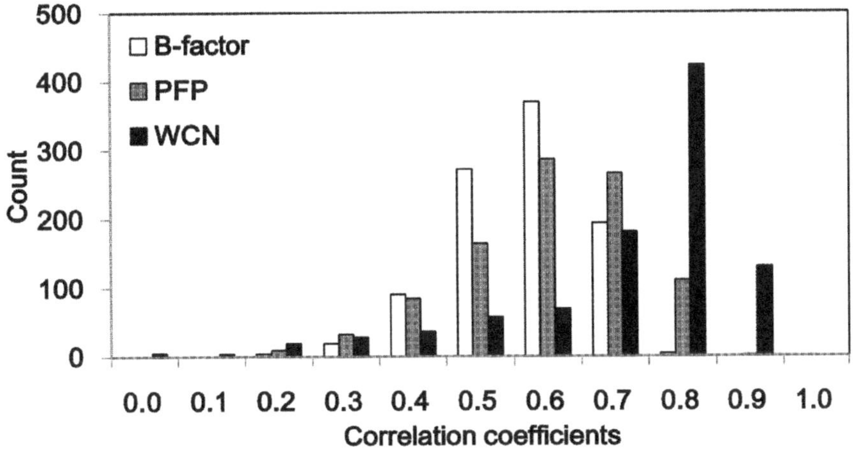

Fig. 7. The distributions of correlation coefficients between sRSA and B-factor (white), PFP (grey), and WCN (black) profiles based on the PR972 dataset

4. Prediction of B-factors from sequences based on predicted RSA profiles

Since the smoothed solvent accessible surface profiles are quite similar to the B-factor profiles, we can indirectly predict B-factors from sequence using the smoothed RSA profile predicted from sequence.

4.1 Methods

4.1.1 Prediction of relative solvent accessibility

The real value prediction of RSA (pRSA) is computed by the support vector machine (SVM) regression model. The inputs to the SVM are position-specific substitution matrix (PSSM) obtained from PSI-BLAST (Altschul et al., 1997), secondary structure profile predicted by PSIPRED (McGuffin et al., 2000) and hydropathy index. The PSSM profile was obtained after three iterations (E-value = 0.003) against the non-redundant protein sequence database. A $N \times 24$ scoring matrix is used as an input to the SVM, which including the PSSM profiles, secondary structure profiles and hydropathy index. Here, the size of the sliding window of the sequence, N, is set to 15. We train the SVM regression model by the commonly used RS126 dataset (Rost & Sander, 1993), a non-homologous data set with pair wise sequence identity less than 25% over a length of more than 80 residues.

4.1.2 The support vector machine

The support vector machine method (SVM) (Vapnik, 1995) has been successfully applied to secondary structure prediction (Hua & Sun, 2001; Kim & Park, 2003; Ward et al., 2003), subcellular localization prediction(Hua & Sun, 2001; Yu et al., 2004), protein fold assignment (Ding & Dubchak, 2001; Yuan et al., 2003) and other biological pattern classification problems (Dobson & Doig, 2003; Chen et al., 2004; Kim & Park, 2004). The original idea of the SVM is to find the separating hyperplane with the largest distance between two classes.

However, because the data to be classified may not always be linearly separable, the SVM overcomes this difficulty by using the kernel function to nonlinearly transform the original input space into a higher dimensional feature space, so that the data may be effectively separated in the higher dimensional space. SVMs perform well compared with other machine-learning methods because of convenient classifier's capacity control and effective avoidance of overfitting. In this work, the software package LIBSVM (Chang & Lin, 2001) was used.

4.2 Prediction results and comparison with other methods

The 2-state model is used to evaluate the performance of RSA prediction. Each residue is assigned buried or exposed by comparing its RSA value with a threshold and the prediction accuracy is defined by the percentage of correctly predicted residues. The prediction accuracies using different thresholds are listed in Table 1. The accuracies of two related prediction methods, SVMpsi and Fuzzy K-NN, are also listed for comparison with our results. The SVMpsi (Kim & Park, 2004) used SVM and the position-specific scoring matrix (PSSM) generated from PSI-BLAST and the Fuzzy K-NN (Sim et al., 2005) used fuzzy k-nearest neighbor method and PSSM as feature vectors.

Method	State threshold			
	25%	16%	5%	0%
SVMpsi	76.8	77.8	79.8	86.2
Fuzzy K-NN	78.3	79.0	82.2	87.2
This work	77.8	78.1	80.0	88.8

Table 1. The accuracies of predicting RSA from sequence in 2-state model based on the RS126 dataset

The correlation coefficients between the B-factor profiles and the predicted RSA (pRSA) and smoothed pRSA (spRSA) profiles are computed for the PR972 data set. The correlation coefficient between the B-factor and pRSA profiles is 0.44 and 31% of the proteins have correlation coefficient ≥ 0.5. After smoothing the pRSA profiles, the correlation coefficient between the B-factor and spRSA profiles increases to 0.53 and 55% of the proteins have correlation coefficient ≥ 0.5. Since spRSA and B-factor are well correlated, we assume that the spRSA of each residue is its predicted B-factor value. We found that the results of utilizing spRSA are comparable to one of the current best B-factor prediction results (Yuan et al., 2005). Yuan used support vector regression approach and PSSM information to predict the B-factor distribution from sequence and reported a correlation coefficient of 0.53 on a dataset of 766 high-resolution proteins. For fair comparison, we tested our method on the same dataset. The average correlation coefficient between spRSA and B-factor is 0.52 which is comparable to theirs (0.53).

Figure 8 shows the B-factor, pRSA and spRSA profiles of human uracil-DNA glycosylase inhibitor (1UGH:I). The correlation coefficient between the B-factor and pRSA profiles of the protein is 0.68 (Figure 8A). After smoothing, the correlation coefficient between the B-factor and spRSA profiles is 0.83. The B-factor profiles are better correlated to the spRSA profiles than the unsmoothed profiles, especially in the N-terminal and the P26-S39 residue regions (Figure 8B).

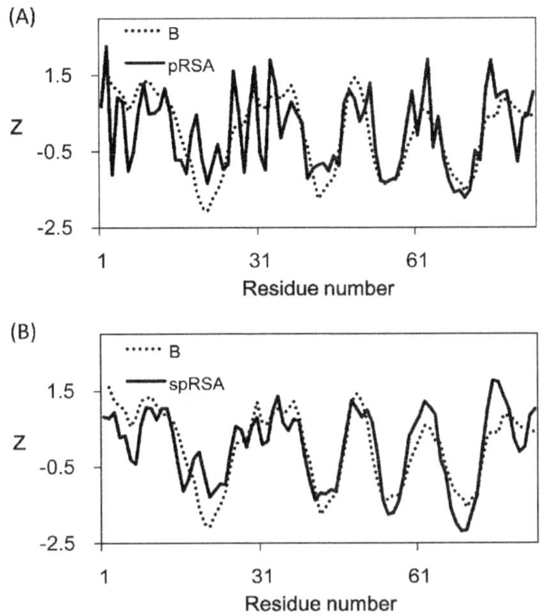

Fig. 8. Comparison of the B-value, pRSA and spRSA profiles of human uracil-DNA glycosylase inhibitor (1UGH:I): (A) the pRSA (solide line) and the B-factors (dotted line) profiles; (B) the spRSA (solid line) and the B-factors (dotted line) profiles.

Several methods have been developed to predict B-factor from sequence using different methodologies and testing datasets. The results of these methods have been compared using a small dataset (290 protein chains) (Radivojac et al., 2004). The average correlation coefficients are 0.32 (Vihinen et al., 1994) by a sliding window averaging technique and B-factor propensities, 0.43 by using neural network and multiple sequence alignment information (Radivojac et al., 2004). There were also structure-based prediction methods, for example, it was reported that the elastic network model has an average correlation coefficient of 0.59 between B-factor and their residue flexibility score (Kundu et al., 2002).

5. Conclusions

Though the dynamic properties of a protein result from a complex network of various interactions, we found that the B-factors of crystallographic structures are closely related to solvent accessibility directly derived from the protein structure. Our results indicate that dynamic properties such as B-factors can be computed directly from the protein's geometrical shape without assuming any mechanical models. Furthermore, we found that the smoothed solvent accessibility profiles are very similar to the B-factor profiles, and in some proteins, these profiles can overlap with each other almost perfectly. In a dataset comprising 972 non-homologous protein sequences, 86% of the proteins have a correlation coefficient between z_B and z_{sRSA} larger than 0.5. The results are consistent with the research by previous work (Zhang et al., 2009) showing the linear relationship between RSA and B-value profiles. In addition, we show that the relationship is not equal for residues in the environments of different rigidity. The correlations are higher for the residues located in the loose regions than those in the rigid environments.

Our results suggest that protein structure and protein dynamics are so closely related that the relative solvent accessibility profile allow one to directly infer the complete B-factor profiles. It will be interesting to investigate further whether there exists similar relationship between B-factors and other structural properties other than solvent accessibility. Recent studies showed a close relationship between the sites of low B-factor values and the active or the binding sites (Yuan et al., 2003; Yang & Bahar, 2005). Our results suggest a potential way to identify active sites from sequence without structure information. Figure 9 shows the spRSA and B-value profiles of type 2 rhinovirus 3C protease (1CQQ). The catalytic residues, H40, E71 and C147 (shown in open circle), are located at the low-mobility regions in the spRSA profile.

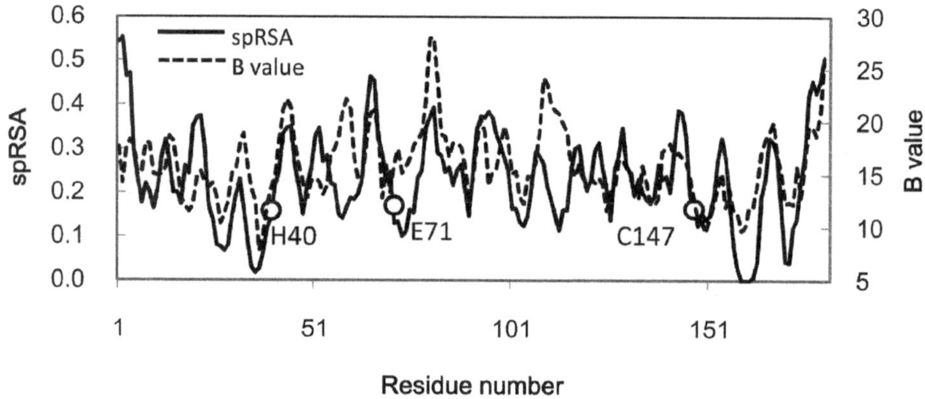

Fig. 9. B-value (dashed line) and spRSA (solid line) profiles of type 2 rhinovirus 3c protease (1CQQ). The catalytic residues, H40, E71 and C147, are labelled as open circle.

Since the prediction of solvent accessibility(Ahmad et al., 2003; Kim & Park, 2004) as well as the prediction of secondary structure elements(Qian & Sejnowski, 1988; Rost & Sander, 1993; Jones, 1999) from protein sequences are among the earliest and the best developed methodologies in computational biology, our findings suggest an indirect way of predicting B-factors from sequences - approaches based on various machine-learning techniques such as the support vector machines or the neural networks have been quite successful in the prediction of real value solvent accessibility, and we can borrow these methods and directly apply them to the prediction of B-factors by the use of the smoothed relatively solvent accessibility. We predict the solvent accessibility from sequence using the support vector machine method. The results are comparable to that of the current best methods, with an average correlation coefficient of 0.53 between the B-factor and spRSA profiles over a data set of 972 proteins.

6. Acknowledgements

Thanks to Tzu-Hui Yang for special supports. This research was supported by the National Science Council and ATU from the Ministry of Education, Taiwan, R.O.C.

7. References

Ahmad, S., Gromiha, M.M. & Sarai, A. (2003). Real value prediction of solvent accessibility from amino acid sequence. *Proteins.*Vol. 50, No. 4, (March 2003), pp 629-635, ISSN 1097-0134

Altschul, S.F., Madden, T.L., Schaffer, A.A., Zhang, J.H., Zhang, Z., Miller, W. & Lipman, D.J. (1997). Gapped BLAST and PSI-BLAST: a new generation of protein database search programs. *Nucleic Acids Research.*Vol. 25, No. 17, (September 1997), pp 3389-3402, ISSN 0305-1048

Bahar, I., Atilgan, A.R. & Erman, B. (1997). Direct evaluation of thermal fluctuations in proteins using a single-parameter harmonic potential. *Folding and Design.*Vol. 2, No. 3, (January 1997), pp 173-181, ISSN 1359-0278

Carugo, O. & Argos, P. (1997). Correlation between side chain mobility and conformation in protein structures. *Protein Engineering.*Vol. 10, No. 7, (July 1997), pp 777-787, ISSN 1741-0126

Chang, C.C. & Lin, C.J. (2001). LIBSVM: a library for support vector machines. Available from: http://www.csie.ntu.edu.tw/~cjlin/libsvm., No.,

Chen, Y.C., Lin, S.C., Lin, C.J. & Hwang, J.K. (2004). Prediction of the bonding states of cysteines using the support vector machines based on multiple feature vectors and cysteine state sequences. *Proteins-Structure Function and Bioinformatics.*Vol. 55, No. 4, (June 2004), pp 1036-1042, ISSN 0887-3585

Connolly, M.L. (1983). Solvent-accessible surfaces of proteins and nucleic acids. *Science.*Vol. 221, No. 4612, (August 1983), pp 709-713, ISSN 1095-9203

Ding, C.H.Q. & Dubchak, I. (2001). Multi-class protein fold recognition using support vector machines and neural networks. *Bioinformatics.*Vol. 17, No. 4, (April 2001), pp 349-358, ISSN 1367-4803

Dobson, P.D. & Doig, A.J. (2003). Distinguishing enzyme structures from non-enzymes without alignments. *Journal of Molecular Biology.*Vol. 330, No. 4, (July 2003), pp 771-783, ISSN 0022-2836

Eisenberg, D. & McLachlan, A.D. (1986). Solvation energy in protein folding and binding. *Nature.*Vol. 319, No. 6050, (January 1986), pp 199-203, ISSN 1476-4687

Fenimore, P.W., Frauenfelder, H., McMahon, B.H. & Young, R.D. (2004). Bulk-solvent and hydration-shell fluctuations, similar to alpha- and beta-fluctuations in glasses, control protein motions and functions. *Proceedings of the National Academy of Sciences of the United States of America.*Vol. 101, No. 40, (October 2004), pp 14408-14413, ISSN 0027-8424

Gromiha, M.M., Oobatake, M., Kono, H., Uedaira, H. & Sarai, A. (1999). Role of structural and sequence information in the prediction of protein stability changes: comparison between buried and partially buried mutations. *Protein Engineering.*Vol. 12, No. 7, (July 1999), pp 549-555, ISSN 0269-2139

Halle, B. (2002). Flexibility and packing in proteins. *Proceedings of the National Academy of Sciences.*Vol. 99, No. 3, (February 2002), pp 1274-1279, ISSN 1091-6490

Hua, S.J. & Sun, Z.R. (2001). A novel method of protein secondary structure prediction with high segment overlap measure: Support vector machine approach. *Journal of Molecular Biology.*Vol. 308, No. 2, (April 2001), pp 397-407, ISSN 0022-2836

Hua, S.J. & Sun, Z.R. (2001). Support vector machine approach for protein subcellular localization prediction. *Bioinformatics.*Vol. 17, No. 8, (August 2001), pp 721-728, ISSN 1367-4803

Huang, S.W., Shih, C.H., Lin, C.P. & Hwang, J.K. (2008). Prediction of NMR order parameters in proteins using weighted protein contact-number model. *Theoretical Chemistry Accounts.*Vol. 121, No. 3-4, (October 2008), pp 197-200, ISSN 1432-881X

Huang, S.W., Yu, S.H., Shih, C.H., Guan, H.W., Huang, T.T. & Hwang, J.K. (2011). On the Relationship Between Catalytic Residues and their Protein Contact Number. *Current Protein and Peptide Science.*Vol. 12, No. 6, (September 2011), pp 574-579, ISSN 1389-2037

Jones, D.T. (1999). Protein secondary structure prediction based on position-specific scoring matrices. *Journal of Molecular Biology.*Vol. 292, No. 2, (September 1999), pp 195-202, ISSN 0022-2836

Kabsch, W. & Sander, C. (1983). Dictionary of protein secondary structure: pattern recognition of hydrogen-bonded and geometrical features. *Biopolymers.*Vol. 22, No. 12, (December 1983), pp 2577-2637, ISSN 1097-0282

Kim, H. & Park, H. (2003). Protein secondary structure prediction based on an improved support vector machines approach. *Protein Engineering.*Vol. 16, No. 8, (August 2003), pp 553-560, ISSN 0269-2139

Kim, H. & Park, H. (2004). Prediction of protein relative solvent accessibility with support vector machines and long-range interaction 3D local descriptor. *Proteins:Structure Function and Genetics.*Vol. 54, No. 3, (February 2004), pp 557-562, ISSN 0887-3585

Kundu, S., Melton, J.S., Sorensen, D.C. & Phillips, G.N. (2002). Dynamics of Proteins in Crystals: Comparison of Experiment with Simple Models. *Biophysical journal.*Vol. 83, No. 2, (August 2002), pp 723-732, ISSN 0006-3495

Levitt, M. & Warshel, A. (1975). Computer simulation of protein folding. *Nature.*Vol. 253, No. 5494, (February 1975), pp 694-8, ISSN 0028-0836

Lin, C.P., Huang, S.W., Lai, Y.L., Yen, S.C., Shih, C.H., Lu, C.H., Huang, C.C. & Hwang, J.K. (2008). Deriving protein dynamical properties from weighted protein contact number. *Proteins-Structure Function and Bioinformatics.*Vol. 72, No. 3, (August 2008), pp 929-935, ISSN 0887-3585

Liu, S., Zhang, C., Liang, S. & Zhou, Y. (2007). Fold recognition by concurrent use of solvent accessibility and residue depth. *Proteins: Structure, Function, and Bioinformatics.*Vol. 68, No. 3, (August 2007), pp 636-645, ISSN 1097-0134

Lu, C.H., Huang, S.W., Lai, Y.L., Lin, C.P., Shih, C.H., Huang, C.C., Hsu, W.L. & Hwang, J.K. (2008). On the relationship between the protein structure and protein dynamics. *Proteins: Structure, Function, and Bioinformatics.*Vol. 72, No. 2, (August 2008), pp 625-634, ISSN 1097-0134

McCammon, J.A., Gelin, B.R. & Karplus, M. (1977). Dynamics of folded proteins. *Nature.*Vol. 267, No. 5612, (June 1977), pp 585-90, ISSN 0028-0836

McGuffin, L.J., Bryson, K. & Jones, D.T. (2000). The PSIPRED protein structure prediction server. *Bioinformatics.*Vol. 16, No. 4, (April 2000), pp 404-405, ISSN 1367-4803

Ming, D., Kong, Y., Lambert, M.A., Huang, Z. & Ma, J. (2002). How to describe protein motion without amino acid sequence and atomic coordinates. *Proceedings of the National Academy of Sciences.*Vol. 99, No. 13, (June 2002), pp 8620-5, ISSN 1091-6490

Noguchi, T. & Akiyama, Y. (2003). PDB-REPRDB: a database of representative protein chains from the Protein Data Bank (PDB) in 2003. *Nucleic Acids Research.*Vol. 31, No. 1, (January 2003), pp 492-493, ISSN 0305-1048

Qian, N. & Sejnowski, T.J. (1988). Predicting the secondary structure of globular proteins using neural network models. *Journal of Molecular Biology.*Vol. 202, No. 4, (August 1988), pp 865-884, ISSN 0022-2836

Radivojac, P., Obradovic, Z., Smith, D.K., Zhu, G., Vucetic, S., Brown, C.J., Lawson, J.D. & Dunker, A.K. (2004). Protein flexibility and intrinsic disorder. *Protein Science.*Vol. 13, No. 1, (January 2004), pp 71-80, ISSN 1469-896X

Rost, B. & Sander, C. (1993). Prediction of protein secondary structure at better than 70% accuracy. *Journal of Molecular Biology.*Vol. 232, No. 2, (July 1993), pp 584-599, ISSN 0022-2836

Sheriff, S., Hendrickson, W.A., Stenkamp, R.E., Sieker, L.C. & Jensen, L.H. (1985). Influence of solvent accessibility and intermolecular contacts on atomic mobilities in hemerythrins. *Proceedings of the National Academy of Sciences.*Vol. 82, No. 4, (February 1985), pp 1104-1107, ISSN 1091-6490

Shih, C.H., Huang, S.W., Yen, S.C., Lai, Y.L., Yu, S.H. & Hwang, J.K. (2007). A simple way to compute protein dynamics without a mechanical model. *Proteins-Structure Function and Bioinformatics.*Vol. 68, No. 1, (July 2007), pp 34-38, ISSN 1097-0134

Sim, J., Kim, S.Y. & Lee, J. (2005). Prediction of protein solvent accessibility using fuzzy k-nearest neighbor method. *Bioinformatics.*Vol. 21, No. 12, (June 15), pp 2844-2849, ISSN 1460-2059

Tirion, M.M. (1996). Large Amplitude Elastic Motions in Proteins from a Single-Parameter, Atomic Analysis. *Physical Review Letters.*Vol. 77, No. 9, (August 1996), pp 1905-1908, ISSN 0031-9007

Vapnik, V. (1995). *The Nature of statistical learning theory.*, Springer, ISBN 0-387-94559-8, New York

Vihinen, M., Torkkila, E. & Riikonen, P. (1994). Accuracy of protein flexibility predictions. *Proteins-Structure Function and Bioinformatics.*Vol. 19, No. 2, (June 1994), pp 141-149, ISSN 0887-3585

Ward, J.J., McGuffin, L.J., Buxton, B.F. & Jones, D.T. (2003). Secondary structure prediction with support vector machines. *Bioinformatics.*Vol. 19, No. 13, (September 2003), pp 1650-1655, ISSN 1367-4803

Warshel, A. (1976). Bicycle-pedal model for the first step in the vision process. *Nature.*Vol. 260, No. 5553, (April 1976), pp 679-683, ISSN 0028-0836

Yang, L.W. & Bahar, I. (2005). Coupling between catalytic site and collective dynamics: A requirement for mechanochemical activity of enzymes. *Structure.*Vol. 13, No. 6, (June 2005), pp 893-904, ISSN 0969-2126

Yu, C.S., Lin, C.J. & Hwang, J.K. (2004). Predicting subcellular localization of proteins for Gram-negative bacteria by support vector machines based on n-peptide compositions. *Protein Science.*Vol. 13, No. 5, (May 2004), pp 1402-1406, ISSN 0961-8368

Yu, C.S., Wang, J.Y., Yang, J.M., Lyu, P.C., Lin, C.J. & Hwang, J.K. (2003). Fine-grained protein fold assignment by support vector machines using generalized npeptide coding schemes and jury voting from multiple-parameter sets. *Proteins-Structure Function and Bioinformatics.*Vol. 50, No. 4, (March 2003), pp 531-536, ISSN 0887-3585

Yuan, Z., Bailey, T.L. & Teasdale, R.D. (2005). Prediction of protein B-factor profiles. *Proteins-Structure Function and Bioinformatics*.Vol. 58, No. 4, (March 2005), pp 905-912, ISSN 0887-3585

Yuan, Z., Zhao, J. & Wang, Z.X. (2003). Flexibility analysis of enzyme active sites by crystallographic temperature factors. *Protein Engineering*.Vol. 16, No. 2, (February 2003), pp 109-114, ISSN 0269-2139

Zhang, H., Zhang, T., Chen, K., Shen, S., Ruan, J. & Kurgan, L. (2009). On the relation between residue flexibility and local solvent accessibility in proteins. *Proteins: Structure, Function, and Bioinformatics*.Vol. 76, No. 3, (August 2009), pp 617-636, ISSN 1097-0134

9

Estimating Hydrogen Bond Energy in Integral Membrane Chromoproteins by High Hydrostatic Pressure Optical Spectroscopy

Liina Kangur[1], John D. Olsen[2],
C. Neil Hunter[2] and Arvi Freiberg[1]
[1]*University of Tartu,*
[2]*University of Sheffield*
[1]*Estonia,*
[2]*United Kingdom*

1. Introduction

Proteins are biological macromolecules that participate in virtually every process in live organisms. By nature they are divided into three categories: globular, membrane, and fibrous proteins. The globular proteins are present in the cytosol of cells and in body fluids such as blood; the membrane proteins are "solubilized" in two-dimensional lipid membranes that organize the macroscopic body space; the, usually, large fibrous proteins reinforce membranes and maintain the structure of cells and tissues. In the cell the membrane proteins are responsible for structural, catalytic, transport, signalling, control, and other crucial life-supporting functions (Palazzo 2006). Protein function is defined by its folded structure (Rose and Wolfenden 1993) and the self-assembly into the folded structure is largely governed by a manifold of generally weak, spatially oriented hydrogen (H) bonds that also confer distinct quantum properties (Li, Walker et al. 2011). Multiple H-bond interactions are responsible for binding the strands of RNA, DNA, and other biopolymers together, as well as for elasticity of skeletal and cardiac muscles. Cooperativity of the H-bond interactions appears to be a defining feature at all levels of biomolecular folding and function (Lu, Isralewitz et al. 1998; Finkelstein and Ptitsyn 2002; Lin, Mohammed et al. 2011). Proteins are only functional if associated with water, but unlike the dynamic network of H-bonds in the bulk solvent, the interactions with biomolecules are short-range, affecting only one or two layers of waters (Ball 2008). Although there is no doubt that H-bonds are extremely important structural elements of proteins, their role in stabilizing proteins is still a matter of debate. Studying H-bonds within proteins, particularly membrane bound proteins, therefore, permits insights into fundamental biological phenomena, indeed all of life on earth.

Hydrogen bond energies in simple model compounds and small peptides have been investigated in great detail (Sheu, Yang et al. 2003; Wendler, Thar et al. 2010). Hydrogen bonds in folded globular proteins, and especially in membrane proteins, are, however, much more difficult to characterize. The most frequently used methods are scanning calorimetry and titration with chemical denaturants such as urea. Both these methods probe unfolding of the whole protein and provide integrated rather than bond-specific information. Moreover, changing the temperature at constant pressure causes simultaneous changes of the system's kinetic energy and its volume/potential energy that are difficult to separate. In this work we show that in certain cases the use of pressure, another fundamental thermodynamic variable, offers attractive advantages. Continuous and reversible tuning of the protein density can be achieved over a wide range without changing its primary structure and chemical potential of the solute. Hydrogen bonds are controversially reported to be either widely insensitive to pressure (Phelps and Hesterberg 2007) or promoted by it (Boonyaratanakornkit, Park et al. 2002). We use native protein co-factors as intrinsic probes and optical spectroscopy as a sensitive tool for identifying specific H-bonds that undergo major changes under external compression as well as for studying the energetics of these bonds.

The light harvesting (LH) antenna pigment-protein complexes of purple photosynthetic bacteria are one of the best-characterized membrane chromoproteins (Blankenship, Madigan et al. 1995; Hunter, Daldal et al. 2008). They absorb solar photons and transfer the resulting excitations to the reaction center (RC) special pair sites, where the excitation energy is transformed into potential chemical energy. In the wild type (wt) purple bacterium *Rhodobacter (Rb.) sphaeroides* the photosynthetic apparatus is organized into spherical chromatophore vesicles of ~50 nm diameter (Hu, Ritz et al. 2002; Cogdell, Gall et al. 2006; Sener, Olsen et al. 2007; Sener, Strümpfer et al. 2010). The reaction center light-harvesting 1 core complex and peripheral light-harvesting 2 protein complexes (RC-LH1 and LH2, respectively) form the majority of the proteins in these vesicles, where the self-assembled photosystems of LH complexes are reminiscent of ordered two-dimensional crystals. Low resolution electron microscope and full atomic resolution crystal structures of the protein complexes have also been known for some time for a number of species (Karrasch, Bullough et al. 1995; McDermott, Prince et al. 1995; Koepke, Hu et al. 1996; Roszak, Howard et al. 2003; Qian, Hunter et al. 2005). The basic building block of these antenna structures is a heterodimer of α-helical polypeptides designated α and β (coloured yellow and magenta in Fig. 1), each non-covalently binding either two (LH1) or three (LH2) bacteriochlorophyll *a* (Bchl) chromophore co-factors along with carotenoid molecules. The number of hetrodimers in the structures of bacterial core complexes vary from 15 in the bacterium *Rhodopseudomonas palustris* (Roszak, Howard et al. 2003) or 16 in *Rhodospirillum rubrum* (Karrasch, Bullough et al. 1995) to 28 in the dimeric core complex of *Rb. sphaeroides* (Qian, Hunter et al. 2005; Bullough, Qian et al. 2009). In contrast, only two forms of the peripheral antenna complex are known, including 8 (as in *Rhodospirillum molichianum* (Koepke, Hu et al. 1996)) or 9 heterodimers (*Rb. sphaeroides, Rhodopseudomonas (Rps.) acidophila,* and *Rubrivivax gelatinosus* (McDermott, Prince et al. 1995; Walz, Jamieson et al. 1998; Ranck, Ruiz et al. 2001)). In LH2 from *Rb. sphaeroides* each polypeptide pair binds two Bchl molecules at the outer membrane surface and one molecule on its cytoplasmic side, forming two concentric pigment circles (B850 and B800, respectively) with C_9 symmetry, see Fig. 1.

The spectroscopic properties of LH1 and LH2 chromoproteins have been extensively studied (Van Amerongen, Valkunas et al. 2000; Hu, Ritz et al. 2002; Cogdell, Gall et al. 2006). While free in organic solvents, the lowest singlet (Q_y) electronic transition of Bchl is located in the near infrared region at ~775 nm (Grimm, Porra et al. 2006; Rätsep, Cai et al. 2011).

Significant red shifts of this transition are observed in the antenna systems, for example, the major absorption band in the LH2 complex from *Rb. sphaeroides* peaks at 850 nm (B850 band, see Fig. 2) (Van Dorssen, Hunter et al. 1988; Freiberg, Rätsep et al. 2011), while the spectroscopic equivalent of the $\alpha\beta$-$(Bchl)_2$ heterodimer subunit called B820, obtained by breaking down the LH1 complex, has a maximum at 820 nm (Loach and Parkes-Loach 2008). These large spectral shifts are primarily related to unique arrangements of the Bchls in the LH proteins, promoting strong inter-pigment (exciton) interactions. The B850 absorption band is the product of the 18 tightly packed and overlapping Bchl molecules on the periplasmic side of the complex, in a waterwheel arrangement, having intermolecular distances of less than 1 nm. The 9 monomeric Bchls of the other ring on the cytoplasmic side of the complex, which are widely separated (~2 nm), give rise to an absorption band at around 800 nm (B800 band). The B800 band shift from the molecular 775-nm absorption is mainly determined by interactions between the chromophores and the surrounding protein. Universal dispersion interactions aside, the factors that contribute most to the solvent/protein shifts are H-bonds to the C_3-acetyl carbonyl of the Bchl chromophores (Fowler, Sockalingum et al. 1994; He, Sundstrom et al. 2002; Uyeda, Williams et al. 2010) and various conformational interactions (Gudowska-Nowak, Newton et al. 1990).

The absorption spectra of Bchl molecules, being sensitive to even minor structural rearrangements in the LH complexes, have been taken advantage of in the present work. According to Lesch et al. (Lesch, Schlichter et al. 2004) the integrity of protein complexes can be monitored with sub-nanometer spatial resolution by the so-called molecular probe method.

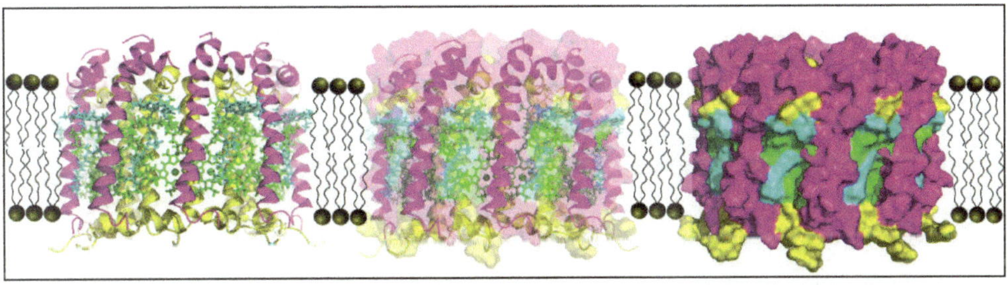

Fig. 1. Different representations of the LH2 pigment-protein complex using the same colour code for structural elements. The space fill model on the right clearly shows how densely packed the molecule actually is; the ribbon representation on left allows easy reference to the individual pigment cofactors; placed in the middle is an overlapped pair. The lipid bilayer covering hydrophobic parts of the complex is also schematically presented. The green B850 and blue B800 Bchl cofactors are sandwiched between the inner yellow ring of α-polypeptides and the outer magenta ring of β- polypeptides. The carotenoid molecules closely attached to the B850 and B800 Bchl rings are omitted for clarity.

To date mostly only water-soluble globular proteins have been studied under high pressures, with the result that functional protein structures cover just a narrow region in the pressure-temperature phase diagram close to physiological temperatures (Silva and Weber 1993; Boonyaratanakornkit, Park et al. 2002; Scharnagl, Reif et al. 2005; Meersman, Dobson et al. 2006). The photosynthetic LH membrane complexes were first investigated under high pressure in (Freiberg, Ellervee et al. 1993). This and subsequent studies (Sturgis, Gall et al. 1998; Gall, Ellervee et al. 2001; Timpmann, Ellervee et al. 2001; Gall, Ellervee et al. 2003) have shown that increasing the pressure at laboratory temperatures causes a smooth red (i.e., lower-energy) shift and broadening of the near-infrared absorption bands of Bchl. In the wt LH2 complexes from *Rb. sphaeroides* and *Rps. acidophila* extracted with mild detergents into a detergent-buffer environment a non-monotonic behaviour was observed (Kangur, Timpmann et al. 2008). With reference to genetic engineering results (Fowler, Visschers et al. 1992; Fowler, Sockalingum et al. 1994) this observation was qualitatively interpreted as due to rupture of the H-bonds at the binding sites of the excitonically coupled B850 chromophores (Kangur, Leiger et al. 2008).

In this chapter, we shall concentrate on a quantitative evaluation of H-bond energies that stabilize the strongly coupled Bchl co-factors in the B850 ring of the peripheral LH2 complex from *Rb. sphaeroides* by means of high-pressure optical spectroscopy. The availability of different genetically-modified preparations is an important advantage of this bacterium, besides its known spatial structure (Walz, Jamieson et al. 1998). The mutants constructed to support and widen our findings on wt complexes include a LH2 complex with a modified carotenoid background (neurosporene instead of a mixture of spheroidene and spheroidenone in the wt complex) and two B850-only LH complexes devoid of the B800 ring of Bchl molecules, one with wt carotenoids and another with neurosporene. To confirm that studies of purified membrane proteins have relevance to the intact membrane, the measurements have been performed both on detergent-isolated and native membrane-bound complexes. As a result, not only pressure-induced breakage of the H-bonds to the B850 chromophores have been demonstrated in all the samples, confirming the previous results on wt complexes (Kangur, Leiger et al. 2008; Kangur, Timpmann et al. 2008), but also their energies have been determined for the first time. The breakage appears to be a cooperative, "all-or-none" type transition, apparently triggered by the rupture of the weakest bond. Furthermore, the energies derived for the membrane-embedded and detergent-isolated complexes match each other within experimental uncertainty.

2. Thermodynamic aspects of protein stability against pressure

The phenomenological basis of protein stability against pressure is well established (Silva and Weber 1993; Boonyaratanakornkit, Park et al. 2002; Scharnagl, Reif et al. 2005; Meersman, Dobson et al. 2006). In a minimalist version of thermodynamic modelling just two global protein states, native (*N*) and denatured (*D*), are assumed. The equilibrium constant of the two-state denaturation reaction is given by Eq. (1), where [*N*] and [*D*] indicate the equilibrium concentrations of native and denatured protein, respectively, *R* is the universal gas constant, *T* is the thermodynamic temperature, and *P* is the pressure.

$$K(P) = [D]/[N] = \exp[-\Delta G(P)/RT] \tag{1}$$

The pressure dependence of the free energy change associated with protein denaturation, in the lowest (linear) order of approximation, can be represented as

$$\Delta G(P) = \Delta G^0 + \Delta V^0 P, \tag{2}$$

where $\Delta G^0 = G_D^0 - G_N^0$ is the standard free energy difference between the denatured and the native state and $\Delta V^0 = V_D - V_N$ is the partial molar volume change in going from the native to the denatured state measured at standard conditions. ΔG^0 has to be positive in order for the protein to be stable at standard conditions. If the volume of the denatured state is smaller than the volume of the native state (i.e., ΔV^0 is negative), the free energy change decreases as pressure is increased. The phase boundary (midpoint) pressure is determined as $P_{1/2} = \left|\Delta G^0 / \Delta V^0\right|$. At pressure $P_{1/2}$, native and denatured structures have equal probabilities of existence. Past this pressure, the denatured state has lower free energy and is stabilized against the native state.

3. Absorption spectra of LH2 complexes under ambient conditions

Figure 2 presents the overview optical absorption spectra of the LH2 complexes in the spectral range from 240 nm to 1000 nm, recorded at ambient temperature and pressure. For solubilization and purification of the complexes from their native membrane environment the detergent dodecyl-dimetyl-amine oxide, commonly known as LDAO, was used. As seen, the spectra of the detergent-isolated and native membrane-bound complexes are similar. Peak positions of the key bands observed in the spectra are presented in Table 1.

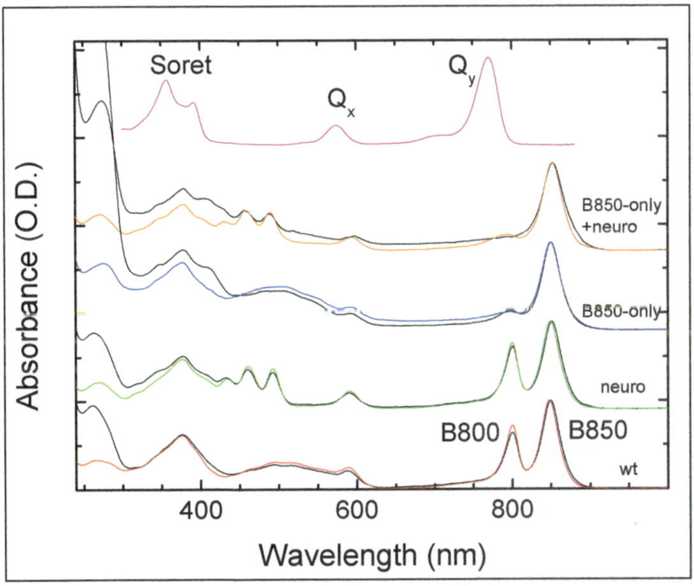

Fig. 2. Absorption spectra of the wt and mutant LH2 complexes studied, recorded at ambient temperature and pressure. The spectra of detergent-isolated (drawn with coloured lines) and native membrane-embedded (black lines) complexes are normalized relative to the B850 absorption band peak. The reference spectrum of Bchl in diethyl ether is shown in magenta. Neuro designates the mutant containing neurosporene, while B850-only+neuro is the double mutant with no B800 molecules and containing neurosporene carotenoids.

Sample		Q_y		Q_x	Soret	Car
		B850	B800			
wt	m	849.4	800.8	587.5	377.4	496.9
	i	847.8	800.8	588.4	376.8	503.1
neuro	m	851.1	801.1	591.3	378.0	492.8 460.9 432.8
	i	849.4	800.7	590.7	373.3	492.9 461.1 434.3
B850-only	m	850.2	-	590.9	378.5	497.2
	i	849.6	-	591.8	377.5	498.9
B850-only+neuro	m	852.4	-	594.7	377.6	489.4 457.2 427.5
	i	852.0	-	594.0	378.7	490.1 458.8 430.8

Table 1. Peak positions in nanometers (±0.5 nm) in the absorption spectra of the membrane-bound (m) and detergent-isolated (i) LH2 complexes recorded at ambient conditions. The bands are classified according to the related Bchl transitions as described in the text.

A comparison with the spectrum of Bchl in diethyl ether implies that the bands of the co-factor chromophores peaking around 850 (B850 band), 800 (B800 band), 590, and 380 nm are related to, respectively, the Q_y, Q_x, and Soret electronic transitions of the Bchl molecule. The bands evident between the Q_x and Soret transitions are from carotenoid molecules, being a mixture of spheroidene and spheroidenone in wt and neurosporene in mutant complexes. Compared with the structure-less spectrum of the native mixture, the neurosporene spectrum is clear-cut and blue shifted, showing three sharp peaks between 430 and 490 nm. Replacement of the wt carotenoids with neurosporene does not significantly influence the electronic states of the Bchl co-factors. The blue-shifted position of the neurosporene spectrum allows easier observation of the Q_x transitions of the protein-bound Bchls. The ultraviolet part of the spectrum, including the peak at 270 nm, is characteristic of the bulk protein scaffold.

The Q_y transitions of LH2 Bchls, which give rise to the B800 and B850 bands, are shifted toward longer (red) wavelengths and are associated with strongly excitonically coupled Bchls in the case of B850. In contrast, the Q_x transitions of the Bchl molecules belonging to the B800 and B850 arrangements apparently overlap. This can be interpreted as arising from the relatively weak oscillator strength of the Q_x transitions, leaving the transitions in all participating molecules almost localized. The stronger B850 exciton coupling compared with B800 clarifies not only the splitting between these bands but also the larger width of the B850 band (Freiberg, Timpmann et al. 1999). A peculiar triple structure in the Soret range of the complexes suggests considerable interactions between the respective strong molecular transitions and related large splitting of the bands. A greater red shift of the B850 band in the B850-only mutant as compared with the wt complex has been noted. It was explained by a somewhat enhanced exciton coupling in this complex, presumably because the missing B800 molecules allow more compact packing of the protein (Timpmann, Trinkunas et al. 2004). A weak shoulder around 795 nm in the spectrum of the B850-only mutant is most probably due to overlapping transitions of the B850 (vibronic) excitons and possibly the presence of some "free" Bchl molecules rather than residual B800 molecules (Rätsep, Hunter et al. 2005). More details about spectral properties of the LH2 complexes and their temperature dependencies can be found in (Freiberg, Rätsep et al. 2011).

In the following sections we shall mainly focus on the lowest-energy B850 absorption band. The particular interest toward this absorption feature is explained by the central role the respective transitions play in native photosynthesis by mediating excitation energy funnelling into the RC. Since Bchl molecules from both the B850 and B800 rings contribute to the Q_x band at 590 nm, we shall briefly characterize the influence of pressure on this band too. The detailed behaviour of the quasi-monomeric B800 band under high pressure will be described elsewhere.

4. Variations of the B850 absorption spectra induced by externally applied high hydrostatic pressure

4.1 Detergent-isolated complexes

As noted above, a non-trivial dependence of the B850 absorption band was observed upon application of high hydrostatic pressure to isolated wt LH2 complexes from *Rb. sphaeroides*. This behaviour was related to the rupture of H-bonds at binding sites of the excitonically coupled B850 chromophores (Kangur, Leiger et al. 2008; Kangur, Timpmann et al. 2008). A comparable situation is demonstrated in Fig. 3 for the mutant LH2 complexes, implying the general applicability of this phenomenon. At low pressures, below ~0.4 GPa, the spectra gradually red shift and broaden. The initial shift and broadening rates of the B850 band, which are similar in wt and mutant species, are a few times greater than that of the B800 band or the Q_y absorption band of isolated Bchl molecules in normal solutions. This is a consequence of the Bchl excited states in antenna complexes having excitonic origin. A contribution of short-range electron overlap effects into the advanced shift rate cannot be excluded (Freiberg, Ellervee et al. 1993; Wu, Rätsep et al. 1997; Timpmann, Ellervee et al. 2001). Toward higher pressures the continuous red shift is interrupted and abruptly reversed to a blue shift, best seen in Fig. 3c, where the B850 band maximum is plotted as a function of pressure. The reverse trend occurs at different pressures in individual samples: ~0.4 GPa in the neurosporene mutant, ~0.5 GPa in the wt, and ~0.6 GPa in the B850-only mutant. Only at about 0.1 GPa higher pressures the red shift is restored, albeit generally with a different slope. The bandwidths (defined as full width at half maximum, FWHM) show correlated changes (see Fig. 3d).

Enhanced stability of isolated complexes was observed (Kangur, Timpmann et al. 2008) at high protein/low detergent concentration, which resulted in aggregation, as well as when a cosolvent (e.g., glycerol) was added into the buffer solution. Aggregation occurred when the molar ratio (D/P) of the detergent (D) and protein (P) dropped below a few hundred (see (Kangur, Timpmann et al. 2008) for determination of this ratio). The combined glycerol-aggregation effect of stabilization is demonstrated in Figs. 3c and 3d on the mutant complexes incorporating neurosporene, where D/P=100 and the complex is solubilized in a buffer that contains 60 % (by volume) glycerol. As seen, the discontinuities of the shift and width dependences move toward significantly higher pressures, and their amplitudes are considerably reduced.

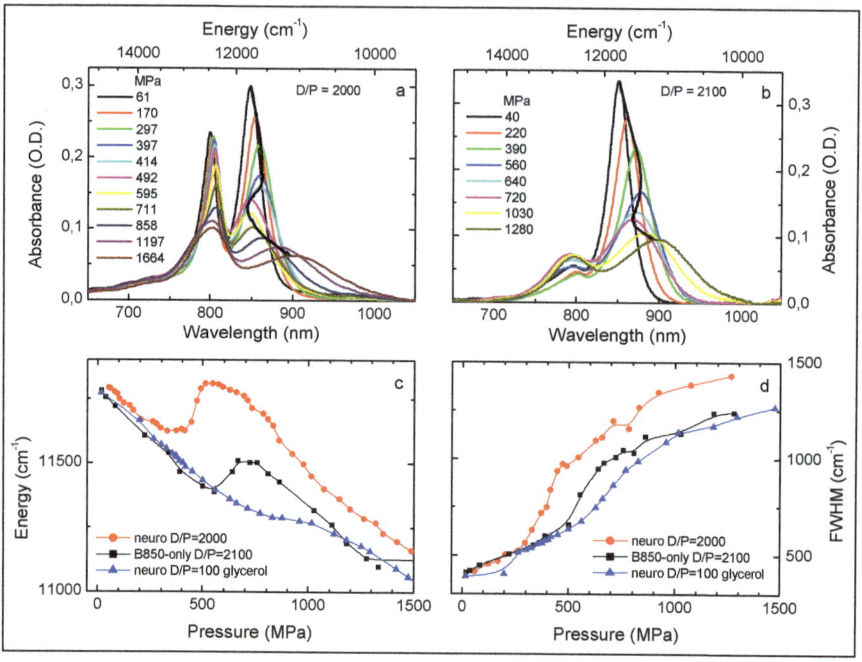

Fig. 3. (a, b) Absorption spectra of the detergent-solubilized neurosporene (a) and B800-only (b) mutant LH2 complexes recorded at the indicated pressures in MPa. Arrowed lines connect the selected absorption maxima. (c, d) Pressure dependence of the B850 absorption peak position (c) and width (d). Solid lines are added as guides to the reader. D/P denotes the ratio of detergent and protein molecules.

Figure 4 demonstrates the recovery of the absorption spectra of isolated complexes, upon pressure release, at the end of the cycle of measurements at elevated pressures. Two major effects following decompression are a change of the spectral shape and overall decrease of intensity. While the latter decrease is unavoidable due to plasticity of the gasket and resulting leakage of the solution (see Experimental section), the transformation of the spectrum is a signature of a partial degradation of the sample. A difference spectrum obtained by subtracting the normalized final spectrum from initial spectrum provides an approximate spectrum of the degradation products. The deformed long red tail aside, this spectrum shown in green in Fig. 4b and peaking at 776 nm is referred to in the literature as the B777 band. It has been identified as a mixture of monomeric α- and β-apoproteins that non-covalently bind just a single Bchl molecule (Loach and Parkes-Loach 2008) (see also Introduction). Reasonable sample homogeneity can be derived from the fact that the spectrum of the surviving complexes almost overlaps with the initial spectrum (not shown). Pressure effects appear to be fully elastic for this sub-population of the 'fittest' complexes, while representatives of the 'less healthy' population(s) fall apart entirely upon decompression.

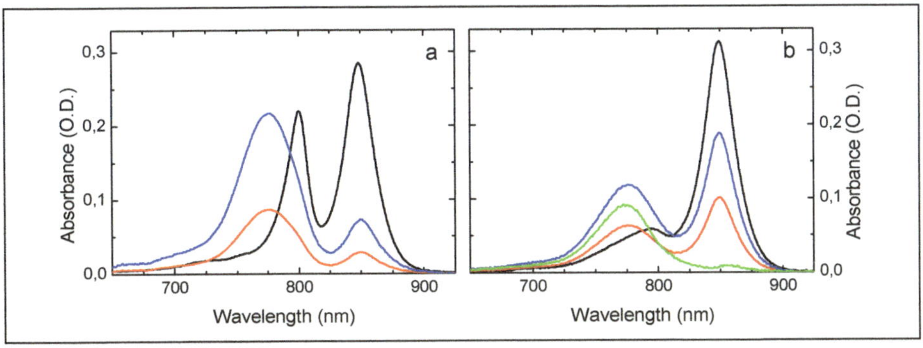

Fig. 4. Recovery of the absorption spectra of the detergent-isolated neurosporene (a) and B850-only (b) mutant complexes. Black, red, and blue lines are, respectively, the initial (ambient pressure) spectrum, the actual spectrum measured after pressure release, and the area-normalized spectrum after pressure release. The evaluated B777 spectrum is shown in green. See text for details.

It is worth noting that of all the samples studied the recovery is best (almost complete) in wt complexes (see (Kangur, Leiger et al. 2008; Kangur, Timpmann et al. 2008)) and worst in purified B850-only+neuro double mutant complexes. Therefore, for the latter samples, pressure dependences are missing.

4.2 Native membrane-bound complexes

Having established that the LH2 complexes purified from the native membrane with mild detergents exhibit clear signs of volatility with respect to high pressure, we undertook a series of experiments on membrane-bound LH2 complexes. It was concluded earlier that the wt membrane-protected complexes are rather resilient to damage by high-pressure compression compared with the isolated complexes (Kangur, Timpmann et al. 2008).

The results of the B850 absorption band shift measurements on wt and mutant membrane-bound LH2 complexes are shown in Fig. 5. Apart from the neurosporene mutant, the remaining two membrane samples in pure buffer solution behave rather uniformly, the main pressure effect being a gradual red shift (and broadening) of the band. These dependences, which demonstrate essential stabilization of the complexes in the native membrane environment, can be reasonably well approximated by two linear components, which have a greater low-pressure slope and a smaller high-pressure slope. The break point between the slopes is at ~0.7 GPa.

In contrast, the neurosporene mutant membrane complexes in buffer solution perform in every sense like their isolated counterparts in Fig. 3c, except that the turning point is moved by ~0.2 GPa toward higher pressures, indicating a degree of stabilization. Since the stability variations of proteins against externally applied pressure generally correlate with the extent of water molecules penetrating into the hydrophobic interior (Chryssomallis, Torgerson et al. 1981; Collins, Hummer et al. 2005; Harano, Yoshidome et al. 2008), it appears plausible that altering the native mixture of spheroidene/spheroidenone carotenoids to neurosporene results in less compact structure, which is more accessible to water than the wt protein.

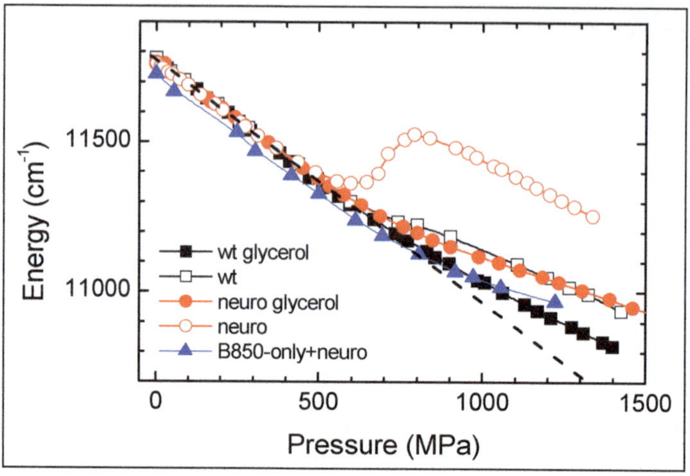

Fig. 5. Pressure dependence of the B850 absorption peak energy for the membrane-bound LH2 complexes. Solid lines are added as guides for the reader; the dashed line represents a linear extrapolation of the data to higher pressures.

Glycerol is known as an effective cryoprotectant; its ability to protect against high-pressure denaturation of detergent-isolated LH2 complexes has also been confirmed (Fig. 3, see also (Kangur, Timpmann et al. 2008)). However, the stabilizing effect of glycerol on native membrane-embedded LH2 complexes demonstrated in Fig. 5 has not been observed before and comes as a surprise. While in the case of the wt membrane adding the glycerol only slightly modifies the dependence, it almost neutralizes the destabilizing effects of neurosporene on the mutant complex in the membrane.

4.3 Comparison of the high-pressure effects on isolated and native membrane-bound LH2 complexes

Figure 6 shows a comprehensive summary of the B850 absorption band relative peak shift and broadening data for the complexes studied, using the peak position as well as the width of wt membrane complexes in glycerol as reference points. In this "free from background" representation we first compensated for the initial, ambient pressure positional differences (see Table 1). Thereafter, the relative shifts were calculated as $\Delta\nu = \nu_i - \nu_r$, where $\nu_{i/r}$ denote the peak frequencies of the studied sample (i) and the reference (r) membrane sample. A blue shift of the studied spectrum with respect to the reference spectrum thus results in a positive-valued $\Delta\nu$, while a red-shift, results in a negative-valued $\Delta\nu$.

The analytical line shape functions greatly simplify bandwidth analysis. We have found that Gaussians reasonably fit the experimental absorption spectra of LH2 complexes at ambient temperatures. In this approximation the relative widths have been evaluated as $\Delta\delta = \left(\Gamma_i^2 - \Gamma_r^2\right)^{1/2}$, where $\Gamma_{i/r}$ are the corresponding FWHM of the B850 absorption bands. The fact that Γ_i is invariably greater than Γ_r implies that the membrane complexes are more ordered than those in detergent micelles.

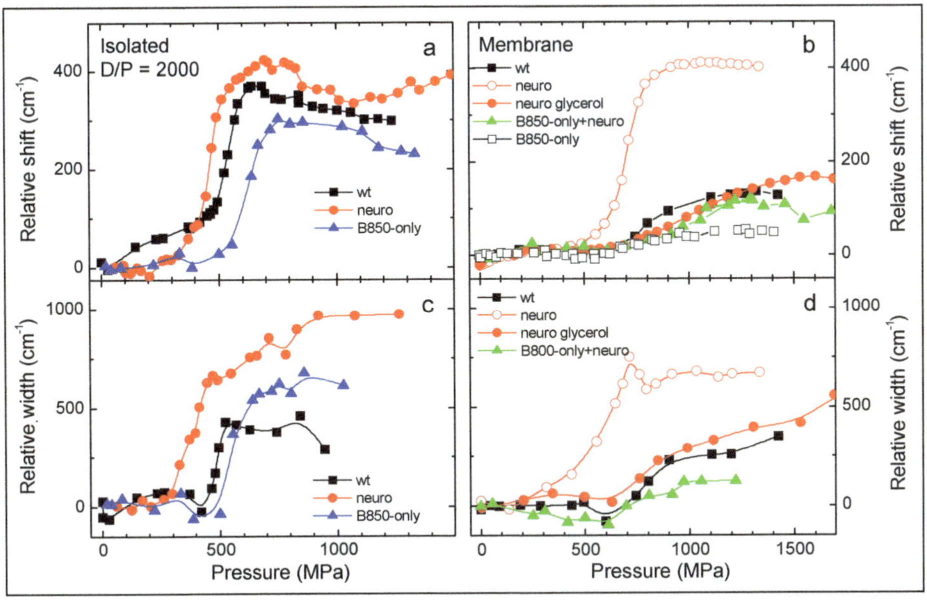

Fig. 6. Relative peak shifts (a, b) and widths (c, d) of the B850 band in various isolated (a, c) and membrane (b, d) LH2 complexes indicated. The shifts and widths are calculated against the wt membrane in glycerol reference values. In isolated samples, similar molar ratio of the detergent and protein molecules (D/P≈2000) has been applied for proper comparison. Solid lines are added as guides to the reader. See text for further details.

Characteristic step-like dependences resembling titration curves of the relative shifts and widths are obtained for all the complexes studied. The data in Fig. 6 that correspond to wt and mutant complexes demonstrate variable thresholds and step heights. In terms of the transition midpoint pressure, explained in paragraph 2, the neurosporene mutant complex appears least resistant to applied pressure, followed by the wt and B850-only mutant complexes. It also demonstrates the largest steps. Similar general trends can be observed for isolated and membrane complexes. However, the midpoint pressures that characterize individual membrane complexes are significantly shifted toward higher pressures and the step heights are generally lower compared to those of their respective isolated complexes. The dependences corresponding to membrane samples are also less noisy than for isolated complexes. This is to be expected as native membranes provide a more protective and less heterogeneous environment for antenna complexes than detergent micelles.

Spectral bandwidths are considered as sensitive reporters of static as well as of dynamic disorder present in antenna complexes. As can be seen in Fig. 6, the steps of relative shift and width correlate with each other, implying a common physical origin. The relative broadening of the spectra indicates an increased freedom of movement of the Bchl molecules in the binding sites of isolated complexes compared with the situation in the glycerol-enhanced membrane environment taken as a reference. Since the pressure-induced denaturation of proteins is driven by a decrease in volume, resulting in penetration of the buffer solvent (essentially of electrically polar water molecules) into the protein interior, rather than swelling of the protein (Chryssomallis, Torgerson et al. 1981; Collins, Hummer et al. 2005; Harano, Yoshidome et al. 2008), the extra freedom of intra-protein movements can only be caused by breaking the bonds that help bind the Bchl molecules in their protein pocket. Given that axial ligation has a minimal effect on the Bchl Q_y band position (Ellervee and Freiberg 2008), the loss of H-bonds in isolated complexes is the most likely explanation of the observed abrupt changes of the absorption bands with pressure.

This interpretation, first provided by Kangur et al. for the wt LH2 complexes from *Rb. sphaeroides* (Kangur, Leiger et al. 2008), is based on the observation that a blue shift of the B850 absorption band can be achieved by genetic manipulation of the H-bonds anchoring the Bchl molecules to the αβ-protein heterodimers (Fowler, Visschers et al. 1992; Fowler, Sockalingum et al. 1994). The α44-Tyr and α45-Tyr residues normally form H-bonds to the C_3-acetyl carbonyls of the α- and β-Bchls, respectively, in the B850 ring. It was shown that removal bonds to the α45-Tyr residues by replacing Tyr to non-H-boding residues in the LH2 complex correlates with the blue shift of the B850 absorption band by 11 nm (or ~150 cm^{-1}). The blue shift corresponding to removal of all H-bonds is 26 nm (~360 cm^{-1}). The latter blue shift of the double H-bond mutant almost coincides with the relative shift obtained in Fig. 5a for the wt isolated complexes (~367 cm^{-1}), implying the high-pressure induced breakage of every C_3-acetyl carbonyl H-bond in the B850 array.

The accompanying broadening of the spectra, an indication of increased freedom of movements of the Bchl probe molecules in their binding sites, supports this explanation. When bond ruptures increase the disorder of the system, they simultaneously widen the spectrum (Fowler, Visschers et al. 1992; Fowler, Sockalingum et al. 1994), thus the broadening effects are to be expected. Due to macroscopic heterogeneity of the samples (see discussion concerning Fig. 4) the broadening generally begins at somewhat lower pressures than the shift.

The step heights of relative shifts measured for the three isolated complexes in Fig. 5a vary between 410 and 292 cm^{-1}. The step height of about 410 cm^{-1} in the neurosporene membrane complex (Fig. 5b) fits the same range. The broad, >100-cm^{-1} variation (410-292=118 cm^{-1}), may be considered as too large to be related to the same shift/broadening mechanism in all these samples. However, taking into consideration the vast diversity of structural adjustments available upon compression, such as rotations of the acetyl carbonyl groups or turning the Bchl planes relative to the vector connecting the centres of the adjacent molecules in the B850 array, this figure is hardly exceptional. Moreover, we will show that the discrepancy between step heights in the LH2 complexes from the same species but having different D/P ratios is almost as great (~90 cm^{-1}). We thus argue that the relative shifts measured for the three (wt, B850-only, and neuro) isolated complexes as well as for the neurosporene membrane complex are not only quantitatively similar but also most probably due to the same mechanism. A similar conclusion, however, cannot be drawn with respect to other membrane complexes (wt, B850-only, B850-only+neuro, and glycerol-enchanced neurosporene), which show step heights that are less than half the size. According to genetic engineering results (Fowler, Visschers et al. 1992; Fowler, Sockalingum et al. 1994), this might imply breakage of every second H-bond rather than all of those to the C_3-acetyl carbonyls of the Bchl molecules in the B850 ring.

5. Evaluation the B850 chromophore-binding hydrogen bond energies

The use of the equilibrium thermodynamic stability model described in paragraph 2 to the present high-pressure experiment is justified by the observed reversibility of the pressure-induced effects. We define the *N* state as corresponding to the protein at ambient pressure, while the *D* state, to its compressed phase with broken H-bonds. The connection with the spectroscopic experiment is established by calculating the pressure-dependent equilibrium constant for a reversible transition between the *N* and *D* phases as

$$K(P)=\left[\Delta\nu(P)-\Delta\nu_i\right]/\left[\Delta\nu_f-\Delta\nu(P)\right], \tag{3}$$

where $\Delta\nu(P)$ is the relative absorption peak frequency shift at pressure P (see Figs. 6a and 6b), and $\Delta\nu_i$ and $\Delta\nu_f$ are the relative shifts of the B850 absorption band measured at initial, pre-transition (i) and final, post-transition (f) pressures, respectively. Inserting Eq. (3) into Eq. (1) provides Eq. (4):

$$-RT\ln K(P)=\Delta G^0+\Delta V^0 P\cdot \tag{4}$$

Assuming $\Delta\nu_i=0$, one gets a still simpler Eq. (5):

$$RT\ln\left[\frac{\Delta\nu_f}{\Delta\nu(P)}-1\right]=\Delta G^0+\Delta V^0 P\cdot \tag{5}$$

The linear pressure functions, Eqs. (4) and (5), allow easy determination of the key model parameters that characterize the N↔D phase equilibrium in LH2 proteins: ΔG^0 as the initial ($P = 0$) value, ΔV^0 as the slope, and $P_{1/2}$ as the crossing point with the y-axis zero.

Experimental plots of $-RT\ln K$ as a function of pressure for the B850 absorption band in all the LH2 samples studied are shown in Fig. 7. The fairly linear dependences obtained around respective phase transition regions justify the applied linear approximation model. The fitting parameters evaluated from these dependences are presented in Table 2.

As we have already noted, the transition midpoint pressures, $P_{1/2}$, are usually larger in membrane bound (where they vary between 641 and 995 MPa) than in isolated complexes (445-621 MPa). The volume changes, ΔV^0, which broadly follow the published volume changes of protein unfolding (Scharnagl, Reif et al. 2005), are generally greater for isolated (-51 to -71 ml/mol) than for membrane (-14 - -39 ml/mol) samples. Both the pressure and volume features agree with the general notion that membranes protect and stabilize integral membrane proteins. The volume changes are negative, meaning that compressed states are stabilized by high pressure. A straightforward explanation of the negative volume effect in connection with H-bonds is as follows: since the states with broken bonds are more compressible, their volume under high pressure is smaller than the volume with intact H-bonds.

Single H-bond energies in proteins between 2 and 25 kJ/mol (167-2090 cm^{-1}) have been reported (Sheu, Yang et al. 2003; Wendler, Thar et al. 2010), depending on the H-bond donor and acceptor as well as their environment. Most of the transition midpoint energies, ΔG^0, in Table 2 fall in this range. Yet a few complexes (wt, B850-only) demonstrate significantly

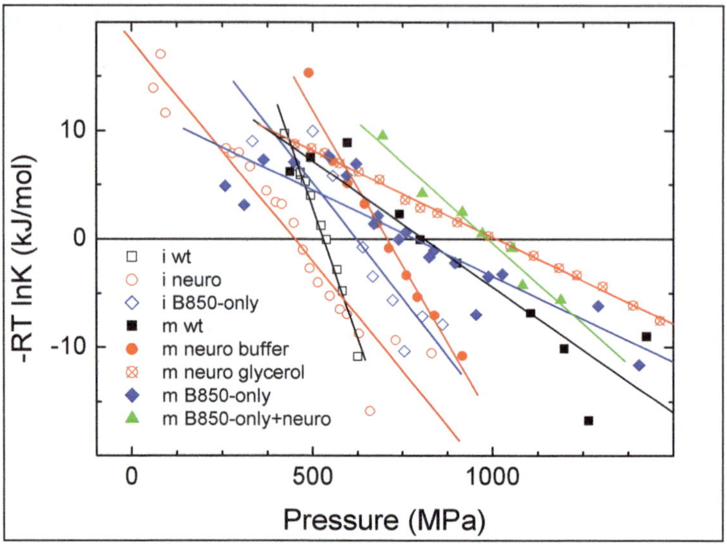

Fig. 7. Pressure dependences of $-RT\ln K(P)$ (see Eq. (4)) for the samples indicated. Continuous lines represent linear fits of the scattered experimental data. The prefix i and m denotes the data for isolated and membrane complexes, respectively.

larger energies, reaching 41 kJ/mol. A closer inspection of the data in Table 2 reveals several notable details about the free energy change related to breakage of the H-bonds in the LH2 complexes: (i) its value for isolated complexes is greater than for membrane/glycerol-enchanced membrane complexes; (ii) it is greater in complexes with native carotenoid background (wt and B850-only) than in complexes, which contain neurosporene (neuro and B850+neuro); (iii) irrespective of the sample, its value in membrane/ glycerol-enhanced membrane complexes is similar (≤19 kJ/mol). The findings (i) and (ii) once again underline the importance of local environment and/or structure of the binding site, specifically the carotenoid background, in determining H-bond energies in the B850 ring.

Sample		ΔG^0 kJ/mol	ΔV^0 ml/mol	$P_{1/2}$ MPa
wt	m	19±3	-23±10	815±15
	i	39±2	-71±10	519±10
B850-only	m	12±5	-16±10	785±10
	i	41±6	-67±10	621±10
neuro	m+g	24±5 14±2	-36±10 -14±10	641±15 995±15
	i	24±5	-51±10	445±10
B850-only+neuro	m	23±7	-23±10	983±10
	i	N/A	N/A	N/A

Table 2. Thermodynamic parameters characterizing phase transitions in isolated (D/P≈2000) and membrane LH2 complexes related to breakage of H-bonds in the B850 ring. m+g indicates the membrane in 60% glycerol; N/A means not available.

The neurosporene mutant complex appears to be unique. Only in this case do the energetic and volume effects that characterize the transitions in isolated and membrane complexes match each other within experimental uncertainty. However, when the neurosporene membrane complexes are stabilized by glycerol, their behaviour becomes similar to other membrane samples (see also Figs. 6 and 7).

6. Dependence on detergent-to-protein ratio

For sake of standardized comparison, a fixed D/P ratio equal to about 2000 has been previously used in the case of isolated complexes. Here, we investigate the possible dependence on this ratio, between 100 and 5000, of the phase transition energetics associated with the breakage of H-bonds in the B850 ring. The results of this study, in terms of the relative absorption peak frequency shift, are presented in Fig. 8 and Table 3. Also included for completeness are mutant membrane data, which effectively correspond to D/P=0.

A systematic decrease of the step height and shift of the midpoint pressure toward higher pressures is observed with decreasing D/P ratio. Nonetheless, the corresponding ΔG^0 values are similar, within experimental uncertainty, due to compensating effects of volume and pressure (see Table 3). In the two membrane samples the step heights differ drastically. While in the m neuro case the height is comparable with that in isolated complexes, in the m neuro glycerol case, it is less than half the size. In terms of the free energy change, m is again similar to the isolated complexes, whereas m+g is very different.

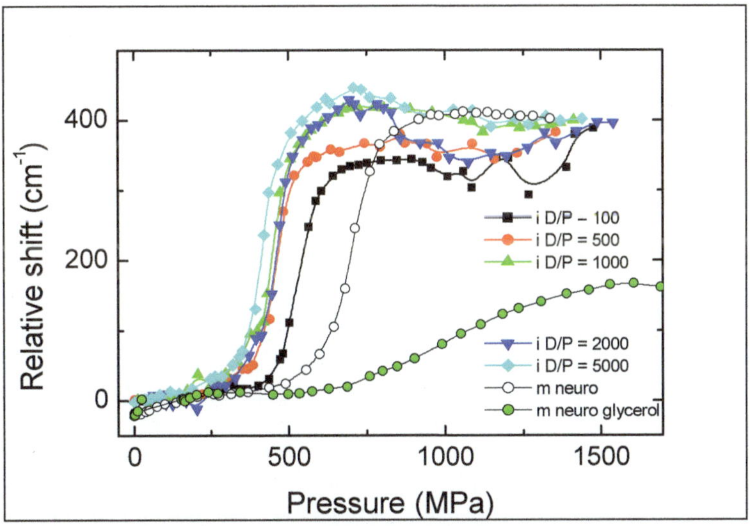

Fig. 8. Dependence on D/P ratio of the relative peak shifts of the B850 band in the spectra of isolated (i) and membrane (m) neurosporene mutant LH2 complexes. The shifts are calculated against the neurosporene mutant membrane in glycerol (in case of isolated complexes) or wt membrane in glycerol (in case of neurosporene mutant membrane in glycerol). Solid lines are added as guides to the reader.

D/P	ΔG^0 kJ/mol	ΔV^0 ml/mol	$P_{1/2}$ MPa
m+g	14±2	-14±5	995±15
m	28±5	-39±10	700±10
100	34±6	-65±10	528±10
500	35±11	-75±25	461±10
1000	24±4	-54±10	445±10
2000	24±5	-51±10	445±10
5000	24±4	-56±10	425±10

Table 3. Thermodynamic parameters characterizing phase transitions in the B850 ring of isolated neurosporene mutant LH2 light-harvesting complexes as a function of D/P ratio. As in Table 2, m+g indicates the membrane in 60% glycerol.

Based on these carotenoid mutant studies, one may conclude that H-bond energies, as determined by high-pressure spectroscopy, do not depend significantly on detergent concentration over a broad (three orders of magitude) D/P ratio. An analysis based on data published in (Kangur, Timpmann et al. 2008) leads to a similar conclusion regarding wt complexes. This is important as it provides evidence that the studies on isolated membrane proteins have direct relevance to the native membrane situation.

It is finally worth noting that the fairly sharp Q_x band at around 590 nm shows no splitting upon high-pressure compression, despite being contributed to by Bchl molecules from both B850 and B800 origin. Yet, as seen in Fig. 9, there is a small blue shift and broadening of this band in correlation with the large blue shift and broadening of the Q_y band, which is related to the B850 Bchl molecules. The Q_x transition is known to be sensitive to axial coordination of chlorophylls (Rätsep, Linnanto et al. 2009; Rätsep, Cai et al. 2011). Pressure-induced formation of an extra (i.e. sixth) coordination bond to the central Mg atom would, however, lead to a spectral red shift (Ellervee, Linnanto et al. 2004; Ellervee and Freiberg 2008). We thus conclude that the small changes in the Q_x spectrum are a reflection of the H-bond breakage in the B850 system.

7. Summary and conclusions

Protein function is governed by its folded structure, whereas denatured states are characterized by disordered conformations. Since the folded state is largely defined by H-bonds, their properties are profoundly important in our understanding of structure, stability, and also function of proteins. In the present chapter, experimental evidence is provided for an externally applied pressure induced rupture of the H-bonds that coordinate the Bchl chromophore cofactors with the protein scaffold in wt and various mutant peripheral LH2 light harvesting complexes from the photosynthetic bacterium *Rb. sphaeroides*. We have taken advantage of the Bchl cofactors which act as sensitive spectroscopic nano-probes of their protein binding sites. The mutant LH2 complexes with a modified carotenoid background and/or with a missing B800 ring of Bchls were constructed to support and extend the relevance of the data obtained on wt complexes. To confirm that the studies of purified membrane proteins have relevance to intact membrane situations, the measurements have been performed both on detergent-isolated and native membrane-bound complexes.

Abrupt spectral blue shifts have been observed in the pressure range of 0.45-1.00 GPa instead of usual gradual pressure-induced red shifts of the absorption and fluorescence emission spectra arising from the lowest-energy optical Q_y and Q_x transitions in the B850 ring of 18 Bchl molecules. The shifts are correlated with similar abrupt broadening of the spectra. For a number of complexes the spectra recover perfectly upon the release of the pressure, demonstrating reversibility of these pressure effects.

The most remarkable quantitative correspondence between the removal of single or double H-bonds in the basic unit of LH2 complexes by genetic engineering, on the one hand, and spectral shifts in membrane or isolated complexes induced by pressure, on the other, has been noted. This suggests a cooperative ("all-or-none" type) rupture of the H-bonds that bind the cofactors to the surrounding protein as the prime source of the observed spectral shifts. The increasing freedom of movements upon removal of the H-bonds to Bchls in their binding sites is the foremost cause of the spectral line broadening. A stepwise ("unit-by-unit") mechanism predicts clear correlations between the energy/ volume changes and the number of H-bonds involved, but these have not been observed. The cooperative process is most probably triggered by significant weakening and finally disruption of the so-called weak-link H-bonds. While concerted disruption of the H-bond network of water under pressure is well documented (Jonas, DeFries et al. 1976; Cunsolo, Formisano et al. 2009), evidence of similar reversible effects in proteins, particularly in membrane proteins has so far been scarce (Kangur, Leiger et al. 2008; Kangur, Timpmann et al. 2008).

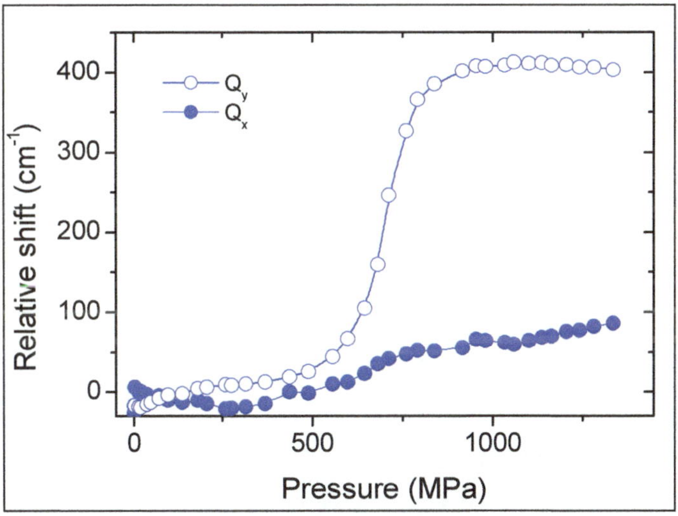

Fig. 9. Relative shifts of the Q_y (B850) and Q_x absorption spectra for neurosporene mutant membrane LH2 complexes. The shifts are calculated against the reference values of wt membranes in glycerol. Solid lines are added as guides for the reader.

The free energy changes related to high pressure-induced rupture of H-bonds have been evaluated by applying a minimalistic two-state thermodynamic model of protein denaturation. The energy required for breaking single H-bond in heterodimeric sub-units of the wt LH2 complexes is 19±3 kJ/mol, and 39±2 kJ/mol for breaking both bonds (see Table

2). The bond ruptures are accompanied by a decrease of the partial molar volume amounting 23±10 ml/mol and 71±10 ml/mol, respectively. Replacement of the native mixture of spheroidene/spheroidenone carotenoids with neurosporene significantly destabilizes the protein, so that the single and double bond energies are reduced to 14±2 kJ/mol and 24±5 kJ/mol, respectively. These results highlight the important role carotenoids play in reinforcement of the photosynthetic light-harvesting protein structures. The H-bond energies determined for the wt photosynthetic bacterial LH complexes are an order of magnitude larger than thermal energy under physiological conditions, explaining the great stability of these proteins both against temperature and pressure.

Comparable data are not available to the best of our knowledge; thus these results validate high-pressure optical spectroscopy as an effective non-invasive tool for the studies of bonding energetic and related structural integrity of integral membrane proteins with individual bond-mapping selectivity.

8. Experimental section

8.1 Samples

The LH2 membranes and complexes were prepared according to (Dawkins, Ferguson et al. 1988; Fowler, Visschers et al. 1992), respectively, and stored at liquid nitrogen temperature. The complexes have been removed from their native membrane environment and solubilized in micelles of the detergent (LDAO) that, above the critical micelle concentration, mimic the embedding of the proteins in native membranes. The detergent concentrations used, which have been elaborated above, resulted in no degradation of the complexes at ambient pressure. The experimental design also ensured that the detergent-solubilized complexes generally remained stable under elevated pressures for at least 10 hours, long enough for the present trials. The samples were thawed before the experiments and diluted with an appropriate buffer to obtain an optical density of about 0.3 of the B850 absorption band maximum in the assembled pressure cell. For complexes in their natural membrane environment no detergent was added. Colloidal properties of membrane proteins depend on the solvent acidity (Palazzo 2006). Therefore, for LH2 samples a 20 mM HEPES buffer, pH 7.5, was utilized since its buffering ability is preserved over a broad pressure range (Samaranayake and Sastry 2010). In some cases glycerol was added to stabilize the samples, to push the solvent solidification limit towards higher pressures, and to improve the solvent hydrostaticity in the solid phase.

8.2 High pressure spectroscopy

The absorption spectra of isolated and membranes complexes under ambient conditions were taken using a Jasco V-570 spectrophotometer (Jasco Corporation). The high-pressure spectroscopy system, based on a commercial diamond anvil cell (DAC D-02, Diacell Products Ltd.) has been previously described (Kangur, Timpmann et al. 2008). It allows absorption, fluorescence, fluorescence excitation, and Raman measurements to be performed in the same set-up. The DAC cell was equipped with a 0.35-mm thick stainless steel gasket, pre-indented between the anvils under small pressure. A ruby-microbead pressure sensor (RSA Le Rubis SA), mounted directly in the sample volume, was used to determine the pressure inside the DAC. The temperature of the cell was maintained at 20 ±

0.5 ^{0}C during the experiment. No essential variations of the results were observedwhen the temperature was varied between 18 and 27 ^{0}C. The accuracy of the pressure measurements (defined as the pressure needed to shift the emission line at the output of the spectrograph by one pixel of the recording CCD camera) was 6.6 MPa. The pressure was increased stepwise with an average rate of 25-30 MPa per minute. Recurrent measurements ensured reproducibility of the data. Reversibility of the measurements was regularly checked by way of releasing pressure from its maximum value down to ambient pressure.

9. Acknowledgments

Estonian Science Foundation (grant No. 8674) and Ministry of Education and Science of Estonia (grant SF0180055s07) supported this work. JDO and CNH gratefully acknowledge financial support from the Biotechnology and Biological Sciences Research Council (UK). The authors are grateful to H. Salujärv for designing the thermally stabilized high-pressure cell.

10. References

Ball, P. (2008). "Water as an active constituent in cell biology." *Chem. Rev.* 108: 74-108.

Blankenship, R. E., M. T. Madigan, et al., Eds. (1995). *Anoxygenic Photosynthetic Bacteria.* Dordrecht, Kluwer Academic Publishers.

Boonyaratanakornkit, B. B., C. B. Park, et al. (2002). "Pressure effects on intra- and intermolecular interactions within proteins." *Biochim. Biophys. Acta* 1595: 235-249.

Bullough, P. A., P. Qian, et al. (2009). Reaction Center-Light-Harvesting Core Complexes of Purple Bacteria. *The Purple Phototrophic Bacteria.* C. N. Hunter, Daldal, F., Thurnauer, M.C., Beatty, J.T. Dordrecht, the Netherlands, Springer.

Chryssomallis, G. S., P. M. Torgerson, et al. (1981). "Effect of hydrostatic pressure on lysozyme and chymotrypsinogen." *Biochemistry* 20: 3955-3959.

Cogdell, R. J., A. Gall, et al. (2006). "The architecture and function of the light-harvesting apparatus of purple bacteria: from single molecules to in vivo membranes." *Quart. Rev. Biophys.* 39: 227-324.

Collins, M. D., G. Hummer, et al. (2005). "Cooperative water filling of a nonpolar protein cavity observed by high-pressure crystallography and simulations." *Proc. Natl. Acad. Sci. USA* 102: 16668-16671.

Cunsolo, A., F. Formisano, et al. (2009). "Pressure dependence of the large-scale structure of water." *J. Chem. Phys.* 131: 194502.

Dawkins, D. J., L. A. Ferguson, et al. (1988). The structure of the 'core' of the purple bacterial photosynthetic unit. *Photosynthetic Light-Harvesting Systems: Organization and Function.* H. Scheer and S. Schneider. Berlin and New York, Walter de Gruyter and Co.: 115-127.

Ellervee, A. and A. Freiberg (2008). "Formation of bacteriochlorophyll a coordination states under external high-pressure." *Chem. Phys. Lett.* 450: 386-390.

Ellervee, A., J. Linnanto, et al. (2004). "Spectroscopic and quantum chemical study of pressure effects on solvated chlorophyll." *Chem. Phys. Lett.* 394: 80-84.

Finkelstein, A. V. and O. Ptitsyn (2002). *Protein Physics.* London, Academic Press.

Fowler, G. J. S., G. D. Sockalingum, et al. (1994). "Blue shifts in bacteriochlorophyll absorbance correlate with changed hydrogen bonding patterns in light-harvesting 2 mutants of *Rhodobacter sphaeroides* with alterations at .alpha.-Tyr-44 and .alpha.-Tyr-45." *Biochem. J.* 299(3): 695-700.

Fowler, G. J. S., R. W. Visschers, et al. (1992). "Genetically modified photosynthetic antenna complexes with blueshifted absorbance bands." *Nature* 355(6363): 848-50.

Freiberg, A., A. Ellervee, et al. (1993). "Pressure effects on spectra of photosynthetic light-harvesting pigment-protein complexes." *Chem. Phys. Lett.* 214(1): 10-16.

Freiberg, A., M. Rätsep, et al. (2011). "A comparative spectroscopic and kinetic study of photoexcitations in detergent-isolated and membrane-embedded LH2 light-harvesting complexes." *Biochem. Biophys. Acta* doi: 10.1016/j.bbabio.2011.11.019.

Freiberg, A., K. Timpmann, et al. (1999). "Disordered exciton analysis of linear and nonlinear absorption spectra of antenna bacteriochlorophyll aggregates: LH2-only mutant chromatophores of *Rhodobacter sphaeroides* at 8 K under spectrally selective excitation." *J. Phys. Chem. B* 103(45): 10032-10041.

Gall, A., A. Ellervee, et al. (2001). "Effect of High Pressure on the Photochemical Reaction Center from Rhodobacter sphaeroides R26.1." *Biophys. J.* 80(3): 1487-1497.

Gall, A., A. Ellervee, et al. (2003). "Membrane protein stability: High pressure effects on the structure and chromophore-binding properties of the light-harvesting complex LH2." *Biochemistry* 42: 13019-13026.

Grimm, B., R. J. Porra, et al., Eds. (2006). *Chlorophylls and Bacteriochlorophylls*. Dordrecht, Springer.

Gudowska-Nowak, E., M. D. Newton, et al. (1990). "Conformational and Environmental Effects on Bacteriochlorophyll Optical Spectra: Correlatlons of Calculated Spectra with Structural Results." *J. Phys. Chem.* 94: 5795.

Harano, Y., T. Yoshidome, et al. (2008). "Molecular mechanism of pressure denaturation of proteins." *J. Chem. Phys.* 129: 145103.

He, Z., V. Sundstrom, et al. (2002). "Influence of the protein binding site on the excited states of bacteriochlorophyll: DFT calculations of B800 in LH2." *Journal of Physical Chemistry B* 106(44): 11606-11612.

Hu, X., T. Ritz, et al. (2002). "Photosynthetic apparatus of purple bacteria." *Quart. Rev. Biophys.* 35: 1-62.

Hunter, C. N., F. Daldal, et al., Eds. (2008). *The Purple Phototrophic Bacteria*. Advances in Photosynthesis and Respiration. Dordrecht, The Netherlands, Springer.

Jonas, J., T. DeFries, et al. (1976). "Molecular motions in compressed liquid water." *J. Chem. Phys.* 65: 582-588.

Kangur, L., K. Leiger, et al. (2008). "Evidence for high-pressure-induced rupture of hydrogen bonds in LH2 photosynthetic antenna pigment-protein complexes." *J. Physics: Conf. Series* 121: 112004.

Kangur, L., K. Timpmann, et al. (2008). "Stability of integral membrane proteins against high hydrostatic pressure: The LH2 and LH3 antenna pigment-protein complexes from photosynthetic bacteria." *J. Phys. Chem. B* 112: 7948-7955.

Karrasch, S., P. A. Bullough, et al. (1995). "The 8.5 A projection map of the light-harvesting complex I from *Rhodospirillum rubrum* reveals a ring composed of 16 subunits." *EMBO J.* 14(4): 631-368.

Koepke, J., X. Hu, et al. (1996). "The crystal structure of the light-harvesting complex II (B800-850) from *Rhodospirillum molischianum*." *Structure* 4(5): 581-597.

Lesch, H., J. Schlichter, et al. (2004). "Molecular probes: What is the range of their interaction with the environment?" *Biophys. J.* 86: 467-472.

Li, X.-Z., B. Walker, et al. (2011). "Quantum nature of the hydrogen bond." *Proc. Natl. Acad. Sci. USA* 108: 6369-6373.

Lin, M. M., O. F. Mohammed, et al. (2011). "Speed limit of protein folding evidenced in secondary structure dynamics." *Proc. Natl. Acad. Sci. USA* 108: 16622-16627.

Loach, P. A. and P. S. Parkes-Loach (2008). Structure-function relationships in bacterial light-harvesting complexes investigated by reconstitution techniques. *The Purple Photosynthetic bacteria*. C. N. Hunter, F. Daldal, M. C. Thurnauer and J. T. Beatty. Dordrecht, Springer: 181-198.

Lu, H., B. Isralewitz, et al. (1998). "Unfolding of titin immunoglobulin domains by steered moldecular dynamics simulations." *Biophys. J.* 75: 662-671.

McDermott, G., S. M. Prince, et al. (1995). "Crystal structure of an integral membrane light-harvesting complex from photosynthetic bacteria." *Nature* 374(6522): 517-521.

Meersman, F., C. M. Dobson, et al. (2006). "Protein unfolding, amyloid fibril formation and configurational energy landscapes under high pressure conditions." *Chem. Soc. Rev.* 35: 908-917.

Palazzo, G. (2006). "Colloidal aspects of photosynthetic membrane proteins." *Curr. Opin. Colloid Interface Sci.* 11: 65-73.

Phelps, D. J. and L. K. Hesterberg (2007). "Protein disaggregation and refolding using high hydrostatic pressure." *J. Chem. Technol. Biotechnol.* 82: 610-613.

Qian, P., C. N. Hunter, et al. (2005). "The 8.5 A projection structure of the core RC-LH1-PufX dimer of Rhodobacter sphaeroides." *J. Mol. Biol.* 349: 948-960.

Ranck, J.-L., T. Ruiz, et al. (2001). "Two-dimensional structure of the native light-harvesting complex LH2 from Rubrivivax gelatinosus and of a truncated form." *Biochem. Biophys. Acta* 1506: 67-78.

Rose, G. D. and R. Wolfenden (1993). "Hydrogen bonding, hydrophobisity, packing, and protein folding." *Annu. Rev. Biophys. Biomol. Struct.* 22: 381-415.

Roszak, A. W., T. D. Howard, et al. (2003). "Crystal structure of the RC-LH1 core complex from *Rhodopseudomonas palustris*." *Science* 302: 1969-1972.

Rätsep, M., Z.-L. Cai, et al. (2011). "Demonstration and interpretation of significant asymmetry in the low-resolution and high-resolution Qy fluorescence and absorption spectra of bacteriochlorophyll a." *J. Chem. Phys.* 134: 024506.

Rätsep, M., C. N. Hunter, et al. (2005). "Band structure and local dynamics of excitons in bacterial light-harvesting complexes revealed by spectrally selective spectroscopy." *Photosynth. Res.* 86: 37-48.

Rätsep, M., J. Linnanto, et al. (2009). "Mirror symmetry and vibrational structure in optical spectra of chlorophyll a." *J. Chem. Phys.* 130: 194501.

Samaranayake, C. P. and S. K. Sastry (2010). "In situ measurement of pH under high pressure." *J. Phys. Chem. B* 114: 13326-13332.

Scharnagl, C., M. Reif, et al. (2005). "Stability of proteins: Temperature, pressure and the role of the solvent." *Biochim. Biophys. Acta* 1749: 187-213.

Sener, M., J. Strümpfer, et al. (2010). "Photosynthetic vesicle architecture and constraints on efficient energy transfer." *Biophys. J.* 99: 67-75.

Sener, M. K., J. D. Olsen, et al. (2007). "Atomic level structural and functional model of a bacterial photosynthetic membrane vesicle." *Proc. Natl. Acad. Sci. USA* 104: 15273-15278.

Sheu, S.-Y., D.-Y. Yang, et al. (2003). "Energetics of hydrogen bonds in peptides " *Proc. Natl. Acad. Sci. USA*(100): 12683-12687.

Silva, J. L. and G. Weber (1993). "Pressure stability of proteins." *Annu. Rev. Phys. Chem.* 44: 89-113.

Sturgis, J. N., A. Gall, et al. (1998). "The effect of pressure on the bacteriochlorophyll *a* binding sites of the core antenna complex from *Rhodospirillum rubrum*." *Biochemistry* 37(42): 14875-14880.

Timpmann, K., A. Ellervee, et al. (2001). "Short-range couplings in LH2 photosynthetic antenna proteins studied by high hydrostatic pressure absorption spectroscopy." *J. Phys. Chem. B* 105: 8436-8444.

Timpmann, K., G. Trinkunas, et al. (2004). "Bandwidth of excitons in LH2 bacterial antenna chromoproteins." *Chem. Phys. Lett.* 398: 384-388.

Uyeda, G., J. C. Williams, et al. (2010). "The influence of hydrogen bonds on the electronic structure of light-harvesting complexes from photosynthetic bacteria." *Biochemistry* 49: 1146-1159.

Walz, T., S. J. Jamieson, et al. (1998). "Projection structures of three photosynthetic complexes from *Rhodobacter sphaeroides*: LH2 at 6 A, LH1 and RC-LH1 at 25 A." *J. Mol. Biol.* 282(4): 833-845.

Van Amerongen, H., L. Valkunas, et al. (2000). *Photosynthetic Excitons*. Singapore, World Scientific.

Van Dorssen, R. J., C. N. Hunter, et al. (1988). "Spectroscopic properties of antenna complexes of Rhodobacter sphaeroides in vivo." *Biochim. Biophys. Acta* 932(2): 179-88.

Wendler, K., J. Thar, et al. (2010). "Estimating the hydrogen bond energy." *J. Phys. Chem. A* 114: 9529-9536.

Wu, H.-M., M. Rätsep, et al. (1997). "Comparison of the LH2 antenna complexes of *Rhodopseudomonas acidophila* (strain 10050) and *Rhodobacter sphaeroides* by high-pressure absorption, high-pressure hole burning, and temperature-dependent absorption spectroscopies." *J Phys. Chem. B* 101(38): 7641-7653.

10

Preserving Proteins Under High Pressure and Low Temperature

Takahiro Takekiyo[1,*], Minoru Kato[2] and Yukihiro Yoshimura[1]
[1]Department of Applied Chemistry, National Defense Academy, Yokosuka,
[2]Department of Pharmacy, Ritsumeikan University, Kusatsu,
Japan

1. Introduction

A promising method of preserving aqueous protein solutions is to subject them to high pressure and low temperature. However, two significant issues can arise under these conditions: (1) ice nucleation in aqueous solution at 273 K and 1.0 GPa, and (2) structural changes to the protein (Kunugi & Tanaka, 2002; Frank, 1995; Heremans & Smeller, 1998; Vandekooi, 1998).

The first issue can be avoided by addition of a freezing-protection agent such as sugar or salt. For example, Miyazaki *et al.* reported that myoglobin in a highly aqueous NaCl solution forms a glassy state during quenching and annealing processes (Miyazaki et al., 1993, 2000).

Our group has a continuing interest in devising new freezing-protection agents. To this end, we have systematically investigated the structure of aqueous electrolyte salt solutions and room-temperature ionic liquids (RTILs), which consist of the cations and anions. We found that such water-mixed solutions remain liquid even at 1.0 GPa and 77 K (achieved by rapid cooling at 500 K/min) (Imai et al., 2010; Takekiyo et al., 2006a, 2006b; Yoshimura et al., 2009, 2011) and that water-soluble RTILs such as 1-butyl-3-methylimidazolium tetrafluoroborate remain liquid even at 2.0 GPa (Imai et al., 2011, Takekiyo et al., 2011). These results suggest that high pressure and low temperature by the use of RTILs-water mixtures might be useful for the freezing preservation of aqueous protein solutions.

The second issue can best be addressed by developing a better understanding of the structural stability of aqueous protein solutions under conditions of high pressure and low temperature.

Pressure–temperature (*P–T*) phase diagrams provide valuable information about the structural stability of proteins. Such diagrams have been obtained by thermodynamic analysis of the pressure and temperature dependencies of structural stability by various experimental methods including NMR, UV–VIS, and vibrational spectroscopy (Brandts et al., 1970; Hawley, 1971; Lassale et al. 2000; Panick et al., 2005; Taniguchi & Suzuli, 1983; Yamaguchi et al., 1995; Zip & Kauzman, 1973). They show that, for many proteins, pressures of >0.5 GPa induce pressure unfolding and temperatures of <253 K induce cold unfolding.

* Corresponding Author

With respect to pressure unfolding, many proteins—such as ribonuclease A(RNase A), lysozyme, myoglobin, and trypsin—pressure unfold at >0.5 GPa (Meersman et al, 2002, 2003;Ruan et al., 1999; Smeller et al., 2006), whereas other proteins such as cytochrome *c* and *trp* repressor do not pressure unfold under high pressure. Dewa *et al.* and Dubins *et al.* showed that cytochrome *c* undergoes pressure-induced recovery (Dewa et al., 1998, Dubins et al., 2003). Moreover, Desai *et al.* showed that the population of the solvated α-helical structure of *trp* repressor increases from approximately 20% to 40% with increasing pressure, whereas the population of the buried α-helical structure, located in the interior of the protein, decreases from approximately 60% to 30% (Desai, et al., 1999).

With respect to cold unfolding, Nash and Jonas suggest that, for some proteins, the cold-unfolded state might resemble an early folding intermediate (Nash & Jonas, 1997). Zhang *et al.* showed that the cold-unfolded state retains a native-like core structure (Zhang et al., 1995). Building on their results, Meersman *et al.* reported that cold unfolding results in the formation of a partially unfolded state in which some secondary contacts are still present (Meersman et al, 2002a).

These results suggest the following: Under high pressure, helical proteins such as myoglobin and cytochrome *c* exhibit different structural stabilities, and α/β- and β-proteins such as RNase A, lysozyme, and trypsin tend to unfold in water. At low temperature, protein denaturation induces partial unfolding.

Thus, two important questions arise: What dominant factor accounts for the different structural stabilities of α-proteins under high pressure? And what is the nature of the structural changes induced in proteins at low temperature? Our goal, in investigating the feasibility of protein preservation under high pressure and low temperature, is to answer these questions. To this end, we report herein the systematic investigation of pressure- and temperature-induced changes in the secondary structures of oligopeptides and helical proteins by means of FTIR and circular dichroism (CD) spectroscopy combined with density functional theory (DFT) and geometric volume calculations.

2. Experimental methodology

2.1 Samples

For peptide samples, Ac-YGAA(KAAAA)$_2$KA-NH$_2$(AK peptide) (Shimizu Laboratory, Soka University) and poly-L-glutamic acid (PLGA) sodium salts (MW = 8000; Peptide Inc.) were used without further purification. For protein samples, bovine serum albumin (BSA) and horse-heart cytochrome *c* (Sigma-Aldrich Co.) were used without further purification. Figure 1 shows the three-dimensional structures of these peptide and proteins.

Sample solutions were prepared as follows. For FTIR spectral measurements, solutions were prepared at concentrations of 20 mg/mL in two different buffer solutions (20 mM MES, pD 5.0; 50 mM tris-DCl buffer, pD 7.6). All exchangeable backbone amide protons in the peptides and proteins were deuterated by incubating in a D_2O solution at 298 K. Completion of hydrogen–deuterium exchange was confirmed by the cessation of shifts in the amide II band. This band, in the frequency region around 1550 cm^{-1}, is known to shift to around 1450 cm^{-1} on deuteration of the backbone amide protons. For CD spectral measurements, solutions were prepared at concentrations of 0.1 mg/mL (~0.06 mM) in a buffer solution (pD 6.7). Sample concentrations were determined by UV absorption at 280 nm of Tyr and Trp (ε = 1197 M^{-1} cm^{-1} for Tyr and 5559 M^{-1} cm^{-1} for Trp).

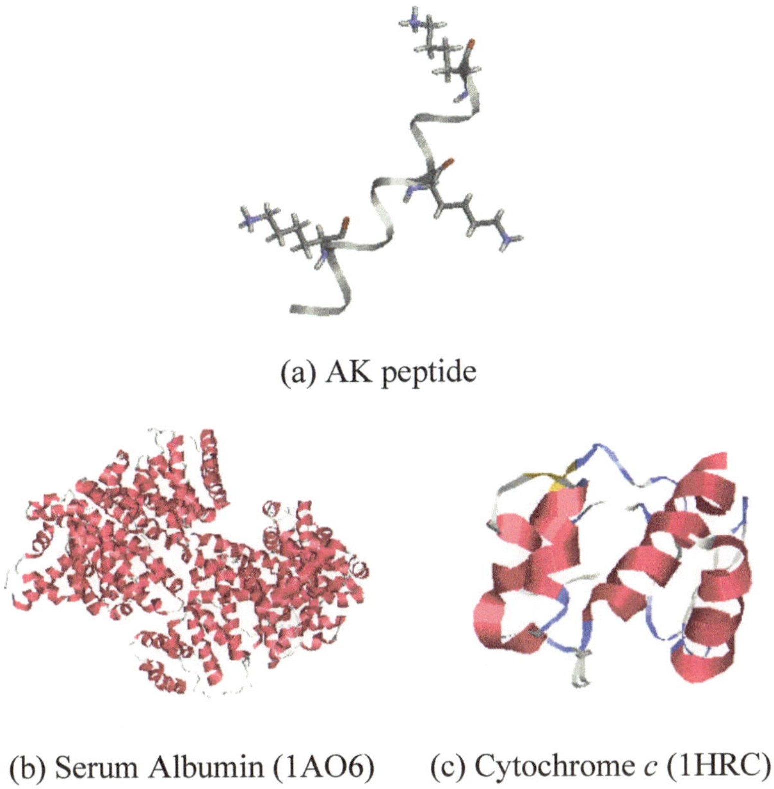

Fig. 1. Three-dimension structures of (a)AK peptide, (b) serum albumin (1AO6), and (c) cytochrome *c* (1HRC). The structure of AK peptide is built using the Gauss view program.

2.2 High-pressure and low-temperature experiments

FTIR and CD spectroscopy are powerful tools for investigating changes in the secondary structures of peptides and proteins. For our purposes, the amide I′ (the deuterium peptide groups) vibrational mode, which consists of C=O stretching, C-N stretching, and C-C-N deformation in the region 1620–1690 cm^{-1} (Figure 2), is highly sensitive to the secondary structures of peptides and proteins, and thus serves as an indicator of α-helical and/or β-sheet structures (Bandekar, 1992; Krimm & Bandekar, 1986). Similarly, CD spectroscopy enables estimation of the helical content of proteins. CD-ellipticity at 208 and 222 nm, which originates from n–π* excitation of the peptide backbone, correlates strongly with helical content (Chen et al., 1974; Greenfield et al., 1969; Kelly et al., 2005). We previously reported the relevant details of FTIR and CD spectroscopy (Takekiyo et al., 2006c, 2009).

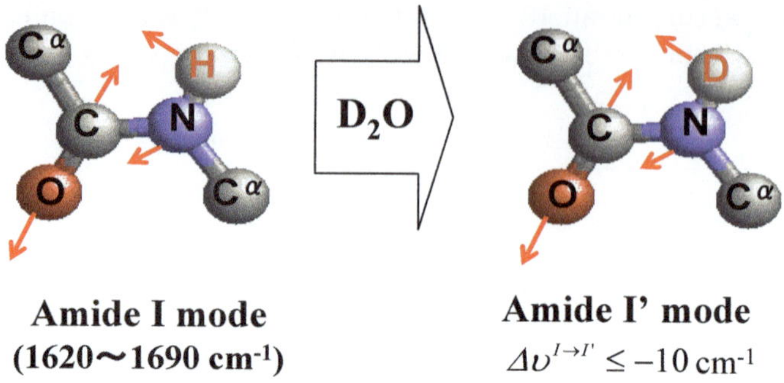

Fig. 2. Relationship between the amide I and I′ modes of protein.

For high-pressure experiments, we used a diamond anvil cell (DAC) (Figure 3a). Sample solution was placed together with a small amount of α-quartz for a pressure marker (Siminovitch, et al., 1987; Wong, et al., 1985) in a Hasteloy C-276 gasket (φ = 1.0 mm, t = 0.05 mm) mounted in the DAC (Figure 3b). Pressure was measured to a precision of ±0.05 GPa. The infrared beam was condensed by a zinc selenide lens system onto the sample in the DAC.

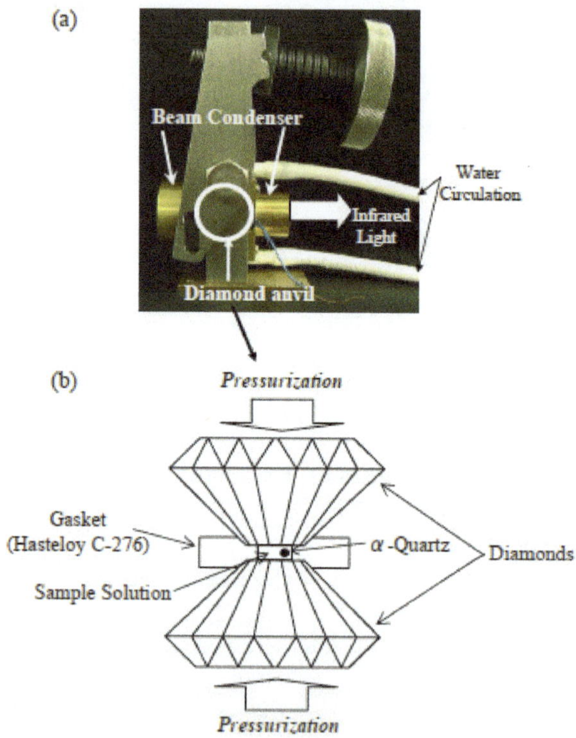

Fig. 3. (a) Schematic picture of diamond anvil cell (DAC) and (b) Schematic diagram of diamond anvils with a gasket and α-quartz.

For low-temperature experiments, we used a cryostat chamber (Figure 4a). The outer windows of the chamber were made of CaF_2. A Teflon spacer (100 μm) was placed between two CaF_2 windows in the transmission cell (Figure 4b), the cell was set in the chamber, and the chamber was filled with liquid N_2 at 0.1 MPa. Temperature was controlled with a mass flow controller. The cooling speed was 30 K/min.

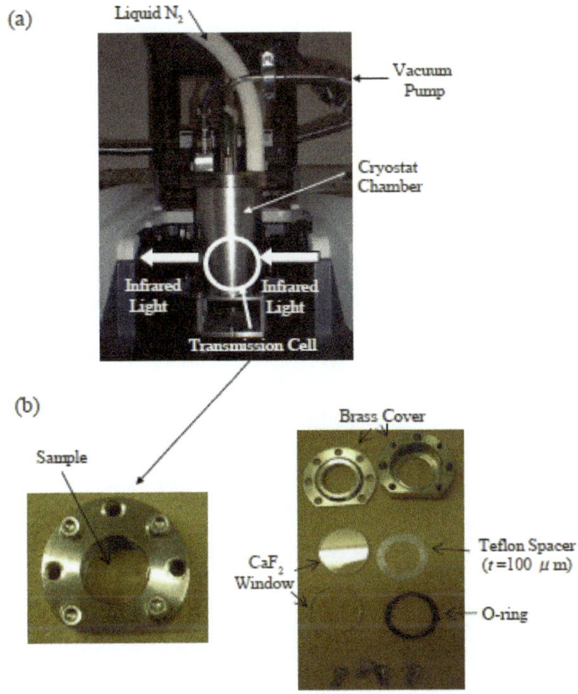

Fig. 4. Schematic picture of (a) cryostat stetted up with FTIR spectrophotometer and (b) transmission cell.

2.3 Density Functional Theory (DFT) and geometric volume calculations

Density functional theory (DFT) calculations have been used as the investigation methodology for the intermolecular interaction between the biomolecules and water molecules (Ham et al., 2003; Kubelka & Keiderling, 2001; Nemukhin et al., 2002). The present DFT calculations were performed using the GAUSSIAN03 program (Frish et al., 2003). For our present calculations, we performed geometry-optimization and frequency calculations at the B3LYP/6-311G++(d,p) level (Becke, 1998, 1998; Lee et al., 1998), which includes the solvent effect, using the polarized continuum model (PCM) (Wiberg & Tomasi, 1982).

Geometric volume calculation has been conducted as the analysis methodology for obtaining the information on a relationship between the partial molar volume (PMV) and the structure of a biomolecule (Imai et al., 2005a, 2005b; Ling et al., 1998). The geometric volume calculations of peptides and proteins were carried out using the alpha-shapes program (Edelsbrunner et al., 1995; Ling et al., 1998). Geometric volume component of peptides and proteins enables the calculation of the solvent accessibility of the molecular surface, assuming that a water molecule is a hard-sphere probe (Edelsbrunner et al., 1995; Imai et al., 2005a, 2005b; Ling et al., 1998). In the present study, we calculated both van der Waals volume (V_W) and molecular volume (V_M). The latter, V_M, consists of V_w plus void volume (V_v), which is the volume of the structural void within the solvent-inaccessible core of the solute molecule. Hence, V_v can be obtained by subtracting V_w from V_M. That is, $V_v = V_M - V_w$. The diameter of water molecule was taken as 2.928 Å, again assuming that a water molecule is a hard-sphere (Imai et al., 2005a). Figure 5 shows schematic definitions of these various volume contributions.

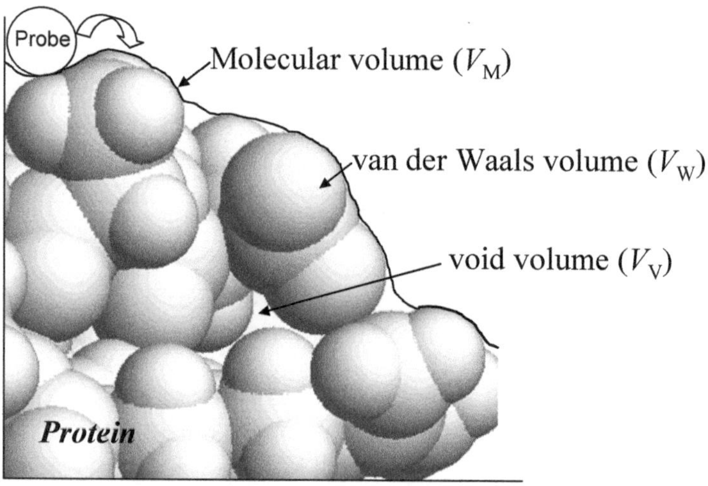

Fig. 5. Definition of geometric volume contributions.

3. Results and discussion

3.1 Pressure-induced structural stability of proteins

Figure 6 shows FTIR and second-derivative spectra as a function of pressure in the amide I′ region for AK peptide and PLGA. The spectra show three peaks at 1635, 1645, and 1672 cm^{-1} for AK peptide and three peaks at 1637, 1648, and 1670 cm^{-1} for PLGA. In a previous FTIR study (Manas et al., 2000; Reisdorf et al., 1996; Silva et al., 2002, Walsh et al., 2003), two peaks at 1635 and 1637 cm^{-1} for AK peptide and PLGA were assigned to the solvated α-helical structure, as shown in the same figure. The peak frequency for a solvated α-helical structure is lower than the characteristic frequency for a buried α-helical structure (~1650 cm^{-1}) due to the hydrogen bond between the α-helix and water molecule as shown in Fig. 7. The peak at ~1645 cm^{-1} for both peptides is thus assigned to a random-coil structure. The peaks at 1672 cm^{-1} for AK peptide are assigned to the asymmetric carboxylate stretching of trifluoroacetic acid ion remained after dialysis (Williams et al., 1996). The peak at 1670 cm^{-1} for PLGA is assigned to a turn structure.

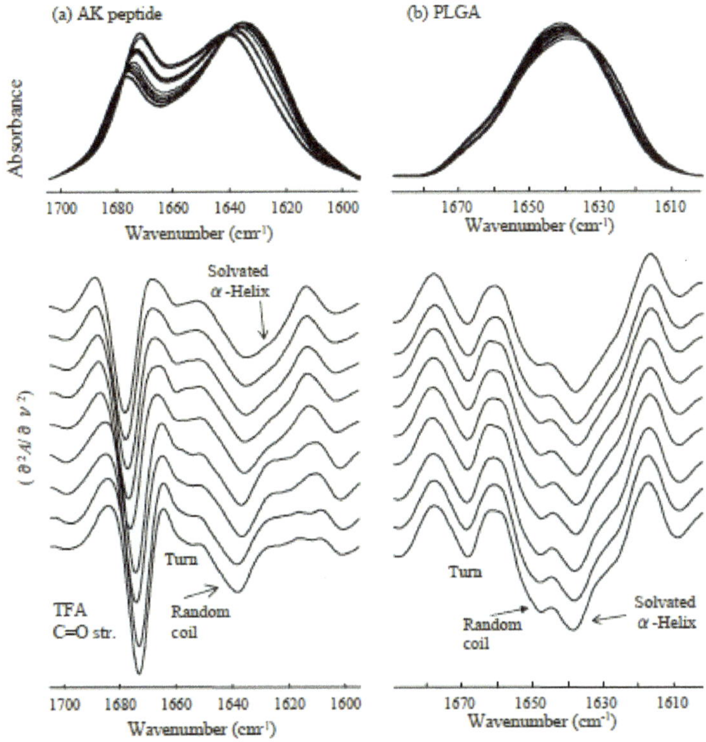

Fig. 6. FTIR (upper) and second derivative (lower) spectra in the amide I′ region of (a) AK peptide and (b) PLGA in aqueous solution as a function of pressure.

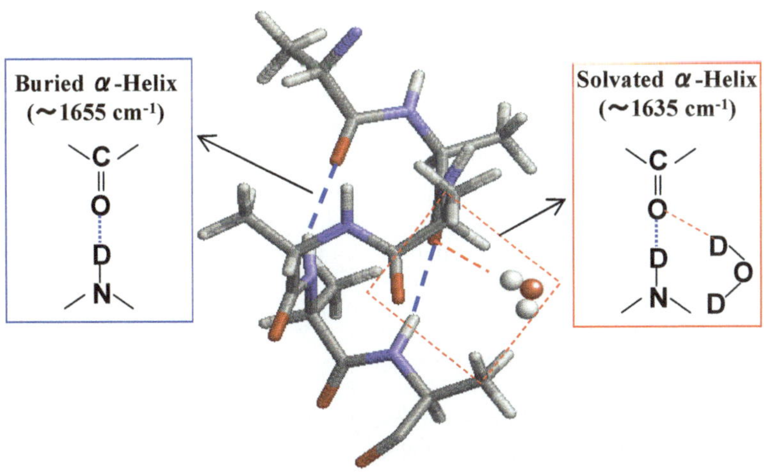

Fig. 7. Scheme of the buried α-helix and solvated α-helix in proteins.

Significantly, for AK peptide and PLGA, the solvated α-helical structures increase with pressure but the random-coil structures decrease from the second derivative spectra. Thus, the solvated α-helical structures of AK peptide and PLGA are clearly stabilized at high pressure. Similar observations have previously been reported. For example, Desai *et al.* reported that the solvated α-helical structure of *trp* repressor is maintained at 0.8 GPa (Desai et al., 1999). A recent FTIR study showed that the solvated α-helical structures of alanine-based peptide (Imamura et al., 2009), four-helix bundle protein $(\alpha\text{-l-}\alpha)_2$ (Takekiyo et al., 2006c) and GCN4-p1 (Imamura et al., 2010) do not unfold even at >1.0 GPa; rather, with increasing pressure, the populations of the buried α-helical structure decrease and those of the solvated α-helical structure increase, implying that the solvated α-helical structures of $(\alpha\text{-l-}\alpha)_2$ and GCN4-p1 hydrate strongly under high pressure. In contrast, however, Meersman *et al.* reported that the α-helical structure of myoglobin begins to break down at >0.5 GPa and denatures completely at 0.7 GPa (Meersman et al., 2002), and Wojciech *et al.* reported that apo and holo types of α-lactalbumin unfold completely at 0.5 GPa (Wojciech et al., 1999). Thus, from the viewpoint of the pressure-induced unfolding of helical proteins, the structural stabilities of even similar helical proteins can differ from one another under high pressure.

3.2 Pressure effect on amide I' frequency shifts of proteins: Interpretation from dft calculations

What, then, is the dominant factor affecting the structural stability of helical proteins under high pressure? We investigated this question by examining the amide I′ frequency shift and geometric volume. The pressure-induced amide I′ frequency shift of a protein is clearly related to environmental changes around the protein.

Recent FTIR studies of helical peptides and proteins have shown that, for $(\alpha\text{-l-}\alpha)_2$ and GCN4-p1, with increasing pressure, the amide I′ frequency of the solvated α-helical structures (~1635 cm^{-1}) shift to lower frequency (Takekiyo et al., 2006c; Imamura et al., 2010). Our current results are consistent with these findings. Figure 8 shows that, for AK peptide and PLGA, with increasing pressure, both amide I′ frequencies of the solvated α-helical structures shift to lower frequency.

To clarify the nature of this pressure-induced amide I′ frequency shift for the α-helical structures of proteins, we focus on the hydrogen bond (H-bond): the intermolecular H-bond between the α-helical structure and water molecules. We then explain the pressure-induced shift to lower frequency by speculating that, with increasing pressure, the intermolecular H-bond between the C=O group of the peptide bond and water molecules strengthens, the C=O force constant weakens, and the contribution of C=O bond unharmonicity increases.

As a model system of solvated α-helical structure, we selected *N*-methylacetamide (NMA; $CH_3CONHCH_3$) dimer + D_2O complex. NMA is the simplest model compound for investigating the structural and physical properties of the peptide group (Ham et al. 2003; Kubelka & Keiderling, 2001). For our present calculations for the NMA dimer + D_2O complex, we examine the two peaks at 1632 and 1656 cm^{-1}, which are amide I′ peaks arising from the NMA–D_2O H-bond and NMA H-bond free, respectively (Figure 9). The former H-bond pattern corresponds to the solvated α-helical structure.

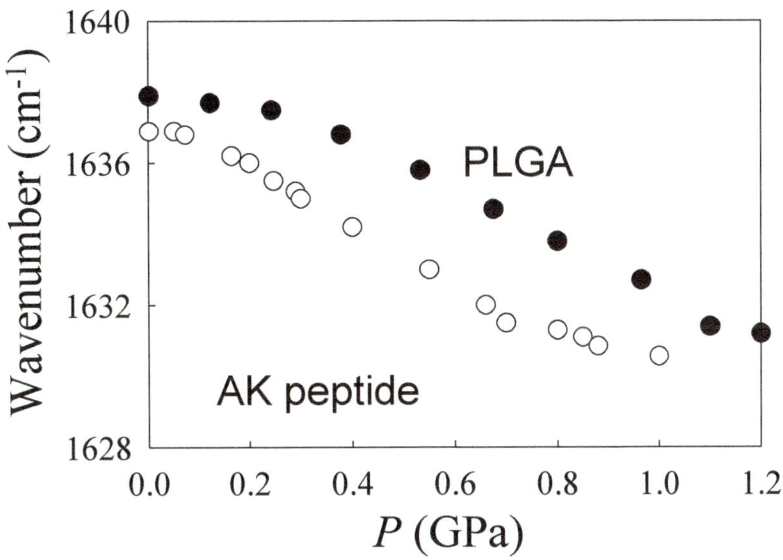

Fig. 8. The peak wavenumber of the solvated α-helical structure of AK peptide and PLGA in aqueous solution as a function of pressure. The closed and open circles represent PLGA and AK peptide, respectively.

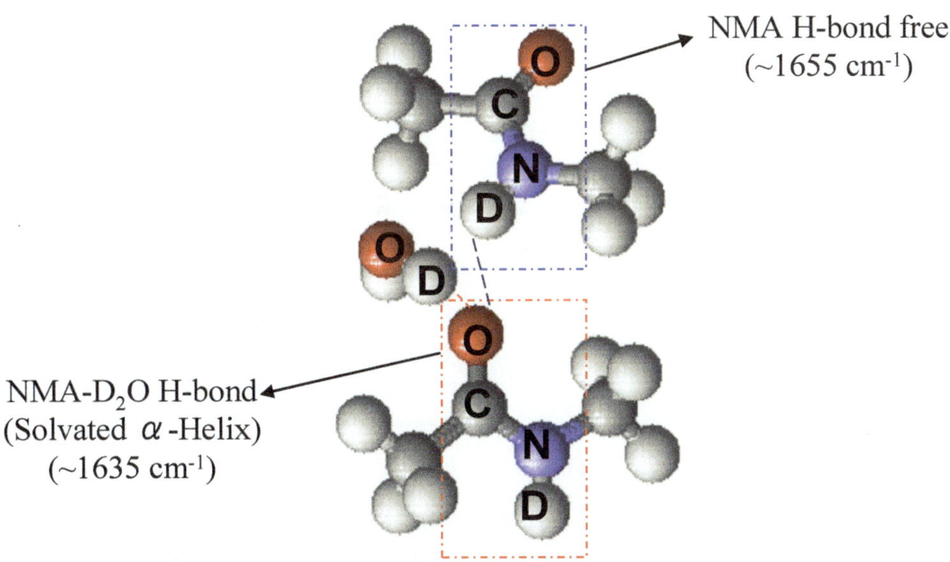

Fig. 9. Amide I′ frequency and hydrogen bond of the peptide bond in the NMA-dimer + D_2O complex. The red and blue long-dashed lines represent the H-bond between the NMA-D_2O and between the NMA-NMA, respectively.

The change in the intermolecular H-bond distance ($d_{O\cdots O}$) between the NMA dimer's C=O group and the water molecule's O–D group is related to pressure-induced changes of the intermolecular H-bond distance in the α-helical structure. We calculated the $d_{O\cdots O}$ dependence of the amide I′ frequencies of NMA dimer + D_2O (Figure 10). With decreasing $d_{O\cdots O}$ (i.e., increasing pressure), the peak at 1632 cm^{-1} (•) shifts to lower frequency. Below $d_{O\cdots O}$ = 2.36 Å, the frequency shift at 1632 cm^{-1} (●) increases significantly. Thus, with decreasing $d_{O\cdots O}$, the intermolecular H-bond between the NMA dimer's C=O group and the water molecule's O–D group strengthens. Because the peak at 1632 cm^{-1} shifts to a slightly extent than that at 1656 cm^{-1}, the average amide I′ frequency shift for NMA dimer + D_2O below $d_{O\cdots O}$ = 2.36 Å is dominated by the peak at 1632 cm^{-1}. This lower amide I′ frequency shift with decreasing $d_{O\cdots O}$ (below $d_{O\cdots O}$ = 2.36 Å) for NMA dimer + D_2O is similar to that for the solvated α-helical structure. Thus, our results indicate that pressure-induced shorting distance of the intermolecular H-bond between the peptide's C=O group and the water molecule's O–D group causes shifts of the amide I′ mode to lower frequency for the solvated α-helical structure. Paschek *et al.*, in a recent IR simulation study (Paschek et al., 2005), proposed that the pressure-induced shift of AK peptide's amide I′ mode to lower frequency is due to hydration effects between the peptide and water molecules rather than structural changes in the peptide. Our results support this suggestion. The amide I′ frequency shifts for the solvated α-helical structure strongly correlate with changes in $d_{C=O\ldots D-O}$.

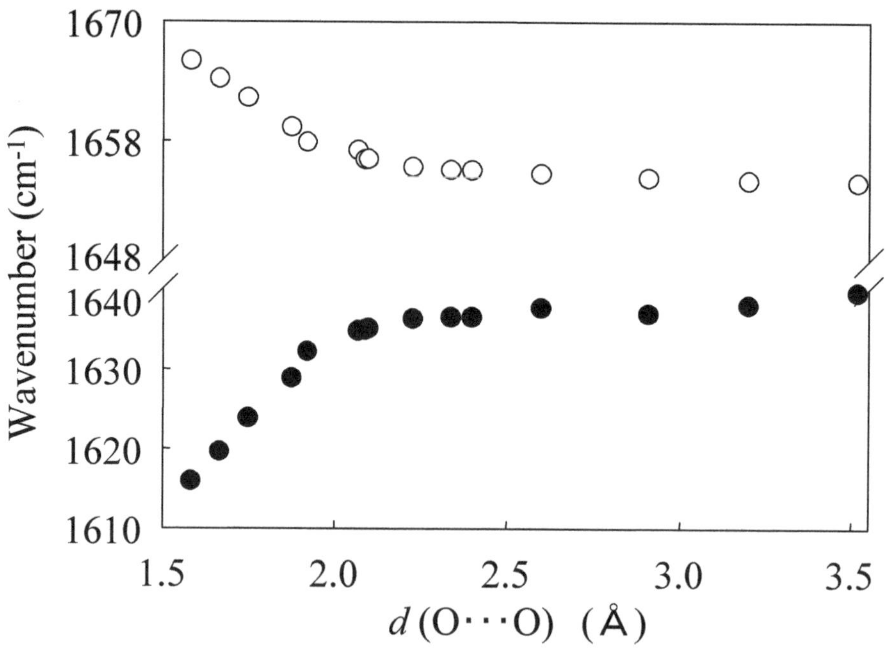

Fig. 10. Changes in the amide I′ frequency of NMA in the NMA-dimer + D_2O complex as a function of the bond distance ($d_{O\cdots O}$) between the C=O group of the peptide bond of NMA dimmer and the O-D group of water calculated at B3LYP/6-311+(d,p) basis set.

3.3 Origin of pressure stability of proteins: Interpretation from geometric volume calculations

We now consider the pressure stability of proteins in terms of geometric volume. PMV is a fundamental thermodynamic quantity that characterizes protein conformation (Chalikian, 2003) and is of principal importance in the analysis of pressure-induced denaturation of protein by Le Chatelier's law (Balny et al., 2002; Royer, 2002). Assuming that the protein unfolding process is a two-state transition (folded (F) ↔ unfolded (U)), the relationship between the pressure dependence of the equilibrium constant (K) and PMV is as follows:

$$(\partial \ln K/\partial p)_T = -\Delta V/RT$$

where K is the equilibrium constant, where ΔV is the difference in PMV between the protein's folded and unfolded states. From this relationship, it is evident that increasing pressure causes the protein structure to change so as to minimize PMV ($\Delta V < 0$). Therefore, analysis of PMV differences associated with pressure unfolding can improve our understanding of the structural stability of proteins under high pressure.

Drawing upon the PMV analysis reported by Chalikian and Breslauer (Chalikian & Breslauer, 1996), we now systematically consider PMV differences between the folded and unfolded conformers of helical peptides and proteins. The PMV difference between the folded and unfolded states can be decomposed into three terms as follows:

$$\Delta V = \Delta V_w + \Delta V_v + \Delta V_{Hyd},$$

where ΔV_w and ΔV_V are the van der Waals and void-volume differences, respectively, and ΔV_{Hyd} is the hydration volume difference caused by changes in intermolecular interactions between the solvent water and solute.

We suggest that protein void space may be the most significant factor affecting the pressure stability of proteins. Lopez *et al.* have speculated similarly (Lopes et al., 2004). They investigated the pressure dependence of the volume contribution of the hydrophobic internal cavity inside three-helix bundle mutant proteins and showed that a large cavity space induces a large change in volume. Therefore, we focus herein on the void volume of various helical peptides and proteins to elucidate the relationship between the pressure stability and volume properties of proteins.

We systematically calculated the geometric volumes of various peptides and proteins (probe size 2.98 Å). Table 1 shows the volume properties of various peptides and proteins. Figure 11(a) shows plots of V_w, V_M, and V_v as a function of molecular weight (MW). V_w, V_M, and V_v all increase linearly with MW. Imai *et al.* recently calculated the MW dependence of the molecular volume of various proteins and showed that V_v correlates well with MW (Imai et al., 2005a); our results are consistent with these. Figure 11(b) shows plots of V_v/MW as a function of MW. V_v/MW increases up to MW = ~20000 and remains constant thereafter.

The pressure stability of a protein structure is thought to correlate with the adiabatic compressibility (β) of the folded protein (Taulier & Chalikian, 2002; Gekko, 2004). That is, in a folded protein, increasing V_v causes an increase in β. The β values for the well-known helical proteins cytochrome *c* (Kamiyama et al., 1999) and myoglobin (Gekko & Noguchi, 1974) are 3.60 and 8.98 $Mbar^{-1}$, respectively. The former is clearly smaller than the latter.

Similarly, the V_v/MW values for cytochrome *c* and myoglobin are 0.153 and 0.223 cm³ mol⁻¹ MW⁻¹, respectively. Thus, for these two proteins, the difference in relationship of V_v/MW values correlates with the difference in β values.

Protein	PDB ID	$N_{residue}$*	MW**	V_w	V_M	V_v	V_v/MW
Helical peptide	1DJF	15	1586	955.6	1067.1	111.6	0.070
GCN4-p1	2TZA	68	7958	4550.2	5900.4	1353.3	0.170
Cytochrome *c*	1HRC	105	11688	7126.1	8913.9	1787.7	0.153
Four helix bundle	1MFT	106	12898	6855.8	8691.9	1836.2	0.142
Myohemerythrin	2MHR	118	13761	9679.0	12495.1	2816.0	0.204
Myoglobin	101M	154	17266	11134.3	14987.6	3853.3	0.223
Phospholipase c	1AH7	245	28373	17438.2	23448.4	6010.2	0.211
Hemerythrin (deoxy)	1HMD	452	53328	31758.1	42452.6	10694.5	0.200
Human Hemoglobin	1A3N	574	61922	39184.1	53289.1	14105.1	0.227

*$N_{residue}$: Amino acid residue
** MW: Molecular weight

Table 1. Volume contributions (cm³/mol) of proteins by geometric volume calculation.

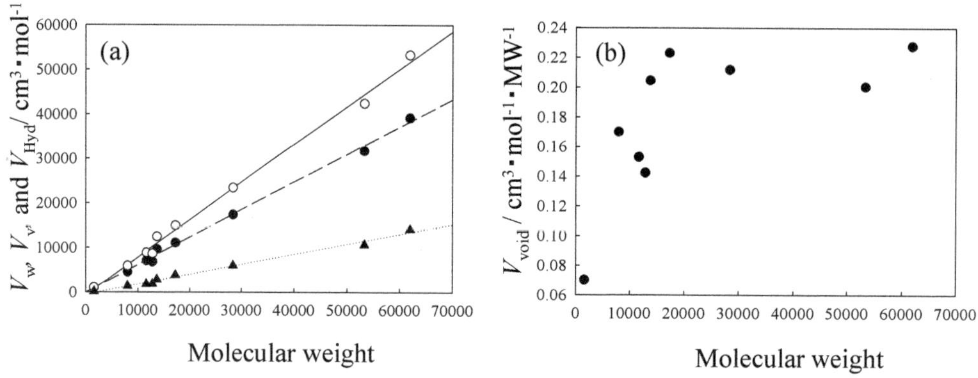

Fig. 11. (a) Fig. 11. (a) The changes in the van der Waals volume (V_w)(●), molecular volume (V_M)(○), and void volume (V_V)(▲) of various helical peptide and proteins as a function of molecular weight. The straight lines are the results of the least-squares analysis. The correlation coefficient values (R) of the least-square fit are 0.9736 for V_w, 0.9769 for V_M, and 0.9398 for V_V, respectively. (b) Change in the V_v per molecular weight of helical peptides and proteins as a function of molecular weight.

Next, we consider the pressure stability of proteins obtained by our present and previous results. The helical structures of AK peptide (15 residues), helix bundle proteins (64 residues) (Takekiyo et al., 2006c), and GCN4-P1 (68 residues) (Imamura et al., 2010) do not unfold even at ~1.0 GPa. The V_v/MW values for these helical peptides and proteins are all <0.2 cm³ mol⁻¹ MW⁻¹. In contrast, the helical structures of myoglobin (153 residues) (Meersman et al., 2002) and α-lactalbumin unfold completely at 0.7 GPa (Wojciech et al., 1999). The V_v/MW value for myoglobin is 0.223 cm³ mol⁻¹ MW⁻¹. From Figure 11 (b), we see that the structures of helical peptides and proteins for which V_v/MW < 0.2 $cm^3 \cdot mol^{-1} \cdot MW^{-1}$ exhibit small pressure stability, and those for which V_v/MW > 0.2 $cm^3 \cdot mol^{-1} \cdot MW^{-1}$ exhibit large pressure stability. Thus, our calculation result is in qualitatively good agreement with previous experimental results for the pressure stability of helical peptides and proteins, and both the examined proteins do indeed exhibit large pressure stability. On the basis of these calculations, we suggest that helical proteins having a small void space are stabilized at high pressure.

Our findings so far suggest that, for successful preservation under high pressure, two conditions must be met: (1) the α-helical structure must be sufficiently hydrated with water molecules, and (2) the protein must have a small void space.

3.4 Comparisons of the structural stability of proteins at low temperature

We have just discussed the structural stability of proteins under high pressure. We now focus on the structural stability of proteins at low temperature (liquid N_2, 77 K).

Figure 12 shows the FTIR (upper) and CD (lower) spectra of PLGA, BSA, and cytochrome *c* in aqueous solution at 298 K and 77 K, both at 0.1 MPa. The FTIR spectrum of PLGA shows an intriguing change: the peak at 1638 cm^{-1} increases in intensity with decreasing temperature from 298 to 77 K. This peak position is close to that assigned to the solvated α-helical structure, despite the fact that bulk water is crystallized at 77 K. If low temperature induces the solvated α-helical structure of PLGA, the associated CD spectrum at low temperature should show double negative bands at 208 and 222 nm, characteristic of an α-helical structure. Indeed, the CD spectra of PLGA show, at 298 K, a negative band at 195 nm, characteristic of a random-coil structure, the CD spectra show the characteristic of α-helical structure. Thus, low temperature induces the transition of PLGA from coil to helix and then stabilizes the resulting solvated α-helical structure.

What is the α-helical structure of proteins at 77 K? In Figure 12, the peaks for BSA and cytochrome *c* at ~1635 cm^{-1} increase with decreasing temperature from 298 to 77 K. The CD spectra show negative bands at 208 nm and 222 nm even at 77 K, albeit at changed intensity. Thus, the solvated α-helical structure of PLGA, BSA, and cytochrome *c* is clearly stabilized at 77 K despite the crystallization of bulk water. Moreover, these spectral changes are completely reversible. Thus, the low-temperature-induced stabilities of the solvated α-helical structure of helical proteins are very similar to the pressure-induced stabilities.

Previous FTIR studies by Manas *et al.* and Walsh *et al.* showed that the solvated α-helical structure of α_3D (a *de novo* designed three-helix bundle protein), GCN4-p1, and parvalbumin in glycerol-D_2O mixed solution does not unfold at low temperature (10 K) but rather is stabilized (Manas et al., 2000; Walsh et al., 2003). Generally, addition of glycerol to a protein solution is known to induce incremental structural stability in the protein (Zelent et al., 2004; Wright et al., 2003). Our present results show that the solvated α-helical structures of PLGA, BSA, and cytochrome *c* are stabilized at 77 K without the addition of glycerol. On the other hand, the secondary structures of RNase A, lysozyme, and trypsin having a β-sheet structure show an increase in the ratio of unfolded structure to β-sheet structure at 77 K (data not shown). Based on these results, we found the difference of the structural stability of proteins at 77 K from the viewpoint of two secondary structure types (the α-helix and β-sheet). A general rule is that the solvated α-helical structure is hydrophilic and the β-sheet

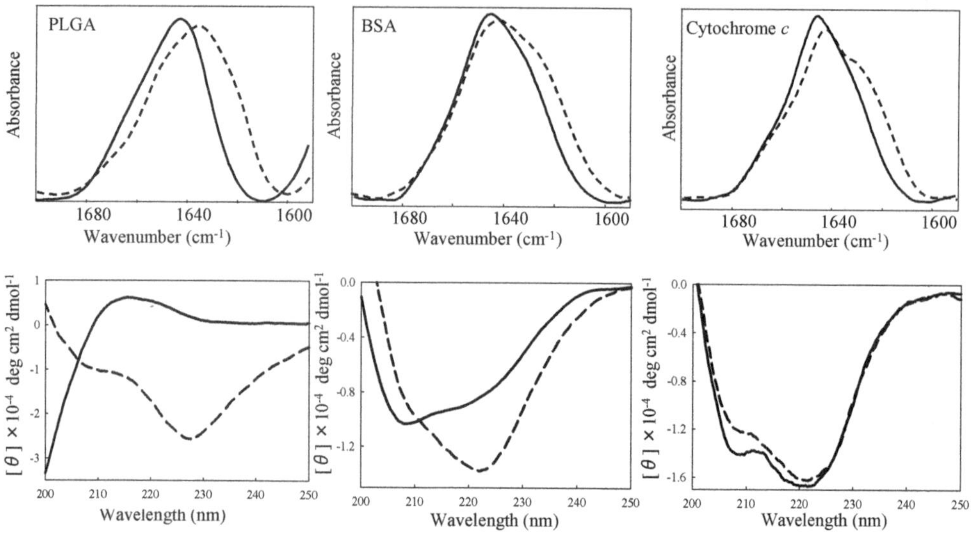

Fig. 12. FTIR (upper) and CD (lower) spectra of PLGA, BSA, and cytochrome *c* in water at 298 (solid line) and 77 K (long-dashed line).

structure is hydrophobic. Typically, the stability of β-sheet cores in proteins requires the exclusion of water since the hydrogen bonding is much sparser between the β-sheets (between individual strands) than between the turn or α-helix. If the bulk water is frozen it is impossible for the water exclusion effects to act and no reduction in free energy is achieved by formation of β-sheet structure. In contrast, as stated above, α-helical structures are formed by H-bonds between successive loops in the helix. These H-bonds would come more stable (relative to thermal fluctuations) at low temperature. Hence, lowering the temperature may even induce α-helix formation such as PLGA and helical proteins.

In this section, our findings so far suggest that, for successful preservation under low pressure, one condition must be met: the α-helical structure must be sufficiently hydrated with water molecules as in the case of high pressure.

3.5 Feasibility of protein preservation under high pressure and low temperature

As mentioned in Section 1, a necessary condition for protein preservation under high pressure and low temperature is that ice formation must be avoided, and this can be accomplished by addition of a freezing-protection agent such as sugar or salt. For example, Yamamoto *et al.* recently reported compression recovery of the secondary structure of RNase A in a sucrose–D_2O mixed solution (Yamamoto, et al. 2010). Manas *et al.* and Walsh *et al.* reported, by FTIR study, that the secondary structures of α_3D, GCN4-p1, and parvalbumin in a glycerol–D_2O mixed solution do not unfold at low temperature (10 K), suggesting that the solvated α-helical structure hydrated with water molecules is stabilized under this condition (Manas et al., 2000; Walsh et al., 2003). Moreover, Thus, under high pressure and low temperature, sugar solutions can induce retention of the secondary structures of proteins.

Given that the protein structure can be retained with addition of a freezing-protection agent, can it also be retained without addition of such an agent? Our present and previous results (Meersman et al, 2002, 2003; Ruan et al., 1999; Smeller et al., 2006) show that proteins having a β-sheet structure unfold at high pressure and show an increase in the ratio of unfolded structure to β-sheet structure at 77 K. Therefore, protein preservation under high pressure and low temperature is difficult without the use of a freezing-protection agent. However, our present results show that the secondary structures of helical proteins having a small void space do not unfold at high pressure and low temperature. Thus, helical proteins may resist low-temperature- and high-pressure-induced protein unfolding, and their preservation under high-pressure and low-temperature conditions may thus be feasible.

4. Summary and conclusions

Protein preservation under high pressure and low temperature can be hampered by two issues: (1) conversion of bulk water to ice, and (2) structural instability of the protein by pressure unfolding (which occurs at pressures of >0.5 GPa) and cold unfolding (which occurs at temperatures of <243 K). The first issue can be solved by use of a freezing-protection agent. Our present results show that the second issue is not a concern for helical proteins having a small void space because these proteins are clearly stabilized at high pressure (~1.0 GPa) or low temperature (77 K). Thus, proteins having a high degree of solvated α-helicity have a high feasibility of preservation under high pressure and low temperature, even without the use of a freezing-protection agent.

Many previous studies have reported the structural stability of proteins in aqueous sugar or salt solutions at low temperature, but only a few have done so at high pressure. However, both pressure and temperature play critical roles in regulating the structures and properties of proteins, and both are important tools for exploring new methods of protein preservation. Our present results demonstrate the feasibility of protein preservation under conditions of both high pressure and low temperature.

5. References

Balny, C., Masson, P., Heremans, K. (2002) High Pressure Effects on Biological Macromolecules: From Structural Changes to Alternation of Cellular Processes. *Biochim. Biophys. Acta* Vol. 1595: 3-10.

Bandekar, J. (1992) Amide modes and protein conformations. *Biochim. Biophys. Acta* Vol. 1120: 123-143.

Becke, A. D. (1988) Density-functional Exchange-energy Approximation with Correct Asymptotic Behavior, *Phys. Rev. A* Vol. 38: 3098-3100.

Brandts, J. F., Oliveira, R. J.,Westort, C. (1970) Thermodynamics of Protein Denaturation : Effect of Pressure on the Denaturation of Ribonuclease A, *Biochemistry* Vol.9: 1038-1047.

Chalikian, T. V. (2003) Volumetric Properties of Proteins, *Annu. Rev. Biophys. Biomol. Struct.* Vol. 32: 207-235.

Chalikian, T. V., & Breslauer, K. (1996) On Volume Changes Accompanying Conformational Transitions of Biopolymers, *Biopolymers* Vol. 39: 619–626.

Chen, Y. H., Yang, J. T., Chau, K. H. (1974) Determination of the Helix and Beta Form of Proteins in Aqueous Solution by Circular Dichroism, *Biochemistry* Vol. 13: 3350-3359.

Desai, G., Panick, G., Zein, M., Winter, R., Royer, C. A. (1999) Pressure-jump Studies of the Folding/Unfolding of *trp*-Repressor, *J. Mol. Biol.* Vol. 288: 461-475.

Dewa, M., Tayauchi, M., Sakurai, M., Nitta, K. (1998) Compression Refolding of Cytochrome c, *Protein Pep. Lett.* Vol. 5: 265-268.

Dubins, D. N., Filfil, R., Magregor, R. B., Chalikian, T. V. (2003) Volume and Compressibility Changes Accompanying Thermallu-induced Native-to Unfolded and Moleten Gloubule-to Unfolded Transitions of Cytochrome *c*: A High Pressure Study, *Biochiemistry* Vol. 42: 8671-8678.

Dzwolak, W., Kato, M., Shimizu, A., Taniguchi, Y. (1999) Fourier-transform Infrared Spectroscopy Study of the Pressure-induced Changes in the Structure of the Bovine α-Lactalbumin: The Stabilizing Role of the Calcium Ion, *Biochim. Biophys. Acta* Vol. 1433: 45-55.

Edelsbrunner, H., Facello, M., Fu, P., & Liang, J. (1995) Measuring Proteins and Voids in Preoteins, *Proceedings of the 28th Annual Hawai International Conference on System Science*; IEEE Computer Society Press: Los Alamos, CA, Vol. 5: 256-264.

Frank, F. (1995) Protein Denaturation at Low Temperatures, *Adv. Protein Chem.* Vol. 46: 105-139.

Frish, M. J., Trucks, G. W., Schlegel, H. B., Scuseria, G. E., Robb, M. A., Cheeseman, J. R., Zakrzewski, V. G., Montogomery, J. A., Daniels, A. D., Kudin, K. N., Strain, M. C., Farkas, O., Tomasi, J., Barone, V., Cossi, M., Cammi, R., Mennucci, B., Pomelli, C., Adamo, C., Clifford, S., Ochterski, J., Petersson, G. A., Ayala, P. Y., Cui, Q., Morokuma, K., Malick, D. K., Rabuck, A. D., Raghavechari, K., Foresman, J. B., Cioslowski, J., Ortiz, J. V., Baboul, A. G., Stefanov, B. B., Liu, G., Liashenko, A., Piskorz, P., Komaromi, I., Gomperts, R., Martin, R. L., Fox, D. J., Kieth, T., Al-Laham, M. A., Peng, C. Y., Nanayaakkara, A., Gonzalez, C., Challacombe, M. P., Gill, M. W., Johnson, B., Chen, W., Wong, M. W., Andres, J. L., Gonzalez, C., Head-Gordon, M., Replogle, E. S., & Pople, J. A. (2003) GAUSSIAN 03, Gaussian, Inc., Pittsburgh, PA.

Gekko, K. (2004) Effect of Hydration on the Volume and Compressibility of Protein Molecules, *Netsu Sokutei* Vol. 31: 186-193 (in Japanese).

Gekko, K., Noguchi, H. (1979) Compressibility of Gloubular Proteins in Water at 25°C, *J. Phys. Chem.* Vol. 83 : 2706-2714.

Greenfield, N., Fasman, D. D. (1969) Computed Circular Dichroism Spectra for the Evaluation of Protein Conformation, *Biochemistry* Vol. 8: 4108-4116.

Ham, S. Kim, J-H. Lee, H. Cho, M. (2003) Correlation between Electronic and Molecular Structure Distortions and Vibrational Properties. II. Amide I modes of NMA–nD_2O complexes. *J. Chem. Phys.* Vol. 118: 3491-3498.

Hawley, S. A. (1971) Reversible Pressure-Temperature Denaturation of Chymotrypsinogen, *Biochemistry* Vol. 10: 2436-2442.

Heremans, K., Smeller, L. (1998) Protein Structure and Dynamics at High Pressure, *Biochim. Biophys. Acta* Vol. 1386: 353-370.

Imai, T., Kovalenko, A., Hirata, F. (2005) Partial Molar Volume of Proteins Studies by the Three-Dimensional Reference Interaction Site Model Theory, *J. Phys. Chem. B* Vol. 109: 6658-6665.

Imai, T., Takekiyo, T., Kovalenko, A., Hirata, F., Kato, M., & Taniguchi, Y. (2005) Theoretical Study of Volume Changes Associated with the Helix-Coil Transition of an Alanine-Rich Peptide in Aqueous Solution, *Biopolymers* Vol. 79: 97-105.

Imai, Y., Abe, H., Goto, T., Michishita, T., Yoshimura, Y. (2010) Pressure-induced Phase Transition of Ionic Liquid [DEME][BF_4]+H_2O Mixtures, *J. Phys. Conf. Ser.* Vol. 215: 012069 (pp.1-4).

Imai, Y., Takekiyo, T., Abe, H., Yoshimura, Y. (2011) Pressure- and Temperature-induced Raman Spectral Changes of 1-Butyl-3-methylimidazolium Tetrafluoroborate, *High. Press. Res.* Vol. 31: 53-57.

Imamura, H., and Kato, M. (2009) Effect of Pressure on Helix-Coil Transition of an Alanine-Based Peptide: an FT-IR Study, *Proteins* Vol. 75: 911-918.

Imamura, H., Isogai, Y., Takekiyo, T., Kato, M. (2010) Effect of Pressure on the Secondary Structure of Coiled Coil Peptide GCN4-p1, *Biochim. Biophys. Acta* Vol. 1804: 193-198.

Kamiyama, T., Sadahide, Y., Nogusa, Y., Gekko, K. (1999) Polyol-induced Molten Globule Cytochrome *c*; An Evidence for Stabilization by Hydrophobic Interaction, *Biochim. Biophys. Acta* Vol. 1434: 44-57.

Kelly, S. M., Jess, T. J., Price. N. C. (2005) How to Study Proteins by Circular Dichroism, *Biochim. Biophys. Acta* Vol. 1751: 119-139.

Krimm, S., Bandekar, J. (1986) Vibrational Spectroscopy and Conformation of Peptides, Polypeptides, and Protein. *Adv. Protein Chem.* Vol. 38: 181-364.

Kubelka, J., Keiderling, T.A. (2001) Ab initio calculation of amide carbonyl stretch vibrational frequencies in solution with modified basis sets. 1. N-methylacetamide. *J. Phys. Chem. A* Vol. 105: 10922-10928.

Kunugi, S., Tanaka, N. (2002) Cold Denaturation of Proteins under High Pressure, *Biochim. Biophys. Acta* Vol. 1595: 329-344.

Lassalle, M. W., Yamada, H., Akasaka, K. (2000) The Pressure-Temperature Free Energy-Landscape of Staphylococcal Nuclease by ^{1}H NMR, *J. Mol. Biol.*, Vol. 298: 293-302.

Lee, C., Yang, W., & Parr, R. G. (1988) Development of the Colle-Salvetti Correlation-Energy Formula into a Functional of the Electron Density, *Phys. Rev. B* Vol. 37: 785-789.

Liang, J., Edelsbrunner, H., Fu, P., Sudlhakar, P. V., & Subramaniam, S. (1998) Analytical Shape Computation of Macromolecules: I. Moleuclar Area and Volume Through Alpha Shape, *Proteins*, Vol. 33: 1-17.

Lopes, D. H. J., Chapeaurouge, A., Manderson, G. A., Johansson, J. S., Ferreira, S. T. (2004) Redesigning the Folding Energetics of a Model Three-Helix Bundle Protein by Site-Directed Mutangesis, *J. Biol. Chem.* Vol. 279: 10991-10996.

Manas, E. S., Getahun, Z., Wright, W. W., DeGrado, W. F., Vanderkooi, J. M. (2000) Infrared Spectra of Amide Groups in α-Helical Proteins: Evidence for Hydrogen bonding between Helices and Water, *J. Am. Chem. Soc.* Vol. 122: 9883-9890.

Meersman, & Heremans, K. (2003) High Pressure induces the Formation of Aggregation-prone States of Proteins under Reducing Conditions, *Biophys. Chem.* Vol. 104: 297-304.

Meersman, F., Smeller, L., Heremans, K. (2002) Comparative Fourier Transform Infrared Spectroscopy Study of Cold-, Pressure-, and Heat-induced Unfolding and Aggregation of Myoglobin, *Biophys. J.* Vol. 82: 2635-2644.

Miyazaki, Y., Matsuo, T., Suga, H. (1993) Glass Transtion of Myoglobin Crystal, *Chem. Phys. Lett.* Vol. 213: 303-308.

Miyazaki, Y., Matuo, T., Suga, H. (2000) Low-Temperature Heat Capacity and Glass Behavior of Lysozyme Crystal, *J. Phys. Chem. B* Vol.104: 8044-8052.

Nash, D. P., Jonas, J. (1997) Structure of the Pressure-assisted Cold Unfolded State of Ubiquitin, *Biochim. Biophys. Res. Commun.* Vol. 238: 289-291.

Nemukhin, A. V., Grigorenko, B. L., Bochenkova, A. V., Topol, I. A., Burt, S. K. (2002) A QM/MM Approach with Effective Fragment Potentials Applied to the Dipole-Water Structures, *J. Mol. Strucut. (THEOCHEM)* Vol.581; 167-175.

Panick, G., Vidugiris, G. J. A., Malessa, R., Rapp, G., Winter, R., Royer, C. A.(1999) Exploring the Temperature-Pressure Phase Diagram of Staphylococcal Nuclease, *Biochemistry* Vol.38; 4157-4164.

Pascheck, D., Gnanakaran, S., Garcia, A. (2005) Simulations of the Pressure and Temperature Unfolding of an α-Helical Peptide, *Proc. Natl. Acad. Sci. USA* Vol. 102: 6765-6770.

Pnkratov, V. V., Friedrich, J., Vanderkooi, J. M., Burin, A. L., Berlin, Y. A. (2004) Physics of Proteins at Low Temperature, *J. Low Temp. Phys.* Vol. 137: 289-317.

Reisdorf, W. C., Krimm. (1996) Infrared Amide I' Band of the Coiled Coil, *Biochemistry* Vol. 35: 13863-13860

Royer, C. A. (2002) Revisiting Volume Changes in Pressure-induced Protein Unfolding, *Biochim. Biophys. Acta* Vol. 1595: 201-209.

Ruan, K., Lange, R., Meersman, F., Heremans, Balny, C. (1999) Fluorescence and FTIR Study of the Pressure-induced Denaturation of Bovine Pancreas Trypsin, *Eur. J. Biochim.* Vol. 265 : 79-85.

Silva, R. A. G. D., Ngyyen, J. Y., Decatur, S. M. (2002) Probing the Effect of the Side Chains on the Conformation and Stability of Helical Peptides via Isotope-edited Infrared Spectroscopy, *Biochemistry* Vol. 41: 15296-15303.

Siminovitch, D. J., Wong, P. T. T., Mantsch, H. H. (1987) Effect of *Cis* and *Trans* Unsaturation on the Structure of Phospholipid Bilayers: A High-Pressure Infrared Spectroscopic Study, *Biochemistry* Vol. 26: 3277-3287.

Smeller, L., Meersman, F., Heremans, K. (2006) Refolding Studies using Pressure: The Folding Landscape of Lysozyme in the Pressure-Temperature Plane, *Biochim. Biophys. Acta* Vol. 1764: 497-505.

Takekiyo, T., Hatano, N., Imai, Y., Abe, H., Yoshimura, Y. (2011) Conformational Preferences of Two Imidazolium-based Ionic Liquids at High Pressures, *Chem. Phys. Lett.* Vol. 511: 241-244.

Takekiyo, T., Ling, W., Yoshimura, Y., Shimizu, A., Keiderling, T. A. (2009) Relationship between the Hydrophobic Interactions and Secondary Structural Stability for Trpzip β-Hairpin Peptides, *Biochemistry* Vol.48: 1543-1552.

Takekiyo, T., Takekda, N., Isogai, Y., Kato, M., Taniguchi, Y. (2006c) Pressure Stability of the α-Helix Structure in a De Novo Designed Protein $(\alpha\text{-l-}\alpha)_2$ Studied by FTIR Spectroscopy, *Biopolymers* Vol. 85 : 185-188.

Takekiyo, T., Yoshimura, Y. (2006a) Raman Spectroscopic Study on the Hydration Structures of Tetraethylammonium Cation in Water, *J. Phys. Chem. A* Vol. 110: 10829-10833.

Takekiyo, T., Yoshimura, Y. (2006b) Drastic Change in the Conformational Equilibria of Tetraalkylammonium Bromide in the Glassy Aqueous Solution, *Chem. Phys. Lett.* Vol. 420 : 1-6.

Taniguchi, Y., Suzuki, K. (1983) Pressure Inactivation of α-Chymotrypsin, *J. Phys. Chem.* Vol.87: 5185-5183.

Taulier, N., Chalikian, T. V (2002) Compressibility of Protein Transitions, *Biochim. Biophys. Acta* Vol. 1595: 48-70.

Vanderkooi, J. M. (1998) The Protein State of Matter, *Biochim. Biophys. Acta* Vol. 1386: 241-253.

Walsh, S. T. R., Cheng, R. P., Wright, W. W., Alonso, D. O. V., Daggett, V., Vanderkooi, J. M., DeGrado, W. F. (2003) The hydration of Amides in Helices; A Comprehensive Picture from Molecular Dynamics, IR, and NMR, *Protein Sci.* Vol.12: 520-531.

Wiberg, K. B., Tomasi, J. (1982) Approximate Evaluations of the Electrostatic Free Energy and Internal Energy Changes in Solution Processes, *Chem. Phys.* Vol. 65: 239-245.

Williams, S., Causgrove, T. P., Gilmanshin, R., Fang, K. S., Callender, R. H., Woodruff, W. H., Dyer, R. B. (1996) Fast Ivents in Protein Folding: Helix Melting and Formation in a Small Peptide, *Biochemistry* Vol. 35: 691-697.

Wong, P. T. T., Moffatt, D. J., Baudais, F. L. (1985) Crystalline quartz as an Internal Pressure Calibrant for High Pressure Infrared Spectroscopy, *Appl. Spectrosc.* Vol. 39: 733-735.

Wright, W. W., Guffanti, G. T., Vanderkooi, J. M. (2003) Protein in Sugar Films and in Glycerol/Water as Examined by Infrared Spectroscopy and by the Fluorescence and Phosphorescence of Tryptophan, *Biophys. J.* Vol. 85: 1980-1995.

Yamaguchi, T., Yamada, H., Akasaka, K. (1995) Thermodynamics of Unfolding of Ribonuclease A under High Pressure; A Study by Proton NMR, *J. Mol. Biol*, Vol.250: 689-694.

Yamamoto, T., Chatani, E., Kato, M. (2010) FTIR Observation of Compression Recovery of the Secondary Structure of Heat Denaturated Ribonuclease A in Sucrose Solution, *J. Phys. Conf. Ser.* Vol.215: 012155 (pp.1-6).

Yoshimura, Y., Goto, T., Abe, H., Imai, Y. (2009) Existence of Nearly-Free Hydrogen Bonds in an Ionic Liquid, *N, N*-Diethyl-*N*-Methyl-*N*-(2-methoxyetyhl) Ammonium Tetrafluoroborate-Water at 77 K, *J. Phys. Chem. B*, Vol.113: 8091-8095.

Yoshimura, Y., Kimura, H., Okamoto, C., Miyashita, T., Imai, Y., Abe, H. (2011) Glass Transition Behavior of Ionic Liquid, 1-Butyl-3-methylimidazolium Tetrafluoroborate-H_2O Mixed Solution, *J. Chem. Thermody.* Vol.43: 410-414.

Zelent, B., Nucci, N. V., Vanderkooi, J. M. (2004) Liquid and Ice Water and Glycerol/Water Glasses Compared by Infrared Spectroscopy from 295 to 12 K. *J. Phys. Chem. A*, Vol. 108: 11141-11150.

Zhang, J., Peng, X., Jonas, A., Jonas, J. (1995) NMR Study of the Cold, Heat, and Pressure Unfolding of Ribonuclease A, *Biochemistry* Vol.34: 8631-8641.

Zip, A., Kauzmann, W. (1973) Pressure Denaturation of Metmyoglobin, *Biochemistry* Vol. 12: 4217-4228.

11

A Stable Protein – CutA1

Azumi Hirata[1,2], Aya Sato[2], Takashi Tadokoro[2], Yuichi Koga[2], Shigenori Kanaya[2] and Kazufumi Takano[1,2]
[1]*Department of Biomolecular Chemistry, Kyoto Prefectural University,*
[2]*Department of Material and Life Science, Osaka University,*
Japan

1. Introduction

CutA1 is a universal protein distributed in bacteria, plants and animals, including humans. This protein was originally isolated and characterized from a gene locus of *Escherichia coli* called *cutA* (Fong et al., 1995). The CutA protein is involved in copper tolerance; moreover, it affects divalent cation tolerance levels of zinc, nickel, cobalt and cadmium salts. The *cutA* locus consists of two operons: one containing a single gene encoding a cytoplasmic protein, CutA1, and the other is composed of two genes encoding the inner-membrane proteins, CutA2 and CutA3. It has been reported that mammalian CutA1 is necessary for anchoring the enzyme acetylcholinesterase (AChE) in neuronal cell membranes (Navaratnam et al., 2000; Perrier et al., 2000). Additionally, CutA1 affects the folding, oligomerization, and secretion of AChE (Falasca et al., 2005; Liang et al., 2009), and co-expression with CutA1 increased the formation and secretion of AChE tetramers (Liang et al., 2009). Therefore, the presence of CutA1 in the secretory pathway affects the processing, probably the folding, and the oligomerization of AChE. However, the precise functions of CutA1 remain to be validated.

The structure of bacterial and mammalian CutA1 has been crystallographically demonstrated in *Pyrococcus horikoshii* (Ph-CutA1) (Tanaka et al., 2004a), *Thermus thermophilus* (Tt-CutA1) (Tanaka et al., 2006), *Oryza sativa* (Os-CutA1) (Sawano et al., 2008), *Escherichia coli* (Ec-CutA1) (Arnesano et al., 2003), *Homo sapiens* (Hs-CutA1) (Bagautdinov et al., 2008), *Thermotoga maritima* (Tm-CutA1) (Savchenko et al., 2004), and *Xanthomonas campestris* (Xc-CutA1) (Lin et al., 2006). In all cases, from bacterial to mammalian, the trimeric structure of CutA1 was conserved. Moreover, the heat-denaturation temperatures of Ph-CutA1, Tt-CutA1, Os-CutA1, Es-CutA1, and Hs-CutA1 are quite high: 150, 113, 97, 90, and 96°C, respectively. The denaturation temperature of Ph-CutA1 (150°C) is among the highest of known proteins. Several factors responsible for the extreme thermostability of proteins have been proposed, such as an increase in the number of ion pairs and hydrogen bonds (Aguilar et al., 1997; Perutz & Raidt, 1975; Tahirov et al., 1998; Tanner et al., 1996; Yamagata et al., 2001), core hydrophobicity (Schumann et al., 1993; Takano et al., 1995), packing density (Russell et al., 1994), as well as the oligomerization of several subunits (Dams & Jaenicke, 1999; Jaenicke et al., 1996; Sterner et al., 1996), and entropic effects due to relatively shorter surface loops and peptide chains (Russell et al., 1994; Yamagata et al., 2001). The difference in stability among CutA1 was explained by the difference in electrostatic interactions (Matsuura et al., 2010; Sawano et al., 2008; Tanaka et al, 2004a; Tanaka et al., 2006) and the conformation of the β2-strand (Bagautdinov et al., 2008; Savchenko et al., 2004; Tanaka et al, 2004a). Though their stabilities fundamentally depended on the optimal growth temperatures of their host organisms, their denaturation temperatures were remarkably higher than the optimal growth temperatures. These results suggested that the unique trimeric structural motif, which enables tightly intertwined subunit interactions among the β-strands, was the critical factor in the unusually high stability of CutA1.

Recently, we cloned the gene encoding CutA1 from *Shewanella* sp. strain SIB1 (SIB1-CutA1), overexpressed it in *E. coli*, purified the recombinant protein and crystallized it (Sato et al., 2011). *Shewanella* sp. SIB1 is a psychrophilic bacterium that grows rapidly at 20°C (Kato et al., 2001). Although this bacterium can grow at temperatures as low as -0.15°C, growth was inhibited at temperatures exceeding 30°C. We demonstrated that the heat-denaturation temperature of SIB1-CutA1 was 95°C, suggesting that SIB1-CutA1 needs to maintain high stability in order to function, even in psychrophilic organisms. However, the precise function of Cut-A1 remains controversial.

In this chapter, we discuss the robustness of CutA1 from the viewpoint of its structure. Herein, we first briefly introduce the structure and stability of Ph-CutA1, Tt-CutA1, Os-CutA1, Tm-CutA1, Es-CutA1, Hs-CutA1, and SIB1-CutA1. Next, we will further discuss why the stability of CutA1 is so structurally remarkable using our novel results of CutA1 from the psychrophilic bacterium *Shewanella oneidensis* MR-1 (So-CutA1).

2. Structure and stability of CutA1 proteins

2.1 CutA1 from *Pyrococcus horikoshii*

Pyrococcus horikoshii OT3, a hyperthermophilic obligate anaerobe, can grow at temperatures between 88 and 104°C with an optimal temperature of 98°C. Differential scanning calorimetry (DSC) analysis revealed that the denaturation peak of Ph-CutA1 appeared near 150°C at pH 7.0, indicating that the folding of Ph-CutA1 was maintained near this temperature and pH (Tanaka et al., 2006). Interestingly, this temperature was 50°C higher than the optimum growth temperature of the organism.

The crystal structure of Ph-CutA1 consisted of a tightly intertwined trimer assembled so as to form a closed circular β-sheet structure (Fig. 1). The monomeric structure consisted of three α-helices and five β-strands. Two short (β1 and β4) and two long (β2 and β3) β-strands formed an askew curved sheet. The trimer was composed of monomers through interactions between the edges of three β-strands. Each trimer had three identical intersubunit interfaces, in which both edges of one strand (β2) interacted with the edges of the β2-strand in the other two subunits, and one short strand (β5) mutually interacted with the β4-strand of a different subunit. This tightly intertwined interaction appeared to contribute to the stabilization of the trimeric structures of the Ph-CutA1 protein.

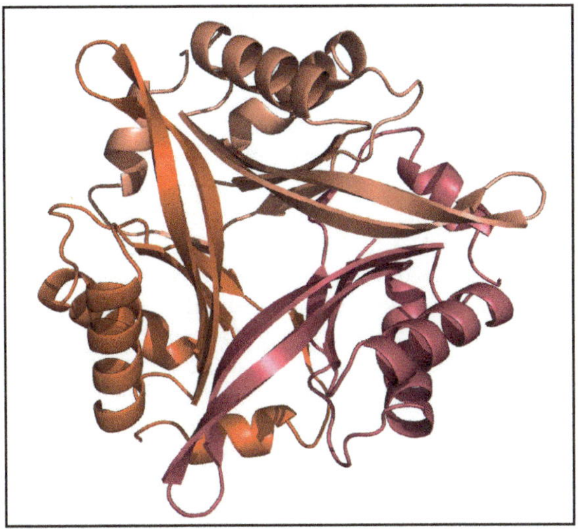

Fig. 1. Ribbon diagram of trimeric structure of Ph-CutA1 (PDB code 1J2V) at 2.00 Å resolution. The figure was prepared using *PyMol* (DeLano, 2004).

The structural characteristics of Ph-CutA1 revealed that Ph-CutA1 had many ionizable residues, though it had few neutral residues. The number of ionizable residues (Asp, Glu, Lys, Arg, and His) and intra-subunit ion pairs in Ph-CutA1 were prominent (43 and 30, respectively) compared to those in Ec-CutA1 (25 and 1, respectively). Importantly, the monomeric structure of Ph-CutA1 was highly stabilized by hydrogen bonds, ion pairs, and hydrophobic interactions (Tanaka et al., 2006). The intra-subunit ion pairs consisted of 14 donor and 16 acceptor residues, which were partially exposed to the solvent. Six donor positively-charged residues formed ion pairs between different secondary segments in the same subunit (Arg82-Glu59, His35-Glu50, Arg36-Glu47, Arg68-Glu24, Arg25-Glu99, and Lys101-Glu64), where the other residues formed pairs within the same segment. All of the donor residues (Lys19, Lys66, Lys70, and Arg36) forming inter-subunit ion pairs shared intra-subunit ion pairs that formed of inter- and intra-subunit ion pair networks.

Chemical denaturation by guanidine HCl (GdnHCl) or urea has been widely used to study protein folding and stability. These denaturants can influence not only the protein stability, but also the ensemble of the native structure. Tanaka et al. (2004b) reported the effects of GdnHCl on Ph-CutA1 through spectroscopic techniques and crystal structure studies. CD spectra results showed no changes in secondary structure of Ph-CutA1 even in 3 to 8 M GdnHCl, indicating that Ph-CutA1 had a rather rigid secondary structure. The crystal structure of Ph-CutA1 in 3 M GdnHCl was determined at 1.6 Å resolution by the molecular replacement method using native Ph-CutA1 as a search model. The crystal structure of native Ph-CutA1 had a large number of intermolecular hydrogen bonds, of which more than 90% were retained in 3 M GdnHCl. The disrupted hydrogen bonds were mainly located on the protein surface. Additionally, Ph-CutA1 in the native state showed that the protein had seven ion pairs per monomer. All ion pairs were present in 3 M GdnHCl, even though they were on the protein surface. These observations indicated that the intermolecular interactions of hydrogen bonds and ion pairs of Ph-CutA1 were extraordinary stable.

2.2 CutA1 from Thermus thermophilus, Oryza sativa, and Thermotoga maritima

Thermus thermophilus, Oryza sativa, and *Thermotoga maritima* is an extreme thermophilic bacterium with an optimum growing temperature of 75°C. The optimum growth temperature of mesophilic *Oryza sativa* from the rice plant is 28°C. Both of the CutA1 proteins from these organisms had a trimeric form (Fig. 2) composed of identical subunits between pH 2.5 and 9.0 (Sawano et al., 2008). DSC analysis has shown that the denaturation temperatures of Tt-CutA1 and Os-CutA1 were 113°C at pH 7.0 and 97°C at pH 7.0, respectively. These results indicated that both CutA1 proteins had high thermostability. On the other hand, Tt-CutA1 showed higher stability to heat, and moreover, resistance to denaturation than Ph-CutA1 at acidic pH (Tanaka et al., 2006). From a structural standpoint, the number of ionizable residues and intra-subunit ion pairs in Tt-CutA1 is not relatively high (29 and 12, respectively) compared to those in Ph-CutA1 (43 and 30, respectively). These phenomena occurred due to protonated ion pairs, which suggested that the ion pairs in Ph-CutA1 maintained its structural integrity at neutral pH. Changes in the cavity volume in the interior of a protein upon its denaturation affected the conformational stability (Eriksson et al., 1992; Funahashi et al., 2001). The total cavity volumes of the tertiary trimeric structure of Tt-CutA1 were smaller than those of Ph-CutA1. In addition, Ph-CutA1 had a large cavity near the center of a trimer along with four smaller cavities. These observations indicated that Tt-CutA1 formed tighter interactions at the trimer interfaces than Ph-CutA1, resulting in the high stability of Tt-CutA1 at acidic pH.

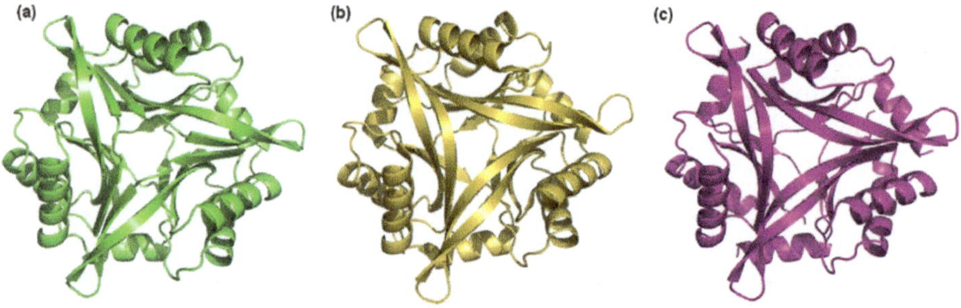

Fig. 2. Ribbon diagrams of trimeric structures of (a) Tt-CutA1 (PDB code 1NZA), (b) Os-CutA1 (PDB code 2Z0M) and (c) Tm-CutA1 (PDB code 1O5J) at 1.70 Å, 3.02 Å, and 1.95 Å resolution, respectively. The figures were generated with *PyMol* (DeLano, 2004).

The CutA1 from *Thermotoga maritima* showed amino acid sequence identities to those of Ec-CutA1 (35%), Hs-CutA1 (32%) and Ph-CutA1 (46%). The structure of Tm-CutA1 was very similar to those of other CutA1 proteins (Fig. 2).

2.3 CutA1 from *Escherichia coli*

The CutA1 protein was originally identified in the *cutA* gene locus of *Escherichia coli,* which was involved in divalent metal tolerance. Although the optimal growth temperature of *Escherichia coli* is 37°C, the denaturation temperature of Ec-CutA1 is greater than 90°C at pH 9.0, which was also significantly higher than other proteins from mesophilic *E. coli.*

The structure of Ec-CutA1 consisted of homotrimers (Arnesano et al., 2003) (Fig. 3). In the crystal asymmetric unit, two homotrimers were present that made extensive contacts leading to a dimer of trimers. One trimer was rotated by 60° around the axis perpendicular to the trimer plane with respect to the other.

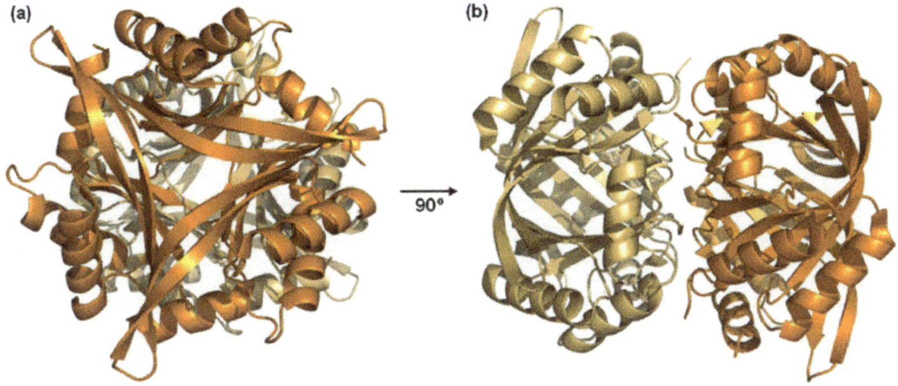

Fig. 3. Ribbon diagram of two trimers of Ec-CutA1 (PDB code 1NAQ) at 1.7 Å resolution. The structures are shown in two different orientations; (a) front view and (b) side view with 90° rotation of (a). The figure was prepared with *PyMol* (DeLano, 2004).

Recently, it was reported that in order to confirm the thermostabilization mechanism, the structure-sequence compatibility between the conformation of Ec-CutA1 and its native sequence was examined using the Stability Profile of Mutant Protein (SPMP) (Matsuura et al., 2010). SPMP analysis is a method to rationally improve the estimation of the conformational stability of a protein (Ota et al., 1995). A pseudo-energy potential derived from a number of Protein Data Bank (http://www.pdb.org/pdb/home/home.do) structures was used in SPMP analysis to predict protein structures and consisted of four elements: packaging of the side-chains, hydration, local structure, and back-bone/side-chain repulsion. From the SPMP scores, seven positions of Ec-CutA1 were mutated to improve protein stability. The mutant Ec-CutA1 proteins were evaluated structurally and for changes in stability by DSC. The crystal structures of these mutant proteins were highly similar to the wild-type Ec-CutA1 structure, while the stabilities of the proteins with mutations at positions 11 and 61 were remarkably improved. The denaturation temperature of single mutant S11V and E61V were 105°C and 103°C, respectively, and that of the S11V/E61V double mutant was 114°C. The values of the denaturation temperatures were improved by 15, 13 and 24°C, respectively, suggesting a cumulative effect for each single mutation. The SPMP evaluations were consistent with the DSC results. These observations suggested that these substitutions resulted in changes in the hydration effect, local structure, and side chain packaging. Positions 11 and 61 were located in the β-sheet and the two substitutions included residues with a higher propensity to form the β-sheet, which indicated enhanced stabilization (Matsuura et al., 2010).

2.4 CutA1 from *Homo sapiens*

The heat-denaturation temperature of CutA1 from *Homo sapiens* (Hs-CutA1) was 96°C, which was remarkably higher than the optimal temperature of *Homo sapiens*. The N-terminal sequence of Hs-CutA1 was hydrophobic and might represent a cleavable secretion signal, a mitochondrial import signal, or a transmembrane anchor (Perrier et al., 2000); whereas, the bacterial CutA1 was devoid of this hydrophobic domain, and Ec-CutA1 was located in the cytoplasm (Arnesano et al., 2003).

Hs-CutA1 was functional as a trimer and the orthorhombic crystal form contained two trimers per asymmetric unit (Bagautdinov et al., 2008). A rigid trimeric core structure appeared to be common to the CutA1 proteins. Similar to other CutA1 structures, the Hs-CutA1 subunit was made up of a double βαβ motif. Strand β2 in Hs-CutA1, however, had a kink at its center that may have been caused by the presence of Pro101 in the middle of the β2-strand. This kink-type deformation of β2 close to Pro101 allowed the formation of hydrogen bond arrangements that stabilized the β2-strand in Hs-CutA1. The conformation of β2 may have a possible role in protein thermostability.

The β2- and β3-strands and the loop between them (the β2-β3 loop) in Hs-CutA1 formed an extended β-hairpin. The six subunits in the asymmetric unit and the two trimers formed by them were very similar. However, superposition of the Hs-CutA1 subunits showed different conformations in the turn region of the β-hairpins, indicating a high degree of conformational flexibility. Actually, superposition of CutA1 trimers from different sources indicated that the main differences were observed in the turn area of the β-hairpins.

The conserved residues were mainly isolated to two regions in the Hs-CutA1 subunit. A loop region between α2 and β4, spatially close to the β-hairpin turn, encompassed the conserved residues His141, Pro142 and Tyr143. The other highly conserved amino acids Cys96 and Tyr107 in β2, Glu118, Lys124 and Thr125 in β3, Trp109 and Gly111 in the β2-β3 loop and Tyr160 and Trp163 in α3 were clustered at the opposite end of the scaffold. The tertiary fold was assembled such that the two conserved regions in the subunits were brought together and formed three potential active sites in the clefts at the trimer interfaces.

In conclusion, the functionally important areas of Hs-CutA1 appeared to be the putative active site clefts and the flexible β-hairpins. Oligomerization of Hs-CutA1 allowed the small protein to form a compact cylinder-shaped structure and offered the three β-hairpins and cleft site for specific interactions with other proteins and molecules. These regions contained conserved residues and the flexible β-hairpin may have changed conformation upon binding of effectors and docking with a receptor. The negatively charged cleft of Hs-CutA1 reflected the positive charge of the substrate ligands and was readily accessible to solvent.

2.5 CutA1 from *Shewanella* sp. SIB1

The CutA1 protein from *Shewanella* sp. SIB1 (SIB1-CutA1) contained 108 amino acid residues and shared 25, 30, and 39% identity with Ph-CutA1, Ec-CutA1, and Hs-CutA1, respectively. Recently, we determined the crystal structure of SIB1-CutA1 and measured its thermal stability (Sato et al., 2011).

The overall structure of SIB1-CutA1 was a trimeric structure, which resembled other homologous CutA1 proteins (Fig. 4). The root-mean-square deviations (r.m.s.d.) of the C^{α} atoms for Ph-CutA1 and Ec-CutA1 against SIB1-CutA1 were 1.08 and 1.00 Å as a monomer, and 1.17 and 1.11 Å as a trimer, respectively. The r.m.s.d. of the C^{α} atoms between any pair of subunit SIB1-CutA1 was 0.50-0.58 Å. Each subunit consisted of three α-helices and six β-strands, although the monomers of any other CutA1 had three α-helices and five β-strands.

The main difference was observed in the β2-strand (Figs. 4 and 5). The β2-strand was divided into two short β2-strands, β2a and β2b, in SIB1-CutA1. Ph-CutA1 and Tm-CutA1 isolated from hyperthermophiles did not exhibit kinks, whereas Tt-CutA1, Os-CutA1, Ec-CutA1, and Hs-CutA1 had a conformational kink in the β2-strand. This kink was caused from an insertion of a Pro residue into the β2-strand. In addition, a Pro deletion variant of Ec-CutA1 increased its stability. A Gln residue (Gln39) was inserted into a β2-strand and was divided into two short β-strands in the SIB1-CutA1 from a psychrotrophic bacterium. So-CutA1 also had an insertion of an Ala residue in the β2-strand. Therefore, these results suggested that the conformation of the β2-strand affected the thermostability of CutA1 since hyperthermophilic CutA1 had no kink, although mesophilic CutA1 contained a kink, and psychrotrophic CutA1 had two short divided β-strands. Moreover, hydrogen bonds between strands may have affected the stability. A kink caused by Pro (Pro58) in the β3-strand was another peculiarity of SIB1-CutA1. The kink in the β3 strand also seemed to be a factor leading the split of the β2-strand in SIB-CutA1, since there was interaction between β2a and β2b of Gln39 and β3 of Pro58.

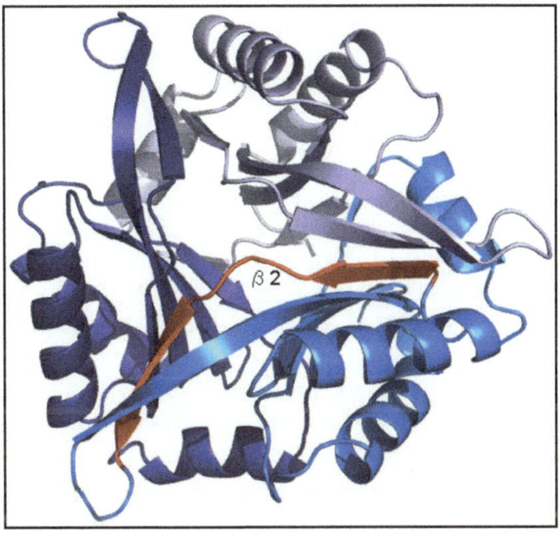

Fig. 4. Ribbon diagram of trimeric structure of SIB1-CutA1 (PDB code 3AHP) at 2.7 Å resolution. The β-strand is shown in red. The figure was prepared using *PyMol* (DeLano, 2004).

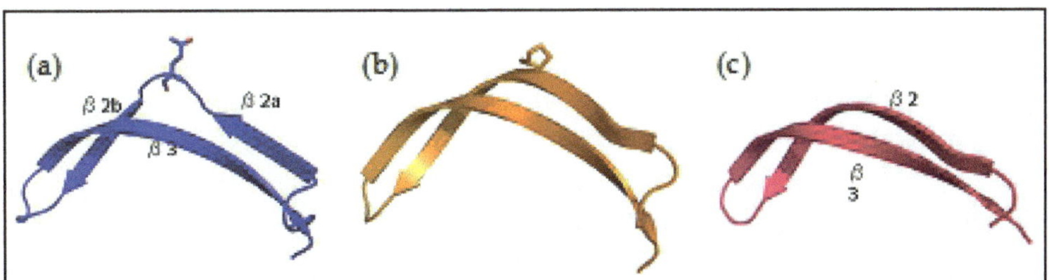

Fig. 5. Close-up views of β2 (β2a and β2b) and β3. (a) SIB1-CutA1. (b) Ec-CutA1. (c) Ph-CutA1. The figure was prepared using *PyMol* (DeLano, 2004).

The SIB1-CutA1 protein began to unfold at 80°C at pH 7.0. However, the unfolding was not completed at 95°C, because the CD value at 95°C and pH 7.0 differed from that at 95°C and pH 5.0. *Shewanella* sp. could not grow at temperatures over 30°C, and other enzymes from *Shewanella* sp. SIB1, FKBP22 and ribonuclease H1 were inactivated at temperatures less than 40°C. These results illustrated the high stability of SIB1-CutA1.

3. Function of CutA1 proteins from the viewpoint of their structure

Despite the similarities in folding of copper chaperones and roles in copper tolerance, CutA1 proteins did not possess the classical CXXC motif that has been shown to bind metal ions via thiol sulphurs in Cys residues (Bull & Cox, 1994; O'Halloran, 1993). In Ec-CutA1, three cysteine residues (16, 39, and 79) were far apart both in sequence and structure, as none were located within a single subunit or trimer. This suggested that potential metal-binding features of CutA1 were different from those of metallochaperones. Moreover, in Tm-CutA1, there was a cavity accessible from the outside via three solvent accessible channels. The cavity of Tm-CutA1 was formed by a number of conserved aromatic residues (Tyr45, Trp47, Tyr81, and Trp101), and charged residues (Asp54, Glu56, and Glu82), which made this cleft a strong candidate for a conserved function. Several conserved residues including Cys35 covered inner surfaces of the cavity that may serve as potential metal binding sites. Though CutA has been known to be involved in metal homeostasis, the metal-binding CXXC sequence motif was not found in Tm-CutA1.

The CutA1 architecture was similar to those of several well-characterized PII signal transducer proteins (Arnesano et al., 2003; Savchenko et al., 2004). PII proteins are trimeric structures and integrate intracellular nitrogen and carbon status signals to the control of enzymes responsible for nitrogen assimilation (Ninfa & Atkinson, 2000). The trimeric PII-like domains were reported to act as signal sensors that regulated unknown catalytic activities of more conserved domains (Godsey et al., 2007; Saikatendu et al., 2006), which might suggest that the conserved trimeric assembly showed a similar mechanism of action as PII proteins and CutA1. On the other hand, the major structural differences included the presence of a C-terminal β-strand in PII proteins, while in CutA1 they replaced the α3-helices. Additionally, a large loop between strands β2 and β3 protruded into the solvent, whereas strands β2 and β3 in CutA1 were longer and formed β-hairpin strands. In PII proteins, the two central β-strands (β3 and β4) were joined by the T-loop, which were composed of 17 amino acids and played a key role in protein-protein interactions. In Tm-CutA1, the β3- and β4-strands were joined by a two amino acid turn to form a hairpin loop. The residues in the C-terminal extension of the Tm-CutA1 β3-strand (Tyr45, Trp46, and Trp47) were conserved. However, the structure of this region in Tm-CutA1 was different from that in PII proteins, suggesting that this region in Tm-CutA1 may not participate in protein-protein interactions (Savchenko et al., 2004).

Recently, it was shown that overexpressed CutA isoform2 sensitized HeLa cells to copper toxicity by promoting copper-induced apoptosis (Yang et al., 2008); moreover, the inhibitory effect of excessive copper on cell proliferation was enhanced by over-expression of CutA isoform2. Based on these results, it was suggested that CutA isoform2 was implicated in an important role in copper toxicity. These studies will be helpful to guide future investigations in understanding the physiological role of the CutA1 protein.

4. High stability of the trimeric structural motif of CutA1 proteins

CutA1 has an extraordinary stability that fundamentally depends on the optimal growth temperatures of their host organisms. The difference in stability among CutA1 proteins was explained by the difference in electrostatic interactions and the conformation of the β2-strand. From a structural point of view, the high stability of CutA1 originated from the common trimer structural motif. The trimeric structure enabled tightly intertwined interactions among the β-strands. It was indicated that hydrophobic interactions in Ph-CutA1 were responsible for stabilizing the trimeric structure. Moreover, it was observed that in SIB1-CutA1, the trimer subunits were solidly locked to each other with highly conserved aromatic residues, such as Tyr45, Trp47, Tyr81, Tyr98 and Trp101 (Fig. 6). These residues were located in the subunit interfaces. Therefore, we suggest that increasing stability of CutA1 was caused by the aromatic cluster. There was no report which provided evidence indicating that the aromatic cluster affected the stability of the CutA1 protein.

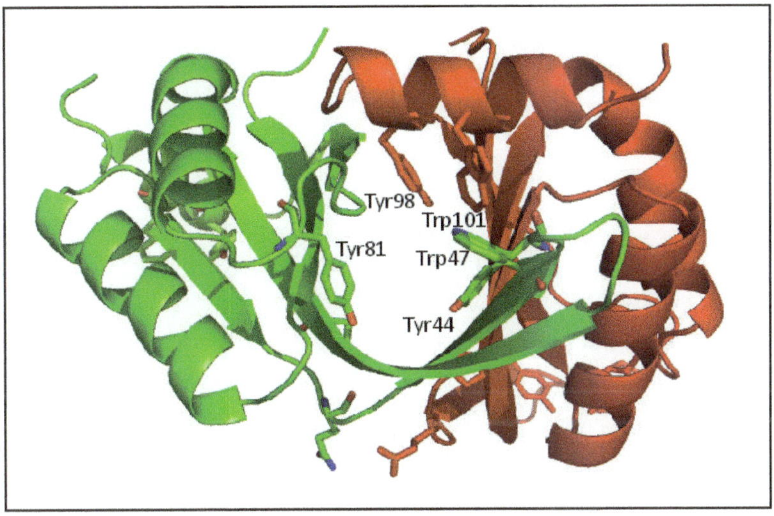

Fig. 6. Close-up view of the subunit interface of SIB1-CutA1. Each monomer is coloured differently. The highly conserved aromatic residues are shown. The figure was prepared using *PyMol* (DeLano, 2004).

SIB1-CutA1 subunit *A* (*B* or *C*) interacted with hydrogen bonds between β2a and β2b of subunit *C* (*A* or *B*), between β2b and β2a of subunit *B* (*C* or *A*) and between β4 and β5 of subunit *B* (*C* or *A*), resulting in a tightly intertwined trimer. The intertwined interactions among β strands of SIB1-CutA1 appeared to stabilize the trimeric structure as in other CutA1 proteins. Furthermore, three α-helices were located on the outside of the trimer and cover β-strands. Highly conserved aromatic residues, Tyr45, Trp47, and Tyr81 from subunit *A* (*B* or *C*) and Tyr98 and Trp101 from subunit *B* (*C* or *A*) existed in the subunit interfaces. These residues formed a hydrophobic core at the trimer interfaces.

In this section, we discussed our novel results in order to demonstrate the effects of the aromatic cluster on So-CutA1 protein stabilization. The highly conserved aromatic residues of So-CutA1 were replaced (Y44A, W46A, Y97A, W100A, and W100F) and the CD spectra measurements and DSC analysis of these mutant proteins were evaluated.

The gene encoding So-CutA1 was amplified by polymerase chain reaction (PCR), where the genomic DNA of *Shewanella oneidensis* MR-1 was used as a template. The resultant DNA fragment was digested with *Nde*I and *Bam*HI. Plasmids for overproduction of So-CutA1 were constructed by ligating the resultant DNA fragment into the *Nde*I-*Bam*HI sites of pET25b. The pET25b derivatives for overproduction of the five mutant proteins were constructed by PCR using the QuikChange II site-directed mutagenesis kit (Stratagene). The pET25b derivative for overproduction of So-CutA1 was used as a template. The mutagenic primers were designed such that the codons for Tyr44, Trp46, Tyr97 and Trp100 were changed to Ala, and the codon for Trp100 was changed to Phe. The resultant DNA fragments were digested with *Nde*I and *Bam*HI and ligated into the *Nde*I-*Bam*HI sites of pET25b. For overproduction of WT and mutant So-CutA1, *E. coli* BL21 (DE3) was transformed with WT or mutant pETCutA1 and grown at 37°C. When D_{600} reached 0.6, 1 mM IPTG was added to the culture medium and cultivation was continued at 37°C for 4 h. The cells were harvested by centrifugation, disrupted by sonication, and heat-treated at 60°C for 10 min. The supernatant was dialyzed against 50 mM Tris-HCl at pH 8.0 and applied to a HiPrep DEAE column (GE Healthcare Life Sciences). The protein was eluted from the column with a linear gradient of 0 to 1.0 M NaCl and the purity was analyzed by SDS-PAGE.

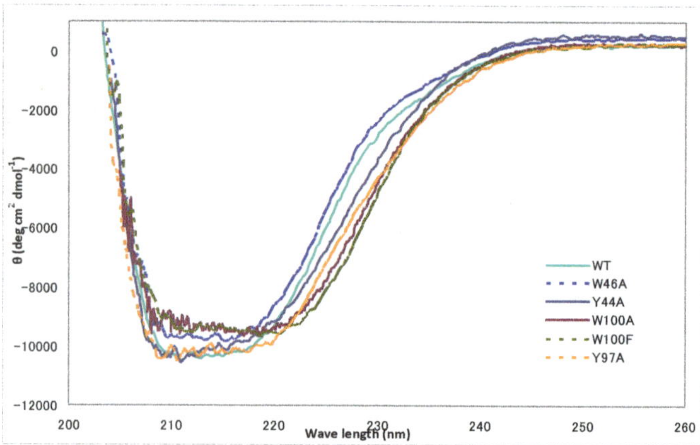

Fig. 7. CD spectra of the WT and mutant So-CutA1 at pH 8.0 and 30°C.

The CD spectra measurements were made on a J-725 automatic spectropolarimeter from Japan Spectroscopic Co., Ltd. (Tokyo, Japan). The mean residue ellipticity, θ, which has units of degrees cm^2 dmol^{-1}, was calculated. For CD spectra measurements, the buffer used consisted of 50 mM Tris-HCl at pH 8.0, 1 mM EDTA, and 1 mM DTT. The protein concentration was 0.2 mg ml^{-1}. DSC measurements were carried out on a high-sensitivity VP-DSC controlled by the VPVIEWER™ software package (Microcal, Inc., Northampton, MA, USA) at a scan rate of 1°C min^{-1}. The buffer for DSC measurements contained 50 mM Glycine-NaOH at pH 9.0, and 1 mM EDTA. The protein concentration during the measurements was ≈ 0.5 mg ml^{-1}. The reversibility of thermal denaturation was verified by reheating the samples.

The far-UV spectra of these five proteins at pH 8.0 and 30°C were nearly similar to those of WT-So-CutA1 as shown in Fig. 7, suggesting they have folded structures. Since far-UV CD spectra of a protein had been affected due to the aromatic residues (Woody, 1994), the differences in far-UV spectra of the six proteins might be caused mainly by differences in the content of aromatic residues. Fig. 8 shows the DSC curves of WT and five mutant proteins at pH 9.0. The peak temperatures of WT-So-CutA1 was 100°C, and those of five mutant proteins were 89 (Y44A), 99 (W46A), 85 (Y97A), 90 (W100A), and 94°C (W100F), respectively. The peak temperatures of mutant proteins were lower than that of WT-So-CutA1, suggesting that the aromatic cluster, especially Tyr44, Tyr97, and Trp100, which were located within the interface of the trimer, contributed to the stabilization of So-CutA1 (Fig. 9).

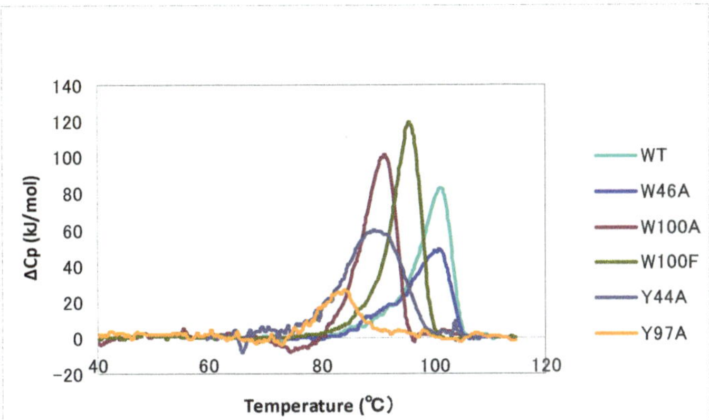

Fig. 8 DSC curves of the WT and mutant So-CutA1 at pH 9.0.

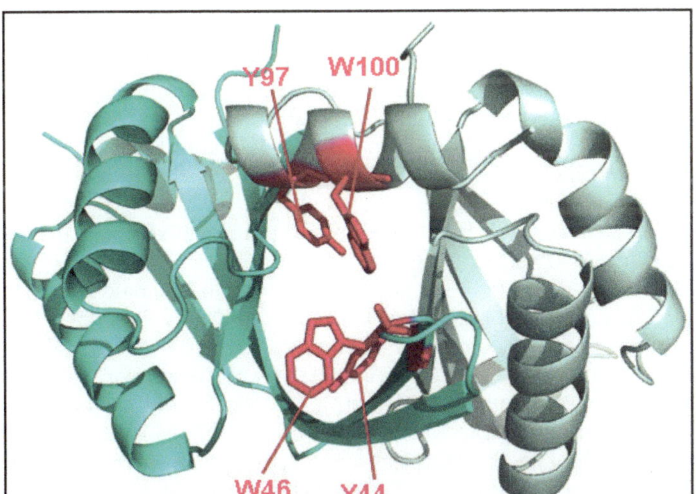

Fig. 9. Close-up view of the subunit interface (model) of So-CutA1. Each monomer is coloured differently. The highly conserved aromatic residues are shown in magenta. The figure was prepared using *PyMol* (DeLano, 2004).

5. Conclusion

In general, high thermostability of proteins correlated with high resistance against proteolysis. CutA1 unfolded at a remarkably higher temperature compared to the growth temperatures of the host organisms. The stability of CutA1 depended upon the growth temperature. Therefore, CutA1 must maintain the robustness for its function. At this present stage, we cannot explain the relationship between the exact functions and stability of CutA1. Nevertheless, the unusual high stability of CutA1 basically originated from a common trimer structure of these proteins. Structurally, the aromatic cluster, which combined the subunits, participated in the stabilization of CutA1. The evolutionarily conserved trimeric structure and thermostability might point to a similar mechanism of CutA1 action.

6. Acknowledgment

This work was supported in part by an Industrial Technology Research Grant Program from the New Energy and Industrial Technology Development Organization (NEDO) of Japan.

7. References

Aguilar, C. F.; Sanderson, I.; Moracci, M.; Ciaramella, M.; Nucci, R.; Rossi, M. & Pearl, L. H. (1997). Crystal Structure of the β-Glycosidase from the Hyperthermophilic archaeon *Sulfolobus solfataricus*: Resilience a Key Factor in Thermostability. *Journal of Molecular Biology*, Vol.271, pp. 789-802, ISSN 0022-2836

Arnesano, F.; Banci, L.; Benvenuti, M.; Bertini, I.; Calderone, V.; Mangani, S. & Viezzoli, M. S. (2003). The Evolutionarily Conserved Trimeric Structure of CutA1 Proteins Suggests a Role in Signal Transduction. *The Journal of Biological Chemistry*, Vol.278, No.46, pp. 45999-46006, ISSN 1083-351X

Bagautdinov, B.; Matsuura, Y.; Bagautdinova S.; Kunishima, N. & Yutani, K. (2008). Structure of Putative CutA1 from *Homo sapiens* Determined at 2.05 Å Resolution. *Acta Crystallographica*, Vol.F64, pp. 351-357, ISSN 1744-3091

Bull, P. C. & Cox, D. W. (1994). Wilson Disease and Menkes Disease: New Handles on Heavy-Metal Transport. *Trends in Genetics*, Vol.10, pp. 246-252, ISSN 0168-9525

Dams, T. & Jaenicke, R. (1999). Stability and Folding of Dihydrofolate Reductase from the Hyperthermophilic Bacterium *Thermotoga maritima*. *Biochemistry*, Vol.38, pp. 9169-9178, ISSN 1520-4995

DeLano, W. L. (2004). The PyMOL Molecular Graphics System. DeLano Scientific, San Carlos, California, USA.

Eriksson, A. E.; Baase, W. A.; Zhang, X. J.; Heinz, D. W.; Blaber, M.; Baldwin, E. P. & Matthews, B. W. (1992). Response of a Protein Structure to Cavity-Creating Mutations and Its Relation to the Hydrophobic Effect. *Science*, Vol.255, pp. 178-183, ISSN 1095-9203

Falasca, C.; Perrier, N.; Massoulié, J. & Bon, S. (2005). Determinants of the t Peptide Involved in Folding, Degradation, and Secretion of Acetylcholinesterase. *The Journal of Biological Chemistry*, Vol.280, No.2, pp. 878-886, ISSN 1083-351X

Fong, S. T.; Camakaris, J. & Lee, B. T. O. (1995). Molecular Genetics of a Chromosomal Locus Involved in Copper Tolerance in *Escherichia coli* K-12. *Molecular Microbiology*, Vol.15, No.6, pp. 1127-1137, ISSN 1365-2958

Funahashi, J.; Takano, K. & Yutani, K. (2001). Are the Parameters of Various Stabilization Factors Estimated from Mutant Human Lysozymes Compatible with Other Proteins? *Protein Engineering*, Vol.14, No.2, pp. 127-134, ISSN 1741-0134

Jaenicke, R.; Schurig, H.; Beaucamp, N. & Ostendorp, R. (1996). Structure and Stability of Hyperstable Proteins: Glycolytic enzymes from Hyperthermophilic Bacterium. *Thermotoga maritima. Advances in Protein Chemistry*, Vol.48, pp. 181-269

Kato, T.; Haruki, M.; Imanaka, T.; Morikawa, M. & Kanaya, S. (2001). Isolation and Characterization of Psychrotrophic Bacteria from Oil-Reservoir Water and Oil sands. *Applied Microbiology and Biotechnology*, Vol.55, No.6, pp. 794-800, ISSN 1432-0614

Liang, D.; Nunes-Tavares, N.; Carvalho, S. ; Bon, S. & Massoulié, J. (2009). Protein CutA Undergoes an Unusual Transfer into the Secretory Pathway and Affects the Folding, Oligomerization, and Secretion of Acetylcholinesterase. *The Journal of Biological Chemistry*, Vol.284, No.8, pp. 5195-5207, ISSN 1083-351X

Lin, C. H.; Chin, K. H.; Gao, F. P.; Lyu, P. C.; Shr, H. L.; Wang, A. H. j. & Chou, S. H. (2006). Cloning, Crystallization and Preliminary X-Ray Studies of XC2981 from *Xanthomonas campestris*, a Putative CutA1 Protein Involved in Copper-Ion Homeostasis. *Acta Crystallographica*, Vol.F62, pp. 1113-1115, ISSN 1744-3091

Matsuura, Y.; Ota, M.; Tanaka, T.; Takehira, M.; Ogasahara, K.; Bagautdinov, B.; Kunishima, N. & Yutani K. (2010). Remarkable Improvement in the Heat Stability of CutA1 from *Escherichia coli* by Rational Protein Design. *Journal of Biochemistry*, Vol.148, No.4, pp. 449-458, ISSN 1756-2651

Navaratnam, D. S.; Fernando, F. S.; Priddle, J. D.; Giles, K.; Clegg, S. M.; Pappin, D. J.; Craig, I. & Smith, A. d. (2000). Hydrophobic Protein that Copurifies with Human Brain Acetylcholinesterase: Amino Acid Sequence, Genomic Organization, and Chromosomal Localization. *Journal of Neurochemistry*, Vol.74, pp. 2146-2153, ISSN 1471-4159

Ninfa, A. J. & Atkinson, M. R. (2000). PII Signal Transduction Proteins. *Trends in Microbiology*, Vol.8, pp. 172-179, ISSN 0966-842X

O'Halloran, T. V. (1993). Transition Metals in Control of Gene Expression. *Science*, Vol.261, pp. 715-725, ISSN 1095-9203

Ota, M.; Kanaya, S. & Nishikawa, K. (1995). Desk-top Analysis of the Structural Stability of Various Point Mutations Introduced into Ribonuclease H. Journal of Molecular Biology, Vol.248, pp. 733-738, ISSN 0022-2836

Perrier, A. L.; Cousin, X.; Boschetti, N.; Haas, R.; Chatel, J. M.; Bon, S.; Roberts, W. L.; Pickett, S. R.; Massoulié, J.; Rosenberry, T. L. & Krejci, E. (2000). Two Distinct Proteins Are Associated with Tetrameric Acetylcholinesterase on the Cell Surface. *The Journal of Biological Chemistry*, Vol.275, No.3, pp. 34260-34265, ISSN 1083-351X

Perutz, M. F. & Raidt, H. (1975). Stereochemical Basis of Heat Stability in Bacterial Ferredoxins and in Haemoglobin A2. *Nature*, Vol.255, pp. 256-259, ISSN 0028-0836

Russell, R. J. M.; Hough, D. W.; Dansos, M. J. & Taylor, G. L. (1994). The Crystal Structure of Citrate Synthase from the thermophilic archaeon *Thermoplasma acidophilum. Structure*, Vol.2, pp. 1157-1167, ISSN 1536-4283

Sato, A.; Yokotani, S.; Tadokoro, T.; Tanaka, S.; Angkawidjaja, C.; Koga, Y.; Takano, K. & Kanaya, S. (2011). Crystal Structure of Stable Protein CutA1 from Psychrotrophic Bacterium *Shewanella* sp. SIB1. *Journal of Synchrotron Radiation*, Vol.18, pp. 6-10, ISSN 0909-0495

Savchenko, A.; Skarina, T.; Evdokimova, E.; Watson, J. D.; Laskowski, R.; Arrowsmith, C. H.; Edwards, A. M.; Joachimiak, A. & Zhang, R. G. (2004). X-Ray Crystal Structure of CutA from *Thermotoga maritima* at 1.4 Å Resolution. *Proteins: Structure, Function, and Bioinformatics*, Vol.54, pp. 162-165, ISSN 1097-0134

Sawano, M.; Yamamoto, H.; Ogasahara, K.; Kidokoro, S.; Katoh, S.; Ohmura, T.; Katoh, E.; Yokoyama, S. & Yutani, K. (2008). Thermodynamic Basis for The Stabilities of Three CutA1s from *Pyrococcus horikoshii, Thermus thermophilus, Oryza sativa,* with Unusually High Denaturation Temperatures. *Biochemistry,* Vol.47, pp. 721-730, ISSN 1520-4995

Schumann, J.; Bohm, G.; Schumacher, G.; Rudolph, R. & Jaenicke, R. (1993). Stabilization of Creatinase from *Pseudomonas putida* by Random mutagenesis. *Protein Science,* Vol.10, pp. 1612-1620, ISSN 0961-8368

Sterner, R.; Kleemann, G. R.; Szadkowski, H., Lustig, A.; Hennig, M. & Kirchner, K. (1996). Phosphoribosyl Anthranilate Isomerase from *Thermotoga maritima* is Extremely Stable and Active Homodimer. *Protein Science,* Vol.5, pp. 2000-2008, ISSN 0961-8368

Tahirov, T. H.; Oki, H.; Tsukihara, T.; Ogasahara, K.; Yutani, K.; Ogata, K.; Izu, Y.; Tsunasawa, S. &Kato, I. (1998). Crystal Structure of Methionine Aminopeptidase from Hyperthermophile, *Pyrococcus furiosus*. *Journal of Molecular Biology,* Vol.284, pp. 101-124, ISSN 0022-2836

Takano, K.; Ogasahara, K.; Kaneda, H.; Yamagata, Y.; Fujii, S.; Kanaya, E.; Kikuchi, M.; Oobatake, M. & Yutani, K. (1995). Contribution of Hydrophobic Residues to the Stability of Human Lysozyme: Calorimetric Studies and X-ray Structural Analysis of the Five Isoleucine to Valine Mutants. Journal of Molecular Biology, Vol.254, pp. 62-76, ISSN 0022-2836

Tanaka, T.; Sawano, M.; Ogasahara, K.; Sakaguchi, Y.; Bagautdinov, B.; Katoh, E.; Kuroishi, C.; Shinkai, A.; Yokoyama, S. &Yutani, K. (2006). Hyper-Thermostability of CutA1 Protein, with a Denaturation Temperature of Nearly 150 °C. *FEBS Letters* Vol.580, pp. 4224-4230, ISSN 0014-5793

Tanaka, Y.; Tsumoto, K.; Nakanishi, T.; Yasutake, Y.; Sakai, N., Yao, M., Tanaka, I. & Kumagai, I. (2004a). Structural Implications for Heavy Metal-Induced Reversible Assembly and Aggregation of a Protein: The Case of *Pyrococcus horikoshii* CutA[1]. *FEBS Letters* Vol.556, pp. 167-174, ISSN 0014-5793

Tanaka, Y.; Tsumoto, K.; Umetsu, M.; Nakanishi, T.; Yasutake, Y.; Sakai, N., Yao, M., Tanaka, I.; Arakawa, T. & Kumagai, I. (2004b). Structural Evidence for Guanidine-Protein Side Chain Interactions: Crystal Structure of CutA from *Pyrococcus horikoshii* in 3M Guanidine Hydrochloride. *Biochemical and Biophysical Research Communications,* Vol.323, pp. 185-191, ISSN 0006-291X

Tanner, J.; Hecht, R. M. & Krause, K. L. (1996). Determinants of Enzyme Thermostability Observed in the Molecular Structure of *Thermus aquaticus* D-Glyceraldehyde-3-Phosphate Dehydrogenase at 2.5 Å Resolutuin. *Biochemistry,* Vol.35, pp. 2597-2609, ISSN 1520-4995

Yamagata, Y.; Ogasahara, K.; Hioki, Y.; Lee, S. J.; Nakagawa, A.; Nakamura, H.; Ishida, M.; Kuramitsu, S. & Yutani, K. (2001). Entropic Stabilization of the Tryptophan Synthase α-Subunit from a Hyperthermophile, *Pyrococcus furiosus*. X-Ray Analysis and Calorimetry. *The Journal of Biological Chemistry,* Vol.276, No.14, pp. 11062-11071, ISSN 1083-351X

Yang, J.; Li, Q.; Yang, H.; Yan, L. ; Yang, L. & Yu, L. (2008). Overexpression of Human CutA1 Isoform2 Enhances the Cytotoxicity of Copper to Hela Cells. *Acta Biochemica Polonica,* Vol.55, No.2, pp. 411-415, ISSN 1734-154X

Woody, R. W. (1994). Contributions of Tryptophan Side Chains to the Far-Ultraviolet Circular Dichroism of Protein. *European Biophysics Journal,* Vol.23, pp. 253-262, ISSN 1432-1017

Section 4

Function and Interaction

12

Functional Difference Between Deuterated and Protonated Macromolecules

Takashi Sugiyama and Tohru Yoshioka
Kaohsiung Medical University,
Taiwan

1. Introduction

Water (H_2O) is very important to living materials, but it is a very strange liquid in the physical sense, because it has an unexpectedly high melting point (0 °C), boiling point (100 °C), density (0.998 g/cm^3), heat of vaporization (10,515 cal/mol), specific heat capacity and temperature of maximum density (4 °C), compared with other typical hydrogen compounds, such as H_2S, H_2Se and H_2Te. Furthermore, it has higher conductivity of proton than that of electron.

To make a water molecule, one pair of bonding hydrogen lose electrons and give them to electro-negative oxygen, and then each hydrogen will have a charge of +0.17e and one pair of electrons near oxygen will give 2x(-0.17e), this asymmetry of the water molecule leads to a dipole moment. The system of hydrogen bond in water determines the key to its biologically significant properties in forming a water network on the surface of proteins.

What is expected when hydrogen (H) is replaced by deuterium (D)? Heavy water (D_2O) has completely the same chemical characteristics as H_2O, but it is quite different in physical aspects, such as melting point (3.82 °C), boiling point (101.72 °C), density (1.017 g/cm^3), temperature of maximum density (11.6 °C) and heat of vaporization (10,864 cal/mol). It has been accepted that two independent physiological effects occur when a living system is exposed to D_2O: i) an isotope exchange effect on functional proteins and ii) a solvent isotope effect on ionic conductivity, meaning that mobility of monovalent cation is reduced to <20% (Bass & Moore, 1973). When H is replaced with D in the biological molecule, the C-D bond is about 10 times stronger than the C-H bond, which means the C-D bond is more resistant than the C-H bond. O-D, N-D and S-D bonds are stronger than the corresponding protonated forms (Katz, 1965; Thomas, 1971). This H/D exchange effect on the functional protein can be observed by using thermal imaging and calcium imaging (Hirakura et al., 2011). In the following sections, we will discuss these two effects from the physical aspect and will propose a thermo-dynamical model for elucidation of water-protein interaction.

2. Effect of D_2O on the ion channel activity

According to the list of ionic conductance appearing in a previous report (Bass & Moore, 1973), the mobility of proton in H_2O is 1.44 times higher than that in D_2O (H/D ratio). The H/D ratio of other cations, K^+ and Na^+, was also higher in mobility as 1.35 and 1.47, respectively. Interestingly, the H/D ratio in proton mobility in ice is enormously high (= 6.25). Therefore, it will be important to study the inside structure of the channel pore by measuring the proton current under D_2O medium.

2.1 Proton channel

The latest data on D_2O effect on proton conductance was obtained by DeCoursey and Cherny (DeCoursey & Cherny, 1997), which showed the H/D ratio in proton conductance to be estimated at 1.9. Simultaneously they obtained some other results, such as (a) D^+ permeated proton channels, (b) the relative permeability of proton channel, however, was 10 times more greater for D^+, (c) D^+ regulated the voltage dependence of the proton channel gating like H^+, (d) D^+ current induced with depolarization was 3 times slower than that for H^+ current, but deactivation was at most 1.5 times slower in D_2O. Both activation and inactivation for D^+ current were found to be slower in general. Does proton go through water in the channel pore or on the surface of channel protein? In order to answer to this question we need to consider how proton flows in the water. We need to consider that proton flow in the hydrated protein appears abruptly when water content exceeds critical concentration, which was predicted by percolation theory (Careri et al., 1990).

Namely, proton is flowing through the hydrated water layer on the protein surface. Since the critical value of water content for proton flow is independent on the pH values, a model for proton flow will be clear by the measurement of pH dependency. Unexpectedly, DeCoursey and Cherny (DeCoursey & Cherny, 1997) found the threshold of applied voltage to generate proton current was pH dependent, while proton conductance was pH independent. What is the role of the proton channel in the cell? We will discuss this problem in the next section.

2.2 Na^+ channel

As was described previously, the H/D ratio in conductance of Na^+ is 1.35 (Schauf & Bullock, 1982). Different from other ionic currents, when the Na^+ channel is abruptly depolarized, a small outward current precedes the Na^+ current. This small displacement current is asymmetric, named "gating current", because it is always associated with the opening of Na^+ channels. The gating current can be isolated by blocking Na^+ current by tetrodotoxin (Armstrong & Bezanilla, 1973). About 30 years later, the gating current was found to be decreased by about 30% when H_2O was changed to D_2O (Landowne, 2000). Na^+ current itself decreased as observed previously by other researchers (Meves, 1974), although these two types of current could be separated by a drug. The interpretation of the gating current by Landowne is that D_2O slowed the rate of the conformational change of channel protein by 30%, thus reducing the amplitude of the gating current and increasing the time required to open the channel due to the high viscosity of D_2O near the channel protein (Landowne, 2000). K^+ channel conductance was also studied by Schauf and Bullock, and H/D ratio in K^+ conductance was reported as 1.47, which was almost the same as that of Na^+ (1.35) (Schauf & Bullock, 1980). Accordingly, the D_2O exchange effect for both channels may be the same as each other. In order to examine the D_2O effect on Na^+ and K^+, high-K^+-induced depolarization of membrane potential was measured using AtT-20 cells (Ikeda et al., 2004). As shown in Fig 1, when the cell is exposed to 30mM KCl, the membrane potential was depolarized from -69mV to -32mV (Fig. 1A). D_2O treatment shifted resting potential slightly

(~ +4.4mV, about 10% of 30mM K^+-induced depolarization). This difference agreed well with the channel conductance difference between Na^+ and K^+ as reported by Shauf and Bullock (~9%) (Schauf & Bullock, 1980).

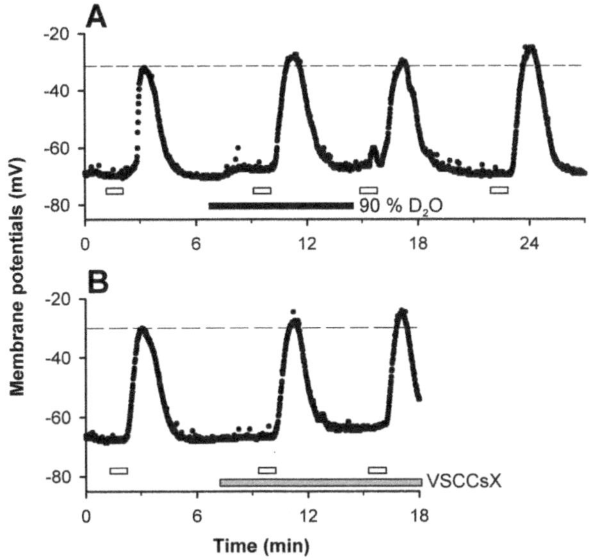

Fig. 1. Effects of 90% D_2O solution on high-K^+-induced membrane depolarization. Open bars = 30mM K^+, black bars = 90% D_2O treatment, grey bar = a mixture of voltage-sensitive Ca^{2+} channel blockers (VSCCsX). (Ikeda et al., 2004)

2.3 Ca^{2+} channel

An initial study on the D_2O exchange effect of the Ca^{2+} channel protein was carried out by Andjus et al. using inter nodal cells of the fresh water alga, which was known as a unique system before the discovery of the patch clump method (Andjus et al., 1994). They reported that there was no D_2O exchange effect on channel gating in the change of the resting potentials made by different concentration of KCl solutions. The single channel conductance was also unchanged, but open channel probability at 10mM KCl was increased irreversibly. The asymmetric distribution of D_2O (extracellular) and H_2O (intracellular) across plasma membrane was able to activate the Ca^{2+} channels of these cells. They suggested that transient osmotic-like stress produced by the rapid trans-membrane diffusion of D_2O may mediate the Ca^{2+} channel activations (Brooks, 1937).

We estimated the kinetics of the voltage-sensitive Ca^{2+} channel by whole cell patch using AtT-20 (Murine anterior pituitary corticotroph tumour) cells. The peak Ca^{2+} current recorded in H_2O-solution-filled electrode was -61.3pA, while the current recorded in D_2O-filled electrode at the same holding potential was significantly reduced to –15.4pA. Here, the H/D ratio of Ca^{2+} conductance was 3.98, which is 2.7 times higher than that of K^+ conductance (1.47) and 2.9 times higher than that of Na^+ conductance (1.35). Since Ca^{2+} is a divalent cation, the effective values will be halved (1.35 times higher than H/D ratio of K^+

conductance and 1.45 times higher than that of Na^+ conductance). Significant difference in H/D ratio of the conductance between Ca^{2+} and Na^+ (or K^+) cannot be explained by a simple model. To elucidate this unique feature of the Ca^{2+} ion in the living cell, we attempted further experiments. The results obtained are shown in Fig. 2.

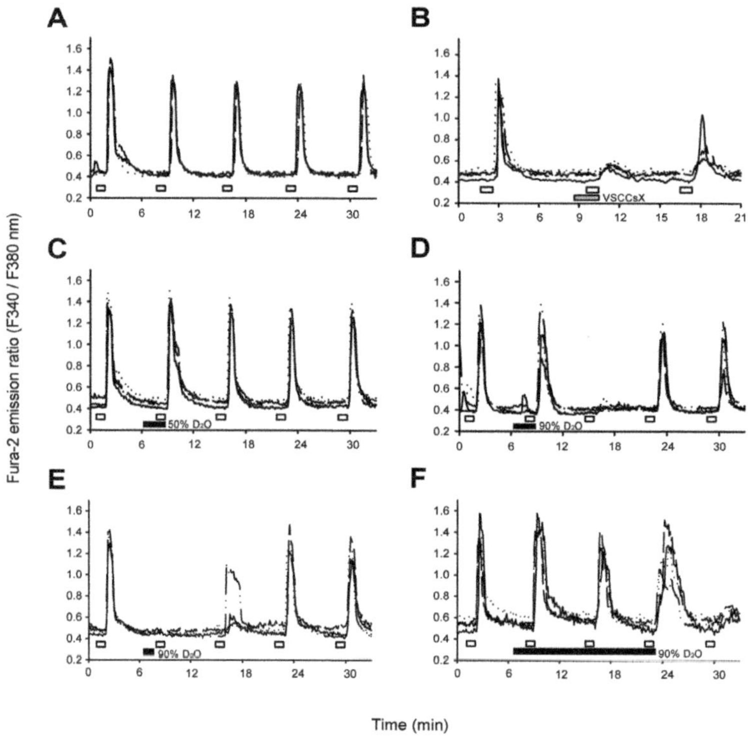

Fig. 2. The effects of D_2O on high-K^+-induced Ca^{2+} influx in AtT-20 cells. Open bars = 30mM K^+, grey bar = voltage-sensitive Ca^{2+} channel blockers (VSCCsX), black bars = 90% D_2O medium. (Ikeda et al., 2004)

As shown in Fig. 2, high-K^+-induced depolarization made a reproducible and rapid increase in the intracellular Ca^{2+} concentration (Fig. 2A). This Ca^{2+} elevation is primarily due to the Ca^{2+} influx through voltage-sensitive Ca^{2+} channels (VSCCs), because typical Ca^{2+} channel blockers, such as nifedipine (10μM) and ω-conotoxin (100nM), significantly reduced (~90%) the amplitude of the high-K^+-induced Ca^{2+} elevation (Fig. 2B). High-K^+-induced Ca^{2+} influx during treatment with D_2O-contained extracellular solution was not changed, whereas Ca^{2+} responses after wash out of D_2O was found to be reduced significantly. But this reduction was recovered gradually within several minutes after D_2O wash out (Fig. 2D). A 1 minute treatment with D_2O solution was sufficient to inhibit such a high-K^+-induced Ca^{2+} entry (~90% reduction; Fig. 2E), whereas longer D_2O treatment (up to 17 min), unexpectedly, showed a slight inhibitory effect (~8% reduction). Based on the data shown in Fig. 1, Fig. 2 and the voltage clamp experiment, it was concluded that the differences between a strong inhibition of Ca^{2+} channel by transient D_2O treatment and a slight inhibition with longer D_2O treatment will be explained by asymmetric distribution of D_2O (inside) and H_2O (outside) in a transient and stable D_2O effect, respectively. Consequently, the H/D ratio on the Ca^{2+} channel is estimated as ~1.08 from Ca^{2+} imaging data and ~1.20 from electrophysiological data.

How can the asymmetric distribution of D_2O and H_2O modulate the Ca^{2+} channel? The involvement of stress-sensitive mechanisms can be excluded, because cell shape change was not observed. The solvent isotope effect was also unlikely to explain the delayed effect of D_2O. Tentatively, it is likely that local unbalanced distribution of D and H in the O-H bonding of the channel protein is the cause. This D/H competition was already observed by Vasdev et al. (1994) with the anti-hypertensive effect of D_2O related to L-type Ca^{2+} channel conductance in myocyte. Proudhon et al. showed that D^+ could compete with H^+ for a single site in the L-type Ca^{2+} channels of guinea pig ventricular myocytes (Proudhon et al., 1994). It was thought that binding and unbinding of protons to this site were essential for Ca^{2+} movement (Kushner et al., 1998). This model can be expanded to the case of Na^+ channels as well as K^+ channels. The mechanism of this phenomenon will be discussed further in section 5.

3. Effect of D_2O on the cytoskeleton

As a solvent, D_2O increases the stability of proteins and other molecules, such as heliozoan microtubule formation, by hydrophobic bond formation (Marsland et al., 1971) and it has been used as a polymerizer of tubulin in a number of systems.

3.1 Tubulin is stabilized by D_2O

A number of studies on the effect of D_2O on protein aggregation have been conducted since the 1970s. It was almost accepted that D_2O stabilized the aggregated form of oligomeric proteins (Baghurst et al., 1972; Bonnete & Zaccai, 1994; Henderson et al., 1970). In 1999, Chakrabarti et al. reported clear results indicating that unstable tubulin protein was stabilized in D_2O and they proposed a mechanism whereby D_2O can have an effect on the conformational step or steps of hydrophobic force disruptions (Chakrabarti et al., 1999). Furthermore, they observed that D_2O stimulated formation of microtubule from tubulin as was observed previously (Ito & Sato, 1984). More interesting, when 8% of DMSO is added in the D_2O solution, tubulin polymerized as a ribbon structure rather than microtubules.

Although classical interpretations of D_2O-induced inhibition of cellular secretion have been based on D_2O-mediated stabilization of microtubules (Hill & Rhoten, 1983; Malaisse-Lagae et al., 1971; Montag & Umanskii, 1976), the effect of D_2O on cytoskeleton appear to be more prevalent and varied. In order to study relationships between cellular function and microtubule aggregation by D_2O, several researchers investigated the effect of some agents on the interaction between D_2O and tubulin. Urata et al. studied the effect of demecolcine (microtubule depolymerizing agent), taxol (microtubule stabilizing agent) and cytochalasin B&D (microfilament blocker) on the IgE-mediated Ca^{2+} influx, arachidonic acid and histamine release in rat basophilic leukaemia cells (Urata et al., 1989). They concluded that microtubule aggregation may be related to the process of secretion. Also, D_2O enhanced secretion of histamine from cultured mouse spleen cells and other mediators from homologous mast cells (Sulowska & Wyczolkowska, 1991). By exposing covalent oligomers of IgE to RBL-2H3 cells (a rat basophilic leukaemia tumour cell), substantial increases in a secretion of histamine was demonstrated by using D_2O in the medium (Maeyama et al., 1986). The differential effects of microtubule-altering agents (vinblastine; VB) on beta-cells during development were shown by comparison with D_2O and they speculated that microtubule is not coupled physico-chemically to other molecules in insulin secretion at day 17 of gestation during development (Hill & Rhoten, 1983). All these data suggest that tubulin microtubule transition is highly regulated by exchange of H/D. That means that the transition may be related to the mass difference between H (= 1.007) and D (= 2.014). The effect of mass difference of water molecules on the cell function will be discussed in section 5.

3.2 Effect of D_2O on the actin structure

Zimmermann et al. reported that D_2O induces the redistribution of filamentous actin (F-actin) and changes the morphology of human neutrophil granulocytes (Zimmermann et al., 1988). More recently, Omori et al. assessed the effect of D_2O on microfilaments and on in vivo actin polymerization using BALB/3T3 cell (Omori et al., 1997). They observed that the cells' stress fibres in the peripheral region became thick and distinct from other regions after being exposed to D_2O (>30%), while the quantity of perinuclear microfilaments was drastically reduced. Cytoplasmic F-actin was found to be increased with the stress fibres. Cell locomotion activity was suppressed in a D_2O concentration. The rate of actin polymerization was accelerated when purified globular actin (G-actin) was polymerized in D_2O. They concluded that alteration of stress fibres in cultured cells may be caused by a direct effect of D_2O on cellular microfilament dynamics (Omori et al., 1997). One possible mechanism underling D_2O-induced actin filament redistribution may involve H to D exchange in globular protein (Hermans & Scheraga, 1959; Scheraga, 1960), which results in a more stable protein structure (Karasz & Gajnos, 1976; Sing & Wood, 1976).

We tested this assumption by Ca^{2+} imaging method with the use of AtT-20 cells. Initially, we examined rhodamine-phalloidin labelling of actin filaments and immunostaining of β-tubulin.

As shown in Fig. 3, actin filaments were observed as filamentous structures that were found to be more highly concentrated in the cell processes than in cell soma (Fig. 3A & 3Ba). Treatment with D_2O-containing solution for 5 minutes immediately increased the amount of actin filaments in the cell soma and eliminated its filamentous structure (Fig. 3A). The relative amount of actin filaments in the cell process was decreased with D_2O treatment (Fig. 3Bb). Longer (15 minutes) treatment with 90% D_2O resulted in an increase in the actin filaments in the cell processes again, however, the filamentous structure was still lacking (Fig. 3A). After D_2O wash out, the filamentous structure of the actin filament recovered, but the shape of the filaments in the cell processes was similar to that of varicose filaments (Fig. 3A & 3Bc). The original shape of the actin filaments did not recover until 30 minutes after the D_2O wash out. Despite the marked changes in the distribution and structure of actin filament, those of β-tubulin (Fig. 3A) or those of neurofilament-M were not changed by D_2O treatment. The results described above indicate that among several cytoskeletal components, D_2O affected actin filaments especially. The effect of D_2O on the amount of actin filament was found to be transient (15 minutes), which is consistent with the results of the previous report (Omori et al., 1974).

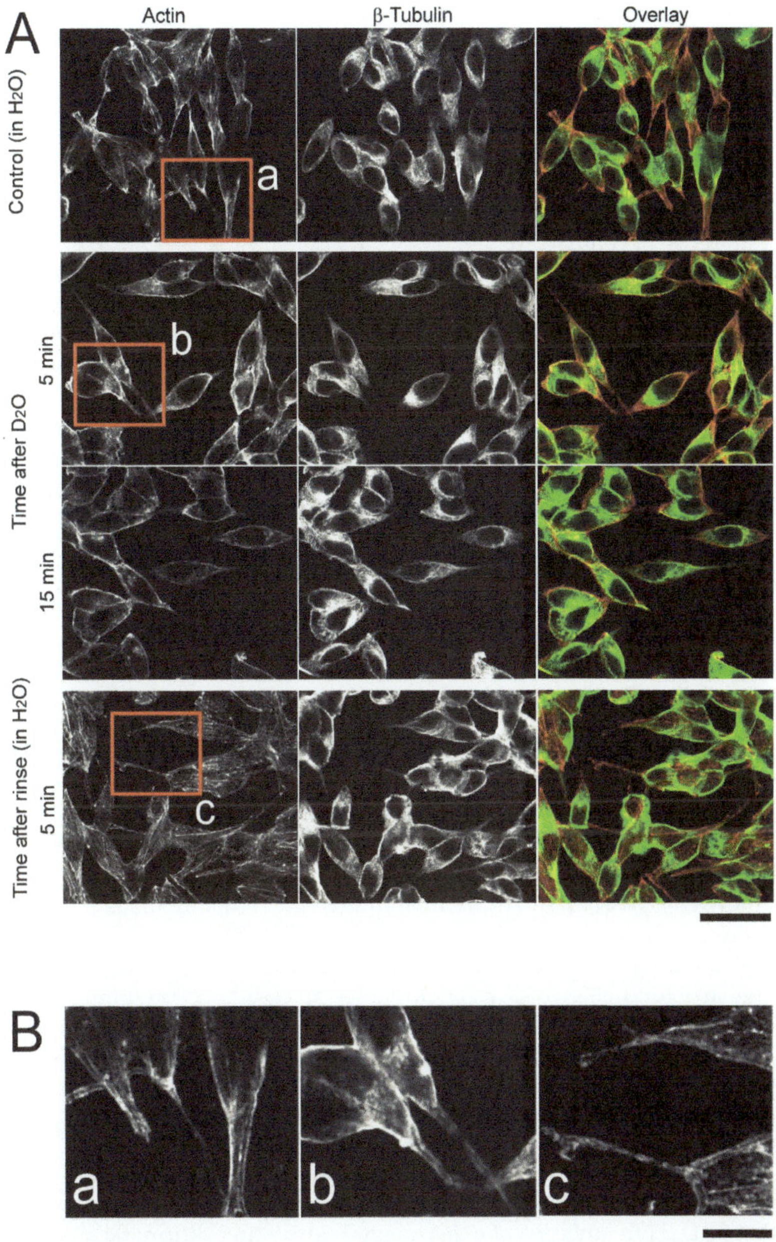

Fig. 3. Effects of 90% D_2O treatment on actin filaments (rhodamine-phalloidin labelling, red) and β-tubulin (FITC-labelling, green) in AtT-20 pituitary cells. The typical changes in the distribution of actin filaments (marked as red frame a-c) are enlarged in (B). Bar = 50μm in (A), 20μm in (B). (Ikeda et al., 2004)

3.3 Novel hypothesis for the molecular mechanism of interaction between channel protein and D_2O

More than 40 years has passed since the beginning of D_2O study, but the molecular mechanism of the interaction between D_2O and cytoskeleton remains unresolved. Therefore, we would like to attempt making a novel model based on the water structure near the cytosol protein proposed by Mentre and Hui Bon Hoa (Mentre & Hui Bon Hoa, 2001). First of all, we have to recall that the H/D ratio of the proton mobility in water is 1.44, while that in ice is 6.25 (4.3 times higher in the ice!). This is the most important numerical value to explain the difference of cellular function between H_2O and D_2O, because, in general, water structure on the surface of protein and lipid membrane is known as bound water, which has a similar structure to ice. According to Kellenberger, water structure on the surface of high polymer is quite different from free water (Kellenberger, 1991). It is called structural water, polarized water, strained water or vicinal water, which form the hydration shell of the surface of high polymer. Water molecules will be structural by hydration with a dissociated group and hydrogen bonding with a polarized group. Attention must be paid so that the structural water can convert chemical energy to mechanical energy, transfer several types of signals and make a biological structure. The structure and characteristics of bound water is very similar to those of ice. It is difficult to be frozen and sublimated. Furthermore, solubility of bound water is changed by hydrated ion, such as Na^+, K^+ and Ca^{2+}, due to exclusion. The order of exclusion follows the Hofmeister series. For example, the exclusion values of Na^+ and Ca^{2+} can be excluded more than that of K^+, which will induce a higher H/D ratio for Ca^{2+} to some extent.

3.4 New hypothesis for interaction between D_2O and cytoskeleton

Albrecht et al. proposed that linear structures in the cell, such as cytoskeleton, transmitted and propagated a signal by changing their structure (Albrecht et al., 1990). Dimer of tubulin α and β can be exchanged in a seesaw action and can make different structures from $\alpha\beta$ to $\beta\alpha$ by the signal initiation, called flipping. This flipping model is available for the explanation of structural change associated with H/D changes. The signal transfer along actin fibre is more complex because of actin binding to trans-membranous fibronectin. When the cell is moving, such as microglia, actin is supported by fibronectin as a fulcrum. Another example is skeletal muscle, where several proteins, such as myosin, dynein and kinesin, can move with the structural change of water induced by Ca^{2+} concentration change near the actin fibre. Thus, cellular signalling will be transferred along actin fibre or microtubule with the assistance of water structure change associated with Ca^{2+} concentration change.

4. Molecular mechanism of hysteresis in the D/H and H/D exchanges in the protein structure

During D/H and H/D exchange experiment, we have noticed that the H/D exchange rate is faster than that of D/H in any type of experiment. In order to confirm the anisotropic effect, a series of new methods was carried out using metabotropic glutamate receptor type 1 (mGluR1)-expressing CHO cells by Ca^{2+} imaging (Hirakura et al., 2011). The first hysteresis experiment was to measure the exchange rate of H_2O to D_2O by utilizing differences in fluorescence intensity of fura-2 in H_2O and D_2O under the same concentration of Ca^{2+}.

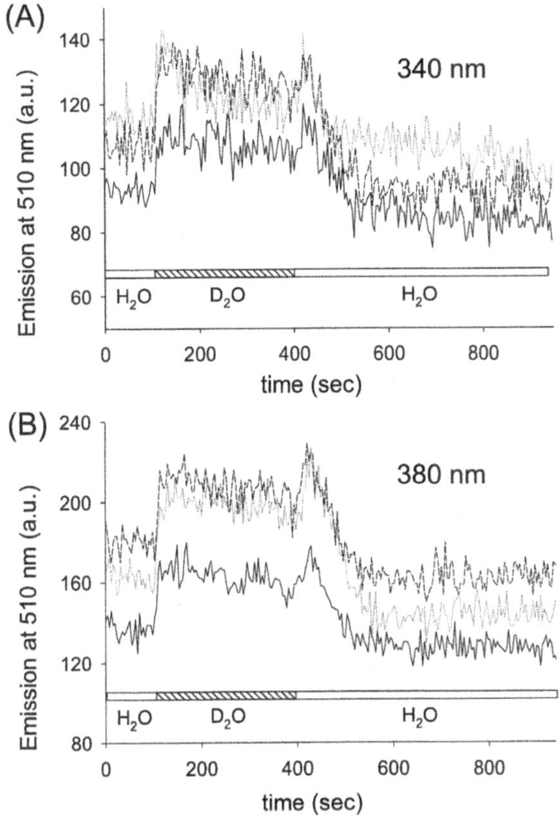

Fig. 4. Measurement of flow rate of D_2O entering the cell and the D/H exchange rate for fura-2 molecules in Ca^{2+}-free media. (A) Ex =340 nm, (B) Ex = 380 nm. (Hirakura et al., 2011)

As shown in Fig. 4, when the external solution was changed from H_2O to D_2O medium, the fluorescence intensity of fura-2 was increased at 380nm excitation. In contrast, when the external solution was changed from D_2O to H_2O, a slight rebound signal appeared and then the intensity declined slowly. The time of fluorescence increase in H to D exchange is about 5 sec, while the decaying time is more than 100 sec. Another example of hysteresis was observed more clearly as shown in Fig. 5.

In these experiments, exchange of extracellular medium was carried out with changes from H_2O to D_2O and back to H_2O at 5 minute intervals and the cells were stimulated with dihydroxyphenylglycine (DHPG; agonist of mGluR1) to induce Ca^{2+} elevation. DHPG-induced Ca^{2+} responses were relatively constant in D_2O, while Ca^{2+} responses were completely blocked during 5 minutes of incubation in H_2O after an incubation in D_2O medium (Fig. 5A). With a 15 minute incubation in H_2O medium after 5 minutes of exposure to D_2O, DHPG-induced Ca^{2+} responses partially recovered (Fig. 5B). When the incubation time in H_2O medium was increased to 30 minutes, Ca^{2+} responses were fully recovered (Fig. 5C & 5D).

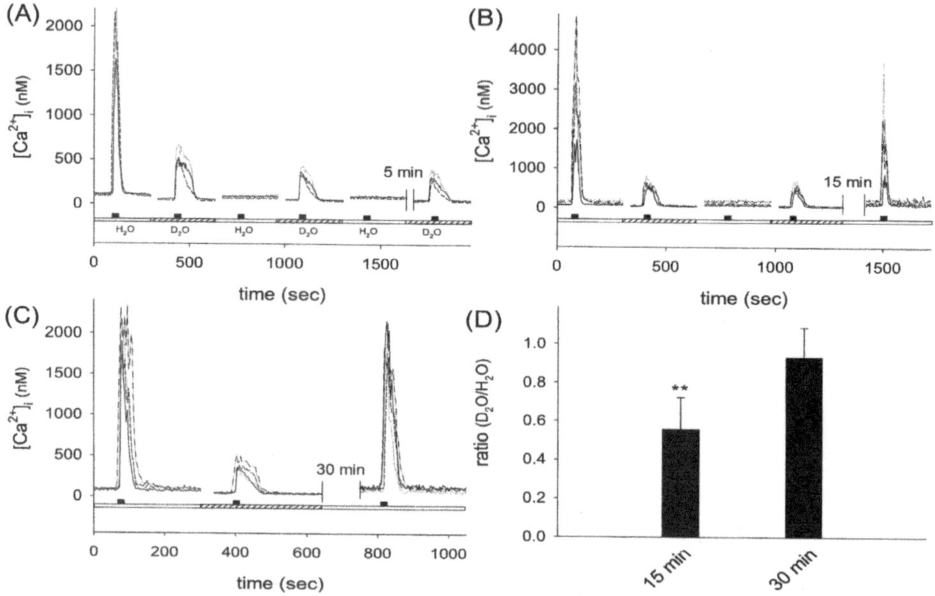

Fig. 5. Hysteresis of DHPG-induced Ca^{2+} responses induced by D/H exchange. Open bars = H_2O, shaded bars = 90% D_2O medium, small black squares = DHPG (30μM). Summary of the Ca^{2+} responses are shown in (D). $^{**}p < 0.01$ (Hirakura et al., 2011)

It is reasonable to explain that these data demonstrate the incubation period in H_2O medium is a critical factor for the recovery after D_2O exposure. When the cells were incubated in D_2O without agonist stimulation, however, DHPG-induced Ca^{2+} responses seemed normal. These results suggest that receptor stimulation is associated with large structural change of receptor that promotes deuteration of the intracellular portion in the receptor protein, which is consistent with a previous report showing that D/H exchange was facilitated during photo-activation of rhodopsin (Rath et al., 1998). The backbone structure of rhodopsin, G-protein-coupled photo-receptor with seven membrane spanning regions, becomes more accessible to D_2O during photo-activation. After stimulation, the receptor protein is more stable and resistant to D/H exchange. The increased stability of the deuterated protein may contribute to the prolonged stable state of the receptor. This hypothesis can be expanded to various types of receptor dynamics, including mGluR1 in the CHO cells. A similar hysteresis effect of H/D exchanges on the voltage-dependent Ca^{2+} channels was also reported previously (Ikeda et al., 2004). Thus we suggest that such hysteresis effect may be due to impairment of D/H exchanges. This may be a typical isotope exchange effect, which is due to a difference in the stability of proteins: deuterated proteins are more stable than protonated protein (Chakrabarti et al., 1999; Omori et al., 1997). Among the many known effects due to the difference in reactivity of D^+ and H^+, the most relevant to the effect observed in this study may be the zero-point energy differences. A quantum chemical calculation indicates that the isotopic substitution of hydrogen by deuterium lowers total energy of hydrophilic amino acids, because of the decrease in zero-point vibration energy

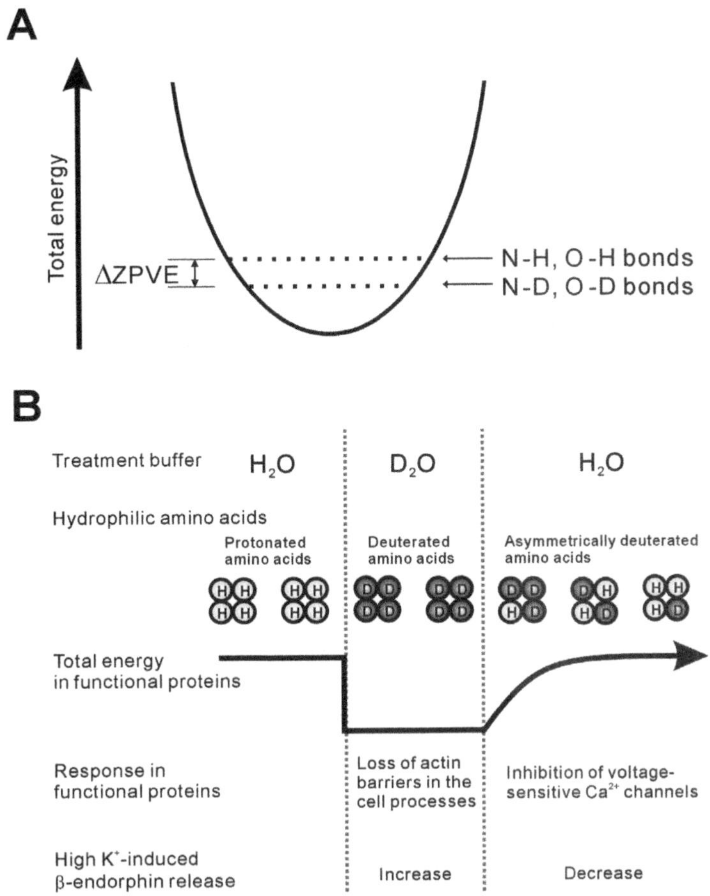

Fig. 6. Schematic illustration of changes in the zero-point vibrational energy and the energy differences between protonated and deuterated amino acids when hydrogen is exchanged by deuterium in hydrophilic amino acid residues of proteins. (A) Solid curve denotes the electronic energy of the system. Single deuterium substitution for hydrogen reduces approximately 2.1 kcal/mol (ΔZPVE) in hydrophilic amino acids. Total vibrational energy reduction in functional proteins depends on the number of deuterated amino acids replaced. (B) For the energy differences between protonated and deuterated amino acids, replacement of H_2O buffer to D_2O buffer immediately exchanges protonated amino acids to deuterated amino acids while replacement of D_2O buffer to H_2O buffer slowly exchanges deuterated amino acids and thus produces long-lasting transient states with asymmetrically deuterated amino acids. The increase in the high K^+-induced β-endorphin release is in the phase of deuterated amino acids and the most probable cause is due to the loss of actin filament barriers in the cell processes, although this effect coincides with many other effects including solvent isotope effects. On the other hand, the decrease in the high K^+-induced β-endorphin release is in the phase of asymmetrically deuterated amino acids and the most probable cause is the complete inhibition of voltage-sensitive Ca^{2+} channels. (Ikeda et al., 2004)

(Ikeda et al., 2004). Therefore, it is likely that cells rapidly reach equilibrium after an exchange of H_2O to D_2O, but take a longer time to reach equilibrium after switching from D_2O back to H_2O. Within the hydrophilic amino acids, a relatively large zero-point energy was found in tyrosine that is known as a critical residue for the voltage sensor of L-type Ca^{2+} channels (Bodi et al., 1997). Thus, this may produce a relatively long lasting transient state in voltage-sensitive Ca^{2+} channels with asymmetric deuterated amino acids after treatment by D_2O medium that will inhibit the Ca^{2+} channel activity (Fig. 6).

5. Role of proton in the regulation of functional proteins

As described above, the H/D exchange experiment was found to be effective to elucidate the role of water in the functional protein. When H is replaced by D, the structure and function of protein was largely changed, although chemical characteristics of D_2O are the same as H_2O. These facts facilitate us to consider carefully the physical properties of H_2O and the physical interaction between H_2O and protein. Here, we will discuss the role of proton in the ion channel and cytoskeleton in the cells, and hysteresis in H/D exchange; it was tentatively proposed that the channel conductivity was regulated by the local unbalanced distribution of D^+ and H^+ in the O-H residue of the channel proteins. Stabilization of cytoskeleton in D_2O will be due to the contribution of structural water and significance of the Hofmeister series. On the hysteresis effect observed in H/D exchange experiment, the significance of zero-point energy was proposed. These three models seem different to each other, but there is a possibility of making a simple model if we accept structural water layer model on the protein surface. Prior to making a novel model, drastic concept for cellular signalling is needed as proposed by Tsong (Tsong, 1989).

5.1 Electro conformational coupling (ECC) model for cellular signalling and energy transduction

In 1989, Tsong insisted that cells can communicate with each other by dispatching and receiving electro-magnetic signals. He observed that Na^+/K^+ ATPase (transmembrane ion transport protein) is sensitive to electric signal and it is most sensitive for alternative electric field of 1000 Hz 20V/cm. Based on these facts, he proposed his theory to explain communication between protein molecules in the cell; all kinds of proteins can respond to electric, electro-magnetic (EM) and acoustic oscillation. This theory was supported by many experimental results. Pulsed EM field was found to promote DNA, RNA and protein biosynthesis (Becker, 1981; Goodman et al., 1983; Pilla et al., 1987). The activity of many membrane enzymes are affected by weak EM signals (Adey, 1986; Blank, 1987). This model postulated that a protein can undergo conformational changes by a coulombic interaction with an oscillating electric field. When the frequency of the electric field matches the kinetic characteristics of the conformational transformation reaction, a phenomenological oscillation among different conformers of the enzyme is induced. He also insisted that this ECC model was consistent with the electric property of the most membrane proteins (Tsong & Astumian, 1987). This model is very attractive for the explanation of the cell-to-cell interaction and intracellular signalling by protein networks. However, some points of modification are needed, because this hypothesis cannot explain completely the functional change by H/D exchange. Therefore, we will try to make a novel model including H/D exchanges.

5.2 Proton signalling model for cellular function

The ECC model described above seems to assume electron transfer occurs in the protein molecule. But it is already accepted by many researchers that the transfer of electron along protein is accomplished by a series of redox centres incorporated into the protein structure or along a chain of conjugated orbital (Tuszynski, 2003). This protein-electron carrier system consists of a protein localized within a lipid bilayer having a single redox centre. The redox centres are usually prosthetic-group-containing non-protein molecules that have conjugated orbital systems and often incorporate metal ion. Furthermore, it has been proposed that protein is insulator at physiological temperature and that electron transport is mediated by proton. In this model, an electron is moving along the protein backbone by protons. Therefore, it is reasonable to assume that such a conformational dynamics of protein is depending on the solvent in *in vitro* experiment. Such a protein–solvent structure should be treated as a single system whose behaviour can be controlled by the environment (Tuszynski, 2003). If that is the case, proton signalling by proton transport will be likely. Accordingly, it is worth noting that structural water in the vicinity of protein molecule has a structure similar to ice. First of all, attention should be given to the fact that ice and water have very low conductivity for electron, while it is a good conductor for proton. In 1962, Grottus proposed an initial model for proton transfer in the water and the model was presented by Klotz in his book (Klotz, 1962). When one signalling proton reaches to the tip of the chain of water molecules, the proton is taken by the first water molecule. The first water molecule transfers one proton of two protons to the second one. Thus, the same type of proton transfer is repeated until the end of a line of water molecules, and then the proton will be released. In this process, the proton itself is not transferred from first water molecule to the last one, but chain-like molecular structure is changed successively. This is a typical type of salutatory conduction of proton in the water (or ice). Since the thickness of structural water layer is very thin (0.2 ~ 1.0 nm), the proton signalling in the water layer was a hypothesis in the 1980s, but it is measureable at present. Let us discuss the water layer inside of the channel pore and surface of cytoskeleton more deeply.

5.2.1 How ion channel permeate ion

According to a typical text book published in 2008, conductivity of the ion channel is formulated based on the diffusion theory (Dale Purves et al., 2008). In famous experiments performed by Hodgkin and Katz using squid giant axon, they found that the value of resting potential became less negative as external K^+ concentration was raised. When the external K^+ concentration was raised high enough to equal the concentration of K^+ inside the neuron, thus K^+ equilibrium potential was 0 mV, the resting membrane potential was also 0mV. In short, the resting membrane potential was not varied exactly 58 mV by ten-fold change in K^+ concentration. They thought the deviation from theoretical value may be due to the contribution of another ion, such as Cl^- and Na^+. But there is clear evidence that the inside of a K^+ channel is covered by a structural water layer and the iceberg structure of water can be disrupted by K^+. Therefore, the resting membrane potential of nerve cell membrane is approaching to Nernst equation because the iceberg structure of the water layer becomes free water at higher concentrations of K^+. If it is the case, ion selectivity of ion channel does not depend on the size of hydrated ion, but on that of metallic ion.

5.2.2 Relationship between shape change of cytoskeleton and iceberg water.

An explanation of the complex behaviour of cytoskeleton is still very difficult. The long fibres of the cytoskeleton are polymers composed of protein subunits. In this article, we will focus on tubulin (microtubules) and actin (microfilament). The tubulin that polymerizes to form microtubule has two isoforms (α-tubulin and β-tubulin) and forms a hetero dimer. The α- and β-tubulin monomers each consist of two β-sheets flanked by α helices to each other. The elementary building block of a microtubule is α and β hetero dimer whose dimensions are 4 by 5 by 8 nm that assemble into a cylindrical structure typically of 13 protofilaments. The outer diameter of a microtubule is 25 nm and the inner diameter is 15 nm. Microtubules are larger and more rigid than actin microfilaments and intermediate filament, and thus serve as major architectural struts of the cytoplasm. Above a certain tubulin concentration threshold, microtubule ensembles show a quasi-periodic regular pattern of damped oscillations (Carlier & Sellier, 1987). It is intriguing to know how the stochastic individual behaviour may change into smooth collective oscillations at high tubulin concentration (Mandelkow & Mandelkow, 1992).

Actin is one of the most abundant proteins in eukaryotic cells. G-actin and F-actin are reversible structures of actin. F-actin is a polymer of G-actin. Actin can be found in non-muscle cells and often is associated with other proteins in the cytoskeletons, where actin works as a dynamic component. G-actin monomers assemble into F-actin filaments in two-stranded geometry. The polymerization of F-actin from G-actin is a largely monotonic process dependent on the concentration of ATP. Once G-actin is assembled, microfilaments have diameters of about 8 nm. Microfilaments are often found with the lattice configuration near the leading edges of growing or motile cells, where they provide greater stability to the newly formed regions. New actin filaments are nucleated at the leading edge of cell growth and trailing microfilaments are disassembled.

These dynamic properties of cytoskeleton can also be explained by structural water theory. The concept of cluster of water molecules proposed by Watterson [Watterson, 1991] is predominant (Watterson, 1991). Water molecules have a resonance due to co-laborational work by hydrogen molecules, and then strong oscillation will be generated. This resonance will not be expanded to the large scale because of disturbance by thermal agitation, but it may be a cause of surface tension of the water molecules and the ordered structure of water, which is called "cluster". The cluster is thought to have the oscillations, whose wave length is calculated as 3.4 nm (~with network [addition of ion]), this impact will be spread out in the network rapidly just like a propagated wave of sound. The wave will decay rapidly in the region with high viscosity and it propagated far away. Therefore, the wave propagation mode is disturbed largely when ion is added to the cluster. Consequently, dynamic shape change can be induced by abrupt application of ion or other molecules.

6. Concluding remarks

In this article we demonstrated the change of activity of ion channels and glutamate receptor, and the structural change of cytoskeleton by H/D changes. These cellular changes were always associated with hysteresis effect of the change of incubation medium from H_2O to D_2O and vice versa. Initially, we tried to approach these phenomena independently, such as zero-point energy difference between protonated and deuterated protein. Finally, we reached the reasonable conclusion that the difference between H_2O and D_2O might be caused by a structural water or proton. If we introduce proton signalling pathway in the structural water in the vicinity of protein, all types of changes in the protein function could be explained without contradiction.

It is worthwhile to note that this model can be interpreted as a docking of the water layer model (Mentrè, 1995) and ECC model (Tsong, 1989), and improved them to some extent to explain reasonably the imaging data recently obtained (Hirakura et al., 2011; Ikeda et al., 2004).

7. Acknowledgement

We gratefully acknowledge the support of Taiwan NSC, Sun's KMU-SMA fund, the Center of Excellence for Environmental Medicine (KMU-EM-97-3.3a), and the Chi-Mei Medical Center and KMU foundation (96CM-KMU-03). This work was also supported by a grant for research for the Future Program (96L00310 to T.Y.) from the Japan Society for the Promotion of Science.

8. References

Adey, W. R. (1986). The sequence and energetics of cell membrane transducing coupling to intracellular enzyme systems. *Bioelectrochemistry and Bioenergetics*, vol.15, pp447-456.

Albrecht, T., Boldogh, I., Fons M., AbuBakar S., & Deng C. Z. (1990). Cell activation signals and the pathogenesis of human cytomegalovirus. *Intervirology*, vol.31(2-4), pp68-75

Andjus, P.R., Kataev, A.A., Alexandrov, A.A., Vucelic, D., & Berestovsky, G.N. (1994). D_2O-induced ion channel activation in Characeae at low ionic strength, *The Journal of membrane biology*, vol.142, No.1, pp43-53

Armstrong, C. M. & Bezanilla F. (1973). Currents related to movement of the gating particles of the sodium channels, *Nature*, vol.242, No.5398, pp459-461

Baghurst, P.A., Nichol, L.W. & Sawyer, W.H. (1972). The effect of D 2 O on the association of -lactoglobulin A. *The Journal of Biological Chemistry*, Vol.247, No.10, pp3199-3204

Bass, L. & Moore, W.J. (1973). A simplified cooperative model of excitable membranes. *The Journal of Membrane Biology*, vol.12, N0o.4, pp361-366

Becker, R. O. ed (1981). *Mechanisms of Growth Control*, Charles C Thomas Pub Ltd, Springfield .

Blank, M. a. F., E. (eds) (1987). *Mechanistic Approaches to Interactions of Electric and Electromagnetic Fields with Living Systems* Plenum Press.

Bodi, I., Yamaguchi, H., Hara, M., He, M., Schwartz, A. & Varadi, G. (1997). Molecular studies on the voltage dependence of dihydropyridine action on L-type Ca2+ channels. Critical involvement of tyrosine residues in motif IIIS6 and IVS6. *The Journal of Biological Chemistry*, vol.272, No.40, pp24952-24960

Bonnete, F. & Zaccai, G. (1994). Small angle neutron scattering, total cross-sections and mass density measurements of concentrated NaCl and KCl solutions in H2O or D2O. *Biophysical Chemistry*, vol.53, No. 1-2, pp69-75

Brooks, B. T. (1937). Robert Kennedy Duncan, *Science*, vol.85, No.2212, pp489-491.

Careri, G., Consolini, G. & Bruni, F. (1990). Proton tunneling in hydrated biological tissues near 200 K. *Biophysical Chemistry*, vol.37, No. 1-3, pp165-170

Carlier, J.P. & Sellier, N. (1987). Identification by gas chromatography-mass spectrometry of short-chain hydroxy acids produced by Fusobacterium species and Clostridium innocuum. *Journal of Chromatography*, vol.420,No.1, pp121-128

Chakrabarti, G., Kim, S., Gupta, M.L., Jr., Barton, J.S. & Himes, R.H. (1999). Stabilization of tubulin by deuterium oxide. *Biochemistry*, vol.38, No.10, pp3067-3072

Dale Purves, G. J. A., David Fitzpatrick, William C. Hall, Anthony-Samuel LaMantia, James O. McNamara & Leonard E. White (2008). *Neuroscience*, Fourth ed., Sinaure Associates, Inc., Massachusetts, U. S. A.

DeCoursey, T. E. & V. V. Cherny (1997). Deuterium isotope effects on permeation and gating of proton channels in rat alveolar epithelium, *J Gen Physiol*, vol.109, No.4, pp415-434.

Goodman, R., C. A. Bassett, and A. S. Henderson (1983). Pulsing electromagnetic fields induce cellular transcription, *Science*, vol.220, No.4603, pp1283-1285.

Henderson, E. J., H. Nagano, H. Zalkin & L. H. Hwang (1970). The anthranilate synthetase-anthranilate 5-phosphoribosylpyrophosphate phosphoribosyltransferase aggregate. Purification of the aggregate and regulatory properties of anthranilate synthetase, *J Biol Chem*, vol.245, No.6, pp1416-1423.

Hermans, J., Jr., & H. A. Scheraga (1959). The thermally induced configurational change of ribonuclease in water and deuterium, *Biochim Biophys Acta*, vol.36, pp534-535.

Hill, R. S. & W. B. Rhoten (1983). Differential effects of microtubule-altering agents on beta-cells during development, *Am J Physiol*, vol.245, No.4, ppE391-400.

Hirakura, Y., T. Sugiyama, M. Takeda, M. Ikeda & T. Yoshioka (2011). Deuteration as a tool in investigating the role of protons in cell signaling, *Biochim Biophys Acta*, vol.1810, No.2, pp218-225.

Ikeda, M., Suzuki, M., Kishio, M., Hirono, M., Sugiyama, T., Matsuura, J., Suzuki, T., Sota, T., Allen, C.N., Konishi, S. & Yoshioka, T. (2004). Hydrogen-deuterium exchange effects on beta-endorphin release from AtT20 murine pituitary tumor cells, *Biophys J*, vol.86, No.1 Pt. 1, pp565-575.

Itoh, T. J. & H. Sato (1984). The effects of deuterium oxide (^2H2O) on the polymerization of tubulin in vitro, *Biochim Biophys Acta*, vol.800, No.1, pp21-27.

Karasz, F.E. & Gajnos, G.E. (1976). Relative stability of the α-helix of deuterated poly(γ-benzyl-L-glutamate). *Biopolymers*, vol.15, pp1939-1950.

Katz, J. J. (1965). *Chemical and biological studies with deuterium*, 1-100 pp., Pennsylvania State University, University Park.

Kellenberger, E. (1991). The potential of cryofixation and freeze substitution: observations and theoretical considerations, *J Microsc*, vol.161, No. 2, pp183-203.

Klotz, I. M. (1962). *Water*, Academic Press, New York and London

Kushner, B. H., N. K. Cheung, K. Kramer, G. Heller, & S. C. Jhanwar (1998). Neuroblastoma and treatment-related myelodysplasia/leukemia: the Memorial Sloan-Kettering experience and a literature review, *J Clin Oncol*, vol. 16, No.12, pp3880-3889.

Landowne, D. (2000). Heavy water (D2O) alters the sodium channel gating current in squid giant axons, *Biol Bull*, vol.199, No.2, pp164-165.

Maeyama, K., R. J. Hohman, H. Metzger & M. A. Beaven (1986). Quantitative relationships between aggregation of IgE receptors, generation of intracellular signals, and histamine secretion in rat basophilic leukemia (2H3) cells. Enhanced responses with heavy water, *J Biol Chem*, vol.261, No.6, pp2583-2592.

Malaisse-Lagae, F., G. R. Brisson & W. J. Malaisse (1971). The stimulus-secretion coupling of glucose-induced insulin release. VI. Analogy between the insulinotropic mechanisms of sugars and amino acids, *Horm Metab Res*, vol.3, No.6, pp374-378.

Mandelkow, E. M. & E. Mandelkow (1992). Microtubule oscillations, *Cell Motil Cytoskeleton*, vol.22, No.4, pp235-244.

Marsland, D., L. G. Tilney & M. Hirshfield. (1971). Stabilizing effects of D2O on the microtubular components and needle-like form of heliozoan axopods: a pressure-temperature analysis. *Journal of Cellular Physiology*, vol.77, pp187-194.

Mentré, P. (1995). *L'eau dans la cellular* Masson, Paris.

Mentre, P. & G. Hui Bon Hoa (2001). Effects of high hydrostatic pressures on living cells: a consequence of the properties of macromolecules and macromolecule-associated water, *Int Rev Cytol*, vol.201, pp1-84.

Meves, H. (1974). The effect of holding potential on the asymmetry currents in squid giant axons, *J Physiol*, vol.243, No.3, pp847-867.

Montag, T. S., and A. Umanskii Iu (1976). Effect of combined use of antilymphocyte serum 7S antibodies and cyclophosphane on the growth of transplanted tumors in rats, *Fiziol Zh*, vol.22, No.5, pp630-634.

Omori, H., M. Kuroda, H. Naora, H. Takeda, Y. Nio, H. Otani & K. Tamura (1997). Deuterium oxide (heavy water) accelerates actin assembly in vitro and changes microfilament distribution in cultured cells, *Eur J Cell Biol*, vol.74, No.3, pp273-280.

Pilla, A. A., Kaufman, J. J. & Ryaby, J. T. (1987). *in Mechanistic Approaches to Interactions of Electric and Electromagnetic Fields with Living Systems* Blank, M. and Findl, E., ed., Plenum Press.

Proudhon, J. F., C. Almange, J. C. Palaric, J. Milon, P. Poulain & J. R. Giraud (1994). Left heart hypoplasia: diagnosis and current obstetrical management. Apropos of 4 cases, *J Gynecol Obstet Biol Reprod (Paris)*. vol.23, No.4, pp425-431.

Rath, P., W. J. DeGrip & K. J. Rothschild (1998). Photoactivation of rhodopsin causes an increased hydrogen-deuterium exchange of buried peptide groups, *Biophys J*, vol.74, No.1, pp192-198.

Schauf, C. L. & J. O. Bullock (1980). Solvent substitution as a probe of channel gating in Myxicola. Differential effects of D2O on some components of membrane conductance, *Biophys J*, vol.30, No.2, pp295-305.

Schauf, C. L. & J. O. Bullock (1982). Solvent substitution as a probe of channel gating in Myxicola. Effects of D2O on kinetic properties of drugs that occlude channels, *Biophys J*, vol.37, No.2, pp441-452.

Scheraga, H. A. (1960). Helix-random coil transformations in deuterated macromolecules, *Ann N Y Acad Sci*, vol.84, pp608-616.

Sing, T.R. & J.L. Wood. 1969. Isotope effect on the hydrogen bound length. *J. Chem. Phys.* vol.50, pp3572-3576.

Sulowska, Z., & J. Wyczolkowska (1991). 5-Hydroxytryptamine releasing activity of the supernatants from cultured mouse spleen cells, *Arch Immunol Ther Exp (Warsz)*, vol.39, No.1-2, pp139-145.

Thomas, A. F. (1971). *Deuterium labeling in organic chemistry*, Appelton-Centry, New York.

Tsong, T. Y. (1989). Deciphering the language of cells, *Trends Biochem Sci*, vol.14, No.3, pp89-92.

Tsong, T. Y. & R. D. Astumian (1987). Electroconformational coupling and membrane protein function, *Prog Biophys Mol Biol*, vol.50, No.1, pp1-45.

Tuszynski, M. H. (2003). Gene therapy for neurological disease, *Expert Opin Biol Ther*, vol.3, No.5, pp815-828.

Urata, C., A. Watanabe, Y. Ogawa, N. Takei, H. Nomoto, M. Mizobe, Y. Abe, K. Mano & S. Urata (1989). Effect of deuterium oxide (D_2O) on the IgE-mediated Ca^{2+} influx, arachidonic acid and histamine release in rat basophilic leukemia. *Arerugi,* vol.38, pp285-295.

Watterson, J. G. (1991). The role of water in cell function, *Biofizika*, vol.36, No.1, pp5-30.

Zimmermann, A., H. U. Keller & H. Cottier (1988). Heavy water (D2O)-induced shape changes, movements and F-actin redistribution in human neutrophil granulocytes, *Eur J Cell Biol*, vol.47, No.2, pp320-326.

13

Ligand-Binding Proteins: Structure, Stability and Practical Application

Olga Stepanenko, Alexander Fonin, Olesya Stepanenko, Irina Kuznetsova and Konstantin Turoverov
Institute of Cytology RAS, Laboratory of Structural Dynamics, Stability and Folding of Proteins, Russia

1. Introduction

A tremendous diversity of ligand binding proteins exists in nature. This undoubtedly creates considerable opportunities for scientific and medicinal applications. In this chapter, we will consider a range of ligand binding proteins, with particular attention to two classes, namely the ligand-binding proteins of the bacterial periplasm and odorant-binding proteins, because these proteins are the building blocks for biosensor development.

2. Diversity of ligand-binding proteins

All functions of living organisms are related to proteins that are present in enormous numbers in the living cell. Usually, proteins are classified by their function. Often, proteins have a range of different functions. Interestingly, the functions of many proteins involved in different biological processes begin with the binding of specific molecules: carbohydrates, amino acids, anions, metal ions, ions, oligo-peptides, proteins, lipids, odorant molecules and others, collectively known as ligands. Vital cell processes, such as DNA replication, gene expression, cell signaling and so on, are initiated by the binding of specific ligands. Trafficking of molecules throughout cellular compartments is possible after the binding of such molecules to a specific carrier protein. To perform their biological functions, enzymes must bind to their cognate substrates. Though performing diverse biological functions, all of these proteins fall into the category of ligand-binding proteins. Such proteins include periplasmic binding proteins, biotin-binding proteins, lipid-binding proteins, lectins, serum albumins, immunoglobulins, and others (De Wolf & Brett, 2000). Given that ligand binding proteins have a high affinity for their ligands, ligand-binding proteins can be used in protein-based controlled delivery systems for bioactive compounds sensitive to environmental factors. Another application of ligand-binding proteins is for biosensing of different disease markers, pathogenic molecules, environmental toxins and chemically or biologically hazardous compounds. Some of these proteins do not alter their structure in response to ligand binding (lipid-binding proteins, lectins). By contrast, other proteins show significant ligand-induced conformational changes. For example, the two-lobed ligand binding proteins of bacterial periplasm switch from their open-form in the absence of ligands to their closed-form in the presence of ligands (Stratton & Loh, 2011). In general, ligand-binding proteins, including lipid-binding proteins, some of the lectins, serum albumins and biotin-binding proteins, recognize a wide array of bound ligands. By contrast, there are proteins with narrow specificity (for example, most of the periplasmic binding proteins). Ligand-binding proteins vary in their overall structure and number of binding sites, but most such proteins have a complex multi-domain structure or exist as multi-mers. As there are excellent reviews on the aforementioned families of ligand-binding proteins, here, we will only briefly discuss ligand-binding proteins and their possible therapeutic and clinical applications.

The biotin-binding proteins, namely, chicken egg-white avidin, bacterial streptavidin and newly discovered tamavidins from basidiomycete fungi, have numerous medical, biological, biochemical and biotechnological applications (Laitinen et al., 2006; Takakura et al., 2010; Wilchek et al., 2006). These tetrameric proteins, consisting of classical β-barrel monomers, bind biotin with exceptionally high affinity. Several peptides having the consensus HPQ tripeptide sequence are reported to be ligands of biotin-binding proteins, though with much lower affinity. The diverse family of lipid-binding proteins (LBP; Banaszak et al., 1994) is made up of extracellular LBPs (eLBPs, which are also known as lipocalins; Grzyb et al., 2006) and intracellular LBPs (iLBPs; Glatz et al., 2002; Haunerland & Spener, 2004). These low-molecular weight proteins share a remarkably similar β-barrel structure, albeit with some differences between iLBPs and lipocalins, and are found in diverse cell types. Individual LBPs can bind a wide range of small hydrophobic ligands, including fatty acids and retinol analogs. A representative member of the lipocalins is β-lactoglobulin (β-LG; Perez & Calvo, 1995), which is shown to be a promising carrier for fatty-acids, as well as a protective agent for bioactive compounds and therapeutically relevant synthetic retinoid derivatives (Liang & Subirade, 2010; Riihimaki-Lampen et al., 2010). An extensive group of lipocalins are the odorant-binding proteins (OBPs) that were successfully adapted to serve as biosensors for dangerous substances, including polyaromatic hydrocarbons (Wei et al., 2008). Other lipocalins, such as neutrophil gelatinase-assotiated lipocalin (NGAL; Taub et al., 2010; Xu & Venge, 2000), are utilized in clinical applications. Importantly, the high structural plasticity of the lipocalin's binding site allows, with the aid of genetic engineering, the generation of artificial lipocalins with novel ligand specificities, that is, the so-called anticalins (Skerra, 2008). Anticalins are immunologically active molecules that bind to small hapten-like compounds and to large protein antigens. Compared with antibodies, they are small (composed of just one polypeptide chain), do not require post-translational modification and exhibit robust biophysical properties. Owing to these properties, anticalins offer many potential applications, not only as reagents for biochemical research but also as a new class of drugs for medical therapy. Other ligand-binding proteins, namely serum albumins, bind to an extremely large number of diverse ligands (more than 70), including fatty acids, amino acids, therapeutic drugs and inorganic ions (Fasano et al., 2005; Varshney et al., 2010). Serum albumins are composed of three structurally homologous domains that are predominantly helical (Carter & Ho, 1994). The binding sites for a variety of ligands are distributed among distinct locations on the protein. The main advantage of serum albumins for in vivo applications is their compatibility with human blood, plasma and body components. Currently, serum albumins are playing an increasing role in the development of drug-delivery systems (Kratz, 2008) and in diverse clinical applications (Caironi & Gattinoni, 2009). A versatile family of periplasmic binding proteins originating in the periplasmic space of bacteria share a characteristic two-lobed structure and traffic different nutrients, such as carbohydrates, amino acids, anions, metal ions, and di- and oligo-peptides (Felder et al., 1999; Fukami-Kobayashi et al., 1999; Tam & Saier, 1993). The members of this family commonly have high specificity for their cognate ligands, though there are exceptions (the case of the di/oligopeptide-binding protein). Their intrinsic ability to undergo a significant, ligand-induced conformational change has been utilized in the engineering of reagentless biosensors to monitor ligand concentration (Dwyer & Hellinga, 2004; de Lorimier et al., 2002). Members of another ubiquitous family of sugar-specific and cell-agglutinating proteins, the lectins, have been found in all kinds of organisms, from viruses to humans (Sharon & Lis, 2004). Lectins function as recognition molecules in ligand-cell and cell-cell interactions in a variety of biological systems. Mature plant lectins are divided into merolectins, hololectins, chimerolectins and superlectins, according to the number of carbohydrate-binding domains (Liu et al., 2010). There are only one or at most two carbohydrate-binding domains in merolectins and hololectins, respectively. It is important to note that different carbohydrate-binding domains of hololectins bind either to the same or structurally similar sugars. Chimerolectins are fusion proteins containing one or more carbohydrate-binding domains and an unrelated domain. Superlectins have at least two carbohydrate-binding domains with specificity for structurally unrelated sugars. Plant lectins have a similar tertiary structure, referred to as the lectin fold. This structural motif consists of a characteristically elaborate jelly roll, derived from antiparallel β-strands, and arranged as two β-sheets. This fold has also been noted in animal lectins. Animal lectins can be divided into 12 groups based on the similarity of their primary structures, including the previously discovered C-type lectins (requiring Ca^{2+} for activity) and the galectins (Kilpatrick, 2002). Animal lectins are often bi-functional, with a carbohydrate-binding domain and an additional domain, which is responsible for the ability of animal lectins to bind to non-carbohydrate ligands via protein-protein, protein-lipid and protein-nucleic acid interactions. Furthermore, there are examples of animal lectins with a carbohydrate-binding domain capable of interaction with non-sugar ligands. Many studies are devoted to the anti-tumor activity of plant lectins against a variety of malignant cells. Lectins have a high potential for development of antineoplastic drugs for cancer therapy and of targeted drug delivery systems (Liu et al., 2010; Robinson et al., 2004). Immunoglobulins, with their typical tetrameric organization consisting of two light and two heavy polypeptide chains, are indispensable in basic research and diagnostics, as they can be adapted for the binding of an incredible variety of ligands (Chester & Hawkins, 1995; Sundberg, 2009). Bispecific antibodies that are capable of simultaneous binding to two different antigens are further improving the prospects for clinical applications of conventional antibodies (Fitzgerald & Lugovskoy, 2011). Some shortcomings of antibodies, such as their lengthy timeframe and high cost of production, and their large size can be overcome by development of small single-domain antibody fragments of high stability with the unique capacity to recognize molecules that are inaccessible to conventional antibodies (de Marco, 2011). Several additional classes of ligand-binding proteins, including inactivated enzymes, penicillin-binding proteins, immunophilins and others, further expand the possibilities for diverse practical applications.

In the next sections, we will focus on two classes of ligand-binding proteins, the two-domain ligand-binding proteins of the bacterial periplasm and odorant-binding proteins for which the use as the sensitive elements in socially important biosensor systems is mostly elaborated. The structure, stability and possible practical applications of these proteins will be discussed in detail.

2.1 Two-domain ligand-binding proteins of bacterial periplasm

Two-domain ligand-binding proteins of the periplasm of gram-negative bacteria (PBPs; Dwyer & Hellinga, 2004; Tam & Saier, 1993) constitute a large bacterial family of proteins serving as primary receptors for a large number of compounds. The ligands of periplasmic binding proteins are represented by carbohydrates such as glucose, maltose, ribose and arabinose; amino acids such as glutamine, leucine, valine and histidine; metal ions such as phosphate, sulfate, iron, zinc and nickel; di- and oligo-peptides and vitamins. Thus, they are involved in the active transport of the soluble molecules inside the bacterial cell. Two-domain ligand-binding proteins are constituents of the ATP-binding cassette in which ligand transport across the membrane is powered by ATP hydrolysis. In general, the ATP-binding cassette consists of two trans-membrane domains, which assist in ligand translocation across the inner membrane, and two nucleotide-binding domains, which provide the energy required for the transport process. The periplasmic binding protein is responsible for trafficking of its ligand across the periplasmic space and release of its ligand near the inner membrane. According to recent studies, the ATP-switch model for the transport function of the ATP-binding cassette has been proposed. In this model, the coupled nucleotide-binding domains switch between an ATP-dependent closed conformation and a nucleotide-free, open conformation to drive the translocation of ligand (Dawson et al., 2007; Linton, 2007). In some cases, the periplasmic binding proteins participate in chemotaxis toward different substances. For example, chemotaxis toward such attractants as some sugars (galactose, ribose and maltose) and amino acids is activated by interaction between a complex of defined binding proteins and attractant molecules recognizing specific chemoreceptors (Felder et al., 1999; Szurmant & Ordal, 2004). Periplasmic binding proteins are involved in bacterial intercellular communication processes, termed quorum sensing (Neiditch et al., 2006; Schauder & Bassler, 2001). In this case, binding of small molecules, the so-called autoinducers, leads to a series of chemical reactions that provide the bacteria with information about the cellular density of the surrounding environment. Some periplasmic binding proteins act as chaperones promoting the proper folding of denatured proteins or their fusion partners (Dalken et al., 2010; Richarme & Caldas, 1997).

The molecular weight of periplasmic binding proteins ranges from 22 to 59 kDa. Despite significant differences in their amino acid sequences, all ligand-binding proteins share the same structural topology of the polypeptide chain. All of them display a secondary structure of the α/β type that is organized at the tertiary level into two domains linked by what is commonly referred to as a hinge region. This hinge region is formed by two or three short flexible peptide segments. The protein ligand-binding site is located in the cleft between the two domains. The periplasmic binding proteins can interconvert through a pronounced (depending on the protein) bending motion around the hinge from a ligand-free open conformation to a ligand-bound closed conformation. Both protein domains adopt a three-layered $\alpha/\beta/\alpha$ sandwich fold, and the distribution of β-strands between the two domains defines the structural subclass of the periplasmic binding protein (figure 1). One of the structural sub-classes (group I), includes *Escherichia coli* D-galactose/D-glucose-binding protein (GGBP; Borrok et al., 2007); and the other subclass (group II) includes *Escherichia coli* glutamine-binding protein (GlnBP; Hsiao et al., 1996; Sun et al., 1998; Fukami-Kobayashi et al., 1999). Both domains possess a CheY-like fold, which gave rise to the hypothesis that the ancestral protein for PBP family members is derived from a one-domain CheY-like protein

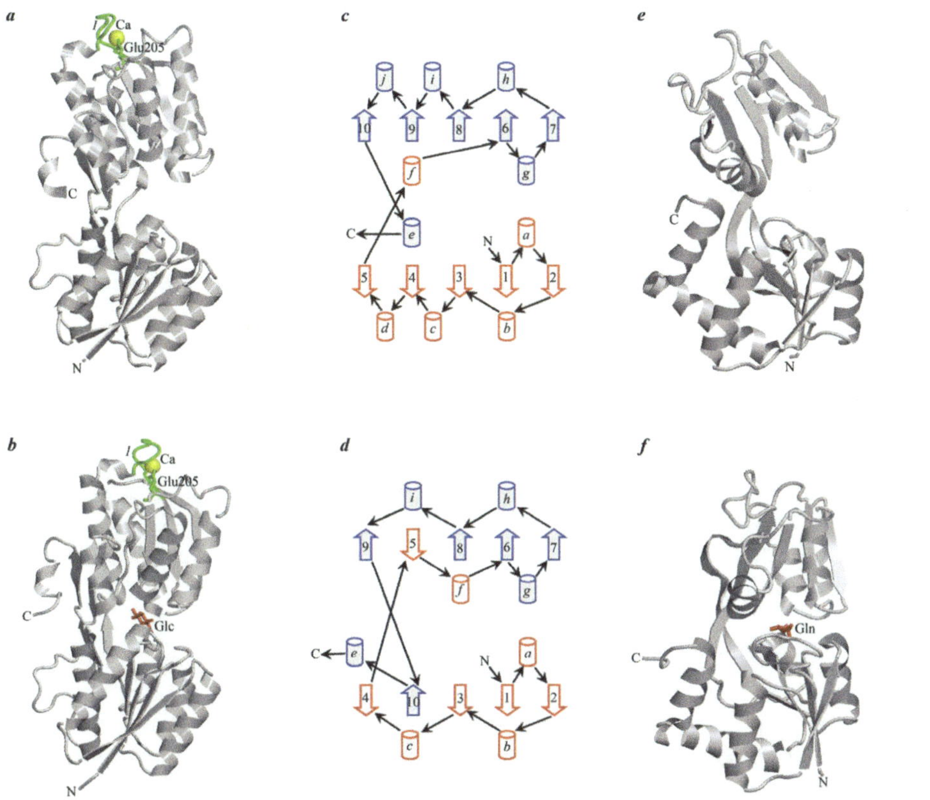

Fig. 1. Spatial pattern of PBP of group I and group II. Tertiary structures of GGBP (***a***) and its complex with glucose GGBP/Glc (***b***), representing group I, are shown. The residues of the Ca-binding site are shown *in green*, including the loop in the protein's C-terminal domain (residues 134-142; *1*) and Glu 205. Calcium is represented as a *yellow* sphere. The structures of group II GlnBP (*e*) and its complex with glutamine GlnBP/Gln (*f*) are shown. In both cases, the ligand (glucose or glutamine) is represented as a *red stick union*. The structures were created based on PDB data (Dutta et al., 2009); PDP codes 2GBP.ent15 (Vyas et al., 1988), 2FWO.ent16 (Borrok et al., 2007), 1GGG (Hsiao et al., 1996) and 1WDN (Sun et al., 1998) using the graphical software VMD (Hsin et al., 2008) and Raster 3D (Merritt & Bacon, 1977). The topology of group I (*c*) and group II (***d***) proteins is drawn with β-strands and α-helices indicated as arrows and cylinders, respectively. Secondary structural elements originating from a monomeric ancestral protein are represented in the same color (red or blue).

through its duplication and subsequent fusion (Lewis et al., 2000). In group I, the domain's β-strands have a 21345 topology and have more regular organization of its secondary structure, while group II proteins, derived later on the evolutionary time scale, are characterized by a more complex topology of their β-strand distribution within each separate domain. The sheet topology of both domains of group II periplasmic binding proteins follows a 213N4 sequence, with β_N becoming the first strand after crossing-over from the N-terminal domain to the C-terminal domain. The group II proteins are supposed to have arisen from the group I proteins through the mutual dislocation of two β-strands in each domain from their original domain to the other: one β-strand in the first domain penetrates into the parallel β-sheet of the second domain, making a new anti-parallel β-sheet, as does the β-strand of the second domain (Fukami-Kobayashi et al., 1999). It is interesting that GGBP has the most regular distribution of βαβ repeats throughout its separate domains. Additionally, there are two extra α-helices in each of the GGBP domains that are absent in all periplasmic binding proteins of group II and in some members of group I. Thus, the spatial structure of GGBP seems to be the closest to that of the ancestral periplasmic binding protein. It is worth noting that di/oligopeptide-binding proteins contain a third domain in addition to the two widely recognized domains of periplasmic binding proteins (Nickitenko et al., 1995; Sleigh et al., 1999). According to the topology of the β-strands typical of periplasmic binding proteins, two domains of the di/oligopeptide-binding proteins can be regarded as group II proteins. The third domain is organized into two hairpins that make only a few contacts with the ligand. Still, some of the periplasmic binding proteins do not fall into any of the two aforementioned structural groups. These include, for example, zinc-binding protein TroA (Lee et al., 2002) and vitamin B12-binding protein BtuF (Karpowich et al., 2003), wherein the protein's N- and C-domains are linked by a single long helix, imposing some rigidity on the overall protein structure. These periplasmic binding proteins can be regarded as group III proteins.

Many solved structures of periplasmic binding proteins, both in the presence and in the absence of ligand, show large-scale motion of the two domains, described as bending and twisting motions around two axes. The degree of the hinge-bending motion varies from 14 to 62 ° for different periplasmic binding proteins (Shilton et al., 1996). The ligand-bound closed state of a periplasmic binding protein in complex with its ligand possesses a protein surface that is quite different from that observed in the open conformation of the protein. This difference in protein conformation is important for recognition of protein-ligand complexes by the trans-membrane proteins of the ATP-binding cassette (Hollenstein et al., 2007). For a long time, these conformational changes were assumed to be triggered by ligand binding. Recent studies revealed that ligand-free forms of periplasmic binding proteins are very flexible compared to ligand-bound forms (Bucher et al., 2011; Pang et al., 2003). This feature is common among all proteins of the PBP family, even among proteins in group III (Kandt et al., 2006; Krewulak et al., 2005). The flexible apo-form of the periplasmic binding proteins tends to oscillate along the modes that lead from the open to closed structure. Further support for this concept is provided by the existence of a dynamic equilibrium between the open and semi-closed conformations of the apo-form of the protein as revealed in the case of maltose-binding protein (MBP; Bucher et al., 2011). Additionally, the fully closed-form of the protein in the absence of ligand is not observed. To the contrary, GlnBP, in its ligand-free state, cannot achieve a partially closed conformation (Bermejo et al., 2010). Possible reasons for this are the stability of the hinge region of GlnBP - unusual in periplasmic binding proteins - and/or instability of the protein's partially closed-unliganded conformation. Whether other members of the PBP family can adopt a semi-closed conformation in the absence of ligand is the subject of future studies.

Generally, the amino acids of the protein ligand-binding site have extensive specific interactions with their cognate ligands, resulting in a high degree of selectivity between anomeric or epimeric carbohydrates, carbohydrates of different sizes or chemically similar anions (Bruns et al., 1997; Cuneo et al., 2009). At the same time, semi-specific ligand-binding takes place. This is the case for the di/oligopeptide-binding protein that binds to peptides ranging in size from two to nine amino acids with little discrimination between the side chains of the peptides (Sleigh et al., 1999). A network of strong hydrogen bonds is formed with the atoms of the main peptide chains, while the peptide's side chains adapt to the binding pocket with the aid of water molecules that donate hydrogen bonds to them and shield the ligand's charge.

The protein family of extra-cellularly bound lipoproteins homologous to the periplasmic binding proteins of gram-negative bacteria exists in gram-positive bacteria, wherein such extra-cellular proteins are covalently bound to outer cell surface (Felder et al., 1999). The spatial fold typical of the two-domain ligand-binding protein has also been found in transcriptional regulators, such as the lac-type repressors. Many eukaryotic receptors contain the PBP fold as a component of multi-domain proteins, such as glutamate/glycine-gated ion channels (for example, the ionotropic glutamate receptor GluR2), G protein-coupled receptors (for example, calcium-sensing receptors CaSRs) and atrial natriuretic peptide receptors. These receptors are regulated by conformational changes of the protein's extracellular domain in response to ligand binding (van den Akker, 2001; Felder et al., 1999).

2.2 Odorant-binding proteins

Odorant-binding proteins in vertebrates (OBPs) belong to the family of lipocalins (Flower, 2000). Lipocalins are a functionally diverse family of small and abundant extracellular proteins that bind mainly hydrophobic molecules, including lipids, odorants, pheromones, retinoids, porphyrins, siderophores, and steroids (Akerstrom et al., 2000; Flower et al., 2000). These proteins have been primarily classified as transport proteins, but it is known currently that lipocalins are involved in many important biological processes (Grzyb et al., 2006). Indeed, many lipocalins have been implicated in the regulation of cell homeostasis: apolipoprotein D, quiescience specific protein, purpurin, alpha-1-microglobulin, and NGAL. Some lipocalins, such as alpha-1-microglobulin, glicodelin and others, participate in the regulation of the immune response (Flower, 2000). Human tear lipocalin is expressed in lacrimal glands of both sexes and released into tears. Human tear lipocalin binds to a broad array of lipophilic substances including fatty acids, fatty alcohols, phospholipids, glycolipids and cholesterol (Breustedt et al., 2005; Glasgow & Gasymov, 2011). Tear lipocalin has a high affinity for retinol, microbial and fungal siderophores and harmful lipophilic compounds. It is proposed that the physiological function of human tear lipocalin is to prevent the corneal surface from desiccating and to stabilize the tear film by binding lipids present in the outer layer of tears. Tear lipocalin is also produced in von Ebner's lingual salivary glands, trachea, prostate, pituitary and sweat. Tear lipocalin in those tissues is supposed to protect epithelia by removing harmful hydrophobic molecules. In addition, tear lipocalin exhibits enzymatic activity such as endonuclease and cystatin-like activity (Glasgow & Gasymov, 2011; Redl, 2000; Yusifov et al., 2008). Tear lipocalin also provides anti-fungal and anti-microbial defenses by capturing siderophores (Fluckinger et al., 2004). Sex-specific pheromone-binding lipocalins and odorant-binding lipocalins (OBPs) are supposed to be associated with chemical communication and olfaction. Pheromone-binding lipocalins include major urinary proteins (MUPs) of male rat and mouse secreted predominantly in the urine and saliva (Beynon & Hurst, 2004), aphrodisin from female hamsters isolated from the vaginal discharge (Briand et al., 2004) and lipocalins of boar saliva (Marchese et al., 1998). These proteins are believed to serve as reservoirs for delivering to the environments and sustained dissemination of pheromones. Chemical information borne by pheromones is perceived by conspecifics and invokes various behavior responses in them (Hurst & Beynon, 2004).

In contrast to pheromone-binding lipocalins, OBPs are secreted in the nasal mucus of the oral sphere epithelia of both sexes. OBPs have been identified in a variety of species, including cow (Pevsner et al., 1985), pig (Dal Monte et al., 1991), rabbit (Garibotti et al., 1997), mouse (Pes et al., 1992), rat (Lobel et al., 2002), elephant (Lazar et al., 2002) and human (Briand et al., 2002). The first OBPs were discovered in the nasal mucus of the cow and several other animals according to their ability to bind pyrazine with a low detection threshold (Baldaccini et al., 1986). The studies of the binding specificity of OBPs toward several common odorants, which were performed mostly with bovine OBP (OBPb) and pig OBP-I (OBPp), have revealed that OBPs can bind a broad spectrum of hydrophobic molecules of medium size (Herent et al., 1995). Among the ligands with the highest affinity for OBPs (dissociation constants in the range of 0.1-1 μM) are heterocyclic derivatives, such as alkyl-substituted pyrazines and thiazoles, terpenoids and their derivatives, such as menthol and thymol, and medium size aliphatic alcohols and aldehyes (Herent et al., 1995). Spherically shaped terpenoids, such as camphor and its analogues, and polar compounds, such as the short chain fatty acids, exhibit poor affinity to OBPs. The low selectivity of OBPs gave rise to the hypothesis that these soluble proteins might function as chaperones for volatile hydrophobic odorants and pheromone molecules crossing the aqueous mucus layer to the olfactory receptors embedded in the membrane of olfactory neurons (Pevsner & Snyder, 1990; Steinbrecht, 1998). Later, it was proposed that OBPs probably play a more specific role in olfaction through their involvement in the first step of odorant discrimination. Indeed, the expression of several sub-types of OBP in the same animal species differing in their primary structure and having different ligand binding patterns has been observed in some animal species (Ganni et al., 1997; Garibotti et al., 1997; Utsumi et al., 1999). Three rat OBPs are specially tuned for distinct chemical classes of odorant molecules (Löbel et al., 2002). Rat OBP-1 preferentially binds heterocyclic compounds such as pyrazine derivatives. Rat OBP-2 is more specific for long-chain aliphatic aldehydes and carboxylic acids, while OBP-3 interacts strongly with odorant molecules having a saturated or unsaturated ring structure (Löbel et al., 2002). Microheterogeneity of OBPs was reported in the pig (Scaloni et al., 2001), mouse (Utsumi et al., 1999), rabbit (Garibotti et al., 1997) and porcupine (Ganni et al., 1997). Recently, it was shown that post-translational modifications can further increase the micro-diversity of OBPs (Le Danvic et al., 2009; Nagnan-Le Meillour et al., 2009). Indeed, porcine OBP can be posttranslationally modified by phosphorylation, resulting in a set of OBPp isoforms with different binding properties (Brimau et al., 2010). Moreover, the binding specificity of the VEG1 isoform, an olfactory binding protein expressed in the vomeronasal organ of the pig, to steroids is linked to the O-N-acetylglucosaminylation of the protein. The porcine VEG2 isoform, which does not undergo this modification, showed specificity for fatty acids rather than steroids (Le Danvic et al., 2009). Thus, it is supposed that phosphorylation and glycosylation could be a mechanism of regulation of OBP binding properties toward odorant and pheromone ligands, and an expanded set of OBP isoforms can serve to preliminary discriminate among ligands prior binding to the olfactory receptor (Le Danvic et al., 2009; Nagnan-Le Meillour et al., 2009).

The similarity in the amino acid sequences of different OBPs is limited with a sequence homology in the range of only 21-26 %. However, few amino acid residues are absolutely conserved within all OBP classes as well as within the lipocalin family (Flower, 2000). They include a GxW motif at the N terminal element (residues 14-16; here and throughout, the section numeration is given according to OBPp), two cysteine residues located on the 4th strand β-barrel and at the C terminal element, respectively (residues 63 and 155), and a Gly (residue 119). Cysteine residues form a disulfide bridge, thus tightening the α- helix domain and the β-barrel. The motif YxxxYxG is also highly conserved (residues 78-84). Bovine OBP is the only protein in the OBP class that has only two of four conserved patches, including GxW and the YxxxYxG motifs (Bianchet et al., 1996). Members of the lipocalin family, including the OBP class, share a common β-barrel structure (figure 2). The nine-stranded β-barrel comprises residues 9-120 (strands 1-8) linked by a turn in the sequence to a short α-helical domain (residues 124-141), followed by the 9th strand of the barrel (residues 146-148) and by a C-terminal tail (residues 149-157). The β-barrel is often preceded by an N-terminal segment, containing a short 3_{10}-like helical moiety (residues 1-8). The barrel of OBPs is markedly deformed to have an elliptical shape in cross-section. The β-barrel encloses a ligand binding site composed of both an internal cavity and an external loop scaffold. Loops L2-L7 connecting strands 2-8 are all typical of short β-hairpins; the exception is loop L1 between the 1st and 2nd strands, which is a large Ω loop. Loop L1 forms a lid, partially closing the internal ligand-binding site at this end of the barrel. The internal ligand-binding cavity is formed of mainly hydrophobic and aromatic amino acids. The cavity is shielded from the solvent, and during the interaction of OBP with the ligand, an opening event takes place, allowing the ligand to enter the binding pocket (Golebiowski et al., 2007; Vincent et al., 2004). A molecular dynamics study shows that these events occur mainly at the junction between the β-strands and loops L1 and L5 with the Tyr-82 residue serving as a gate in the OBPs (Golebiowski et al., 2007). The tyrosine residue at this position is highly conserved

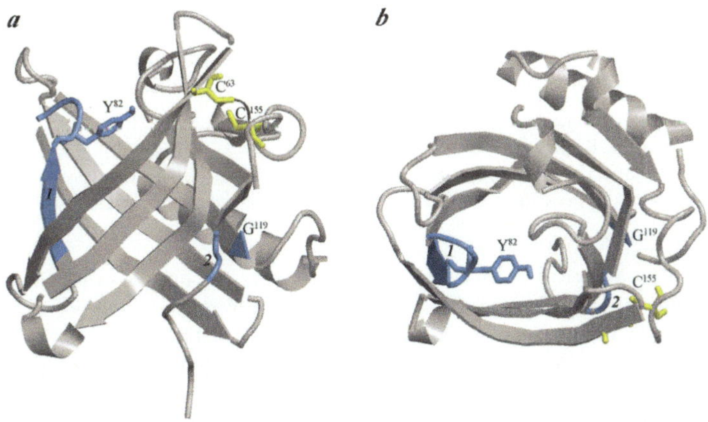

Fig. 2. Structure of OBPp (PDB code 1A3Y, Spinelly et al., 1998) in two projections. Conserved patches are shown *in blue*, including the $Y^{78}xxxYxG^{84}$ (*1*), $G^{14}xW^{16}$ (2) motifs and Gly-119 residue. Residues Cys-63 and Cys-155 are shown as *yellow stick unions*. The drawing was generated by the graphic programs VMD (Hsin et al., 2008) and Raster3D (Merritt & Bacon, 1977).

within the OBP class and tolerates substitution only with a phenylalanine residue, which preserves the structure of "the door". Investigation of the crystallographic structure of OBPb and OBPp complexed with different ligands revealed that the orientation of the ligands inside the cavity appeared to be opportunistic with no specific target patches for aromatic or charged groups and no correlation between the number of contacts and the affinity measured in solution (Vincent et al., 2004).

In contrast to classical lipocalin, dimeric bovine OBP, OBPb, is characterized by a unique folding pattern that involves the crossing of the α-helical domain of each monomer over the β-barrel of the other (Bianchet et al., 1996). In addition to the ligand binding site formed by the internal cavity of the β-barrel in each monomer, OBPb has a putative third binding site called the "central pocket", which is located at the dimer interface in communication with the solvent (Bianchet et al., 1996; Pevsner et al., 1985). The function of the central pocket is not well understood. It has been recently proposed that the holo-form of OBPb captures the first encountered odorant molecules at the central pocket irrespective of their affinity to OBPb (Ikematsu et al., 2005). The initial ligand binding is supposed to be a prerequisite for stabilization of the OBPb structure and for the adjustment of internal binding sites for interaction with ligands. OBPb can further bind the odorant with high affinity at its internal cavity, releasing the pre-bound ligand at the central odorant-binding pocket. The internal cavity-bound odorant can be released by the binding of other odorants at the internal cavity of the second OBPb subunit or at the central pocket, depending on the nature of the odorant. It is believed that such interactions of OBPb with its ligands make it more reactive than other monomeric OBPs and enable a quick recognition of a change in the environment that is highly desirable for ruminant animals, thus permitting them to escape from danger.

Careful examination of the structure of OBPb and OBPp, which exhibits a classical lipocalin fold, revealed the presence of a single insertion of a glycine residue at the hinge between the α-helical moiety and β-barrel domain in OBPp with respect to OBPb (the stretch $L^{117}LGKG^{121}TDIED^{126}$ in OBPp; Spinelli et al., 1998). The Gly-121 residue is located exactly at the position where domain swapping occurs in OBPb, and its presence induces a shift in the alignment of the structural elements joining the swappable domains. The longer and more flexible linker segment between the α-helical and β-barrel domains in OBPp is proposed to be sufficient for preventing domain swapping in the protein (Spinelli et al., 1998), while a conserved disulfide bridge between Cys-63 and Cys-155 stabilizes this non-swapped conformation. The fact that the mutant variant of OBPb, with an insertion of a glycine residue after position 121, became monomeric has further confirmed these findings (Ramoni et al., 2002).

Studies of OBPs over a period of more than 20 years were not able to completely clarify their role. A broad range of ligands and the low number of OBPs with respect to the number of olfactory receptors in mammalian species argue in favor of OBP function as a non-specific carrier of hydrophobic molecules (Pevsner & Snyder, 1990). A role of OBPs in the termination of the olfactory signal has also been proposed, which entails the "removal" of odorants from the olfactory receptors once they have been stimulated to keep them in their active state. The low specificity of OBPs led to the hypothesis that OBPs function as a scavenger of excess odorants to prevent olfactory receptor saturation (Burchell, 1991). Recent findings imply a more active role for OBPs in olfaction, which could involve the preliminary discrimination of odorant molecules or even a direct interaction with olfactory receptors. As already mentioned, the spectrum of OBPs in animal species can be expanded through phosphorylation and glycosylation. Further, it has been shown with the chip based surface plasmon resonance technique that the OBP is able to modulate the activity of olfactory receptors (Vidic et al., 2008). Some evidence for subtle conformational changes in rat and porcine OBPs after ligand binding has been obtained in molecular dynamics studies wherein opening of the β-strand pair was observed (Hajjar et al., 2006). These structural dynamics of OBPs might be essential for recognition of the OBP-ligand complex by the olfactory receptor. OBPs, at least in ruminants, might fulfill a protective role. The natural ligand of the bovine OBP, 1-octen-3-ol, which is produced by endogenous ruminal microflora of ruminants, is an attractant for many insect species (Ramoni et al., 2001). Thus, the OBP can be used to capture 1-octen-3-ol and decrease the risk of infections carried by insects. OBPs have been shown to bind to high affinity aldehydes derived from lipid peroxidation. This observation gave rise to the proposal that OBPs might be used to scavenge toxic substances from nasal epithelia to protect them against oxidative stress (Grolli et al., 2006).

3. Structure and stability of ligand-binding proteins

One of the most intriguing questions in modern molecular and cell biology is how a globular protein folds into a unique, compact, highly organized and functionally active state. In the past decade, our knowledge about protein folding into the native state and even the notion of the native state itself has undergone considerable changes. At the turn of the century, publications appeared that showed that the polypeptide chains of many proteins could not, in principle, fold into a compact globular state. Although these proteins are intrinsically disordered, they are functionally active and notably are in their native state. These proteins form a compact globular state only upon interaction with their specific binding partners, such as low-molecular weight ligands, other proteins, or nucleic acids. As a consequence, understanding the effects of ligand binding on proteins is of great interest (Turoverov et al., 2010). From this point of view, ligand-binding proteins can be convenient models to investigate the role of the ligand in the structure and in the stabilization of proteins in their native state.

3.1 The role of ligands in the process of folding and stability of two-domain ligand-binding proteins

The division of periplasmic binding proteins into two different structural sub-classes according to their structural topology makes it tempting to attribute this fact to differences in folding pathways of these protein sub-classes. In fact, group I proteins in general are characterized by simple two-state folding processes (Kashiwagi et al., 2003). More complex folding, accompanied by intermediate state accumulation, is observed for periplasmic binding proteins of group II (Chun et al., 1993; Staiano et al., 2005). For example, the folding of the *Escherichia coli* glutamine-binding protein, as studied by protein intrinsic fluorescence, 1-anilinonaphthalene-8-sulfonic acid (ANS) emission fluorescence, far- and near-UV circular dichroism spectroscopy, and the parametric presentation of fluorescence data (Kuznetsova et al., 2004), proceeds through the formation of two intermediate states I_1 and I_2 (Staiano et al., 2005). Glutamine binding makes the GlnBP structure more resistant to the denaturing action of the chemical agent, guanidine hydrochloride (GdnHCl); thus, unfolding of the GlnBP complex with Gln (GlnBP/Gln) begins at higher GdnHCl concentrations, and the denaturing process becomes more cooperative. This creates the illusion of a single stage of GlnBP/Gln folding. However, GlnBP/Gln folding successively passes through the same intermediate states as GlnBP but in a narrower range of denaturant concentrations so that the three stages of the folding process essentially overlap. Thus, the ligand serves as an agent to protect the entire protein structure from denaturation (figure 3, *a* – *c*).

The folding mechanism for the *Escherichia coli* D-galactose/D-glucose-binding protein and its complex with glucose (GGBP/Glc), representing group I proteins, is shown to be a one-stage process, with GGBP/Glc being more stable to denaturation than GGBP alone (Stepanenko et al., 2011a). This is supported by the sigmoidal contour of all recorded characteristics, such as fluorescence and far- and near-CD. The linear shape of the parametrically represented fluorescent data and the absence of an increase in ANS fluorescence during GGBP and GGBP/Glc complex unfolding also argue for the lack of any intermediate states in protein folding (figure 3, *d*; Stepanenko et al., 2011a). It is noteworthy that GGBP has an extra ligand, the calcium ion, located at a distinct position from the glucose-binding site (Vyas et al., 1988). The Ca-binding site consists of the loop of the protein's C-terminal domain (residues 134-142), and its structure resembles the "EF-hand" motif, which is typical of intracellular Ca-binding proteins (figure 1). The calcium ion forms coordination bonds with the oxygen atoms of every second residue of this loop and with the Glu 205 residue. The role of calcium in GGBP stability and folding has been evaluated. Although producing little effect on protein structure both in the absence and in the presence of glucose, calcium removal results in pronounced destabilization of GGBP even at small denaturing actions. Meanwhile, calcium depletion has practically no effect on GGBP/Glc stability (figure 3, *d*). Thus, the calcium ion serves as a guardian of the protein's structure in the absence of glucose. These results indicate that the role of calcium consists of maintaining the native structure of GGBP in its open form. Such a stabilizing effect of calcium was previously observed for other calcium-containing proteins (Turoverov et al., 2010). Recently, the folding mechanism of GGBP has been shown to be more complex than expected for the folding pathways of group I proteins (Piszczek et al., 2004; Stepanenko et al., 2011b). An extensive analysis of the experimental data reveals that the two domains of GGBP have a slightly different thermal stability, which is more marked for mutant variants of GGBP carrying point amino acid substitutions in the ligand-binding site of the protein (figure 3, *g* – *i*; Stepanenko et al., 2011b). These results suggest that more careful interpretation of accumulated data are needed.

While studying GdnHCl-induced unfolding – refolding of GGBP in the presence of glucose an interesting effect was observed (figure 3, *e* – *f*; Stepanenko et al., 2011a). The equilibrium curves for complex refolding-unfolding are attained only after 10 days of incubation of GGBP/Glc with GdnHCl. This effect is not revealed upon GGBP/Glc heating. Slow equilibrium acquisition between the native protein in the form of the GGBP/Glc complex and the unfolded state of the protein in the presence of GdnHCl is connected with increased viscosity of the solution at moderate and high GdnHCl concentrations, which interferes with diffusion of glucose molecules. Before equilibrium is established for an appreciable period of time, an excess concentration exists (in comparison with equilibrium) of the native complex $(GGBP/Glc)_N$ in the unfolding pathway or of the unfolded protein $(GGBP)_U$ in the refolding pathway. These imbalances are caused by the activation barrier, which must be overcome in both cases. In the unfolding pathway, the elementary process of complex dissociation does not bring about a disturbance in the configuration fit of the interacting GGBP and Glc molecules, so that the probability of the inverse reaction remains high. In the refolding pathway, $(GGBP/Glc)_N$ is formed because of the coincidence of two conditions: the formation of the native molecule $(GGBP)_N$ and the appearance of a configuration fit between $(GGBP)_N$ and Glc. Thus, the rate-limiting step in the unfolding-refolding process for the GGBP/Glc complex is the disruption/tuning of the configuration fit between the protein in its native state and the ligand.

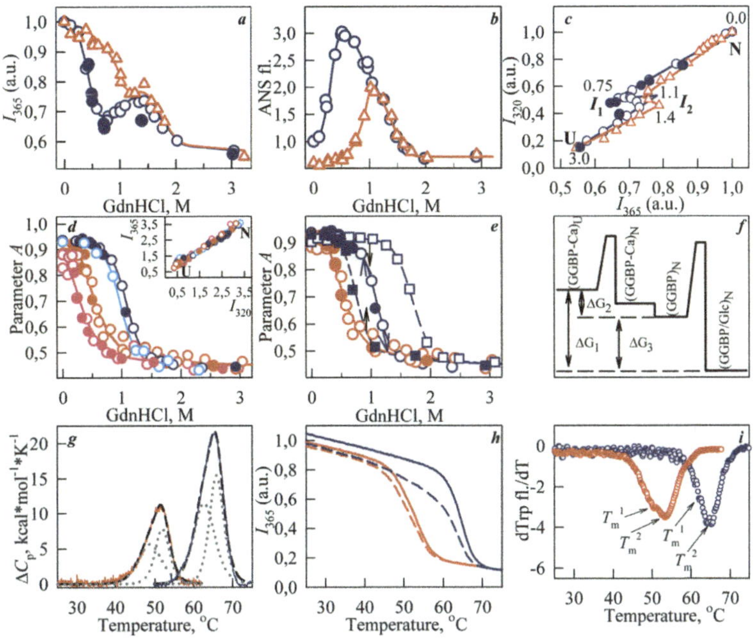

Fig. 3. Conformational changes of GlnBP (panels ***a*** - ***c***) and GGBP (panels ***d*** - ***i***). GdnHCl-induced transitions of GlnBP (blue circles) and complex with glutamine GlnBP/Gln (red triangles) were recorded by tryptophan fluorescence intensity at 365 nm (***a***, λ_{ex}=297 nm), ANS fluorescence intensity (***b***) and parametric presentation of fluorescence intensities at 320 and 365 nm (***c***). GdnHCl-induced transitions of GGBP (red circles) and complex with glucose GGBP/Glc (blue circles) and their calcium-depleted forms GGBP-Ca (pink circles) and GGBP-Ca/Glc (light blue circles) were characterized by equilibrium changes in parameter $A = I_{320}/I_{365}$, λ_{ex}=297 nm (***d***) and parametric presentation of fluorescence intensities at 320 and 365 nm (inset in ***d***). In panels ***a*** - ***d***, data for unfolding are depicted with empty symbols and data for refolding, with filled symbols. Equilibration (***e***) of unfolding - refolding curves as recorded by parameter A was achieved for GGBP after an incubation of less than 24 h in the presence of GdnHCl (red circles) and for GGBP/Glc, only after an incubation of 10 days (blue circles). The renaturation (closed blue squares) and denaturation (open blue squares) curves for GGBP/Glc do not coincide after 24 h incubation, but the curves tend to approach each other. The kinetics scheme characterizes the GGBP and GGBP/Glc unfolding —refolding processes (***f***). Heat-induced denaturation of GGBP (red) and GGBP/Glc (blue) was recorded by differential scanning calorimetry (***g***) and by the fluorescence intensity at 365 nm (***h***) and the first derivative of fluorescence intensity (***i***). The deconvolution of the calorimetric traces into two separate thermal transitions in panel g is shown in gray. Two sequential scans (solid and dashed lines, respectively) of the temperature dependency of the fluorescence intensity at 365 nm of the studied proteins are shown to characterize the reversibility of the thermal transitions (***h***).

All of the data reveal a common effect of the ligand on the folding and stability of all periplasmic binding proteins. The presence of the bound ligand does not change the folding pattern of the individual ligand-binding protein, but makes the protein more cooperative. Additionally, the structure of these proteins becomes more stable to various denaturing actions on the binding of their cognate ligands. The protein preserves its native structure up to the point of ligand dissociation induced by increased denaturing effects.

3.2 Beta-barrel scaffold of odor-binding proteins

Investigations into heat-induced and chemical denaturation of a series of OBPs revealed their high structural stability. Indeed, in the case of denaturation of porcine OBPp by the chemical denaturing agent GdnHCl, the difference in free energy between native and unfolded states of OBPp is as high as -5.95 kcal/mol (Staiano et al., 2007). The native structure of OBPp is highly resistant to heating with a transition mid-point of approximately 70 °C (Burova et al., 1999; Paolini et al., 1999; Stepanenko et al., 2008). In the absence of GdnHCl, OBPp preserves some residual structure under heating up to 80 °C as indicated by parameter *A* measurements (figure 4, *a*; Stepanenko et al., 2008). It noteworthy that parameter *A* being calculated as $A = I_{320}/I_{365}$ (where I_{320} and I_{365} are fluorescence intensities measured at the emission wavelengths of 320 and 365 nm, respectively), characterizes the

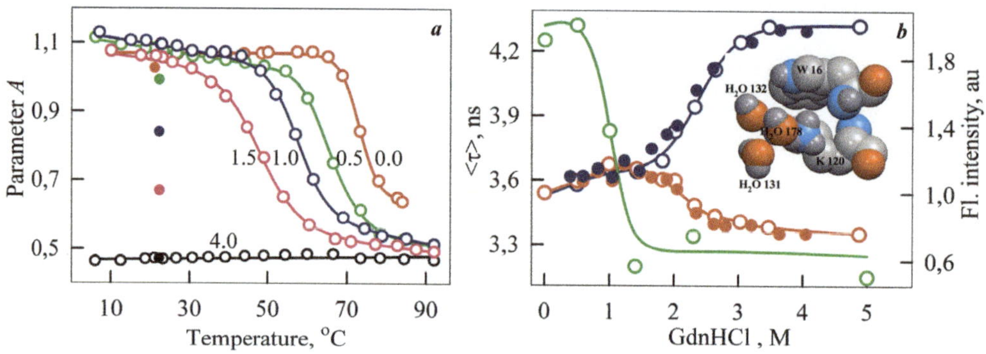

Fig. 4. Conformational changes of OBPp.

Heat-induced denaturation of OBPp in the absence and in the presence of different concentrations of GdnHCl recorded as a change in the parameter *A* (***a***). Numbers near the curves are GdnHCl concentrations. GdnHCl-induced conformational transitions of OBPp as revealed by changes in the fluorescence lifetime (green circles) and fluorescence intensities at 320 (red circles) and 365 (blue circles) nm (***b***). λ_{ex} = 297 nm. The microenvironment for Trp 16 of OBPp is shown (insert to ***b***). Open circles represent the protein denaturation processes, while closed circles show the protein renaturation processes.

shape and position of protein's fluorescence spectra (Turoverov & Kuznetsova, 2003). Changes of the parameter *A* can provides information about even subtle perturbation of spatial structure of proteins. Complete temperature denaturation of the protein can be achieved only in the presence of GdnHCl. Aggregation of OBPp that occurs at high temperatures is explained as the result of increased conformational mobility of the Ω loop with a loss of an ion pair connecting Glu 31 and Arg 152 (Stepanenko et al., 2008). As shown by studies of the double mutant variant of OBPp with C63A and C155A substitutions, the conserved disulfide bridge between Cys-63 and Cys-155 plays an important role in the stability and function of this protein (Parisi et al., 2005).

Interestingly, the study of GdnHCl-induced denaturation of OBPp revealed unusual characteristics of the fluorescence of the single Trp residue (Trp 16) in this protein (Staiano et al., 2007). In OBPp, the only polar group in the vicinity of Trp 16 is Lys 120. The side chain of Lys 120 is virtually parallel to the indole ring of Trp 16 with its positive charge located just over the center of the indole ring of Trp 16. This is the most favorable position for the formation of a complex between this group and the indole ring that could be responsible for the low fluorescence intensity of the native OBPp. Pre-denaturating GdnHCl concentrations result in an increase in the fluorescence quantum yield and a decrease in the fluorescence lifetime of Trp16 in OBPp (figure 4, *b*). These effects are not accompanied by noticeable changes in protein secondary and tertiary structure. The accessibility of Trp16 to the solvent remains unchanged, and near UV CD spectra become more pronounced, indicating disturbances to the microenvironment of the tryptophan residue. This likely leads to the disruption of the complex between Trp 16 and the positively charged NZ atom of Lys 120 and consequently to the enhancement of fluorescence intensity. The possibility of a distance change between atom NZ of Lys 120 and the indole ring of the tryptophan residue of OBPp in water was confirmed by molecular dynamic simulations. At the same time, the increase in the asymmetry of the tryptophan residue microenvironment can also promote the formation of exciplexes of Trp 16 with bound water molecules in close vicinity to the indole ring and can explain the decrease in the fluorescence lifetime. Thus, pre-denaturating GdnHCl concentrations induce local changes in the microenvironment of Trp16 in OBPp.

Increased stability of OBPs can be attributed to their β-barrel fold. The other group of proteins with β-barrel topology is the family of fluorescent proteins (FPs) that are widely used in a variety of applications in modern biology. For FPs, an extremely high resistance to environmental conditions has also been observed (Stepanenko et al., 2010).

4. Ligand-binding proteins as the sensitive element of socially important biosensors

The selective interaction of ligand-binding proteins with their partners is the major prerequisite for their use as sensitive elements in biosensor systems for definite analytes (Deuchle et al., 2005). In some cases, binding of a ligand to a ligand-binding protein is followed by a change in total charge, refractive index or molecular mass that can be detected (Stratton and Loh, 2011). These approaches in vivo are complicated by a significant noise level, e.g., in serum blood, there are approximately 3000 proteins, and the total protein concentration is approximately 70 mg/mL (Tang et al., 2005). Two-domain ligand-binding proteins sharing the intrinsic feature of ligand-induced conformational changes and targeting a diversity of natural ligands have been exploited in protein engineering to construct different biosensor systems (Dwyer and Hellinga, 2004; Tolosa et al., 2010). The large protein domains motion about the hinge has been utilized to transduce the ligand binding event to a variety of physical signals. Though the most preferable approaches for analyte registration are fluorescent methods, intrinsic protein fluorescence is not suitable as cells and living tissues are completely impervious to UV-light. Rational introduction of FRET-compatible protein pairs and environmentally sensitive dyes into different sites on the ligand-binding protein, as well as use of surface plasmon resonance methods, have been successful approaches to the design of biosensor systems (Ge et al., 2004; Khan et al., 2008, 2010; de Lorimier et al., 2002; Okada et al., 2009; Thomas et al., 2006). Changing the affinity and specificity of ligand-binding proteins has been achieved by the re-engineering of the protein's ligand-binding site. Much work has been performed with PBP-based biosensors for sugar detection (e.g., glucose, ribose and other) (Amiss et al., 2007; Medintz & Deschamps, 2006; Vercillo et al., 2007). Utilization of the *Escherichia coli* D-galactose/D-glucose-binding protein as the sensitive element in the glucose biosensor is one of the most promising directions for continuous glucose monitoring (Deuchle et al., 2005; Shilton et al., 1996). In this case, the necessity to lower the affinity of GGBP to glucose should be taken into account in the development of methods to monitor the glucose level in human blood (Amiss et al., 2007). It is worth mentioning that the ligand-binding proteins represent a target for the development of antimicrobial agents. These small molecules that are antagonists to native PBP ligands act as inhibitors of the conformational changes in proteins and thus block key processes in bacteria driven by PBP (Borrok et al., 2009).

Odorant binding proteins have been proposed as promising building blocks for construction of optical biosensors for dangerous substances such as toxic, explosive molecules and so on. It has been shown that the ligand-binding site of lipocalins can be optimized to bind molecules that are structurally different from their cognate ligands. For example, mutant OBPp has been successfully utilized for the monitoring of polyaromatic hydrocarbons that are among the most dangerous pollutants in water and atmosphere (Wei et al., 2007).

Thus, the construction of biosensor systems with sensors derived from ligand-binding proteins is one of the most promising areas in modern science. The development of an increasing number of such biosensor systems is dictated by the requirements of our daily lives, as they can be applied to the control of foodstuff quality and drug purity, environmental safety, detection of explosive and dangerous substances, monitoring of clinically relevant molecules and narcotics control (Tolosa, 2010).

5. Conclusion

The ligand-binding proteins, which are diverse in their structure and function, have numerous applications. Still, there is another way to enhance their practical importance. In particular, proteins from extremophilic sources, e.g., from hyperthermophilic bacteria (Staiano et al., 2010), could broaden the exploitation of ligand-binding proteins owing to their high stability under extreme conditions. In addition, ligand-binding proteins are convenient subjects of inquiry to elucidate one of the most intriguing and perplexing questions in structural and molecular biology, namely, the problem of protein folding into a unique, compact, highly ordered and functionally active form and, especially, the role of ligands in the structure and in the stabilization of proteins in their native states.

6. Acknowledgment

Financial support from the Ministry of Education and Science (Contracts 02.740.11.5141 and 16.512.11.2114), the Program MCB RAS and from the SPb government (OVS, 2011) is gratefully acknowledged.

7. References

Akerstrom, B.; Flower, D.R. & Salier, J.-P. (2000). Lipocalins: unity in diversity. *Biochimica et biophysica acta*, Vol.1482, No.1-2, (October 2000), pp. 1-8, ISSN 0006-3002

van den Akker, F. (2001). Structural insights into the ligand binding domains of membrane bound guanylyl cyclases and natriuretic peptide receptors. *Journal of molecular biology*, Vol.311, No.5, (August 2001), pp. 923-937, ISSN 0022-2836

Amiss, T.J.; Sherman, D.B.; Nycz, C.M.; Andaluz, S.A. & Pitner, J.B. (2007). Engineering and rapid selection of a low-affinity glucose/galactose-binding protein for a glucose biosensor. *Protein science: a publication of the Protein Society*, Vol.16, No.11, (November 2007), pp. 2350-2359, ISSN 0961-8368

Baldaccini, N.E.; Gagliardo, A.; Pelosi, P. & Topazzini, A. (1986). Occurrence of a pyrazine binding protein in the nasal mucosa of some vertebrates. *Comparative biochemistry and physiology. B*, Vol.84, No.3, pp. 249-253, ISSN 0305-0491

Banaszak, L.; Winter, N.; Xu, Z.; Bernlohr, D.A.; Cowan, S. & Jones, T.A. (1994). Lipid-binding proteins: a family of fatty acid and retinoid transport proteins. *Advances in Protein Chemistry*, Vol.45, pp 89-151, ISSN 0065-3233

Bermejo, G.A.; Strub, M.P.; Ho, C. & Tjandra, N. (2010). Ligand-free open-closed transitions of periplasmic binding proteins: the case of glutamine-binding protein. *Biochemistry*, Vol.49, No.9, (March 2010), pp. 1893-1902, ISSN 0006-2960

Beynon, R.J. & Hurst, J.L. (2004). Urinary proteins and the modulation of chemical scents in mice and rats. *Peptides*, Vol.25, No.9, (September 2004), pp. 1553-1563, ISSN 0196-9781

Bianchet, M.A.; Bains, G.; Pelosi, P.; Pevsner, J.; Snyder, S.H.; Monaco, H.L. & Amzel, L.M. (1996). The three-dimensional structure of bovine odorant binding protein and its mechanism of odor recognition. *Nature structural biology*, Vol.3, No.11, (November 1996), pp. 934-939, ISSN 1072-8368

Borrok, M.J.; Kiessling, L.L. & Forest, K.T. (2007). Conformational changes of glucose/galactose-binding protein illuminated by open, unliganded, and ultra-high-resolution ligand-bound structures. *Protein science*, Vol.16, No.6, (June 2007), pp. 1032-1041, ISSN 0961-8368

Borrok, M.J.; Zhu, Y.; Forest, K.T. & Kiessling, L.L. (2009). Structure-based design of a periplasmic binding protein antagonist that prevents domain closure. *American Chemical Society chemical biology*, Vol.4, No.6, (June 2009), pp. 447-456, ISSN 1554-8929

Breustedt, D.A.; Korndörfer, I.P.; Redl, B. & Skerra, A. (2005). The 1.8-A crystal structure of human tear lipocalin reveals an extended branched cavity with capacity for multiple ligands. *The Journal of biological chemistry*, Vol.280, No.1, (October 2004), pp. 484-493, ISSN 0021-9258

Briand, L.; Blon, F.; Trotier, D. & Pernollet, J.C. (2004). Natural ligands of hamster aphrodisin. *Chemical senses*, Vol.29, No.5, (June 2004), pp. 425-430, ISSN 0379-864X

Briand, L.; Eloit, C.; Nespoulous, C.; Bezirard, V.; Huet, J.C.; Henry, C.; Blon, F.; Trotier, D. & Pernollet, J.C. (2002). Evidence of an odorant-binding protein in the human olfactory mucus: location, structural characterization, and odorant-binding properties. *Biochemistry*, Vol.41, No.23 (June 2002), pp. 7241-7252, ISSN 0006-2960

Brimau, F.; Cornard, J.P.; Le Danvic, C.; Lagant, P.; Vergoten, G.; Grebert, D.; Pajot, E. & Nagnan-Le Meillour, P. (2010), Binding specificity of recombinant odorant-binding protein isoforms is driven by phosphorylation. *Journal of chemical ecology*, Vol.36, No.8, (June 2010), pp. 801-813, ISSN 0098-0331

Bruns, C.M.; Nowalk, A.J.; Arvai, A.S.; McTigue, M.A.; Vaughan, K.G.; Mietzner, T.A. & McRee, D.E. (1997). Structure of Haemophilus influenzae Fe(+3)-binding protein reveals convergent evolution within a superfamily. *Nature structural biology*, Vol.4, No.11, (November 1997), pp. 919-924, ISSN 1072-8368

Bucher, D.; Grant, B.J.; Markwick, P.R. & McCammon, J.A. (2011). Accessing a hidden conformation of the maltose binding protein using accelerated molecular dynamics. *PLoS computational biology*, Vol.7, No. 4, (April 2011), e1002034, ISSN 1553-734X

Burchell, B. (1991). Turning on and turning off the sense of smell. *Nature*, Vol.350, No. 6313, (March 1991), pp. 16-17, ISSN 0028-0836

Burova, T.V.; Choiset, Y.; Jankowski, C.K. & Haertle, T. (1999). Conformational stability and binding properties of porcine odorant binding protein. *Biochemistry*, Vol.38, No.45, (November 1999), pp. 15043-15051, ISSN 0006-2960

Caironi, P. & Gattinoni, L. (2009). The clinical use of albumin: the point of view of a specialist in intensive care. *Blood Transfusion*, Vol.7, No.4, (October 2009), pp. 259-267, ISSN 1723-2007

Carter, D.C. & Ho, J.X. (1994). Structure of serum albumin. *Advances in Protein Chemistry*, Vol.45, pp.153-203, ISSN 0065-3233

Chester, K.A. & Hawkins, R.E. (1995). Clinical issues in antibody design. *Trends in biotechnology*, Vol. 13, No. 8, (August 1995), pp. 294-300, ISSN 0167-7799

Cuneo, M.J.; Changela, A.; Beese, L.S. & Hellinga, H.W. (2009). Structural adaptations that modulate monosaccharide, disaccharide, and trisaccharide specificities in periplasmic maltose-binding proteins. *Journal of molecular biology*, Vol. 389, No.1, (May 2009), pp. 157-166, ISSN 0022-2836

Chun, S.Y.; Strobel, S.; Bassford, P.Jr. & Randall, L.L. (1993). Folding of maltose-binding protein. Evidence for the identity of the rate-determining step in vivo and in vitro. *The Journal of biological chemistry*, Vol.268, No.28, (October 1993), pp. 20855-20862, ISSN 0021-9258

Dalken, B.; Jabulowsky, R.A.; Oberoi, P.; Benhar, I. & Wels, W.S. (2010). Maltose-binding protein enhances secretion of recombinant human granzyme B accompanied by in vivo processing of a precursor MBP fusion protein. *PLoS One*, Vol.5, No.12, (December 2010), e14404, ISSN 1932-6203

Dal Monte, M. Andreini, I.; Revoltella, R. & Pelosi, P. (1991). Purification and characterization of two odorant-binding proteins from nasal tissue of rabbit and pig. *Comparative biochemistry and physiology. B*, Vol.99, No.2, pp. 445-451, ISSN 0305-0491

Dawson, R.J.; Hollenstein, K. & Locher, K.P. (2007). Uptake or extrusion: crystal structures of full ABC transporters suggest a common mechanism. *Molecular microbiology*, Vol.65, No.2, (July 2007), pp. 250-257, ISSN 0950-382X

Deuschle, K.; Okumoto, S.; Fehr, M.; Looger, L.L.; Kozhukh, L. & Frommer, W.B. (2005). Construction and optimization of a family of genetically encoded metabolite sensors by semirational protein engineering. *Protein science : a publication of the Protein Society*, Vol.14, No.9, (September 2005), pp. 2304-2314, ISSN 0961-8368

Dutta, S.; Burkhardt, K.; Young, J.; Swaminathan, G.J.; Matsuura, T.; Henrick, K.; Nakamura, H. & Berman, H.M. (2009). Data deposition and annotation at the worldwide protein data bank. *Molecular biotechnology*, Vol.42, No.1, (May 2009), pp. 1-13, ISSN 1073-6085

Dwyer, M.A. & Hellinga, H.W. (2004). Periplasmic binding proteins: a versatile superfamily for protein engineering. *Current Opinion in Structural Biology*, Vol.14, No.4, (August 2004), pp. 495-504, ISSN 0959-440X

Fasano, M.; Curry, S.; Terreno, E.; Galliano, M.; Fanali, G.; Narciso, P.; Notari, S. & Ascenzi. P. (2005). The extraordinary ligand binding properties of human serum albumin. IUBMB Life, Vol.57, No.12, (December 2005), pp. 787–796, ISSN 1521-6543

Felder, C.B.; Graul, R.C.; Lee, A.Y.; Merkle, H.P. & Sadee, W. (1999). The Venus flytrap of periplasmic binding proteins: an ancient protein module present in multiple drug receptors. AAPS Pharm Scie., Vol.1, No. 2, p. E2, ISSN 1522-1059

Fitzgerald, J. & Lugovskoy, A. (2011). Rational engineering of antibody therapeutics targeting multiple oncogene pathways. *Monoclonal antibodies*, Vol.3, No.3, (May 2011), pp. 299-309, ISSN 1942-0862

Flower, D.R. (2000). Experimentally determined lipocalin structures. *Biochimica et biophysica acta*, Vol.1482, No.1-2, (October 2000), pp. 46-56, ISSN 0006-3002

Flower, D.R.; North, A.C. & Sansom, C.E. (2000). The lipocalin protein family: structural and sequence overview. *Biochimica et biophysica acta*, Vol.1482, No.1-2, (October 2000), pp. 9-24, ISSN 0006-3002

Fluckinger, M.; Haas, H.; Merschak, P.; Glasgow, B.J. & Redl, B. (2004). Human tear lipocalin exhibits antimicrobial activity by scavenging microbial siderophores. *Antimicrobial agents and chemotherapy*, Vol.48, No.10, (September 2004), pp. 3367-3372, ISSN 0066-4804

Fukami-Kobayashi, K.; Tateno, Y. & Nishikawa, K. (1999). Domain dislocation: A change of core structure in periplasmic binding proteins in their evolutionary history. *Journal of Molecular Biology*, Vol.286, No.1, (February 1999), pp. 279-290, ISSN 0022-2836

Ganni, M.; Garibotti, M.; Scaloni, A.; Pucci, P. & Pelosi, P. (1997). Microheterogeneity of odorant-binding proteins in the porcupine revealed by N-terminal sequencing and mass spectrometry. *Comparative biochemistry and physiology. B*, Vol.117, No.2, (June 1997), pp. 287-291, ISSN 0305-0491

Garibotti, M.; Navarrini, A.; Pisanelli, A.M. & Pelosi, P. (1997). Three odorant-binding proteins from rabbit nasal mucosa. *Chemical senses*, Vol.22, No.4, (August 1997), pp. 383-390, ISSN 0379-864X

Glasgow, B.J. & Gasymov O.K. (2011). Focus on molecules: tear lipocalin. *Experimental eye research*, Vol.92, No.4, (August 2010), pp. 242-243, ISSN 0014-4835

Glatz, J.F.; Luiken, J.J.; van Bilsen, M. & van der Vusse, G.J. (2002). Cellular lipid binding proteins as facilitators and regulators of lipid metabolism. *Molecelar and Cellular Biochemistry*, Vol.239, No.1-2, (October 2002), pp 3-7, ISSN 0300-8177

Ge, X.; Tolosa, L. & Rao, G. (2004). Dual-labeled glucose binding protein for ratiometric measurements of glucose. *Analytical chemistry*, Vol.76, No.5, (March 2004), pp. 1403-1410, ISSN 0003-2700

Golebiowski, J.; Antonczak, S.; Fiorucci, S. & Cabrol-Bass, D. (2007). Mechanistic Events Underlying Odorant Binding Protein Chemoreception. *Proteins: Structure, Function, and Bioinformatics*, Vol.67, No.2, (May 2007), pp. 448–458, ISSN 0887-3585

Grolli, S.; Merli, E.; Conti, V.; Scaltriti, E. & Ramoni, R. (2006). Odorant binding protein has the biochemical properties of a scavenger for 4-hydroxy-2-nonenal in mammalian nasal mucosa. *The FEBS journal*, Vol.273, No.22, (October 2006), pp. 5131-5142, ISSN 1742-464X

Grzyb, J.; Latowski, D. & Strzałka, K.J. (2006). Lipocalins - a family portrait. *Journal of Plant Physiology*, Vol.163, No.9, (September 2006), pp 895-915, ISSN 0176-1617

Hajjar, E.; Perahia, D.; Debat, H.; Nespoulous, C. & Robert, C.H. (2006). Odorant Binding and Conformational Dynamicsin the Odorant-binding Protein. *The Journal of biological chemistry*, Vol.281, No.40, (July 2006), pp. 29929–29937, ISSN 0021-9258

Haunerland, N.H. & Spener, F. (2004). Fatty acid-binding proteins--insights from genetic manipulations. *Progress in Lipid Research*, Vol.43, No. 4, (July 2004), pp 328-349, ISSN 0163-7827

Herent, M.F.; Collin, S. & Pelosi, P. (1995). Affinities of nutty and green-smelling compounds to odorant-binding proteins. *Chemical senses*, Vol.20, No.6, (December 1995), pp. 601-610, ISSN 0379-864X

Hollenstein, K.; Frei, D.C. & Locher, K.P. (2007). Structure of an ABC transporter in complex with its binding protein. *Nature*, Vol.446, No.7132, (March 2007), pp. 213-216, ISSN 0028-0836

Hsiao, C.D.; Sun, Y.J.; Rose, J. & Wang, B.C. (1996). The crystal structure of glutamine-binding protein from Escherichia coli. *Journal of molecular biology*, Vol.262, No.2, (September 1996), pp. 225-242, ISSN 0022-2836

Hsin, J.; Arkhipov, A.; Yin, Y.; Stone, J.E. & Schulten, K. (2008). Using VMD: an introductory tutorial. *Current protocols in bioinformatics*, Chapter 5, Unit 5.7, ISSN 1934-3396

Hurst, J.L. & Beynon, R.J. (2004). Scent wars: the chemobiology of competitive signalling in mice. *BioEssays: news and reviews in molecular, cellular and developmental biology*, Vol.26, No.12, (December 2004), pp. 1288-1298, ISSN 0265-9247

Ikematsu, M.; Takaoka, D. & Yasuda, M. (2005). Odorant binding initially occurring at the central pocket in bovine odorant-binding protein. *Biochemical and biophysical research communications*, Vol.333, No.4, (August 2005), pp 1227-1233, ISSN 0006-291X

Kandt, C.; Xu, Z. & Tieleman, D.P. (2006). Opening and closing motions in the periplasmic vitamin B12 binding protein BtuF. *Biochemistry*, Vol.45, No.44, (November 2006), pp. 13284-13292, ISSN 0006-2960

Karpowich, N.K.; Huang, H.H.; Smith, P.C. & Hunt, J.F. (2003). Crystal structures of the BtuF periplasmic-binding protein for vitamin B12 suggest a functionally important reduction in protein mobility upon ligand binding. *The Journal of biological chemistry*, Vol.278, No.10, (March 2003), pp. 8429-8434, ISSN 0021-9258

Kashiwagi, K.; Shiba, K.; Fukami-Kobayashi, K.; Noda, T.; Nishikawa, K. & Noguchi, H. (2003). Characterization of folding pathways of the type-1 and type-2 periplasmic binding proteins MglB and ArgT. *Journal of biochemistry*, Vol133, No.3, (March 2003), pp. 371-376, ISSN 0021-924X

Khan, F.; Gnudi, L. & Pickup, J.C. (2008). Fluorescence-based sensing of glucose using engineered glucose/galactose-binding protein: a comparison of fluorescence resonance energy transfer and environmentally sensitive dye labelling strategies. *Biochemical and biophysical research communications*, Vol.365, No.1, (January 2008), pp. 102-106, ISSN 0006-291X

Khan, F.; Saxl, T.E. & Pickup, J.C. (2010). Fluorescence intensity- and lifetime-based glucose sensing using an engineered high-Kd mutant of glucose/galactose-binding protein. *Analytical biochemistry*, Vol.399, No.1, (April 2010), pp. 39-43, ISSN 0003-2697

Kilpatrick, D.C. (2002). Animal lectins: a historical introduction and overview. *Biochimica et biophysica acta,* Vol.1572, No.2-3, (September 2002), pp. 187-97, ISSN 0006-3002

Kratz, K. (2008). Albumin as a drug carrier: Design of prodrugs, drug conjugates and nanoparticles. *Journal of Controlled Release,* Vol.132, No.3, (December 2008), pp. 171-183, ISSN 0168-3659

Krewulak, K.D.; Shepherd, C.M. & Vogel, H.J. (2005). Molecular dynamics simulations of the periplasmic ferric-hydroxamate binding protein FhuD. *Biometals,* Vol.18, No.4, (August 2005), pp. 375-386, ISSN 0966-0844

Kuznetsova, I.M.; Turoverov, K.K. & Uversky, V.N. (2004). Use of the phase diagram method to analyze the protein unfolding-refolding reactions: fishing out the "invisible" intermediates. *Journal of proteome research,* Vol.3, No.3, (May-June 2004), pp. 485-494, ISSN 1535-3893

Laitinen, O.H.; Hytonen, V.P.; Nordlund, H.R. & Kulomaa, M.S. (2006). Genetically engineered avidins and streptavidins. *Cellular and Molecular Life Science,* Vol.63, No.24, (December 2006), pp 2992-3017, ISSN 1420-682X

Lazar, J.; Greenwood, D.R.; Rasmussen, L.E.L. & Prestwich, G.D. (2002). Molecular and functional characterization of an odorant binding protein of the asian elephant, elephas maximus: implications for the role of lipocalins in mammalian olfaction. *Biochemistry,* Vol.41, No.39, (October 2002), pp. 11786-11794, ISSN 0098-0331

Le Danvic, C.; Guiraudie-Capraz, G.; Abderrahmani, D.; Zanetta, J.P. & Nagnan-Le Meillour, P. (2009). Natural ligands of porcine olfactory binding proteins. *Journal of chemical ecology,* Vol.35, No.7, (May 2009), pp. 741-751, ISSN 0098-0331

Lee, Y.H.; Dorwart, M.R.; Hazlett, K.R.; Deka, R.K.; Norgard, M.V.; Radolf, J.D. & Hasemann, C.A. (2002). The crystal structure of Zn(II)-free Treponema pallidum TroA, a periplasmic metal-binding protein, reveals a closed conformation. *Journal of bacteriology,* Vol.184, No.8, (April 2002), pp. 2300-2304, ISSN 0021-9193

Lewis, R.J.; Muchova, K.; Brannigan, J.A.; Barak, I.; Leonard, G. & Wilkinson, A.J. (2000). Domain swapping in the sporulation response regulator Spo0A. *Journal of molecular biology,* Vol.297, No.3, (March 2000), pp. 757-770, ISSN 0022-2836

Liang, L. & Subirade, M. (2010). β-lactoglobulin/folic acid complexes: formation, characterization, and biological implication. *Journal of Physical Chemistry B,* Vol.114, No.19, (May 2010), pp. 6707–6712, ISSN 1520-6106

Linton, K.J. (2007). Structure and function of ABC transporters. *Physiology (Bethesda),* Vol.22, (April 2007), pp. 122-130, ISSN 1548-9213

Liu, B.; Bian, H.J. & Bao, J.K. (2010). Plant lectins: potential antineoplastic drugs from bench to clinic. *Cancer Letters,* Vol.287, No.1, (January 2010), pp. 1-12, ISSN 0304-3835

Löbel, D.; Jacob, M.; Völkner, M. & Breer, H. (2002). Odorants of Different Chemical Classes Interact with Distinct Odorant Binding Protein Subtypes. *Chemical senses,* Vol.27, No.1, (January 2002), pp. 39-44, ISSN 0379-864X

de Lorimier, R.M.; Smith, J.J.; Dwyer, M.A.; Looger, L.L.; Sali, K.M.; Paavola, C.D.; Rizk, S.S.; Sadigov, S.; Conrad, D.W.; Loew, L. & Hellinga, H.W. (2002). Construction of a fluorescent biosensor family. *Protein Science,* Vol.11, No.11, (November 2002), pp. 2655-2675, ISSN 0961-8368

Marchese, S.; Pes, D.; Scaloni, A.; Carbone, V. & Pelosi, P. (1998). Lipocalins of boar salivary glands binding odours and pheromones. *European journal of biochemistry,* Vol.252, No.3, (March 1998), pp. 563-568, ISSN 0014-2956 de Marco, A. (2011). Biotechnological applications of recombinant single-domain antibody fragments. *Microbial cell factories*, Vol.10 (June 2011), p. 44, ISSN 1475-2859

Medintz, I.L. & Deschamps, J.R. (2006). Maltose-binding protein: a versatile platform for prototyping biosensing. *Current opinion in biotechnology*, Vol.17, No.1, (February 2006), pp. 17-27, ISSN 0958-1669

Merritt, E.A. & Bacon, D.J. (1977). Raster3D: photorealistic molecular graphics. *Methods in enzymology*, Vol.277, pp. 505-524, ISSN 0076-6879

Nagnan-Le Meillour, P.; Le Danvic, C.; Brimau, F.; Chemineau, P. & Michalski, J.C. (2009). Phosphorylation of native porcine olfactory binding proteins. *Journal of chemical ecology*, Vol.35, No.7, (July 2009), pp. 752-760, ISSN 0098-0331

Neiditch, M.B.; Federle, M.J.; Pompeani, A.J.; Kelly, R.C.; Swem, D.L.; Jeffrey, P.D.; Bassler, B.L. & Hughson, F.M. (2006). Ligand-induced asymmetry in histidine sensor kinase complex regulates quorum sensing. *Cell*, Vol.126, No.6, (September 2006), pp. 1095-10108, ISSN 0092-8674

Nickitenko, A.V.; Trakhanov, S. & Quiocho, F.A. (1995). 2 A resolution structure of DppA, a periplasmic dipeptide transport/chemosensory receptor. *Biochemistry*, Vol.34, No.51, (December 1995), pp. 16585-16595, ISSN 0006-2960

Okada, S.; Ota, K. & Ito, T. (2009). Circular permutation of ligand-binding module improves dynamic range of genetically encoded FRET-based nanosensor. *Protein science: a publication of the Protein Society*, Vol.18, No.12, (December 2009), pp. 2518-2527, ISSN 0961-8368

Pang, A.; Arinaminpathy, Y.; Sansom, M.S. & Biggin, P.C. (2003). Interdomain dynamics and ligand binding: molecular dynamics simulations of glutamine binding protein. *FEBS letters*, Vol.550, No.1-3, (August 2003), pp. 168-174, ISSN 0014-5793

Paolini, S.; Tanfani, F.; Fini, C.; Bertoli, E. & Pelosi, P. (1999). Porcine odorant-binding protein: structural stability and ligand affinities measured by fourier-transform infrared spectroscopy and fluorescence spectroscopy. *Biochimica et biophysica acta*, Vol.1431, No.1, (April 1999), pp. 179-188, ISSN 0006-3002

Parisi, M.; Mazzini, A.; Sorbi, R.T.; Ramoni, R.; Grolli, S. & Favilla, R. (2005). Role of the disulphide bridge in folding, stability and function of porcine odorant binding protein: spectroscopic equilibrium studies on C63A/C155A double mutant. *Biochimica et biophysica acta*, Vol.1750, No.1, (June 2005), pp. 30-39, ISSN 0006-3002

Perez, M.D. & Calvo, M. (1995). Interaction of β-lactoglobulin with retinol and fatty acids and its role as a possible biological function for this protein. *Journal of Dairy Science*, Vol.78, No.5, (May 1995), pp. 978-988, ISSN 0022-0302

Pes, D.; Dal Monte, M.; Ganni, M. & Pelosi, P. (1992). Isolation of two odorant-binding proteins from mouse nasal tissue. *Comparative biochemistry and physiology. B*, Vol.103, No.4, (December 1992), pp. 1011-1017, ISSN 0305-0491

Pevsner, J. & Snyder, S.H. (1990). Odorant-binding protein: odorant transport function in the vertebrate nasal epithelium. *Chemical senses*, Vol.15, No.2, (April 1990), pp. 217-222, ISSN 0379-864X

Pevsner, J.; Trifiletti, R.R.; Strittmatter, S.M. & Snyder, S.H. (1985). Isolation and characterization of an olfactory receptor protein for odorant pyrazines. *Proceedings of the National Academy of Sciences of the United States of America*, Vol.82, No.9, (May 1985), pp. 3050-3054, ISSN 0027-8424

Piszczek, G.; D'Auria, S.; Staiano, M.; Rossi, M. & Ginsburg, A. (2004). Conformational stability and domain coupling in D-glucose/D-galactose-binding protein from Escherichia coli. *The Biochemical journal*, Vol.381, (July 2004), pp. 97-103, ISSN 0264-6021

Ramoni, R.; Vincent, F.; Ashcroft, A.E.; Accornero, P.; Grolli, S.; Valencia, C.; Tegoni, M. & Cambillau, C. (2002). Control of domain swapping in bovine odorant-binding protein. *The Biochemical journal*, Vol.365, No.Pt 3, (August 2002), pp. 739-748, ISSN 0264-6021

Ramoni, R.; Vincent, F.; Grolli, S.; Conti, V.; Malosse, C.; Boyer, F.D.; Nagnan-Le Meillour, P.; Spinelli, S.; Cambillau, C. & Tegoni, M. (2001). The insect attractant 1-octen-3-ol is the natural ligand of bovine odorant-binding protein. *The Journal of biological chemistry*, Vol.276, No.10, (December 2000), pp. 7150–7155, ISSN 0021-9258

Redl, B. (2000). Human tear lipocalin. *Biochimica et biophysica acta*, Vol.1482, No.1-2, (October 2000), pp. 241–248, ISSN 0006-3002

Richarme, G. & Caldas, T.D. (1997). Chaperone properties of the bacterial periplasmic substrate-binding proteins. *The Journal of biological chemistry*, Vol.272, No.25, (June 1997), pp. 15607-15612, ISSN 0021-9258

Riihimaki-Lampen, L.H.; Vainio, M.J.; Vahermo, M.; Pohjala, L.L.; Heikura, J.M.S.; Valkonen, K.H.; Virtanen, V.T.; Yli-Kauhaluoma, J.T. & Vuorela, P.M. (2010). The Binding of Synthetic Retinoids to Lipocalin β-Lactoglobulins. *Journal of Medicinal Chemistry*, (November 2009), Vol.53, No. 1, pp. 514–518, ISSN 0022-2623

Robinson, M.A.; Charlton, S.T.; Garnier, P.; Wang, X.T.; Davis, S.S.; Perkins, A.C.; Frier, M.; Duncan, R.; Savage, T.J.; Wyatt, D.A.; Watson, S.A. & Davis, B.G. (2004). LEAPT: lectin-directed enzyme-activated prodrug therapy. *Proceedings of the National Academy of Sciences of the USA*, Vol.101, No.40, (October 2004), pp. 14527-14532, ISSN 0027-8424

Schauder, S. & Bassler, B.L. (2001). The languages of bacteria. *Genes and development*, Vol.15, No.12, (June 2001), pp. 1468-1480, ISSN 0890-9369

Scaloni, A.; Paolini, S.; Brandazza, A.; Fantacci, M.; Bottiglieri, C.; Marchese, S.; Navarrini, A.; Fini, C.; Ferrara, L. & Pelosi, P. (2001). Purification, cloning and characterisation of odorant- and pheromone-binding proteins from pig nasal epithelium. *Cellular and molecular life sciences*, Vol.58, No.5-6, (May 2001), pp. 823-834, ISSN 1420-682X

Sharon, N. & Lis, H. (2004). History of lectins: from hemagglutinins to biological recognition molecules. *Glycobiology*, Vol.14, No.11, (November 2004), pp. 53R-62R, ISSN 0959-6658

Shilton, B.H.; Flocco, M.M.; Nilsson, M. & Mowbray, S.L. (1996). Conformational changes of three periplasmic receptors for bacterial chemotaxis and transport: the maltose-, glucose/galactose- and ribose-binding proteins. *Journal of molecular biology*, Vol.264, No.2, (November 1996), pp. 350-363, ISSN 0022-2836

Skerra A. (2008). Alternative binding proteins: anticalins - harnessing the structural plasticity of the lipocalin ligand pocket to engineer novel binding activities. *FEBS Journal*, Vol.275, No.11, (June 2008), pp. 2677-2683, ISSN 1742-464X

Sleigh, S.H.; Seavers, P.R.; Wilkinson, A.J.; Ladbury, J.E. & Tame, J.R. (1999). Crystallographic and calorimetric analysis of peptide binding to OppA protein. *Journal of molecular biology*, Vol.291, No.2, (August 1999), pp. 393-415, ISSN 0022-2836

Spinelli, S.; Ramoni, R.; Grolli, S.; Bonicel, J.; Cambillau, C. & Tegoni, M. (1998). The structure of the monomeric porcine odorant binding protein sheds light on the domain swapping mechanism. *Biochemistry*, Vol.37, No.22, (June 1998), pp. 7913-7918, ISSN 0098-0331

Staiano, M.; Scognamiglio, V.; Rossi, M.; D'Auria, S.; Stepanenko, O.V.; Kuznetsova, I.M. & Turoverov, K.K. (2005). Unfolding and refolding of the glutamine-binding protein from Escherichia coli and its complex with glutamine induced by guanidine hydrochloride. *Biochemistry*, Vol.44, No.15, (April 2005), pp. 5625-5633, ISSN 0006-2960

Staiano, M.; D'Auria, S.; Varriale, A.; Rossi, M.; Marabotti, A.; Fini, C.; Stepanenko, O.V.; Kuznetsova, I.M. & Turoverov, K.K. (2007). Stability and dynamics of the porcine odorant-binding protein. *Biochemistry*, Vol.46, No.39, (October 2007), pp. 1120-1127, ISSN 0006-2960

Staiano, M.; Baldassarre, M.; Esposito, M.; Apicella, E.; Vitale, R.; Aurilia, V. & D'Auria, S. (2010). New trends in bio/nanotechnology: stable proteins as advanced molecular tools for health and environment. *Environmental technology*, Vol.31, No.8-9, (July-August 2010), pp. 935-942, ISSN 0959-3330

Steinbrecht, R.A. (1998). Odorant-binding proteins: expression and function. *Annals of the New York Academy of Sciences*, Vol.855, (November 1998), pp. 323-332, ISSN 0077-8923

Stepanenko, O.V.; Marabotti, A.; Kuznetsova, I.M.; Turoverov, K.K.; Fini, C.; Varriale, A.; Staiano, M.; Rossi, M. & D'Auria, S. (2008). Hydrophobic interactions and ionic networks play an important role in thermal stability and denaturation mechanism of the porcine odorant-binding protein. *Proteins*, Vol.71, No.1, (April 2008), pp. 35-44, ISSN 0887-3585

Stepanenko, O.V.; Kuznetsova, I.M.; Verkhusha, V.V.; Staiano, M.; D'Auria, S. & Turoverov, K.K. (2010). Denaturation of proteins with beta-barrel topology induced by guanidine hydrochloride. *Spectroscopy: Biomedical Applications*, Vol.24, No.3-4, (July 2010), pp. 367-373, ISSN 0712-4813

Stepanenko, O.V.; Stepanenko, O.V.; Povarova, O.I.; Fonin, A.V.; Kuznetsova, I.M.; Turoverov, K.K.; Staiano, M.; Varriale, A. & D'Auria S. (2011a). New insight into protein-ligand interactions. The case of the D-galactose/D-glucose-binding protein from Escherichia coli. *The journal of physical chemistry B*, Vol.115, No.12, (March 2011), pp. 2765-2773, ISSN 1520-6106

Stepanenko, O.V.; Fonin, A.V.; Stepanenko, O.V.; Morozova, K.S.; Verkhusha, V.V.; Kuznetsova, I.M.; Turoverov, K.K.; Staiano, M. & D'Auria, S. (2011b). New insight in protein-ligand interactions. 2. Stability and properties of two mutant forms of the D-galactose/D-glucose-binding protein from E. coli. *The journal of physical chemistry B*, Vol.115, No.29, (July 2011), pp. 9022-9032, ISSN 1520-6106

Stratton, M.M. & Loh, S.N. (2011). Converting a protein into a switch for biosensing and functional regulation. *Protein Science*, Vol.20, No.1, (January 2011), pp. 19-29, ISSN 0961-8368

Sun, Y.J.; Rose, J.; Wang, B.C. & Hsiao, C.D. (1998). The structure of glutamine-binding protein complexed with glutamine at 1.94 A resolution: comparisons with other amino acid binding proteins. *Journal of molecular biology*, Vol.278, No.1, (April 1998), pp. 219-229, ISSN 0022-2836

Sundberg, E.J. (2009). Structural basis of antibody-antigen interactions. *Methods in molecular biology*, Vol.524, No. 4, pp. 23-36, ISSN 1064-3745

Szurmant, H. & Ordal, G.W. (2004). Diversity in chemotaxis mechanisms among the bacteria and archaea. *Microbiology and molecular biology reviews*, Vol.68, No.2, (June 2004), pp. 301-319, ISSN 1092-2172

Takakura, Y.; Oka, N.; Kajiwara, H.; Tsunashima, M.; Usami, S.; Tsukamoto, H.; Ishida, Y. & Yamamoto, T. (2010). Tamavidin, a versatile affinity tag for protein purification and immobilization. *Journal of Biotechnology*, Vol.145, No.4, (February 2010), pp 317-322, ISSN 0168-1656

Tam, R. & Saier, M.H.Jr. (1993). Structural, functional, and evolutionary relationships among extracellular solute-binding receptors of bacteria. *Microbiological Reviews*, Vol.57, No.2, (June 1993), 320-346, ISSN 0146-0749

Tang, H.Y.; Ali-Khan, N.; Echan, L.A.; Levenkova, N.; Rux, J.J. & Speicher, D.W. (2005). A novel four-dimensional strategy combining protein and peptide separation methods enables detection of low-abundance proteins in human plasma and serum proteomes. *Proteomics*, Vol.5, No.13, (August 2005), pp. 3329-3342, ISSN 1615-9853

Taub, P.R.; Gabbai-Saldate, P. & Maisel, A. (2010). Biomarkers of heart failure. *Congestive Heart Failue*, Vol.16, No.4(suppl 1), (July 2010), pp. S19–S24, ISSN 1527-5299

Thomas, K.J.; Sherman, D.B.; Amiss, T.J.; Andaluz, S.A. & Pitner, J.B. (2006). A long-wavelength fluorescent glucose biosensor based on bioconjugates of galactose/glucose binding protein and Nile Red derivatives. *Diabetes technology & therapeutics*, Vol.8, No.3, (June 2006), pp. 261-268, ISSN 1520-9156

Tolosa, L. (2010). On the design of low-cost fluorescent protein biosensors. *Advances in biochemical engineering/biotechnology*, Vol.116, pp. 143-157, ISSN 0724-6145

Turoverov, K.K. & Kuznetsova, I.M. (2003). Intrinsic Fluorescence of Actin. *Journal of fluorescence*, Vol.13, No.1, pp. 41-57, ISSN 1053-0509

Turoverov, K.K.; Kuznetsova, I.M. & Uversky, V.N. (2010). The protein kingdom extended: ordered and intrinsically disordered proteins, their folding, supramolecular complex formation, and aggregation. *Progress in biophysics and molecular biology*, Vol.102, No.2-3, (June-July 2010), pp. 73-84, ISSN 0079-6107

Utsumi, M.; Ohno, K.; Kawasaki, Y.; Tamura, M.; Kubo, T. & Tohyama, M. (1999). Expression of major urinary protein genes in the nasal glands associated with general olfaction. *Journal of neurobiology*, Vol.39, No.2, (May 1999), pp. 227-236, ISSN 0022-3034

Varshney, A.; Sen, P.; Ahmad, E.; Rehan, M.; Subbarao, N. & Khan, R.H. (2010). Ligand binding strategies of human serum albumin: how can the cargo be utilized? *Chirality*, Vol.22, No.1, (January 2010), pp. 77-87, ISSN 0899-0042

Vercillo, N.C.; Herald, K.J.; Fox, J.M.; Der, B.S. & Dattelbaum, J.D. (2007). Analysis of ligand binding to a ribose biosensor using site-directed mutagenesis and fluorescence spectroscopy. *Protein science: a publication of the Protein Society*, Vol.16, No.3, (March 2007), pp. 362-368, ISSN 0961-8368

Vidic, J.; Grosclaude, J.; Monnerie, R.; Persuy, M.-A.; Badonnel, K.; Baly, C.; Caillol, M.; Briand, L.; Salesse, R. & Pajot-Augy, E. (2008). On a chip demonstration of a functional role for odorant binding protein in the preservation of olfactory receptor activity at high odorant concentration. *Lab on a chip*, Vol.8, No.5, (March 2008), pp. 678-688, ISSN 1473-0197

Vincent, F.; Ramoni, R.; Spinelli, S.; Grolli, S.; Tegoni, M. & Cambillau, C. (2004). Crystal structures of bovine odorant-binding protein in complex with odorant molecules. *European journal of biochemistry*, Vol.271, No.19, (October 2004), pp. 3832-3842, ISSN 0014-2956

Vyas, N.K.; Vyas, M.N. & Quiocho, F.A. (1988). Sugar and signal-transducer binding sites of the Escherichia coli galactose chemoreceptor protein. *Science*, Vol.242, No.4883, (December 1988), pp. 1290-1295, ISSN 0036-8075

Wei, Y.; Brandazza, A. & Pelosi, P. (2008). Binding of polycyclic aromatic hydrocarbons to mutants of odorant-binding protein: A first step towards biosensors for environmental monitoring. *Biochimica et Biophysica Acta*, Vol.1784, No.4, (April 2008), pp. 666-671, ISSN 0006-3002

Wilchek, M.; Bayer, E.A. & Livnah, O. (2006). Essentials of biorecognition: the (strept)avidin-biotin system as a model for protein-protein and protein-ligand interaction. *Immunology Letters*, Vol.103, No.1, (February 2006), pp. 27-32, ISSN 0165-2478

De Wolf, F.A. & Brett, G.M. (2000). Ligand-binding proteins: their potential for application in systems for controlled delivery and uptake of ligands. *Pharmacological Reviews*, Vol.52, No.2, pp. 207-236, ISSN 0031-6997

Xu, S. & Venge, P. (2000). Lipocalins as biochemical markers of disease. *Biochimica et Biophysica Acta*, Vol.1482, No.1, (October 2000), pp. 298-307, ISSN 0006-3002

Yusifov, T.N.; Abduragimov, A.R.; Narsinh, K.; Gasymov, O.K. & Glasgow, B.J. (2008). Tear lipocalin is the major endonuclease in tears. *Molecular vision*, Vol.14, (January 2008), pp. 180-188, ISSN 1090-0535

Ye, K. & Schultz, J.S. (2003). Genetic engineering of an allosterically based glucose indicator protein for continuous glucose monitoring by fluorescence resonance energy transfer. *Analytical chemistry*, Vol.75, No.14, (July 2003), pp. 3451-3459, ISSN 0003-2700

14

Slit/Robo Signaling: Inhibition of Directional Leukocyte Migration

Ilya M. Mukovozov and Lisa A. Robinson
The University of Toronto,
The Hospital for Sick Children,
Canada

1. Introduction

Localized inflammation and the associated influx of leukocytes is a hallmark of the pathogenesis of many diseases. The ability to target the recruitment of leukocytes holds vast therapeutic potential in inflammatory diseases where there is excessive cell recruitment due to an overactive immune response, or the improper resolution of the initial response resulting in chronic leukocyte infiltration.

1.1 The neutrophil

Polymorphonuclear leukocytes, or neutrophils, are a critical component of the innate immune system, participating in host defence against bacterial and fungal infections. Not surprisingly, neutropenias can lead to severe infections and sepsis. During an inflammatory response, neutrophils are recruited to the sites of infection and/or injury by chemoattractants, including the chemokine family of proteins. Once in the tissue, neutrophils fight infections by ingesting microorganisms and producing reactive oxygen intermediates (ROI) as well as other anti-microbial substances, such as defensins (Ganz, T., 2003). Neutrophils can also produce and/or exacerbate inflammatory disease states as a result of the potent systems that have evolved in these cells for microbial killing. Inappropriate or excessive activation of these systems results in tissue damage (Fujishima et al., 1995). To better understand the role of the neutrophil in the fine balance between host defence and tissue injury, the mechanisms underlying neutrophil recruitment will be discussed.

1.2 Neutrophils and tissue injury

Neutrophils have been implicated in the pathogenesis of several inflammatory conditions, including: ischemia reperfusion injury (following coronary artery occlusion) (Frangogiannis et al., 2002), idiopathic pulmonary fibrosis (Haslam et al., 1980), arthritis (Weissmann et al., 1984), asthma (Lemanske et al., 1983), vasculitis (Fauci et al., 1978), glomerulonephritis (Holdsworth et al., 1984) and acute respiratory distress syndrome (ARDS) (Wieland et al., 1999). Neutrophil-mediated tissue injury results from the release of neutrophil anti-microbial factors such as ROI and proteases, and other mediators that amplify cell recruitment into the extracellular milieu (Frangogiannis et al., 2001). This can occur in two ways: 1) activation of neutrophils leads to fusion of antimicrobial granules to the plasma membrane and subsequent release of granule contents, and 2) attempts to ingest large particles result in a large open vacuole, and subsequent granule fusion and release of granule contents into the extracellular space (Weissmann et al., 1971).

ROI are strong oxidizing and reducing agents that damage the integrity of cell membranes by lipid peroxidation (Li et al., 2002). ROI also promote arachidonic acid synthesis by activating phospholipase A2. Arachidonic acid is an important precursor of eicosanoids and prostaglandins, including thromboxane A_2 and leukotriene B_4 (Toyokuni et al., 1999). Increased production of these pro-inflammatory molecules enhances recruitment of leukocytes. ROI also induce activation of transcription factors such as nuclear factor κB (NF-κB) and activator protein1 (AP1) (Toyokuni et al., 1999), leading to increased expression of adhesion molecules, including P-selectin, and chemokines (such as IL-8) thereby facilitating leukocyte arrest and recruitment from the circulation (Eltzschig et al., 2004).

Activated neutrophils also secrete matrix metalloproteases (MMPs), including collagenase and gelatinase. These enzymes are structurally specialized to digest basement membranes and interstitial structural proteins to facilitate neutrophil extravasation and subsequent migration through the interstitium (Kang et al., 2001). MMPs degrade several major structural components of the extracellular matrix (ECM), including collagen, fibronectin, proteoglycans, laminin and gelatin. MMPs are antagonized by tissue inhibitors of metalloproteases (TIMPs; Own et al., 1999). It has been shown that the imbalance between TIMPs and neutrophil-derived MMPs is a key feature of inflammatory conditions, including ARDS and asthma (Cederqvist et al., 2001). Neutrophil derived elastase is another bactericidal protease that is also associated with tissue damage. Like the MMPs, elastase displays proteolytic activity against structural components of the ECM. Elevated levels of neutrophil derived elastase and collagenase have been detected in patients with chronic inflammatory conditions, such as rheumatoid arthritis (Garcia et al., 1987). Increased neutrophil-derived protease activity has also been linked to cartilage destruction (Mohr et al., 1981). In ARDS, elastase activity has been associated with degradation of surfactant proteins in the lung (Hirche et al., 2004; Rubio et al., 2004). These proteins increase bacterial opsonization and clearance of apoptotic neutrophils (Vandivier et al., 2002). Therefore, increased elastase activity could indirectly increase susceptibility to infection and delay resolution of inflammation in the lung.

Commonly prescribed anti-inflammatory drugs, such as aspirin and glucocorticoids, have shown some success in reducing neutrophil-mediated tissue damage. However, these drugs generally attenuate activation of transcription factors such as NF-κB, thereby non-specifically reducing expression of cytokines and leukocyte adhesion molecules (Panes et al., 1999). One alternative method to prevent neutrophil-mediated tissue injury is to selectively block neutrophil recruitment to inflammatory foci. However, the redundancy in chemoattractant pathways means that interruption of a particular chemoattractant pathway may result in another pathway assuming its function. In principle, localized general chemoattractant blockade could be a useful strategy. Unique strategies to achieve this may be gained from studying central nervous system (CNS) development, in which positive and negative guidance cues for neuronal migration and axonal pathfinding have been defined.

1.3 Leukocyte trafficking and the adhesion cascade

The purpose of the inflammatory response is to selectively recruit the appropriate subsets of leukocytes to a site of inflammation. Inflammatory cytokines, such as interleukin 1 (IL-1) and tumour necrosis factor α (TNF-α), and soluble chemoattractants, are released within the local inflammatory environment. This results in local vasodilation, increased volume of blood perfusing the inflamed area, and a simultaneous decrease in the flow velocity within the vessel, facilitating extravasation of circulating leukocytes. Leukocytes are recruited to sites of inflammation in a series of coordinated interactions with endothelial cells lining the vascular wall. The classical leukocyte adhesion cascade involves these main steps: i) leukocyte capture and rolling, ii) activation and arrest, and iii) transendothelial migration (Fig. 1). Failure in any one of these steps can result in severe immunodeficiencies (Beutler, B., 2004). However, there exists a substantial therapeutic potential for the localized blockade of leukocyte adhesion and diapedesis.

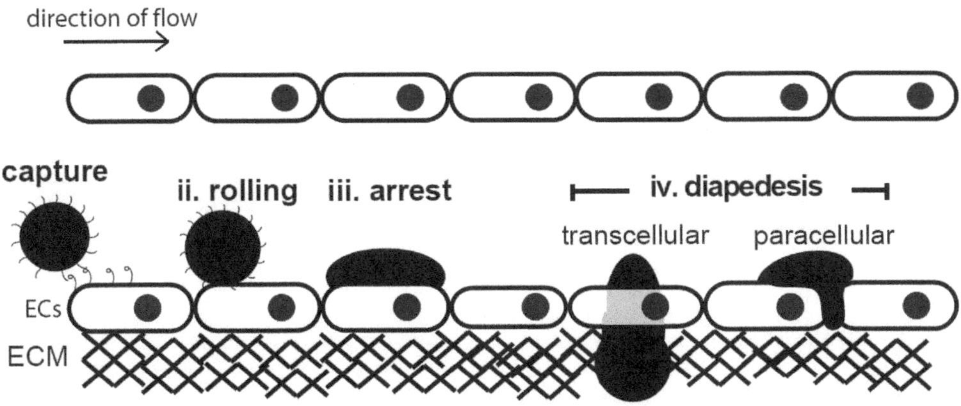

Fig. 1. Endothelial–leukocyte interactions leading to transmigration across the vascular wall. (i) Capture and (ii)rolling: The initial tethering of leukocytes to the endothelial cells lining the vessel wall is mediated by the selectins. These structural interactions enable the leukocyte to roll along the venular wall and to 'sample' the endothelial surface for activating factors (iii) Arrest: These interactions lead to leukocyte integrin activation. Firm adhesion of the leukocyte is mediated through binding of integrins to members of the immunoglobulin superfamily expressed in stimulated endothelial cells.(iv) Diapedesis: Following firm adhesion, the cell changes shape in response to local chemoattractant gradients and transmigrates across the endothelial barrier.

Selectins are a family of adhesion molecules that are structurally specialized for the initial capture of circulating leukocytes. Rolling is mediated by E-selectin and P-selectin, expressed by endothelial cells, and by L-selectin expressed on the majority of circulating neutrophils, monocytes, eosinophils, and T and B lymphocytes (Kansas, G., 1996). The broad expression pattern of L-selectin allows for nonspecific recruitment of all leukocyte lineages. P-selectin is constitutively found in Weibel-Palade bodies of endothelial cells, and mobilized to the cell surface within minutes following activation by inflammatory mediators (Frangogiannis et al., 2002). All of the selectins interact with P-selectin glycoprotein ligand 1 (PSGL1), although other glycoprotein ligands exist, such as CD34 and MadCAM-1 (McEver et al., 1997; Puri et al., 1995). Following initial leukocyte capture, the binding of leukocyte L-selectin to PSGL1 facilitates secondary leukocyte capture, where adherent leukocytes assist in the recruitment of additional cells (Eriksson et al., 2001). Interactions of selectins with their ligands allow leukocytes to roll on inflamed endothelium under the rapid flow of the bloodstream (Alon et al., 1995). In fact, shear stress is required to support L-selectin and P-selectin dependent adhesion, and rolling cells detach when flow is stopped (Finger et al., 1996; Lawrence et al., 1997). This selectin-mediated slow rolling allows the leukocyte to 'sample' the repertoire of chemokines and other activation signals presented on the luminal surface of endothelial cells.

In addition to selectins, various integrins participate in rolling. Integrins bind members of the immunoglobulin superfamily, including vascular cell-adhesion molecule 1 (VCAM-1) and intercellular adhesion molecule 1 (ICAM-1). Neutrophils roll on immobilized VCAM-1 by engaging the leukocyte integrin receptor, very late antigen 4 (VLA-4; $\alpha_4\beta_1$-integrin). β2-integrins also support rolling (Sigal et al., 2000). Resting mouse neutrophils roll on surfaces coated with E-selectin ligand and ICAM-1. Ligation of endothelial E-selectin induces a structural conformational change in leukocyte lymphocyte function-associated antigen 1 (LFA-1; $\alpha_L\beta_2$-integrin) allowing it to bind to its endothelial ligand, ICAM-1 (Salas et al., 2004). In addition, it has recently been demonstrated that the mechanochemical design of LFA-1 allows shear stress to induce and maintain a state of high ligand-binding affinity (Astrof et al., 2006). Rolling *in vivo* requires E-selectin (Kunkel et al., 1996), engagement of the β2-integrins (Jung et al., 1998), LFA-1 and macrophage antigen-1 (MAC1; Dunne et al., 2002).

Although leukocytes (particularly neutrophils) roll under normal conditions, during inflammation leukocytes undergo integrin-dependent arrest. Arrest of leukocytes on endothelial cells is rapidly triggered by the binding of chemokines and other chemoattractants (Campbell et al., 1998). These chemoattractants are secreted by activated endothelial cells and platelets. In fact, platelets can deposit chemokines, such as CC-chemokine ligand 5 (CCL5), CXC-chemokine ligand 4 (CXCL4), and CXCL5 onto the inflamed endothelial lumen to trigger leukocyte arrest (von Hundelshausen et al., 2001; Huo et al., 2003).

Following firm arrest, leukocytes migrate, by a process called diapedesis, across the endothelial cell barrier, its associated basement membrane, and the pericyte sheath. Leukocyte diapedesis and chemotaxis is triggered by chemokines (such as IL-8) presented to rolling leukocytes on the luminal surface of endothelial cells. Leukocytes can cross the endothelium between adjacent endothelial cells (paracellular route) or directly through an endothelial cell (transcellular route). Transcellular migration generally occurs in 'thin' parts of the endothelium where there is less distance for the leukocyte to migrate (Ley et al., 2007). In addition, caveolae containing ICAM-1 link together to form vesiculo-vacuolar organelles (VVOs), providing shortcuts for transcellular leukocyte diapedesis (Dvorak et al., 2001). This creates a channel inside the cell through which leukocytes can migrate. During paracellular migration, ligation of endothelial-cell adhesion molecules results in reduced interendothelial contacts, facilitating the migration of leukocytes through endothelial cell junctions (Ley et al., 2007). Transendothelial migration requires an increase in intracellular endothelial calcium, which promotes opening of endothelial cell junctions via the activation of myosin light chain kinase and endothelial cell contraction. The route of leukocyte migration is determined by both the surface density of ICAM-1 and the shape of endothelial cells (Yang et al., 2005). Both a high density of ICAM-1 and endothelial cells with a polygonal morphology promote transcellular migration (Yang et al., 2005). Many endothelial junctional molecules, such as platelet/endothelial-cell adhesion molecule 1 (PECAM-1), ICAM -1,

ICAM-2, junctional adhesion molecule A (JAM-A), (JAM-B), (JAM-C), endothelial cell-selective adhesion molecule (ESAM), and CD99, play a role in leukocyte transmigration. Although the leukocyte adhesion cascade has been divided into several steps, these are not temporally exclusive, but instead synergistically promote leukocyte arrest and diapedesis. Leukocyte diapedesis was described almost 200 years ago, but its molecular mechanisms are only now beginning to be more fully understood (Imhof et al., 2004). In the past decade, new insights have been gained into the signaling events that underlie integrin activation, post-adhesion strengthening of leukocyte attachment and the structural significance of molecules involved in diapedesis (Muller, W., 2003).

1.4 Chemotaxis

Following extravasation, leukocytes migrate through the interstitial ECM, following a chemoattractant gradient, to reach the site of inflammation. Chemotaxis, directed cell migration towards external chemical gradients, occurs in many eukaryotic cells including: free-living organisms, leukocytes (during inflammation), endothelial cells (angiogenesis), spermatocytes (fertilization) and neurons (neurogenesis) (Singer et al., 1986). Upon exposure to a chemoattractant the cell orients itself in the direction of locomotion along the chemoattractant gradient. Polarization results from preferential pseudopod extension towards areas of higher chemoattractant concentration (Zigmond, S., 1974). Efficient chemotaxis requires coordination between pseudopod formation at the leading edge of the cell, and uropod retraction at the trailing edge. During chemotaxis, neutrophils extend short surface protrusions called filopodia, or microspikes, which are membrane extensions of approximately 0.1-0.2 μm in diameter and up to 20 μm in length. These structures act as cellular tentacles and are supported by a core bundle of actin microfilaments (Mattila et al., 2008). In neutrophils, filopodia support thin sheets of membrane-enclosed cytoplasm, called lamellipodia. Lamellipodia contain actin filaments and a meshwork of myosin II-associated microfilaments. In neutrophils, the actin network within the lamellipodia, together with other structural and regulatory proteins, comprises the molecular motor which drives cell locomotion (Jones et al., 1998). This locomotory apparatus works against cell-to-substratum adhesions called focal contacts or focal adhesions. Focal adhesions are molecular structures that utilize integrins to link the myosin II-containing bundles of cytoplasmic microfilaments (called stress fibers) to proteins in the extracellular matrix (ECM) (Critchley et al., 1999). In neutrophils, integrin-mediated contacts to the ECM take two forms: focal complexes and podosomes. Focal complexes are structurally similar to focal adhesions but lack stress fibers (Allen et al., 1997), while podosomes are distinct circular structures that are only observed in cells of the myeloid lineage (DeFife et al., 1999; Correia et al., 1999; Linder et al., 2003). In this way, cytoskeletal rearrangement permits leukocytes to migrate toward chemoattractant gradients.

1.5 Chemoattractants

Many types of chemoattractant recruit leukocytes to inflammatory foci. These include bacterial components, leukotrienes, complement factors and chemokines. C5a, the first chemoattractant identified, is a cleaved product derived from complement component C5 (Shin et al., 1968). Bacterial products such as fMLP (N-formyl-methionyl-leucyl-phenylalanine) and other N-formylpeptides also act as chemoattractants that non-specifically recruit leukocyte subsets to inflammatory foci. An important family of chemoattractants involved in leukocyte recruitment to inflammatory foci is a family of chemoattractant cytokines called chemokines. Chemokines constitute a large family of small peptides that are structurally similar and that bind to a family of seven transmembrane G-protein coupled receptors (Rossi et al., 2000). The specific expression, regulation, and receptor binding patterns of each chemokine determine their functional diversity. Most chemokines are structurally conserved to bind to glycosaminoglycans (GAGs) on the luminal surface of endothelial cells. This binding is required for leukocyte recruitment *in vivo*. Indeed, chemokines with mutations in their GAG binding domains can induce *in vitro* chemotaxis, but are unable to recruit leukocytes to the peritoneal cavity *in vivo* (Johnson et al., 2005).

The binding of chemoattractants to their receptors activates leukocyte integrins instantaneously by inside-out signalling mechanisms (Shamri et al., 2005). They rapidly regulate integrin avidity by increasing both integrin affinity (by a conformational change that results in increased ligand binding energy and a decreased ligand dissociation rate), and valency (the density of integrins per area of plasma membrane involved in adhesion, determined by expression levels and lateral mobility) (Laudanna et al., 2002; Constantin et al., 2000). Through these signaling mechanisms, chemokines work as powerful activators of integrin-mediated adhesion and leukocyte recruitment.

1.6 Intracellular signaling of chemoattractant receptors

Several neutrophil chemoattractants, particularly chemokines, interact with specific receptors on the plasma membrane, transducing signals by coupling to heterotrimeric G proteins. Heterotrimeric G proteins are composed of an α, β, and γ subunit. The α subunit is the GDP/GTP binding element. When bound to GDP, the α subunit interacts with the β and γ subunits to form an inactive heterotrimer complex. Chemoattractant binding induces a conformational change in the receptor, exchanging GDP for GTP on the α subunit. The α subunit then dissociates from the receptor, releasing the Gβγ complex. The free Gα and Gβγ subunits are then available to bind and activate target enzymes such as phosphatidylinositol 3-kinase (PI3K), phospholipase C (PLC), or adenyl cyclase (Fig. 2). These enzymes generate secondary intracellular messengers that initiate a cascade of signaling events that ultimately culminate in cytoskeletal rearrangement and leukocyte migration.

Ligation of chemoattractant receptors leads to the activation of four major signaling pathways (Fig. 2): PLC, PI3K, mitogen-activated protein kinases (MAPKs) and Rho guanosine triphosphatases (GTPases). Once the Gα subunit dissociates, the Gβγ complex activates PLC, which cleaves phosphatidylinositol (4,5)-bisphosphate (PI(4,5)P_2) to generate inositol (1,4,5)-triphosphate (IP_3) and diacylglycerol (DAG). Generation of IP_3 leads to the mobilization of intracellular calcium stores from the endoplasmic reticulum, and together with DAG, activates protein kinase C (PKC) (Li et al., 2000). The activation and recruitment of PKC to the plasma membrane promotes changes in the actin cytoskeleton that facilitate and/or drive cell spreading and migration (Fig. 2).

A convincing role for PI3K in chemoattractant receptor signaling and chemotaxis has been established (Li et al., 2000; Sasaki et al., 2000; Hirsch et al., 2000; Servant et al., 2000; Jin et al., 2000). Although there are at least four Class I PI3K isoforms in mammalian cells (Vanhaesebroeck et al., 1999), only a single Class IB variant has been shown to interact with chemoattractant receptors in leukocytes. The outcome of Class I PI3K activation is phosphorylation of membrane $PI(4,5)P_2$ by activated PI3K, generating $PI(3,4,5)P_3$ at the plasmalemma. The Gβγ complex also activates PI3Kγ, activating Src-family kinases and generating $PI(3,4,5)P_3$ from membrane $PI(4,5)P_2$ (Krugmann et al., 1999), resulting in the recruitment of Ras GTPases and subsequent activation of MAPK pathways (Fig. 2) (Kintscher et al., 2000). Although MAPK signaling pathways are involved in chemotaxis and adhesion, the most important biochemical events for cell polarization are the production of PIP_3 and activation of Rho GTPases at the leading edge of the cell.

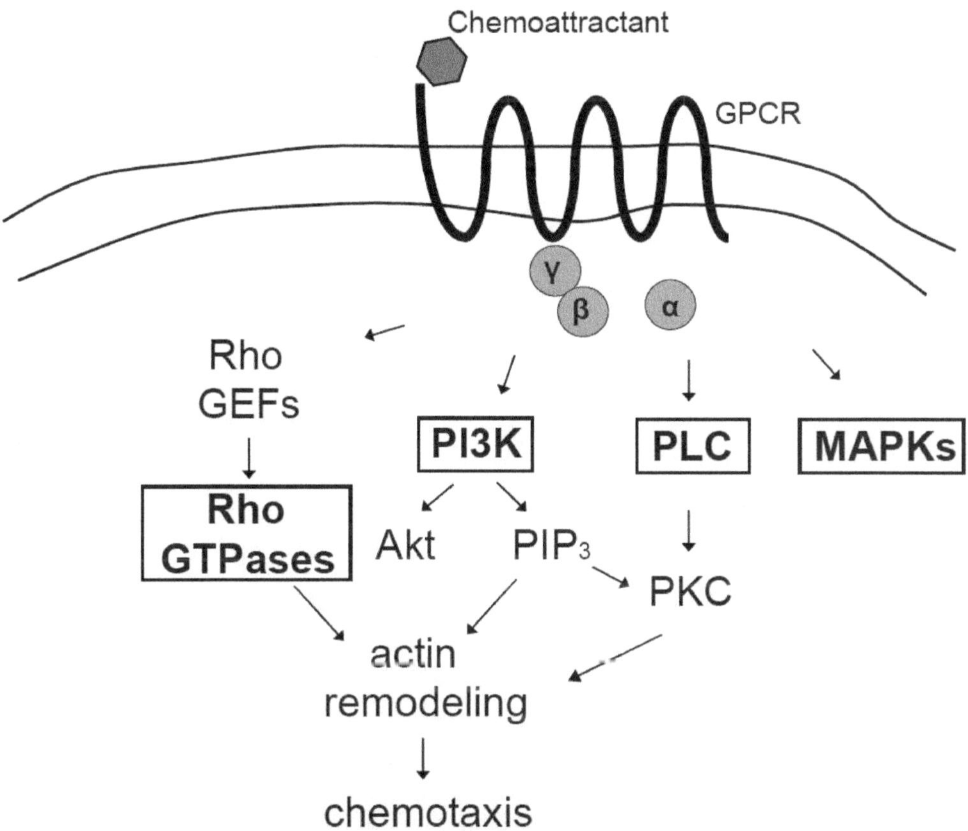

Fig. 2. Intracellular signaling cascade upon ligation of chemoattractant receptors. Chemoattractant binding to GPCRs induces a conformational change that results in the dissociation of Gα subunits from the Gβγ complex. This leads to rapid outside-in signaling resulting in the activation of four major signaling pathways that contribute to the generation of cell polarity and chemotaxis: Rho GTPases, PI3K, PLC, and MAPKs.

The PI3K dependent production of PIP_3 at the cell membrane allows for the recruitment of the Rho-family GTPases, Rac and Cdc42, to the cell membrane. The localization of PIP_3, Rac and Cdc42 then stimulate polymerization of actin, a process necessary for the formation of filopodia and lamellipodia at the front of the cell. At the back of the cell, Rho-kinase phosphorylation results in inactivation of myosin light chain phosphatase, leading to increased myosin light-chain kinase (MLK) dependent activation of myosin (Nguyen et al., 1999). These biochemical conditions favour the formation of actomyosin bundles, contraction, de-adhesion from the substratum and tail retraction (Ridley, A., 2001; Bokoch, G., 2005). Interestingly, signals at the leading edge inhibit signals at the trailing edge, allowing for the maintenance of cell polarity (Fenteany et al., 2004). To prevent the accumulation of PIP_3 at the trailing edge, PTEN dephosphorylates $PI(3,4,5)P_3$ to $PI(4,5)P_2$. The lack of PIP_3 in the back of the cell prevents activation and recruitment of Rho GTPases and subsequent actin polymerization, allowing the formation of actomyosin bundles and tail retraction (Worthylake et al., 2001). Actin polymerization at the leading edge coupled with tail retraction in the back allows for directed leukocyte chemotaxis.

1.7 Rho-family GTPases: Rac, Cdc42, and Rho

Small GTPases of the Rho family are a part of the Ras superfamily of small GTP-binding proteins. They are pivotal regulators of many signaling networks that are activated by a diverse variety of receptor types. To date, over 20 mammalian Rho GTPases have been characterized, and these can be grouped into 6 different classes: Rac (Rac1, Rac2, Rac3, RhoG), Rho (RhoA, RhoB, RhoC), Cdc42 (Cdc42Hs, G25K, TC10), Rnd (Rnd3/RhoE, Rnd1/Rho6, Rnd2/Rho7), RhoD, and RhoH/translocation three four (TTF) (Aspenström, P., 1999; Kjoller et al., 1999). When activated, Rho GTPases regulate many important processes in all eukaryotic cells, including actin cytoskeleton dynamics, transcription, cell cycle progression, and membrane trafficking. The activity of Rho GTPases is regulated by outside-in signals from a variety of receptor types, including GPCR, tyrosine kinase receptors, cytokine receptors and adhesion receptors. Rho-family GTPases play a critical role in regulating leukocyte chemotaxis, adhesion and phagocytosis.

1.7.1 Rho GTPases: Structure and regulation

All Rho GTPases contain two main structural domains, the C-terminal 'CAAX' motif and a catalytic GTP domain. The 'CAAX' motif undergoes post-translational processing, involving carboxy-terminal proteolysis of the AAX residues followed by carboxyl-methylation. The modified C-terminal domain can then attach to membrane lipids and facilitates membrane association and subcellular localization of Rho GTPases (Gutierrez et al., 1989; Casey et al., 1989; Fujiyama et al., 1990). The catalytic domain contains two regions, switch I and switch II. These domains correspond to different structural conformations in the GTP-bound and GDP-bound forms. Rho GTPases function as molecular switches by cycling between GDP-bound and GTP-bound forms. When bound to GDP, Rho GTPases are inactive. Binding of ligands to cell surface receptors, results in exchange of GDP for GTP, switching the protein to an active state. The active form interacts with downstream effector molecules. The intrinsic GTPase activity of Rho GTPases completes this cycle by hydrolyzing GTP, returning the GTPase to its inactive GDP-bound state.

Three classes of molecules interact with Rho GTPases and regulate their activation state: guanine nucleotide exchange factors (GEFs), GTPase-activating proteins (GAPs), and guanine nucleotide dissociation inhibitors (GDIs). GEFs catalyze the exchange of GDP for GTP, leading to the activation of Rho GTPases. To date, over 69 mammalian GEFs for Rho GTPases have been identified (Rossman et al., 2005). They are characterized by the presence of a Dbl homology domain (DH), which interacts with both the switch I and switch II regions and catalyses the exchange of GDP for GTP. In addition, many of these DH-domain containing proteins, such as Vav, contain a Pleckstrin homology (PH) domain which allows GEFs to bind phosphoinositides, such as PIP_3. This localizes GEFs to the plasma membrane where they can bind other Rho-family GTPase-interacting proteins. GAPs enhance the intrinsic GTPase activity of Rho GTPases, and thus suppress their activity. Although GTPases possess intrinsic GTPase activity, the actual rate of GTP hydrolysis is relatively slow. Therefore, the interaction with a GAP is required for efficient GTP hydrolysis, as this accelerates the cleavage step by several orders of magnitude (Vetter et al., 2001). To date, more than 70 eukaryotic RhoGAPs have been discovered, of which 35 are found in humans (Tcherkezian et al., 2007). There exists a large diversity in the primary sequences of the various GAPs. However, each one contains a Rho GAP domain with a conserved tertiary structure composed of α helices and a catalytically critical 'arginine finger' which stabilizes the formation of the transition state during GTP hydrolysis (Nassar et al., 1998). In addition, the Rho GAP domain interacts with both the switch I and switch II regions on the GTPase domain (Gamblin et al., 1998), allowing GAPs to facilitate the intrinsic hydrolysis of GTP, resulting in the inactivation of Rho GTPases.

Finally, GDIs associate with Rho GTPases in their inactive GDP-bound state and inhibit their activation by GEFs. GDIs also bind to GTP-bound GTPases, and suppress their activity (Oloffson, B., 1999). There is evidence that GDIs can bind to isoprenyl moieties on the C-terminus of GTPases in order to sequester them in the cytosol (Keep et al., 1997). The role of GDIs in partitioning GTPases between the membrane and cytosol may be physiologically more important than the inhibition of their activation, as this may provide a storage pool of Rho GTPases that is readily utilized upon cell activation. Overall, GDIs prevent the activation of Rho GTPases, prevent their interaction with membranes, and inhibit downstream signaling networks.

1.7.2 Rho GTPases and the actin cytoskeleton

The movement of eukaryotic cells relies on the coordinated extension of actin-rich lamellipodia in the leading edge and retraction of the uropod at the rear of the cell. The extension of lamellae in the leading edge involves rapid turnover of actin filaments (Symons et al., 1991; Wang, Y., 1985). More stable actin-myosin cables can be found in more established protrusions and in the middle and rear of the cell (DeBiasio et al., 1988). Recycling of the plasma membrane and integrin-mediated adhesion to the substratum and/or ECM are also important for cell motility (Bretscher, M., 1996; Martenson et al., 1993; Yamada et al., 1995; Mitra et al., 2005). Coordinated mobilization of the actin cytoskeleton is regulated by deployment of actin-binding proteins by activated Rho-family GTPases. Rho-family GTPases control cell motility and morphological changes in response to extracellular chemoattractants. Activation of Rho in fibroblasts results in the assembly of stress fibers and focal adhesions (Ridley et al., 1992). The activation of Rac causes extension of lamellipodia and assembly of small focal complexes (Nobes et al., 1995; Ridley et al., 1992). In contrast, activation of the Cdc42 Rho-family GTPase leads to the formation of filopodial extensions (Nobes et al., 1995).

As discussed above, the influx of neutrophils and other leukocytes to inflammatory foci relies on activation of Rho-family GTPases and dynamic actin turnover. In principle, one method to prevent neutrophil-mediated tissue injury would involve blocking neutrophil recruitment. However, the redundancy in chemoattractant pathways means that interruption of a particular chemoattractant may result in another assuming its function. Thus, a localized general chemoattractant blockade could be a useful strategy. Unique strategies to target neutrophil recruitment may be gained from studying central nervous system (CNS) development, in which structurally distinct positive and negative guidance cues for migration and axonal pathfinding have been defined.

2. Slit2: A guidance cue for cell migration

During the development of the CNS, neurons must migrate and project axons over long distances. Most axons emanating from the CNS must cross the midline and then project longitudinally towards their synaptic targets. The molecular mechanisms that guide this pathfinding include contact attraction, chemoattraction, contact repulsion and chemorepulsion. Guidance cues selectively promote or repress migration of neurons and axonal projection. For example, netrins are diffusible chemotropic factors that attract commissural axons to the midline (Kennedy et al., 1994). The Slit family of secreted proteins, together with their cell-surface receptor Roundabout (Robo), repel neurons during CNS development. Once commissural axons have crossed the midline, midline glial cells express Slit to prevent axons from re-crossing the midline. Mutant Drosophila lacking Slit proteins exhibit midline defects, such as collapse of the regular scaffold of commissural and longitudinal axon tracts in the embryonic CNS (Rothberg et al., 1988; Rothberg et al., 1990). A similar defect is observed in mutant Drosophila lacking Robo, where projecting axon tracts cross the midline repeatedly (Kidd et al., 1998).

2.1 Slit and robo: Structure

The Slit family of proteins contain an N-terminal signal peptide, four leucine-rich repeats (LRRs), nine epidermal growth factor (EGF) repeats and a C-terminal cysteine knot (Fig. 3) (Rothberg et al., 1988; Rothberg et al., 1990; Rothberg et al., 1992). The EGF repeats and LRR allow Slit proteins to interact with ECM components, such as glypican-1, enabling them to act as localized, non-diffusible, signaling molecules (Ronca et al., 2001). Furthermore, Slit2 can be proteolytically cleaved after the fifth EGF repeat to form N-terminal (Slit2-N) and C-terminal (Slit2-C) fragments (Brose et al., 1999; Wang et al., 1999). Slit2-N includes the first 1118 amino acids and contains the four LRRs and the first five EGF repeats, while Slit2-C contains the remaining residues (Brose et al., 1999). Importantly, only the second LRR of human Slit2 is required to bind with the first Ig domain of Robo and initiate downstream signaling (Morlot et al., 2007). Therefore, both full length Slit2 and Slit2-N bind Robo receptors to repel migrating cells and projecting axons (Nguyen Ba-Charvet et al., 2001). Although the cleavage of Slit2 does not eliminate its activity, it may play a role in its diffusion since Slit-N appears to be more tightly associated with the cell membrane. In rat neural tissue both Slit2-N and Slit2-C were shown to bind heparan sulfate proteoglycan glypican-1 (Liang et al., 1999), although Slit2-C bound with higher affinity, suggesting a possible regulatory mechanism for its diffusion.

Robo, a member of the immunoglobulin superfamily, is a single-pass type-1 receptor for the Slit proteins. The extracellular region of human Robo-1 contains five immunoglobulin (Ig) repeats and three fibronectin type III domains. The cytoplasmic region of Robo-1 contains four conserved cytoplasmic signaling motifs, CC0, CC1, CC2 and CC3 (Kidd et al., 1998; Zallen et al., 1998). Only the first Ig domain of Robo is required to bind to the second LRR domain in Slit2 and Slit2-N (Battye et al., 2001; Chen et al., 2001; Nguyen Ba-Charvet et al., 2001). The cytoplasmic CC motifs of Robo are required for response to Slit (Bashaw et al., 2000).

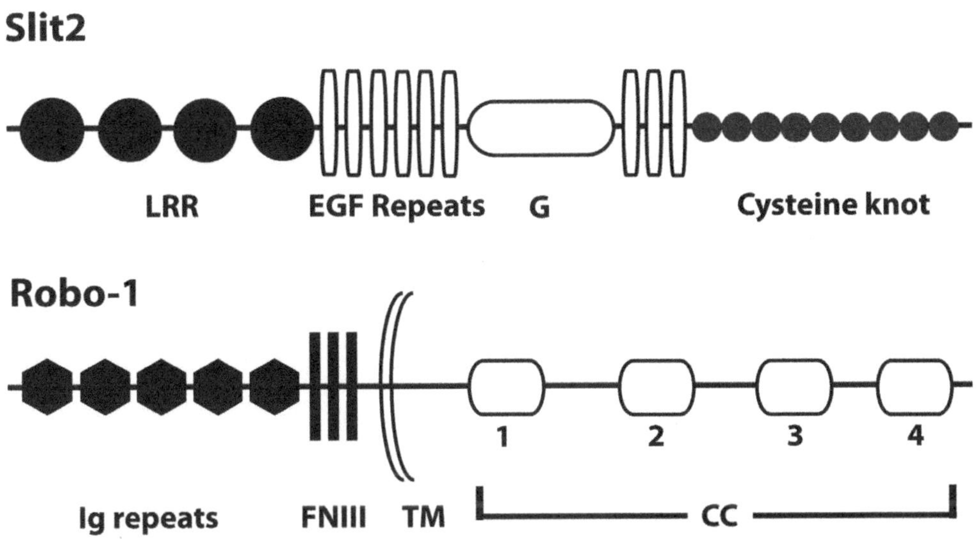

Fig. 3. Primary Protein Structure of Mammalian Slit2 and Robo-1 Proteins. Mammalian Slit2 contains four leucine rich repeats (LRRs), nine epidermal growth factor (EGF) repeats, a laminin G (G) domain, and a cysteine rich C terminus. The Robo-1 receptor contains five immunoglobulin (Ig) repeats, three fibronectin (FN) type III, a transmembrane Domain (TM) and four conserved cytoplasmic (CC) signaling motifs.

The detection of an amino-terminal fragment of Robo-1 (Robo-1-NTF) in the conditioned medium of cancer cell lines and in the serum of patients with hepatocellular carcinoma suggests that Robo-1 may undergo proteolytic cleavage (Ito et al., 2006). The cleavage site was recently shown to be between Glu852 and Glu853, only 10 residues away from the plane of the plasma membrane (Seki et al., 2010). Following cleavage of transmembrane Robo-1 by MMPs, a soluble Robo-1-NTF is generated. The remaining carboxy-terminal fragment (Robo-1-CTF1) is subsequently cleaved by γ-secretase to form Robo-1-CTF2 (Fig. 4). Robo-1-CTF2 translocates to the nucleus, although its function is unknown (Seki et al., 2010).

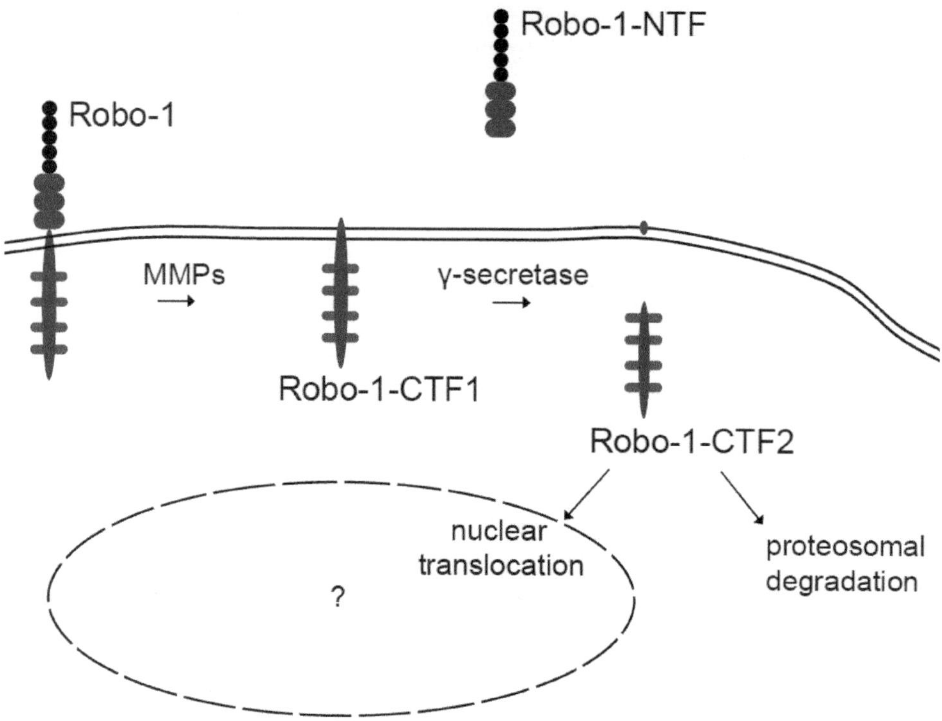

Fig. 4. Successive cleavage of the Robo-1 receptor. Full-length Robo-1 is first cleaved by MMPs to form Robo-1-NTF and Robo-1-CTF1. The second cleavage, mediated by γ-secretase, releases Robo-1-CTF2 which translocates to the nucleus. The function of Robo-1-CTF2 at this location is unknown.

2.2 Slit and robo: Expression

Expression of the Slit genes has been demonstrated in many organisms, including Drosophila (Battye et al., 1999), *Caenorhabditis elegans* (Hao et al., 2001), Xenopus (Chen et al., 2000), *Gallus gallus domesticus* (Holmes et al., 2001; Vargesson et al., 2001), mice (Holmes et al., 1998; Piper et al., 2000), rats (Marillat et al., 2002) and humans (Itoh et al., 1998). In mammals there are three members of the Slit family. Although Slit1 is predominantly expressed in the developing CNS (Yuan et al., 1999), Slit2 and Slit3 are expressed outside the CNS, particularly in lung, kidney, and heart (Wu et al., 2001). Importantly, Slit expression persists in the adult organism, suggesting a role for Slit proteins beyond embryogenesis.

Expression of Robo has been demonstrated in Drosophila (Kidd et al., 1998), mice (Yuan et al., 1999) and humans (Kidd et al., 1998). There are four isoforms of Robo in mammals. Robo-1 is most highly expressed in tissues outside the CNS, including human leukocytes (Wu et al., 2001). Robo-2 is expressed during vertebrate limb development (Vargesson et al., 2001). Robo-3 is expressed following cerebellar and spinal cord lesions (Wehrle et al., 2005). Robo-4 is expressed in the adult organism by primary human endothelial cells, including umbilical vein endothelial cells and microvascular endothelial cells (Suchting et al., 2005). Interestingly, the tissue expression of Slit and Robo is relatively complementary, suggesting a synergistic relationship (Yuan et al., 1999).

2.3 Slit and robo: Function

Recent studies demonstrate a role for Slit and Robo as repellents outside the CNS. For example, in mesoderm migration in Drosophila, myocyte precursors migrate away from the midline towards peripheral target sites where they fuse to form muscle fibers. In Slit and Robo mutants, these cells do not migrate away from the midline and instead fuse across it (Rothberg et al., 1990). Interestingly, this defect can be reversed by expressing Slit protein in midline cells (Kramer et al., 2001). Slit and Robo signaling also plays a role in nephrogenesis. During renal development, formation of a ureteric bud requires secretion of glial cell derived neurotrophic factor (GDNF) by nearby mesenchymal cells. Slit2 and Robo-2 knockout mice display abnormal patterns of GDNF secretion and develop multiple ureteric buds and multiple urinary collecting systems (Ray, L., 2004). Furthermore, polymorphisms in the human *Robo2* gene are associated with familial vesicoureteral reflux (Bertoli-Avella et al., 2008), a condition involving improper insertion of ureters into the bladder resulting in retrograde flow of urine from the bladder to the kidney. Slit2 also acts as a repellent in the mature organism. A recent study demonstrated that Slit2 inhibits vascular smooth muscle cell migration toward a gradient of platelet-derived growth factor (PDGF) (Liu et al., 2006). This inhibition occurred by suppression of activation of the small GTPase, Rac1. Slit2 has been shown to prevent cancer cell metastasis. The chemokine receptor, CXCR4, is expressed by some human breast cancer cells, allowing them to migrate towards gradients of the CXCR4 ligand, stromal cell-derived factor-1 (SDF-1α), and promoting their metastasis to the lung. Slit2 inhibited chemotaxis, adhesion and chemoinvasion of these breast cancer cells (Prasad et al., 2004). Several other studies have demonstrated a role for Slit2 as a tumor suppressor. Slit2 was shown to inhibit colony formation in lung, colorectal and breast cancer cell lines (Dallol et al., 2002). *Slit2* has also been shown to be epigenetically silenced in more aggressive forms of these and other cancers (Dallol et al., 2003; Dallol et al., 2003; Dickinson et al., 2004). Collectively, these studies demonstrate a repellent role for Slit and Robo in the adult organism and in cancer biology.

The role of Robo-4 signaling in endothelial cells is controversial. Kaur et al. (2006) showed that Robo-4 signaling mediates attractive guidance mechanisms by activating Cdc42 and Rac1 in endothelial cells and inducing actin-mediated cell protrusions, including filopodia and lamellipodia. In fact, Robo-4-induced phenotypic effects in endothelial cells are rescued by dominant negative constructs of Cdc42. Thus, Robo-4 may mediate attractive signaling via activation of Rho-family GTPases, Cdc42 and Rac1. However, in 2008, Jones et al. showed that Slit inhibits endothelial cell migration and angiogenesis. In fact, Robo-4 signaling was shown to stabilize endothelial cell barriers (Jones et al., 2009). Thus, the precise role of Slit/Robo signaling in endothelial cells is yet to be determined.

2.4 Slit2/Robo-1 intracellular signal transduction

Studies of neuronal tissue have demonstrated that Robo-1 signals through two pathways that lead to remodeling of the cytoskeleton: Enabled (Ena) protein and Rho GTPases. Both of these pathways require the CC motifs in the cytoplasmic domain of Robo. Ena and its mammalian homologue (Mena) are members of a family of proteins that link signal transduction to localized remodeling of the actin cytoskeleton by binding to profilin, an actin binding protein which regulates actin polymerization (Lanier et al., 1999; Wills et al., 1999). Ena is a substrate for Abelson kinase (Gertler et al., 1989). Ena and Abelson both bind to Robo. Ena binds to the CC1 motifs while Abelson binds to the CC3 motif (Fig. 5) (Bashaw et al., 2000). Impairing Ena binding reduces Robo function, while mutations in Abelson result in Robo hyperactivity (Bashaw et al., 2000).

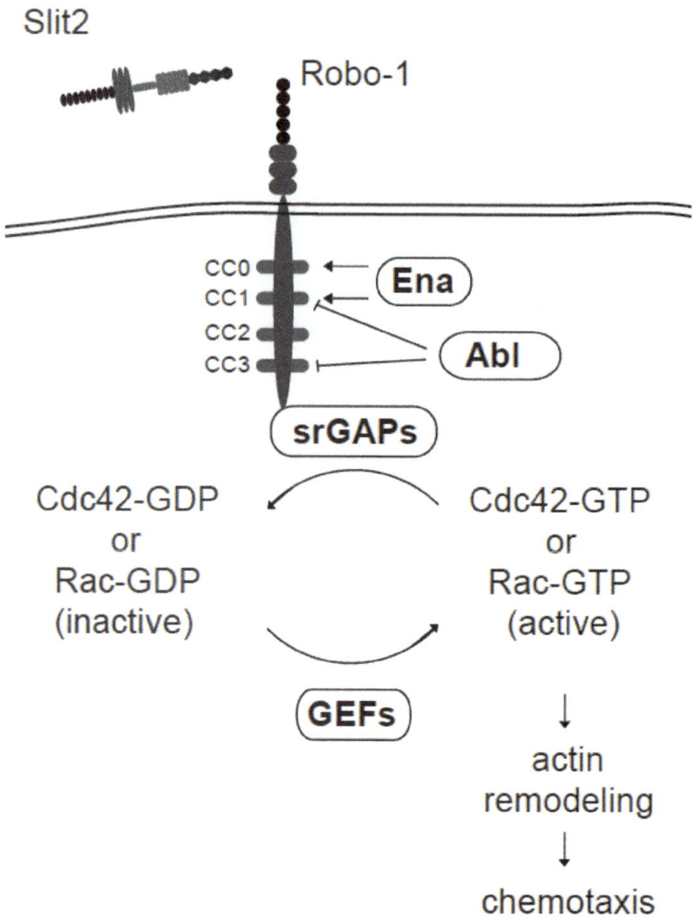

Fig. 5. Intracellular signaling downstream of the Robo-1 receptor. Enabled protein bind to Robo-1 and may contribute to Slit-mediated repulsion. Abelson kinase phosphorylates intracellular domains of Robo and antagonizes Robo function. Ligation of Robo-1 by Slit2 results in the recruitment of srGAPs to the plasma membrane. srGAPs convert active GTP-bound forms of Cdc42 and Rac to their inactive, GDP-bound counterparts, thereby inhibiting the dynamic actin polymerization required for chemotaxis and preventing cell migration.

Slit/Robo also mediate cell repulsion through modulation of Rho GTPase activity. A family of GTPase activating proteins, Slit Robo GTPase activating proteins (srGAPs), were shown to bind Robo (Fig. 5) (Wong et al., 2001). The SH3 domain of srGAP binds the CC3 motif of Robo, while the GAP domain has activity for the Rho GTPases, Rac, Cdc42 and Rho (Wong et al., 2001). Ligation of Robo by Slit induces the recruitment of srGAP, thereby inactivating Rho-family GTPases and inhibiting actin remodeling and cell motility (Wong et al., 2001).

2.5 Slit/Robo in cell trafficking

Both neuronal and leukocyte chemotaxis require recognition of guidance cues, polarization of the cell, and mobilization of the actin cytoskeleton. In addition to repelling developing axons, Slit2 also inhibits chemotaxis of other cell types including vascular smooth muscle cells (Liu et al., 2006). However, the first study to demonstrate that Slit2 inhibits leukocyte chemotaxis, in 2001, utilized transwell migration assays to show that Slit2 inhibits chemotaxis of rat lymph node cells and neutrophil-like HL-60 cells towards MCP-1 and fMLP respectively (Wu et al., 2001). Subsequently, Kanellis et al. (2004) demonstrated that Slit2 inhibits chemotaxis of rat-derived macrophages towards MCP-1 and fMLP. Another study showed that Slit2 inhibited migration of dendritic cells (DCs) (Guan et al., 2003). In 2007, Prasad et al. demonstrated that Slit2 inhibits chemotaxis and transendothelial migration of primary CD4+ T lymphocytes toward SDF-1. Recently, Slit2 was shown to promote chemotaxis of eosinophils towards the chemokine, eotaxin, and to exacerbate allergic airway inflammation (Ye et al., 2010). Thus, Slit2 can negatively or positively regulate directional migration of individual leukocyte subsets.

3. Slit2/Robo-1 signaling inhibits neutrophil migration

Using immunoblotting, we previously demonstrated Robo-1 protein in human and mouse neutrophils (Tole et al., 2009). Immunofluorescence microscopy and flow cytometry revealed that Robo-1 was on the surface of cells.

We used Transwell migration assays to study the effects of Slit2 on chemotaxis of primary human neutrophils. In the presence of Slit2, fMLP-induced migration of neutrophils was inhibited in a dose-dependent manner. In fact, we observed that Slit2 is a potent inhibitor of neutrophil migration toward diverse types of chemotactic cues, including IL-8 and C5a (Tole et al., 2009).

Neutrophil exposure to chemoattractants results in the activation of the Rho GTPases, Rac and Cdc42 and the subsequent reorganization of actin filaments. (Sun et al., 2004; Srinivasan et al., 2003). Since the predominant isoform of Rac in human neutrophils is Rac2, not Rac1, the activation of Rac2 was studied. Following stimulation with fMLP, levels of activated Cdc42 and Rac2 in the presence of Slit2 were less than half of those observed in untreated control cells.. We found that Slit2 inhibits neutrophil chemotaxis and actin polymerization by preventing cell polarization and disrupting generation and recruitment of activated Rac2 and Cdc42.

We examined the effects of Slit2 on the activation of kinase signaling pathways associated with neutrophil chemotaxis, namely, PI3K, Akt, Erk, and p38 MAPK. Stimulation of neutrophils with fMLP resulted in levels of activated Akt that were comparable in the presence or absence of Slit2, indicating that Slit2 does not impair the ability of neutrophils to generate membrane PIP_3. Similarly, Slit2 treatment had no effect on fMLP-induced phosphorylation of Erk and p38 MAPK. Thus, Slit2 inhibits neutrophil chemotaxis by specifically preventing activation of Cdc42 and Rac2 but not activation of Akt, Erk, or p38 MAPK (Tole et al., 2009).

We studied the effect of Slit2 on neutrophil recruitment *in vivo* using a mouse model of chemical irritant peritonitis (Glogauer et al., 2003). The administration of Slit2 prior to induction of peritonitis with sodium periodate, resulted in a significant decrease in neutrophil recruitment to the peritioneum.. Slit2 also prevented neutrophil recruitment to the peritoneal cavity in response to other chemoattractant factors tested, including C5a and MIP-2. These data demonstrate that Slit2 acts as a potent inhibitor of chemotaxis for circulating neutrophils toward diverse inflammatory stimuli. Slit2 also inhibited infiltration of other leukocyte subsets, especially monocytes/macrophages (Tole et al., 2009).

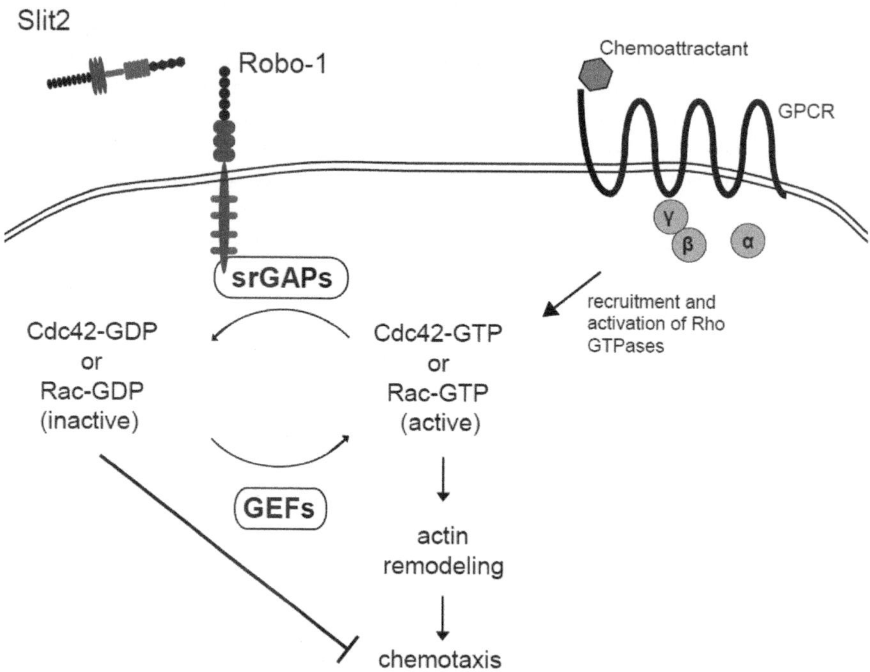

Fig. 6. Slit2/Robo-1 signaling inhibits actin remodelling required for chemotaxis. Chemoattractant signaling induces the activation of Rho GTPases Cdc42 and Rac, allowing for actin remodeling and chemotaxis. The binding of the LRRs on Slit2 to Robo-1 recruits srGAPs to the membrane, converting active Rho GTPases to their inactive, GDP-bound, forms. Inactivation of Rho GTPases abolishes actin remodeling and prevents cell chemotaxis.

4. Conclusion and discussion

The LRRs contained in Slit proteins can inhibit the migration of diverse cells, including neuronal cells and vascular smooth muscle cells. The conserved structure of Slit proteins also allows them to inhibit the migration of several subsets of leukocytes, including DCs and lymphocytes (Guan et al., 2003; Kanellis et al., 2004; Prasad et al., 2007). We have recently shown that Slit2 inhibits migration of neutrophils to diverse inflammatory attractants, *in vitro* and *in vivo*. Furthermore, we have demonstrated that this inhibition is mediated by inactivation of Rho-family GTPases, Rac and Cdc42 (Fig. 5). Excessive infiltration of leukocytes, particularly neutrophils, is associated with local tissue damage seen in inflammatory conditions such as rheumatoid arthritis and ischemia reperfusion injury (Weissmann et al., 1984; Kaminski et al., 2002). Thus, the protein structure of the conserved LRR regions contained in Slit proteins may be utilized as a novel therapeutic strategy to locally inhibit leukocyte recruitment.

Extensive glycosylation makes Slit2 a large and relatively "sticky" protein, potentially allowing it to maintain a high local concentration through adherence to extracellular matrix proteins such as glypican-1 (Ronca et al., 2001). Thus, after regional administration, Slit2 may be retained at sites of inflammation, such as joints and transplanted organs, thereby alleviating neutrophil-inflicted tissue injury associated with rheumatoid arthritis and ischemia reperfusion injury. As Slit2 blocks migration of several types of inflammatory cells, including neutrophils, T lymphocytes, macrophages, and dendritic cells, toward diverse chemoattractant signals, it could act as a highly effective anti-inflammatory agent (Guan et al., 2003; Kanellis et al., 2004; Prasad et al., 2007; Wu et al., 2001). Further studies are required to explore the therapeutic use of Slit2, or of a Slit-like compound containing the structurally critical LRRs, for prevention and treatment of localized inflammation.

5. Acknowledgment

This work was supported by the Canadian Institute of Health Research (L. A. R.). L. A. R. holds a Canada Research Chair, Tier 2.

6. References

Allen, W. E., Jones, G. E., Pollard, J. W., & Ridley, A. J. (1997). Rho, rac and Cdc42 regulate actin organization and cell adhesion in macrophages. *Journal of Cell Science, 110*(6), 707-720.

Alon, R., Hammer, D. A., & Springer, T. A. (1995). Lifetime of the P-selectin-carbohydrate bond and its response to tensile force in hydrodynamic flow. *Nature, 374*(6522), 539.

Aspenström, P. (1999). Effectors for the rho GTPases. *Current Opinion in Cell Biology, 11*(1), 95.

Astrof, N., Salas, A., Shimaoka, M., Chen, J., & Springer, T. (2006). Importance of force linkage in mechanochemistry of adhesion receptors. *Biochemistry, 45*(50), 15020.

Bashaw, G. J., Kidd, T., Murray, D., Pawson, T., & Goodman, C. S. (2000). Repulsive axon guidance: Abelson and enabled play opposing roles downstream of the roundabout receptor. *Cell, 101*(7), 703-715.

Battye, R., Stevens, A., & Jacobs, J. R. (1999). Axon repulsion from the midline of the drosophila CNS requires slit function. *Development, 126*(11), 2475-2481.

Battye, R., Stevens, A., Perry, R. L., & Jacobs, J. R. (2001). Repellent signaling by slit requires the leucine-rich repeats. *The Journal of Neuroscience, 21*(12), 4290-4298.

Benard, V., Bohl, B. P., Bokoch, G. M. (1999). Characterization of Rac and Cdc42 activation in chemoattractant-stimulated human neutrophils using a novel assay for active GTPases. *J. Biol. Chem.* 274, 13198–13204.

Bertoli-Avella, A., Conte, M., Punzo, F., de Graaf, B., Lama, G., La Manna, A., et al. (2008). ROBO2 gene variants are associated with familial vesicoureteral reflux. *Journal of the American Society of Nephrology, 19*(4), 825-831.

Beutler, B. (2004). Innate immunity: An overview. *Molecular Immunology, 40*(12), 845-859.

Bokoch, G. M. (2005). Regulation of innate immunity by rho GTPases. *Trends in Cell Biology, 15*(3), 163.

Bretscher, M. S. (1996). Getting membrane flow and the cytoskeleton to cooperate in moving cells. *Cell, 87*(4), 601.

Brose, K., Bland, K. S., Wang, K. H., Arnott, D., Henzel, W., Goodman, C. S., et al. (1999). Slit proteins bind robo receptors and have an evolutionarily conserved role in repulsive axon guidance. *Cell, 96*(6), 795-806.

Calderwood, D. A., Shattil, S. J., & Ginsberg, M. H. (2000). Integrins and actin filaments: Reciprocal regulation of cell adhesion and signaling. *The Journal of Biological Chemistry, 275*(30), 22607-22610.

Campbell, J. J., Hedrick, J., Zlotnik, A., Siani, M. A., Thompson, D. A., & Butcher, E. C. (1998). Chemokines and the arrest of lymphocytes rolling under flow conditions. *Science, 279*(5349), 381.

Casey, P. J., Solski, P. A., Der, C. J., & Buss, J. E. (1989). p21ras is modified by a farnesyl isoprenoid. *Proceedings of the National Academy of Sciences of the United States of America, 86*(21), 8323-8327.

Cederqvist, K., T. Sorsa, et al. (2001). "Matrix metalloproteinases-2, -8, and -9 and TIMP-2 in tracheal aspirates from preterm infants with respiratory distress." Pediatrics 108(3): 686-92.

Chen, J. H., Wen, L., Dupuis, S., Wu, J. Y., & Rao, Y. (2001). The N-terminal leucine-rich regions in slit are sufficient to repel olfactory bulb axons and subventricular zone neurons. *The Journal of Neuroscience, 21*(5), 1548-1556.

Chen, J., Wu, W., Li, H., Fagaly, T., Zhou, L., Wu, J., et al. (2000). Embryonic expression and extracellular secretion of xenopus slit. *Neuroscience, 96*(1), 231.

Clark, R. A., Volpp, B. D., Leidal, K. G., Nauseef, W. M. (1990). Two cytosolic components of the human neutrophil respiratory burst oxidase translocate to the plasma membrane during cell activation. *J. Clin. Invest.* 85, 714–721.

Constantin, G., Majeed, M., Giagulli, C., Piccio, L., Kim, J. Y., Butcher, E. C., et al. (2000). Chemokines trigger immediate beta2 integrin affinity and mobility changes: Differential regulation and roles in lymphocyte arrest under flow. *Immunity, 13*(6), 759.

Correia, I., Chu, D., Chou, Y., Goldman, R., & Matsudaira, P. (1999). Integrating the actin and vimentin cytoskeletons: Adhesion-dependent formation of fimbrin-vimentin complexes in macrophages. *The Journal of Cell Biology, 146*(4), 831.

Critchley, D. R., Holt, M. R., Barry, S. T., Priddle, H., Hemmings, L., & Norman, J. (1999). Integrin-mediated cell adhesion: The cytoskeletal connection. *Biochemical Society Symposia, 65*, 79-99.

Dallol, A., Da Silva, N., Viacava, P., Minna, J., Bieche, I., Maher, E., et al. (2002). SLIT2, a human homologue of the drosophila Slit2 gene, has tumor suppressor activity and is frequently inactivated in lung and breast cancers. *Cancer Research, 62*(20), 5874-5880.

Dallol, A., Krex, D., Hesson, L., Eng, C., Maher, E., & Latif, F. (2003). Frequent epigenetic inactivation of the SLIT2 gene in gliomas. *Oncogene, 22*(29), 4611-4616.

Dallol, A., Morton, D., Maher, E., & Latif, F. (2003). SLIT2 axon guidance molecule is frequently inactivated in colorectal cancer and suppresses growth of colorectal carcinoma cells. *Cancer Research, 63*(5), 1054-1058.

DeBiasio, R. L., Wang, L. L., Fisher, G. W., & Taylor, D. L. (1988). The dynamic distribution of fluorescent analogues of actin and myosin in protrusions at the leading edge of migrating swiss 3T3 fibroblasts. *The Journal of Cell Biology, 107*(6), 2631-2645.

Defacque, H., Egeberg, M., Habermann, A., Diakonova, M., Roy, C., Mangeat, P., et al. (2000). Involvement of ezrin/moesin in de novo actin assembly on phagosomal membranes. *EMBO Journal, 19*(2), 199-212.

DeFife, K. M., Jenney, C. R., Colton, E., & Anderson, J. M. (1999). Cytoskeletal and adhesive structural polarizations accompany IL-13-induced human macrophage fusion. *The Journal of Histochemistry and Cytochemistry, 47*(1), 65-74.

Dickinson, R., Dallol, A., Bieche, I., Krex, D., Morton, D., Maher, E., et al. (2004). Epigenetic inactivation of SLIT3 and SLIT1 genes in human cancers. *The British Journal of Cancer, 91*(12), 2071.

Dunne, J., Ballantyne, C., Beaudet, A., & Ley, K. (2002). Control of leukocyte rolling velocity in TNF-alpha-induced inflammation by LFA-1 and mac-1. *Blood, 99*(1), 336.

Dvorak, Ann., Feng, Dian. (2001). The vesiculo–vacuolar organelle (VVO): a new endothelial cell permeability organelle. *The Journal of Histochemistry & Cytochemistry, 49*(4), 419-431.

Eltzschig, H. K. and C. D. Collard (2004). "Vascular ischaemia and reperfusion injury." Br Med Bull 70: 71-86.

Eriksson, E. E., Xie, X., Werr, J., Thoren, P., & Lindbom, L. (2001). Importance of primary capture and L-selectin-dependent secondary capture in leukocyte accumulation in inflammation and atherosclerosis in vivo. *The Journal of Experimental Medicine, 194*(2), 205.

Fauci, A. S., B. Haynes, et al. (1978). "The spectrum of vasculitis: clinical, pathologic, immunologic and therapeutic considerations." Ann Intern Med 89(5 Pt 1): 660-76.

Fenteany, G., & Glogauer, M. (2004). Cytoskeletal remodeling in leukocyte function. *Current Opinion in Hematology, 11*(1), 15-24.

Finger, E. B., Puri, K. D., Alon, R., Lawrence, M. B., von Andrian, U. H., & Springer, T. A. (1996). Adhesion through L-selectin requires a threshold hydrodynamic shear. *Nature, 379*(6562), 266.

Frangogiannis, Nikolaos., Smith, Wayne., Entman, Mark. (2002). The inflammatory response in myocardial infarction. *Cardiovascular Research, 53*, 31-47.

Fujishima, S., Aikawa, N. (1995). Neutrophil-mediated tissue injury and its modulation. *Intensive Care Medicine, 21*(3), 277-285.

Fujiyama, A., & Tamanoi, F. (1990). RAS2 protein of saccharomyces cerevisiae undergoes removal of methionine at N terminus and removal of three amino acids at C terminus. *The Journal of Biological Chemistry, 265*(6), 3362-3368.

Gamblin, S. J., & Smerdon, S. J. (1998). GTPase-activating proteins and their complexes. *Current Opinion in Structural Biology, 8*(2), 195-201.

Ganz, T. (2003). Defensins: Antimicrobial peptides of innate immunity. *Nature Reviews.Immunology, 3*(9), 710-720.

Guan, H., Zu, G., Tang, H., Johnson, M., Xu, X., Kevil, C., Xiong, W-C., Elmets, C., Rao, Y., Wu, J. Y., Xu, H. (2003) Neuronal repellent Slit2 inhibits dendritic cell migration and the development of immune responses. *J. Immunol.* 171, 6519–6526.

Garcia, J. G., H. L. James, et al. (1987). "Lower respiratory tract abnormalities in rheumatoid interstitial lung disease. Potential role of neutrophils in lung injury." Am Rev Respir Dis 136(4): 811-7.

Gertler, F. B., Bennett, R. L., Clark, M. J., & Hoffmann, F. M. (1989). Drosophila abl tyrosine kinase in embryonic CNS axons: A role in axonogenesis is revealed through dosage-sensitive interactions with disabled. *Cell, 58*(1), 103-113.

Glogauer, M., Marchal, C. C., Zhu, F., Worku, A., Clausen, B. E., Foerster, I., Marks, P., Downey, G. P., Dinauer, M., Kwiakowski, D. J. (2003). Rac1 deletion in mouse neutrophils has selective effects on neutrophil function. *J. Immunol.* 170, 5652–5657.

Guan, H., Zu, G., Xie, Y., Tang, H., Johnson, M., Xu, X., et al. (2003). Neuronal repellent Slit2 inhibits dendritic cell migration and the development of immune responses. *The Journal of Immunology, 171*(12), 6519-6526.

Gu, Y., Filippi, M-D., Cancelas, J. A., Siefring, J. E., Williams, E. P., Jasti, A. C., Harris, C. E., Lee, A. W., Prabhakar, R., Atkinson, S. J., Kwiakowski, D. J., Williams, D. A. (2003) Hematopoietic cell regulation by Rac1 and Rac2 guanosine triphosphatases. *Science* 302, 445–449.

Gutierrez, L., Magee, A. I., Marshall, C. J., & Hancock, J. F. (1989). Post-translational processing of p21ras is two-step and involves carboxyl-methylation and carboxy-terminal proteolysis. *EMBO Journal, 8*(4), 1093-1098.

Hao, J. C., Yu, T. W., Fujisawa, K., Culotti, J. G., Gengyo-Ando, K., Mitani, S., et al. (2001). C. elegans slit acts in midline, dorsal-ventral, and anterior-posterior guidance via the SAX-3/Robo receptor. *Neuron, 32*(1), 25-38.

Haslam, P. L., C. W. Turton, et al. (1980). "Bronchoalveolar lavage in pulmonary fibrosis: comparison of cells obtained with lung biopsy and clinical features." Thorax 35(1): 9-18.

Hirche, T. O., E. C. Crouch, et al. (2004). "Neutrophil serine proteinases inactivate surfactant protein D by cleaving within a conserved subregion of the carbohydrate recognition domain." J Biol Chem 279(26): 27688-98.

Hirsch, E., Katanaev, V. L., Garlanda, C., Azzolino, O., Pirola, L., Silengo, L., et al. (2000). Central role for G protein-coupled phosphoinositide 3-kinase gamma in inflammation. *Science, 287*(5455), 1049-1053.

Holdsworth, S. R. and R. Bellomo (1984). "Differential effects of steroids on leukocytemediated glomerulonephritis in the rabbit." Kidney Int 26(2): 162-9.

Holmes, G. P., Negus, K., Burridge, L., Raman, S., Algar, E., Yamada, T., et al. (1998). Distinct but overlapping expression patterns of two vertebrate slit homologs implies functional roles in CNS development and organogenesis. *Mechanisms of Development, 79*(1-2), 57-72.

Holmes, G., & Niswander, L. (2001). Expression of slit-2 and slit-3 during chick development. *Developmental Dynamics, 222*(2), 301-307.

Huo, Y., Schober, A., Forlow, S. B., Smith, D., Hyman, M., Jung, S., et al. (2003). Circulating activated platelets exacerbate atherosclerosis in mice deficient in apolipoprotein E. *Nature Medicine, 9*(1), 61.

Imhof, B., & Aurrand-Lions, M. (2004). Adhesion mechanisms regulating the migration of monocytes. *Nature Reviews.Immunology, 4*(6), 432.

Ito, H. et al. (2006) Identification of ROBO1 as a novel hepatocellular carcinoma antigen and a potential therapeutic and diagnostic target. Clin. Cancer Res. 12, 3257–3264.

Itoh, A., Miyabayashi, T., Ohno, M., & Sakano, S. (1998). Cloning and expressions of three mammalian homologues of drosophila slit suggest possible roles for slit in the formation and maintenance of the nervous system. *Molecular Brain Research, 62*(2), 175-186.

Jin, T., Zhang, N., Long, Y., Parent, C. A., & Devreotes, P. N. (2000). Localization of the G protein betagamma complex in living cells during chemotaxis. *Science, 287*(5455), 1034-1036.

Johnson, Z., Proudfoot, A. E., & Handel, T. M. (2005). Interaction of chemokines and glycosaminoglycans: A new twist in the regulation of chemokine function with opportunities for therapeutic intervention. *Cytokine Growth Factor Reviews, 16*(6), 625.

Jones, G. E., Allen, W. E., & Ridley, A. J. (1998). The rho GTPases in macrophage motility and chemotaxis. *Cell Adhesion Communication, 6*(2-3), 237-245.

Jones, C. A., Londin, N. R., Chen, H., Park, K. W., Sauvaget, D., Stockton, R. A., Wythe, J. D., Suh, W., Larrieu-Lahargue, F., Mukoutama, Y-S., Lindblom, P., Seth, P., Frias, A., Nishiya, N., Ginsberg, M. H., Gerhardt, H., Zhang, K., Li, D. Y. (2008) Robo4 stabilizes the vascular network by inhibiting pathologic angiogenesis and endothelial hyperpermeability. *Nat. Med.* 14, 448–453.

Jones, C., Nishiya, N., London, N., Zhu, W., Sorensen, L., Chan, A., et al. (2009). Slit2-Robo4 signalling promotes vascular stability by blocking Arf6 activity. *Nature Cell Biology, 11*(11), 1325-1331.

Jung, U., Norman, K. E., Scharffetter-Kochanek, K., Beaudet, A. L., & Ley, K. (1998). Transit time of leukocytes rolling through venules controls cytokine-induced inflammatory cell recruitment in vivo. *Journal of Clinical Investigation, 102*(8), 1526.

Kaminski, K. A., Bonda, T. A., Korecki, J., Musial, W. J. (2002) Oxidative stress and neutrophil activation—the two keystones of ischemia/reperfusion injury. *Int. J. Cardiol.* 86, 41–59.

Kanellis, J., Garcia, G., Ping, L., Parra, G., Wilson, C., Rao, Y., et al. (2004). Modulation of inflammation by slit protein in vivo in experimental crescentic glomerulonephritis. *The American Journal of Pathology, 165*(1), 341.

Kang, T., J. Yi, et al. (2001). "Subcellular distribution and cytokine- and chemokine regulated secretion of leukolysin/MT6-MMP/MMP-25 in neutrophils." J Biol Chem 276(24): 21960-8.

Kansas, G. S. (1996). Selectins and their ligands: Current concepts and controversies. *Blood, 88*(9), 3259.

Kaur, S., Castellone, MD., Bedell, V., Konar, M., Gutkind, J., Ramchandran, R. (2006). Robo4 signaling in endothelial cells implies attraction guidance mechanisms. *Journal of Biological Chemistry, 281*(16), 11347.

Keep, N. H., Barnes, M., Barsukov, I., Badii, R., Lian, L. Y., Segal, A. W., et al. (1997). A modulator of rho family G proteins, rhoGDI, binds these G proteins via an immunoglobulin-like domain and a flexible N-terminal arm. *Structure, 5*(5), 623-633.

Kennedy, J., Kelner, G. S., Kleyensteuber, S., Schall, T. J., Weiss, M. C., Yssel, H., et al. (1995). Molecular cloning and functional characterization of human lymphotactin. *The Journal of Immunology, 155*(1), 203.

Kennedy, T. E., Serafini, T., de la Torre, J. R., & Tessier-Lavigne, M. (1994). Netrins are diffusible chemotropic factors for commissural axons in the embryonic spinal cord. *Cell, 78*(3), 425-435.

Kidd, T., Brose, K., Mitchell, K. J., Fetter, R. D., Tessier-Lavigne, M., Goodman, C. S., et al. (1998). Roundabout controls axon crossing of the CNS midline and defines a novel subfamily of evolutionarily conserved guidance receptors. *Cell, 92*(2), 205-215.

Kintscher, U., Goetze, S., Wakino, S., Kim, S., Nagpal, S., Chandraratna, R. A., et al. (2000). Peroxisome proliferator-activated receptor and retinoid X receptor ligands inhibit monocyte chemotactic protein-1-directed migration of monocytes. *European Journal of Pharmacology, 401*(3), 259-270.

Kjoller, L., & Hall, A. (1999). Signaling to rho GTPases. *Experimental Cell Research, 253*(1), 166-179.

Kramer, S. G., Kidd, T., Simpson, J. H., & Goodman, C. S. (2001). Switching repulsion to attraction: Changing responses to slit during transition in mesoderm migration. *Science, 292*(5517), 737-740.

Krugmann, S., Hawkins, P. T., Pryer, N., & Braselmann, S. (1999). Characterizing the interactions between the two subunits of the p101/p110gamma phosphoinositide 3-kinase and their role in the activation of this enzyme by G beta gamma subunits. *The Journal of Biological Chemistry, 274*(24), 17152-17158.

Kunkel, E. J., & Ley, K. (1996). Distinct phenotype of E-selectin-deficient mice. E-selectin is required for slow leukocyte rolling in vivo. *Circulation Research, 79*(6), 1196.

Lanier, L. M., Gates, M. A., Witke, W., Menzies, A. S., Wehman, A. M., Macklis, J. D., et al. (1999). Mena is required for neurulation and commissure formation. *Neuron, 22*(2), 313-325.

Laudanna, C., Kim, J., Constantin, G., & Butcher, E. (2002). Rapid leukocyte integrin activation by chemokines. *Immunological Reviews, 186*, 37.

Lawrence, M. B., Kansas, G. S., Kunkel, E. J., & Ley, K. (1997). Threshold levels of fluid shear promote leukocyte adhesion through selectins (CD62L,P,E). *The Journal of Cell Biology, 136*(3), 717.

Lemanske, R. F., Jr., D. A. Guthman, et al. (1983). "The biologic activity of mast cell granules. VII. The effect of anti-neutrophil antibody-induced neutropenia on rat cutaneous late phase reactions." J Immunol 131(2): 929-33.

Li, C. and R. M. Jackson (2002). "Reactive species mechanisms of cellular hypoxiareoxygenation injury." Am J Physiol Cell Physiol 282(2): C227-41.

Li, Z., Jiang, H., Xie, W., Zhang, W., Smrcka, A., & Wu, D. (2000). Roles of PLC-2 and-3 and PI3K in chemoattractant-mediated signal transduction. *Science's STKE, 287*(5455), 1046.

Liang, Y., Annan, R. S., Carr, S. A., Popp, S., Mevissen, M., Margolis, R. K., et al. (1999). Mammalian homologues of the drosophila slit protein are ligands of the heparan sulfate proteoglycan glypican-1 in brain. *The Journal of Biological Chemistry, 274*(25), 17885-17892.

Liu, D., Hou, J., Hu, X., Wang, X., Xiao, Y., Mou, Y., et al. (2006). Neuronal chemorepellent Slit2 inhibits vascular smooth muscle cell migration by suppressing small GTPase Rac1 activation. *Circulation Research, 98*(4), 480-489.

Linder, S., & Aepfelbacher, M. (2003). Podosomes: Adhesion hot-spots of invasive cells. *Trends in Cell Biology, 13*(7), 376-385.

Luster, A. D. (1998). Chemokines--chemotactic cytokines that mediate inflammation. *New England Journal of Medicine, the, 338*(7), 436.

Martenson, C., Stone, K., Reedy, M., & Sheetz, M. (1993). Fast axonal transport is required for growth cone advance. *Nature, 366*(6450), 66-69.

Mattila, P., & Lappalainen, P. (2008). Filopodia: Molecular architecture and cellular functions. *Nature Reviews.Molecular Cell Biology, 9*(6), 446-454.

Marillat, V., Cases, O., Nguyen-Ba-Charvet, K. T., Tessier-Lavigne, M., Sotelo, C., & Chdotal, A. (2002). Spatiotemporal expression patterns of slit and robo genes in the rat brain. *Journal of Comparative Neurology, 442*(2), 130-155.

McDermott, David., Fong, Alan., Yang, Qiong., Sechler, Joan., Cupples, Adrienne., Merrell, Maya., Wilson, Peter., D'Agostino, Ralph., O'Donnell, Christopher., Patel, Dhavalkumar., and Murphy, Philip. (2003). Chemokine receptor mutant CX_3CR1-M280 has impaired adhesive function and correlates with protection from cardiovascular disease in humans. *Journal of Clinical Investigation, 111*(8), 1241-1250.

McEver, R. P., & Cummings, R. D. (1997). Perspectives series: Cell adhesion in vascular biology. role of PSGL-1 binding to selectins in leukocyte recruitment. *Journal of Clinical Investigation, 100*(3), 485.

Mitra, S., Hanson, D., & Schlaepfer, D. (2005). Focal adhesion kinase: In command and control of cell motility. *Nature Reviews.Molecular Cell Biology, 6*(1), 56-68.

Mohr, W., H. Westerhellweg, et al. (1981). "Polymorphonuclear granulocytes in rheumatic tissue destruction. III. an electron microscopic study of PMNs at the pannus-cartilage junction in rheumatoid arthritis." Ann Rheum Dis 40(4): 396-9.

Moore, K. J., Andersson, L. P., Ingalls, R. R., Monks, B. G., Li, R., Arnaout, M. A., et al. (2000). Divergent response to LPS and bacteria in CD14-deficient murine macrophages. *The Journal of Immunology, 165*(8), 4272-4280.

Morlot, C., Thielens, N., Ravelli, R. B. G., Hemrika, W., Romijn, R., Gros, P., et al. (2007). Structural insights into the slit-robo complex. *Proceedings of the National Academy of Sciences of the United States of America, 104*(38), 14923-14928.

Muller, W. A. (2003). Leukocyte–endothelial-cell interactions in leukocyte transmigration and the inflammatory response. *Trends in Immunology, 24*(6), 326-333.

Nassar, N., Hoffman, G. R., Manor, D., Clardy, J. C., & Cerione, R. A. (1998). Structures of Cdc42 bound to the active and catalytically compromised forms of Cdc42GAP. *Nature Structural Biology, 5*(12), 1047-1052.

Nguyen Ba-Charvet, K. T. N., Brose, K., Ma, L., Wang KH., Marillat, V., Sotelo, C., et al. (2001). Diversity and specificity of actions of Slit2 proteolytic fragments in axon guidance. *The Journal of Neuroscience, 21*(12), 4281.

Nguyen, D. H., Catling, A. D., Webb, D. J., Sankovic, M., Walker, L. A., Somlyo, A. V., et al. (1999). Myosin light chain kinase functions downstream of Ras/ERK to promote migration of urokinase-type plasminogen activator-stimulated cells in an integrin-selective manner. *The Journal of Cell Biology, 146*(1), 149-164.

Nobes, C. D., & Hall, A. (1995). Rho, rac, and cdc42 GTPases regulate the assembly of multimolecular focal complexes associated with actin stress fibers, lamellipodia, and filopodia. *Cell, 81*(1), 53-62.

Oloffson, B. (1999). Rho Guanine Dissociation Inhibitors - Pivotal Molecules in Cellular Signalling . *Cellular Signalling, 11*(8), 545-554.

Owen, C. A. and E. J. Campbell (1999). "The cell biology of leukocyte-mediated proteolysis." J Leukoc Biol 65(2): 137-50.

Panes, J., M. Perry, et al. (1999). "Leukocyte-endothelial cell adhesion: avenues for therapeutic intervention." Br J Pharmacol 126(3): 537-50.

Piper, M., Georgas, K., Yamada, T., & Little, M. (2000). Expression of the vertebrate slit gene family and their putative receptors, the robo genes, in the developing murine kidney. *Mechanisms of Development, 94*(1-2), 213-217.

Prasad, A., Fernandis, A., Rao, Y., & Ganju, R. (2004). Slit protein-mediated inhibition of CXCR4-induced chemotactic and chemoinvasive signaling pathways in breast cancer cells. *The Journal of Biological Chemistry, 279*(10), 9115-9124.

Prasad, A., Qamri, Z., Wu, J., & Ganju, R. (2007). Slit-2/Robo-1 modulates the CXCL12/CXCR4-induced chemotaxis of T cells. *Journal of Leukocyte Biology, 82*(3), 465-476.

Puri, D., Finger, B., Springer, A. (1995 Sialomucin CD34 is the major L-selectin ligand in human tonsil high endothelial venules. The *Journal of Cell Biology, 131*(1), 261-270.

Ray, L. B. (2004). STKE: Slit and robo in kidney formation. *Science, 304*(5675), 1215c.

Ridley, A. J. (2001). Rho proteins, PI 3-kinases, and monocyte/macrophage motility. *FEBS Letters, 498*(2-3), 168.

Ridley, A. J., & Hall, A. (1992). The small GTP-binding protein rho regulates the assembly of focal adhesions and actin stress fibers in response to growth factors. *Cell, 70*(3), 389-399.

Robinson, L. A., Nataraj, C., Thomas, D. W., Howell, D. N., Griffiths, R., Bautch, V., et al. (2000). A role for fractalkine and its receptor (CX3CR1) in cardiac allograft rejection. *The Journal of Immunology, 165*(11), 6067.

Ronca, F., Andersen, J. S., Paech, V., & Margolis, R. U. (2001). Characterization of slit protein interactions with glypican-1. *The Journal of Biological Chemistry, 276*(31), 29141-29147.

Rossi, D., & Zlotnik, A. (2000). The biology of chemokines and their receptors. *Annual Review of Immunology, 18*, 217-242.

Rossman, K., Der, C., & Sondek, J. (2005). GEF means go: Turning on RHO GTPases with guanine nucleotide-exchange factors. *Nature Reviews.Molecular Cell Biology, 6*(2), 167-180.

Rothberg, J. M., & Artavanis-Tsakonas, S. (1992). Modularity of the slit protein. characterization of a conserved carboxy-terminal sequence in secreted proteins and a motif implicated in extracellular protein interactions. *Journal of Molecular Biology, 227*(2), 367-370.

Rothberg, J. M., Hartley, D. A., Walther, Z., & Artavanis-Tsakonas, S. (1988). Slit: An EGF-homologous locus of D. melanogaster involved in the development of the embryonic central nervous system. *Cell, 55*(6), 1047-1059.

Rothberg, J. M., Jacobs, J. R., Goodman, C. S., & Artavanis-Tsakonas, S. (1990). Slit: An extracellular protein necessary for development of midline glia and commissural axon pathways contains both EGF and LRR domains. *Genes Development*, 4(12A), 2169-2187.

Rubio, F., J. Cooley, et al. (2004). "Linkage of neutrophil serine proteases and decreased surfactant protein-A (SP-A) levels in inflammatory lung disease." Thorax 59(4): 318-23.

Salas, A., Shimaoka, M., Kogan, A., Harwood, C., von Andrian, U., & Springer, T. (2004). Rolling adhesion through an extended conformation of integrin alphaLbeta2 and relation to alpha I and beta I-like domain interaction. *Immunity, 20*(4), 393.

Sasaki, T., Irie-Sasaki, J., Jones, R. G., Oliveira-dos-Santos, A. J., Stanford, W. L., Bolon, B., et al. (2000). Function of PI3Kgamma in thymocyte development, T cell activation, and neutrophil migration. *Science, 287*(5455), 1040-1046.

Sehr, P., Joseph, G., Genth, H., Just, I., Pick, E., Aktories, K. (1998) Glucosylation and ADP ribosylation of Rho proteins: effects of nucleotide binding, GTPase activity, and effector coupling. *Biochemistry* 37, 5296–5304.

Seki, M., Watanabe, A., Enomoto, S., Kawamura, T., Ito, H., Kodama, T., et al. (2010). Human ROBO1 is cleaved by metalloproteinases and γ-secretase and migrates to the nucleus in cancer cells. *FEBS Letters, 584*(13), 2909-2915.

Servant, G., Weiner, O. D., Herzmark, P., Balla, T., Sedat, J. W., & Bourne, H. R. (2000). Polarization of chemoattractant receptor signaling during neutrophil chemotaxis. *Science, 287*(5455), 1037-1040.

Shamri, R., Grabovsky, V., Gauguet, J., Feigelson, S., Manevich, E., Kolanus, W., et al. (2005). Lymphocyte arrest requires instantaneous induction of an extended LFA-1 conformation mediated by endothelium-bound chemokines. *Nature Immunology, 6*(5), 497.

Shin, H. S., Snyderman, R., Friedman, E., Mellors, A., & Mayer, M. M. (1968). Chemotactic and anaphylatoxic fragment cleaved from the fifth component of guinea pig complement. *Science, 162*(851), 361.

Sigal, A., Bleijs, D. A., Grabovsky, V., van Vliet, S. J., Dwir, O., Figdor, C. G., et al. (2000). The LFA-1 integrin supports rolling adhesions on ICAM-1 under physiological shear flow in a permissive cellular environment. *The Journal of Immunology, 165*(1), 442-452.

Singer, S. J., & Kupfer, A. (1986). The directed migration of eukaryotic cells. *Annual Review of Cell Biology, 2*, 337-365.

Srinivasan, S., Wang, F., Glavas, S., Ott, A., Hofmann, F., Aktories, K., Kalman, D., Bourne, H. R. (2003). Rac and Cdc42 play distinct roles in regulating PI(3,4,5)P3 and polarity during neutrophil chemotaxis. *J. Cell Biol.* 160, 375–385.

Suchting, S., Heal, P., Tahtis, K., Stewart L., and Bicknell, R. (2005). Soluble Robo4 receptor inhibits in vivo angiogenesis and endothelial cell migration. *The FASEB Journal, 19*(1), 121-123.

Sun, C. X., Downey, G. P., Zhu, F., Koh, A. L. Y., Thang, H., Glogauer, M. (2004). Rac1 is the small GTPase responsible for regulating the neutrophil chemotaxis compass. *Blood* 104, 3758–3765.

Symons, M. H., & Mitchison, T. J. (1991). Control of actin polymerization in live and permeabilized fibroblasts. *The Journal of Cell Biology, 114*(3), 503-513.

Tcherkezian, J., & Lamarche-Vane, N. (2007). Current knowledge of the large RhoGAP family of proteins. *Biology of the Cell, 99*(2), 67-86.

Vanhaesebroeck, B., & Waterfield, M. D. (1999). Signaling by distinct classes of phosphoinositide 3-kinases. *Experimental Cell Research, 253*(1), 239-254.

Vandivier, R. W., V. A. Fadok, et al. (2002). "Elastase-mediated phosphatidylserine receptor cleavage impairs apoptotic cell clearance in cystic fibrosis and bronchiectasis." J Clin Invest 109(5): 661-70.

Vargesson, N., Luria, V., Messina, I., Erskine, L., & Laufer, E. (2001). Expression patterns of slit and robo family members during vertebrate limb development. *Mechanisms of Development, 106*(1-2), 175-180.

Vetter, I. R., & Wittinghofer, A. (2001). The guanine nucleotide-binding switch in three dimensions. *Science, 294*(5545), 1299-1304.

von Hundelshausen, P., Weber, K. S., Huo, Y., Proudfoot, A. E., Nelson, P. J., Ley, K., et al. (2001). RANTES deposition by platelets triggers monocyte arrest on inflamed and atherosclerotic endothelium. *Circulation, 103*(13), 1772.

Wang, K. H., Brose, K., Arnott, D., Kidd, T., Goodman, C. S., Henzel, W., et al. (1999). Biochemical purification of a mammalian slit protein as a positive regulator of sensory axon elongation and branching. *Cell, 96*(6), 771-784.

Wang, Y. L. (1985). Exchange of actin subunits at the leading edge of living fibroblasts: Possible role of treadmilling. *The Journal of Cell Biology, 101*(2), 597.

Wehrle, R., Camand, E., Chedotal, A., Sotelo, C., & Dusart, I. (2005). Expression of netrin-1, slit-1 and slit-3 but not of slit-2 after cerebellar and spinal cord lesions. *European Journal of Neuroscience, 22*(9), 2134-2144.

Weissmann, G. and H. Korchak (1984). "Rheumatoid arthritis. The role of neutrophil activation." Inflammation 8 Suppl: S3-14.

Weissmann, G., R. B. Zurier, et al. (1971). "Mechanisms of lysosomal enzyme release from leukocytes exposed to immune complexes and other particles." J Exp Med 134(3 Pt2): 149s-165s.

Wieland, T. and C. K. Chen (1999). "Regulators of G-protein signalling: a novel protein family involved in timely deactivation and desensitization of signalling via heterotrimeric G proteins." Naunyn Schmiedebergs Arch Pharmacol 360(1): 14-26.

Wills, Z., Marr, L., Zinn, K., Goodman, C. S., & Van Vactor, D. (1999). Profilin and the abl tyrosine kinase are required for motor axon outgrowth in the drosophila embryo. *Neuron, 22*(2), 291-299.

Wong, K., Ren, X. R., Huang, Y. Z., Xie, Y., Liu, G., Saito, H., et al. (2001). Signal transduction in neuronal migration: Roles of GTPase activating proteins and the small GTPase Cdc42 in the slit-robo pathway. *Cell, 107*(2), 209-221.

Worthylake, R. A., Lemoine, S., Watson, J. M., & Burridge, K. (2001). RhoA is required for monocyte tail retraction during transendothelial migration. *The Journal of Cell Biology, 154*(1), 147-160.

Wu, J. Y., Feng, L., Park, H. T., Havlioglu, N., Wen, L., Tang, H., et al. (2001). The neuronal repellent slit inhibits leukocyte chemotaxis induced by chemotactic factors. *Nature, 410*(6831), 948-952.

Yamada, K. M., & Miyamoto, S. (1995). Integrin transmembrane signaling and cytoskeletal control. *Current Opinion in Cell Biology, 7*, 681.

Yang, L., Froio, R., Sciuto, T., Dvorak, A., Alon, R., Luscinskas, F. (2005). ICAM-1 regulates neutrophil adhesion and transcellular migration of TNF-α-activated vascular endothelium under flow. *Blood, 106*(2), 584.

Ye, BQ., Geng, Z., Ma, L., & Geng, J. (2010). Slit2 regulates attractive eosinophil and repulsive neutrophil chemotaxis through differential srGAP1 expression during lung inflammation. *The Journal of Immunology, 185*(10), 6294.

Yuan, W., Zhou, L., Chen, J. H., Wu, J. Y., Rao, Y., & Ornitz, D. M. (1999). The mouse SLIT family: Secreted ligands for ROBO expressed in patterns that suggest a role in morphogenesis and axon guidance. *Developmental Biology, 212*(2), 290-306.

Zallen, J. A., Yi, B. A., & Bargmann, C. I. (1998). The conserved immunoglobulin superfamily member SAX-3/Robo directs multiple aspects of axon guidance in C. elegans. *Cell, 92*(2), 217-227.

Zigmond, S. H. (1974). Mechanisms of sensing chemical gradients by polymorphonuclear leukocytes. *Nature, 249*, 450.

Section 5

Applications

15

Fibrinolytic Enzymes from Medicinal Mushrooms

Chung-Lun Lu and Shiu-Nan Chen
College of Life Science, National Taiwan University, Taiwan

1. Introduction

Mushrooms, a special group of macrofungi, are not plants and thus do not use photosynthesis for nutrient acquisition. Mushrooms are fleshy, have the spore-bearing fruiting body of fungi, and typically grow above ground on soil, or other food sources. Most mushrooms are Basidiomycota or Agaricomycetes. The fruiting body of mushrooms is an important food and is used in many cuisines worldwide. Additionally, several species have been consumed extensively as a crude drug or folk tonics; in East Asia, these mushrooms are considered as medicinal mushrooms. Some medicinal mushrooms can be cultured and are an abundant source of natural proteins and polysaccharides. These mushrooms have garnered considerable attention for clinical research and for modern scientific and medicinal researches investigating their biological functions (Kino et al., 1989).

1.1 Functional mushrooms

In recent decades, medicinal mushrooms have been used to improve human health and strengthen the immune system, which have become significant issues in medical and pharmacological researchers (Guillamon et al., 2010; Kwok et al., 2005). The activities that have been identified include immunomodulatory and hypocholesterolemic actions, and antitumor, anti-inflammatory, anti-allergic, anticoagulation, and antithrombin activity as well as fibrino(geno)lysis stimulation (Lu et al., 2010b; Wang et al., 1995). Based on the medicinal potentials of edible mushrooms, scientists and businesses have devoted considerable effort to increase the quality of cultivated mushroom. It's estimated that approximately 14,000 species of mushrooms exist (Miles & Chang, 2004); as edible and medicinal mushrooms account for only a small proportion. More than 10 million metric tons of edible and medicinal mushrooms are cultivated annually worldwide. Products from cultivated mushrooms have been used extensively as food additives, health foods, and for medicinal purposes to treat cancers and circulatory disorders (Kim et al., 2003; Mao et al., 2005; Park et al., 2001). Improvements in mushroom cultivation technology have brought countless benefits to commercial production and laboratory applications of mushroom.

1.2 Bioactive compounds in mushrooms

Many bioactive compounds, such as cordycepin, polysaccharides, polysaccharide-peptide complex, ergosterol, mannitol, peptides and protein/protease, in various mushrooms have chemotherapeutic or medicinal activity (Das et al., 2010). The sources of these bioactive compounds include fruiting body, mycelia, cultivation broth, submerged cultivation mycelia and fermentation derivatives (Cheung, 1996; Fiore & Kakkar, 2003; Kwok et al., 2005; Sugimoto et al., 2007; Wang et al., 1995; Wong et al., 2011; Wu et al., 2010, Yamamoto et al., 2005; Yoon et al., 2003). This chapter focuses on the fibrinolytic enzymes derived from medicinal mushrooms. The function and characterization of these fibrinolytic enzymes are described in detail. Technologies for purification and characterization of fibrinolytic enzyme from *Schizophyllum commune* are discussed.

2. Haemostasis and antithrombotic studies

Cardiovascular diseases are the leading cause of death worldwide. A common cause of cardiovascular diseases is abnormal fibrin accumulation in the blood vessels or a fibrin clot adhering to the unbroken vessel walls of the endoepithelium. An abnormal clot formation is called a "thrombus". Thrombosis can stop blood circulation in vessels (arteries or veins), and may cause a hypoxiation syndrome such as acute myocardial infarction, high blood pressure, ischemic heart, and stroke (Mine et al., 2005). In response to the high mortality rates associated with thrombosis, antithrombotic studies and clinical therapies are progressing rapidly.

2.1 Mechanisms of coagulation

The balance of circulation blood in a liquid or clotted state is called haemostasis, which includes two complementary mechanisms: blood coagulation and fibrinolysis. Coagulation limits blood loss from a damaged vessel via clot formation. After the rehabilitation of the vessel's endoepithelium, fibrinolysis system processes dissolve clots and recover circulating blood (Takada et al., 1994). When a blood vessel is damaged by external force (e.g. a cut or scrape), this damage induces the platelet activation and aggregation of clotting plasma proteins. Platelets adhere to the subendothelium and simultaneously activate a coagulation cascade that induces fibrin production (Heemskerk et al., 2002). This coagulation cascade is the outcome of multiple interdependent interactions among plasma proteins (tissue factors), platelets, prothrombin, thrombin, fibrinogen, and fibrin. Via the intrinsic and extrinsic pathways, soluble fibrinogen is converted into insoluble fibrin. Fibrins construct a mesh structure over the platelet plug, sealing the injury site (Gentry, 2004; Norris, 2003; Wolberg, 2007).

2.2 Fibrinolysis

Under normal conditions, fibrin clots formed in blood vessels should be disassembled rapidly and removed by the fibrinolysis system effectively. The key enzyme in the fibrinolysis system is plasmin, a serine protease, which is activated from proenzyme plasminogen via a tissue-type plasminogen activator (tPA) or a urinary plasminogen activator (u-PA) trigger. The fibrinolysis system is regulated by a number of orchestrated interactions between fibrin, specific inhibitors, and plasminogen; plasminogen activators are indeed necessary that generate clot degradation (Medved & Nieuwenhuizen, 2003). Collen (1999) demonstrated that a fibrin formation can trigger activation of fibrinolytic system and generate of active plasmin; the latter substance may then degrade fibrin into soluble fibrin degradation products (FDPs), followed by clot disintegration (Collen, 1999).

2.3 Clinical thrombolytic agents

Pharmacologic dissolution of an established thrombus is now an accepted therapeutic approach for thrombotic occlusive disease. Intravenous infusion of commercial thrombolytic agents - plasminogen activators (PA), including recombinant tPA (r-tPA), u-PA, rokinase, streptokinase, and anisoylated plasminogen streptokinase-activator complex is effective in restoring blood flow in occluded arteries and veins (Liu et al., 2005). However, these agents are expensive and have a number of drawbacks, such as rapid degradation, uncontrollable acceleration of fibrinolysis and haemorrhage. Widespread systemic activation of fibrinolysis leads to potentially life-threatening side effects. To overcome these risks, a safer thrombolytic agent is need for treating thrombolytic processes.

3. Overview of fibrinolytic enzymes

Fibrinolytic enzymes have been found in natural sources. Their activity resembles that of plasmin, which can degrade fibrin and inhibit fibrin clot formation (Chen et al., 1991; Mihara et al., 1991; Sumi et al., 1987). Moreover, some fibrinolytic enzymes exhibit activity similar to that of a PA. These enzymes may have great potential for antithrombotic therapy. Their specific characteristics, such as fibrinolysis, fibrinogenolysis and the proteolytic effect, alter the balance between coagulation and anticoagulation, resulting in wide-ranging therapeutic applications.

3.1 Fibrinolytic enzymes from animals

Earthworms have been used for their antithrombotic effect in East Asia traditional fork medicine for a thousand years. However, their precise physiological and biochemical mechanism remains unclear. In this century, applications of earthworms have been investigated intensively. In 1991, lumbrokinase (LK), a proteolytic enzyme, was first extracted from the *Lumbricus rubellus* by Mihara et al. (Mihara et al., 1991). Studies of LK demonstrated that these enzymes, which have a molecular weight of 25-32 kDa, found in the earthworm's body cavity and digestive organs, perform PA and plasmin activities. Recent studies have shown that LK enzymes can dissolve blood fibrin clots and inhibit platelets activation and aggregation, such that LK enzymes can be administered to treat stroke patients as well those with cardiovascular diseases (Nakajima et al., 1993; Tang et al., 2002). In 2008, absorption and efficacy of earthworm fibrinolytic enzyme d (EFE-d) was enhanced when delivered in water-in-oil (w/o) microemulsions to rats (Cheng et al., 2008). Intestinal absorption experiments for LK enzymes have also showed that these heterologous proteolytic enzymes have potent properties, facilitating their development as anti-thrombotic drugs (Fan et al., 2001).

Many snake venoms, which consist of a multitude of biologically active proteins and peptides, are lethal to humans by adversely altering haemostasis. These biological molecules have been classified as serine proteases, metalloproteinases, C-type lectins, disintegrins and phospholipases. Each may act selectively on different blood coagulation factors, blood cells, and tissues (Clemetson et al., 2007). Venom proteases may involve activation or inactivation of each factor related to coagulation and fibrinolysis. Notably, a thrombin-like function, which stimulates fibrinogen forward clotting processes, and a fibrino(geno)lytic function which can digest fibrin and fibrinogen, also exist in snake venoms. Further study has showed that venom contains two groups of fibrino(geno)lytic enzyme, with molecular masses of approximately 25 and 60 kDa, respectively. Fibrino(geno)lytic enzymes have been isolated in snake venom, including those of *Agkistrodon acutus, A. contortrix, A. rhodostoma, A. halys brevicaudus, A. piscivorus piscivorus, A. piscivorus conami,* and *Crotalus atrox.* The most significant characteristic of the amino acid composition of an enzyme is very high levels of Asx and Glx residues (Hahn et al., 1995). Swenson and Markland (2005) characterized venom fibrino(geno)lytic metalloproteinases and serine proteinases. Two sub-classes of proteinases that have distinct sensibility to enzymatic inhibitors, such as EDTA, a metalloproteinase inhibitor, and phenylmethanesulfonylfluoride (PMSF), a serine proteinase inhibitor. The mechanism of action of venom fibrin(ogen)olytic metalloproteinases and serine proteinases differs, and they target different amino acid sequences in fibrin(ogen). The α-chain and β-chain fibrinogenases can be defined as venom enzymes degrading preferentially (although not exclusively) either the α- or β-chain of fibrinogen, respectively (Swenson & Markland, 2005). Recently, several venom fibrin(ogen)olytic enzymes and genetically recombined venom fibrin(ogen)olytic enzymes have been examined using animal models and a promising result was obtained (Gasmi et al., 1997; Marsh & Fyffe, 1996; Moise & Kashyap, 2008; Toombs, 2001). Based on the high fibrinolytic activity of the fibrinolytic enzyme process in venom, these enzymes are currently examined under examination in preclinical and clinical experiments for their thrombolytic efficacy and haemostatic safety.

3.2 Fibrinolytic enzymes from microbial

Peng et al. (2005) presented an overview of microbial fibrinolytic enzymes. Microbial fibrinolytic enzymes are derivatives from bacteria (e. g. streptomyces, actinomyces, and bacilli), fungi, and algae (Peng et al., 2005). Two well-known plasminogen activators, streptokinase and staphylokinase from *Streptococcus hemolyticus* and *Streptococcus aureus,* were demonstrated to be effective in thrombolytic therapy. In 1997, the streptokinase gene was cloned and its expression in the non-pathogenic *Escherichia coli* was characterized. The recombinant 47.5 kDa protein corrected to native streptokinase has a peptide sequence that was successfully used to treat Thrombus (Avilan et al., 1997). Unlike native streptokinase, the recombinant forms of streptokinase have low antigenicity and high fibrin-selective activity in human circulatory system (Collen & Lijnen, 1994). The recombinant fibrinolytic enzyme produced by recombinant technology is now mass-produced. Additionally, the side effect of the recombinant fibrinolytic enzyme is reduced.

Studies of fermented foods showed that fibrinolytic enzymes may be purified from such sources as Japanese Natto, Korean Chungkook-Jang soy sauce, dochi, fermented shrimp paste, salt-fermented fish, fermented vegetables (e.g., Kimchi), and Indonesia soy products (e.g., Tempeh). Most fermented foods are derived from raw materials such as beans, grains, fish, meat, vegetables, and dairy products. Fermentation is carried out by edible bacteria or fungi (Kim et al., 1996; Sugimoto et al., 2007; Sumi et al., 1987; Sumi et al., 1995; Wong & Mine, 2004). For example, the Bacillus genus contains microbial species that usually are used for fermentation during fermented-foods production. Natto a popular soybean food in Japan, is fermented by the microorganism *Bacillus subtilis* natto. The first commercial fibrinolytic enzyme, Nattokinase (NK), was purified and characterized from Natto. Nattokinases, which were found with fibrin and plasmin substrate H-D-Val-Leu-Lys-pNA (S-2251) digestion activity (Sumi et al., 1987), have been investigated extensively worldwide.

In 1993, NK was characterized as a substilin-like serine protease, base on its high sensitivity to protein substrate Suc-Ala-Ala-Pro-Phe-pNA for substilin (Fujita et al., 1993). In 2005, a novel NK protein (NKCP) with both antithrombotic and fibrinolytic effects was discovered. Dose-dependent prolongations of both prothrombin time (PT) and active partial thromboplastin time (APTT) were observed in rats administered NKCP intraduodenally (Omura et al., 2005). Two fibrinolytic enzymes, QK-1 and QK-2, from supernatant of *B. subtilis* QK02 culture broth were purified and characterized in 2004; QK-1 is a plasmin-like serine protease and QK-2 is a subtilisin family serine protease (Ko et al., 2004). A strong fibrinolytic subtilisin doenjang (DJ)-4 was purified from doenjang, a traditional Korean fermented soybean paste (Kim & Choi, 2000). Fibrinolytic recombinant full-subtilisin DJ-4 (rf-subDJ-4) and mature-subtilisin DJ-4 (rm-subDJ-4) were expressed by a pET29 vector system (Choi et al., 2004). The rf-subDJ-4 had higher heat- and acid-resisting (pH 3.0–4.0) properties than native subtilisin DJ-4. Notably, rf-subDJ-4 has the same abilities as the hydrolyzed α-, β-, and γ-chains of fibrinogen; however, rm-subDJ-4 does not. Sumi et al. (1990) further demonstrated that Natto or NK capsules administered orally enhance fibrinolysis in canine plasma. NK has great potential as a drug candidate for treating and preventing thrombus in animal models (Fujita et al., 1995; Sumi et al., 1990). Moreover, as the fibrinolytic enzyme was purified from common edible sources, oral administration is likely safe.

4. Mushroom fibrinolytic enzymes

The fruiting body of mushrooms can produce and disperse a large number of spores within a short period. Spores may create new individuals when in an environment suitable for growth. Previously, the fruiting body was cultured in growing mediums comprising woodchips, straw, sawdust, coffee grounds, logs, and similar organic items. However, after seeding spores or the mycelium of fungi on artificial solid or liquid culture medium under controlled temperature and moisture, now desired mushroom cultures can be successfully created. The products of mushroom cultivation include hyphae, mycelium, fruiting body, culture broth, and derivatives that dissolve in broth. Many fibrinolytic enzymes were discovered from these products recently. Table 1 lists fibrinolytic enzymes derived from mushrooms.

4.1 Fibrin(ogen)olytic enzyme of mushroom

The proteolytic complex from fungus *Flammulina velutipes* was studied by gel chromatography and the activities of the enzyme complex were compared with those of *Aspergillus terricola* and *Streptomyces griseus* proteinases (Morozova et al., 1982). This was the first study of fibrinolytic enzyme from mushrooms and their application as therapeutic agents. Thereafter, fibrin(ogen)olytic proteases were discovered in the fruiting of *Pleurotus ostreatus, Armillaria mellea, Tricholoma saponaceum,* and *Cordyceps militaris* (Choi & Shin, 1998, Kim & Kim, 1999, 2001; Kim et al., 2006). The mushroom fruiting body can be collected from nature and cultured via a sterilized growth medium made of organic substances for saprophytic utilization. Since air, ground, plants (e.g., trees) and ground water can be pollute, non-polluted substances for growing mediums are becoming rare. Moreover, cultivation of mushroom fruiting body is space occupied and labor-intensive. In the last decade, technology for mushroom submerged path culture has been established in the laboratory and at an industrial level. Mushrooms (fungi) may generate a new generation

Mushroom	Protease	Fibrinogen Degradation	Native Mol. Wt.	Reference
From fruiting body				
Flammulina velutipes	Two protease (no name)	N*	N	(Morozova et al., 1982)
Pleurotus ostreatus	Metalloprotease	+	24 kDa	(Choi & Shin, 1998)
Armillaria mellea	Metalloprotease	+	18.5 kDa	(Kim & Kim, 1999)
Tricholoma saponaceum	TSMEP1	+	18.1 kDa	(Kim & Kim, 2001)
Cordyceps militaris	no name	+	52 kDa	(Kim et al., 2006)
From mycelium				
Ganoderma lucidum	metalloprotease	N	100 kDa	(Choi & Sa, 2000)
Armillaria mellea	AMMP	+	21 kDa	(Lee et al., 2005)
Flammulina velutipes	FVP-1	+	37 kDa	(Park et al., 2007)
Perenniporia fraxinea	metalloprotease	+	42 kDa	(Kim et al., 2008)
From culture broth				
Fomitella fraxinea	FFP1	+	32 kDa	(Lee et al., 2006)
	FFP2	+	42 kDa	
Cordyceps sinensis	CSP	+	31 kDa	(Li et al., 2007)
Fusarium sp. BLB	FP	N	27 kDa	(Ueda et al., 2007)
Schizophyllum commune	no name	+	21.32 kDa	(Lu et al., 2010a)
From recombinant source				
Fusarium sp. BLB	FP	N	28.5 kDa	(Sugimoto et al., 2007)

* N means undetermined

Table 1. Fibrinolytic enzymes derived from mushrooms

when a fresh environment and nutrients are provided. The mycelium may grow stably without pollutants. Choi and Sa (2000) isolated a metalloprotease with fibrinolytic activity from cultured mycelium of *Ganoderma lucidum* (Choi & Sa, 2000). In the following decade, fibrin(ogen)olytic proteases from mushroom mycelium, including metalloprotease from *A. mellea* (AMMP), metalloprotease from *F. velutipes* FVP-1, and metalloprotease from *Perenniporia fraxinea*, were identified (Kim et al., 2008; Lee et al., 2005; Park et al., 2007). Fibrinolytic metalloproteases are sensitive to metalloprotease inhibitors (e.g., EDTA and 1,10-phenanthroline) and metal ions.

Mushrooms grown in medium or submerged broth may differ morphologically. Nutrients and differences between growth mediums and submerged broth contribute to the particular metabolite of mushrooms. Fukushima et al. (1991) reported that when soy sauce oil was the carbon source for *Asperyillus oryzae*, protease secretion was increased significantly during submerged cultivation. Furthermore, specific protease production was stimulated selectively by the oils (Fukushima et al., 1991). Changing culture medium parameters, such as osmotic pressure, salt concentrations, protein content, and the carbon source, markedly alter the extracellular performance of fungi mycelium (Archer et al., 1995; Archer & Peberdy, 1997; Bobowicz-Lassociska & Grajek, 1995; Kadimaliev et al., 2008). Broth for submerged cultivation is rich in mushroom derivatives. Over the last 50 years, mushroom derivatives with biological activities for humans have been investigated extensively. Several extracellular fibrinolytic enzymes have been purified from submerged broth of *Fomitella fraxinea* (i.e., FFP1 and FFP2),

Cordyceps sinensis (i.e., CSP) and *S. commune* (Lee et al., 2006; Li et al., 2007; Lu et al., 2010a; Ueda et al., 2007). Previous studies have demonstrated that the fibrinolytic enzymes may be available in the fruiting bodies of some mushrooms but not in the cultured mycelium (Kim et al., 2008). The purification stratagem for these extracellular fibrinolytic enzymes differs from those of fruiting bodies or mycelium. Fungi derivatives in submerged broth are biological molecules considered important for human health (Papagianni, 2004).

Ueda et al. (2007) first purified a fibrinolytic protease from *Fusarium* sp. BLB (FP), and identified the N-terminal amino acid (Ueda et al., 2007). Sugimoto et al. (2007) cloned an FP gene encoding a novel protease derived from *Fusarium* sp. BLB. The hydrolytic activity of FP toward synthetic peptide subtracts is higher than that of proteases from *Bacillus subtilis* natto, *Aspergillus oryzae*, *Streptomyces griseus* and commercial plasmin (Sugimoto et al., 2007). This development of FP from *Fusarium* sp. BLB may demonstrate that manufacturing mushroom fibrinolytic enzymes may be possible in the future. Extracellular secreted protease of fungi is in some cases the fungi metabolite that may reflect cultivation conditions. Nutrition and the physical environment may be the dominant factor for growth of fungi and the rate at which extracellular proteases are produced in a submerged culture system. Base on the critical influence that produce mushroom extracellular secreted protease in submerged culture system, we expect that the high production efficiency of mushroom fibrin(ogen)olytic enzymes can be achieved by well-designed cultivation processes.

4.2 Characteristics of medicinal mushroom fibrin(ogen)olytic enzymes

Fibrin(ogen)olytic enzymes including those from *P. ostreatus*, *A. mellea*, *T. saponaceum*, *C. militaris*, *G. lucidum*, *P. fraxinea*, *F. fraxinea*, *C. sinensis*, *F. velutipes*, *Fusarium* sp. BLB, and *S. commune*, have been identified. Table 2 lists their biochemical properties, including molecular weight, optimal pH, thermal stability, inhibitors, and substrate specificity. The N-terminal sequences of most fibrinolytic enzymes have been determined. These fibrin(ogen)olytic enzymes, except for metalloprotease derived from *A. mella* fruiting body and mycelia, have markedly different N-terminal sequences. The optimal pH for these mushroom fibrin(ogen)olytic enzymes is 5-10; the optimal temperature is 20-60°C. An overview of microbial fibrinolytic enzymes (Table 2) showed that mushroom fibrinolytic enzymes may be classified as serine protease (i.e., inhibited by serine protease inhibitors) and metalloprotease (i.e., inhibited by metalloprotease inhibitors) according to protease inhibitor specificity.

The protease activity of mushroom serine fibrinolytic enzymes can be irreversibly inhibited by PMSF but no other protease inhibitors. Previous studies have showed that most fibrinolytic serine protease from traditional fermented foods belong to subtilisin of Bacillus origin such as nattokinase, subtilisin DFE and subtilisin QK-1 (Peng et al., 2005). Via study of N-terminal sequence alignment; a chymotrypsin-like serine protease from mushroom *C. militaris* had high sequence identify with subtilisin PR1J (GeneBank, CAC95048) (Kim et al., 2006); FFP1 from *F. fraxinea* also has 20% identify with Nattokinase and subtilisin E (Lee et al., 2006). Serine fibrinolytic enzymes from mushrooms of *F. fraxinea* (i.e., FFP1), *Fusarium* sp. BLB (i.e., FP), and *C. sinensis* (i.e., CSP) were found to have a broad substrate specificity for synthetic substrates, which including fibrin, fibrinogen, casein, substrates for subtilisin and substrates for plasmin. In laboratory works, fibrinolytic protease from *Fusarium* sp. BLB had higher fibrin degradation and plasminogen activation than Nattokinase (Sugimoto et al., 2007).

Mushroom Protease	Optimal pH and temperature	Ion induced	Ion inhibit	Fibrinogen degrade	Protein inhibitor	N-Terminal sequence	Reference
Serine protease type-							
Cordyceps militaris (52 kDa)	pH 7.4, 37°C	Ca, Mg	Cu, Co	YES	PMSF, PMASF	Yes	(Kim et al., 2006)
Fomitella fraxinea (32 kDa, FFP1)	pH 10, 40°C	N*	N	Yes	PMSF, aprotinin	Yes	(Lee et al., 2006)
Cordyceps sinensis (31 kD, CSP)	pH 7, 40°C	N	Mn, Cu, Hg	Yes	PMSF	Yes	(Li et al., 2007)
Fusarium sp. BLB (27 kDa, FP)	pH 9.5, 50°C	N	N	N	DFP, PMSF	Yes	(Ueda et al., 2007)
Fusarium sp. BLB (28.5 kDa, FP)	N	N	N	N	N	Yes	(Sugimoto et al., 2007)
Metalloprotease type--							
Flammulina velutipes (metalloprotease)	N	N	N	N	EDTA	N	(Morozova et al., 1982)
Pleurotus ostreatus (24 kDa, metalloprotease)	pH 7.5~8.0, up to 50°C	Zn, Co	N	Yes	1,10-phenanthroline	N	(Choi & Shin, 1998)
Armillaria mellea (18.5 kDa, metalloprotease)	pH 7, 45~55°C	Zn, Co, Mg	Hg	Yes	1,10-phenanthroline, EDTA	Yes	(Kim & Kim, 1999)
Ganoderma lucidum (100 kDa, metalloprotease)	pH 7.0~7.5, up to 60°C	Zn, Co	N	Yes (assume)	1,10-phenanthroline, EDTA	N	(Choi & Sa, 2000)
Tricholoma saponaceum (18.1 kDa, TSMEP1)	pH 7.5~9, 30~40°C	Zn, Co, Mg,Fe	Hg, Cu	Yes	1,10-phenanthroline, EDTA	Yes	(Kim & Kim, 2001)
Armillaria mellea (21 kDa, AMMP)	pH 6, 33°C	Ca, Mg	Cu, Co	Yes	EDTA	Yes	(Lee et al., 2005)
Fomitella fraxinea (42 kDa, FFP2)	pH 5, 40°C	Zn, Co	Cu, Ni	Yes	1,10-phenanthroline, EDTA	Yes	(Lee et al., 2006)
Flammulina velutipes (37 kDa, FVP-1)	pH 6, 20~30°C	Mn,Mg	Cu, Fe2+, Fe3+	Yes	EDTA, EGTA	Yes	(Park et al. 2007)
Perenniporia fraxinea (42 kDa, metalloprotease)	pH 6, 35~40°C	Mn, Mg	Cu,Fe,Zn	Yes	EDTA	Yes	(Kim et al., 2008)
Schizophyllum commune (21.3 kDa, metalloprotease)	pH 5, 45°C	Mg	Hg, Cu, Co	Yes	EDTA	Yes	(Lu et al., 2010)

*N means undetermined.

Table 2. Review of medicinal mushroom fibrin(ogen)olytic enzymes

Mushroom fibrinolytic metalloproteases were discovered in *F. velutipes*, *P. ostreatus*, *A. mellea*, *G. lucidum*, *T. saponaceum*, *F. fraxinea*, *P. fraxinea* and *S. commune* (Choi & Shin, 1998; Kim & Kim, 1999, 2001; Kim et al., 2008; Lee et al., 2005, Lee et al., 2006; Lu et al., 2010; Morozova et al., 1982; Park et al., 2007). Proteases were purified in the fruiting body, or mycelium, or culture supernatant. The presence of Zn^{2+} was detected in metalloprotease from *P. ostreatus* and *G. lucidum* by mass spectrometry; both of these metalloproteases have Zn^{2+}-dependent protease activity. Additionally, Zn^{2+}-dependent protease activity exists in metalloprotease derived from *F. fraxinea* (FFP2), *A. mellea* and *T. saponaceum* (TSMEP1), with an undefined ion structure. Mg^{2+}-dependent protease activity was found in metalloproteases from *F. velutipes* (FVP-1), *A. mellea* (AMMP), *P. fraxinea* and *S. commune*. The activity of mushroom fibrinolytic metalloproteases can be inhibited by EDTA and 1,10-phenanthroline predominantly . The metal ions dependent activity is not only a character of mushroom fibrin(ogen)olytic enzymes but also can be considered as a critical point for proteases under the clinical application for thrombolytic therapy (Lu & Chen, 2010).

5. Production, purification and characterization of mushroom fibrinolytic enzymes

Streptokinase (SK) from streptococci is a widely used therapeutic agent for acute myocardial infarction. However, the manufacturing capacity of SK from haemolytic streptococci is limited with a high price tag. Pharmacological uses of SK take risks of causing potential myocardium and liver damages due to the residual bacteriohemolysin in the manufacturing process (Zhang et al., 1999). Peng et al. (2005) illustrated that microbial fibrinolytic enzymes, especially those from food-grade microorganisms, have potential to be developed as functional food additives and drugs to prevent or cure thrombotic disease (Peng et al., 2005). Therefore, fibrin(ogen)olytic enzymes from non-toxic mushrooms has gradually become the centre of attention for investigators in thrombolytic therapy. Mushroom of *S. commune* is a ubiquitous white rot and widespread fungus in existence. This medicinal mushroom has been an additive in traditional folk medicine and the subject of genetic analysis. The production, purification and characterization of an fibrin(ogen)olytic protease from *S. commune* were discussed in this section (Lu et al., 2010a; Lu et al., 2010b; Lu & Chen, 2010). The fast growing and easy cultivation raise the universal uses for *S. commune* in science researches and medicinal applications.

5.1 Submerged cultivation of *S. commune*

Schizophyllum commune derivatives have been suspected with antithrombotic effect for human beings by scientists (Okamura-Matsui et al., 2001), yet the detail mechanism is unclear. Base on the high protease and extracellular biological substance production ratio in submerged cultivation (Desrochers et al., 1981), *S. commune* is a good biological resource for scientists whom study the functional substance. In our study, *S. commune* was cultured in YM agar plate (contains peptone, malt extract, dextrose, yeast extract and 0.2% agar) at 25°C for one week. The mycelia grew and covered plate surface, was like white flannelette. Surface of plate agar was pure colour and without contamination. As a seed culture, a piece of plate agar covered with grown mycelia, was transferred to 50ml YM broth for 5 days at 25°C. For mass production, the seed culture then transferred to 15L YM broth for additional

7 days with shaking platform and air exchangeable cover. Colonies of fungi in submerged cultivation grow exponentially, that is different from yeast cells grow at a constant rate. The growth rate increases with time so that the logarithm of the amount of fungus mycelium increases with time.

5.2 Purification of fibrinolytic enzyme from *S. commune*

Mycelia in submerged culture broth were removed by centrifugation at 4°C, 8,200 ×g, for 30min. The supernatant was filtered by 0.45μm membrane and fractionized by cross-flow filtration, to collect the <100-kDa fraction, which was further concentrated by 3-kDa ceramic column filtration. Due to the extracellular polysaccharide with huge molecular weight present in broth, that will obstruct the protein purification. The ceramic cross-flow filtration is done before protein precipitation. Protease purification was then performed by fast performance liquid chromatography. The protein concentration of each eluted fraction was detected with A280 in real time by ÄKTA purifier 10 (GE Healthcare). Measuring activity to azocasein is a general selected detection method to identify target protease fractions. Combination of liquid chromatography and protease activity screening, the purified fraction containing target protease is available.

The concentrated fraction (molecular weight 3 -100 kDa) was precipitated by saturated ammonium sulfate at 4°C for 8 hour followed by centrifugation at 10,000×g, 30min at 4°C. The protein pellet was dissolved in ddH_2O and followed by dialysis with FPLC column buffer containing 50mM sodium phosphate (pH 8.0) and 1M $(NH_4)_2SO_4$ as 1st dialysis substrate. In step 1, a hydrophobic interaction column, Phenyl Sepharose™ High Performance beaded packing column was preequilibrated with column buffer. Equilibrated 1st dialysis substrate was loaded on column at flow rate of 1 ml/min. Figure 1A illustrated the result of elution processes with column buffer without $(NH_4)_2SO_4$ in a stepwise manner of 60 min interval at flow rate of 1 ml/min. Eluted fractions were analyzed for protease activity to azocasein and recorded (Fig. 1A). The active fractions were pooled; protein precipitated and dialyzed against 20mM Tris (pH 8.0) as 2nd dialysis substrate for next step of chromatography.

In step 2, an ion exchange column, Mono Q™ 5/50 GL column was preequilibrated with 20mM Tris buffer (pH 8.0). The 2nd dialysis substrate was loaded on Mono Q™ 5/50 GL column at flow rate of 1 ml/min. Elution was carried out with the same buffer but containing 1M NaCl, by linear gradient of 20-fold column volume to 40% 1M NaCl. After examination of protease activity to azocasein (see in Fig. 1B), the active fractions were pooled; protein precipitated and dialyzed against 50mM sodium phosphate buffer (pH 8.0) as 3rd dialysis substrate for next step of chromatography.

In step 3, a size exclusion column, Superdex 75 10/300 GL column was preequilibrated with 50mM sodium phosphate buffer. The 3rd dialysis substrate was loaded with same buffer at flow rate of 1.2 ml/min. Eluted fractions were analyzed for protease activity to azocasein (see in Fig. 1C). Figure 2 showed the purification purity by SDS-PAGE (Fig. 2). The purified protease then can be applied for more.

Fibrinolytic enzyme purification from mushroom may divide into two processes. One is the pre-treatment of resource, including the crude protein extraction from fruiting body, or mycelium, or culture supernatant. Another is the chromatography stratagem, which including the combination of ion exchange, hydrophobic interaction and size exclusion chromatography. Techniques with selectivity are highly independent of protease resources and the enzyme properties.

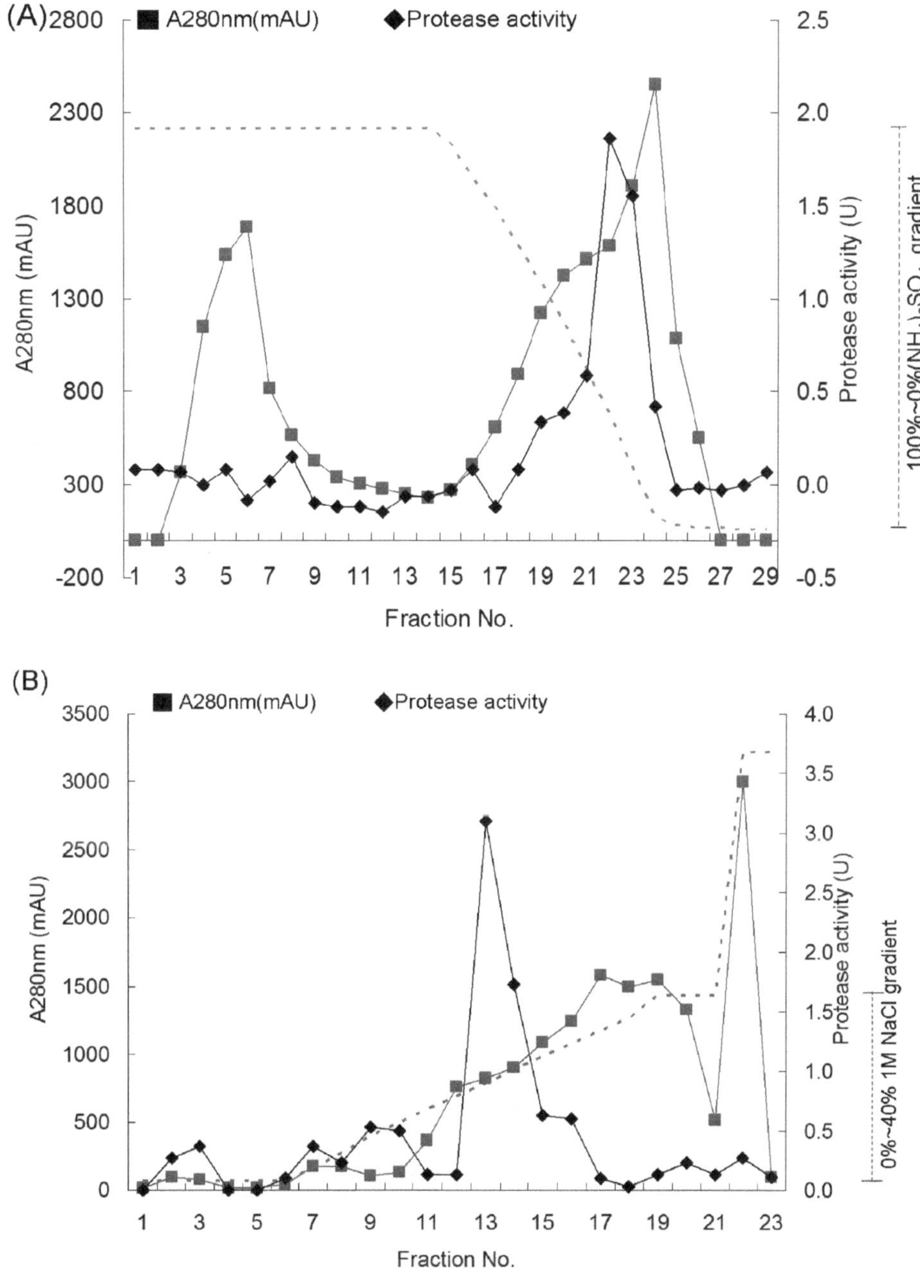

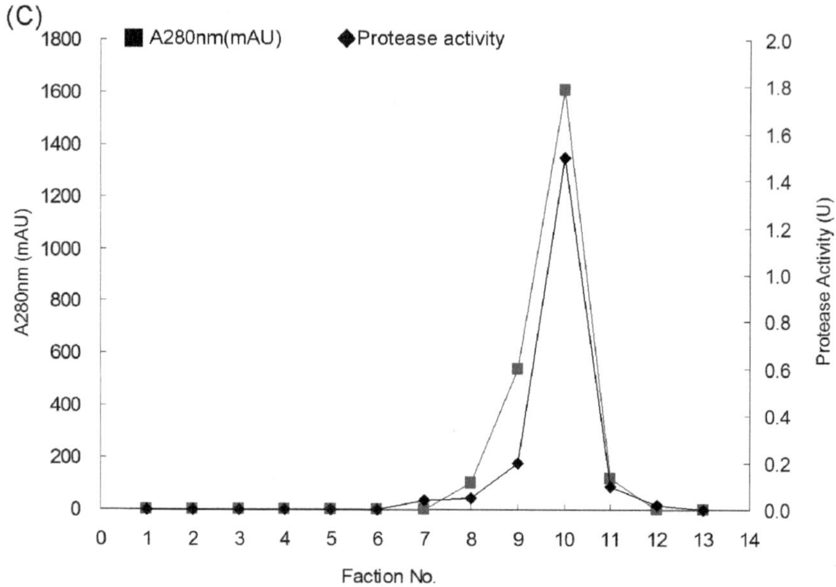

Fig. 1. Chromatography purification of *S. commune* fibrinolytic enzyme (A) Eluted fractions of Phenyl Sepharose™ High Performance column; (B) Eluted fractions of Mono Q column; (C) Eluted fractions of Superdex 75 10/300 GL column.

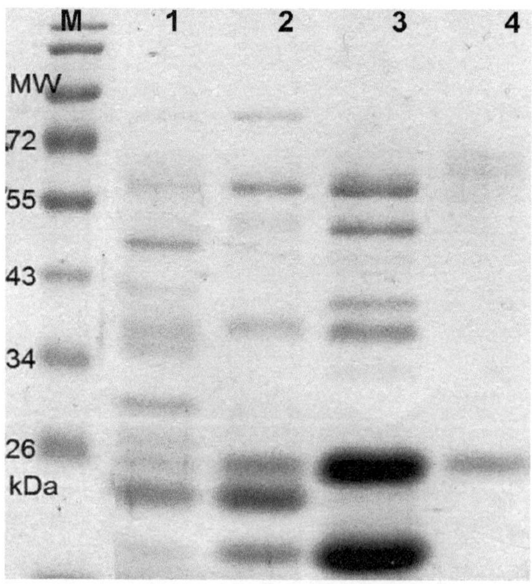

Fig. 2. SDS-PAGE result of chromatographic purification M: protein markers; Lane 1: 1st dialysis substrate; Lane 2: 2nd dialysis substrate; Lane 3: 3rd dialysis substrate; Lane 4: Protein with highest protease activity eluted from Superdex 75 10/300 GL column.

5.3 Characterization of fibrinolytic enzyme from *S. commune*

To study the effects of pH and temperature to fibrinolytic enzyme from *S. commune*, purified enzyme was incubated in various pH and temperature respectively, and the protease activity to azocasein then examined. Optimal protease activity reveal at pH 5.0 and 45°C, the data was illustrated in Fig. 3A and 3B.

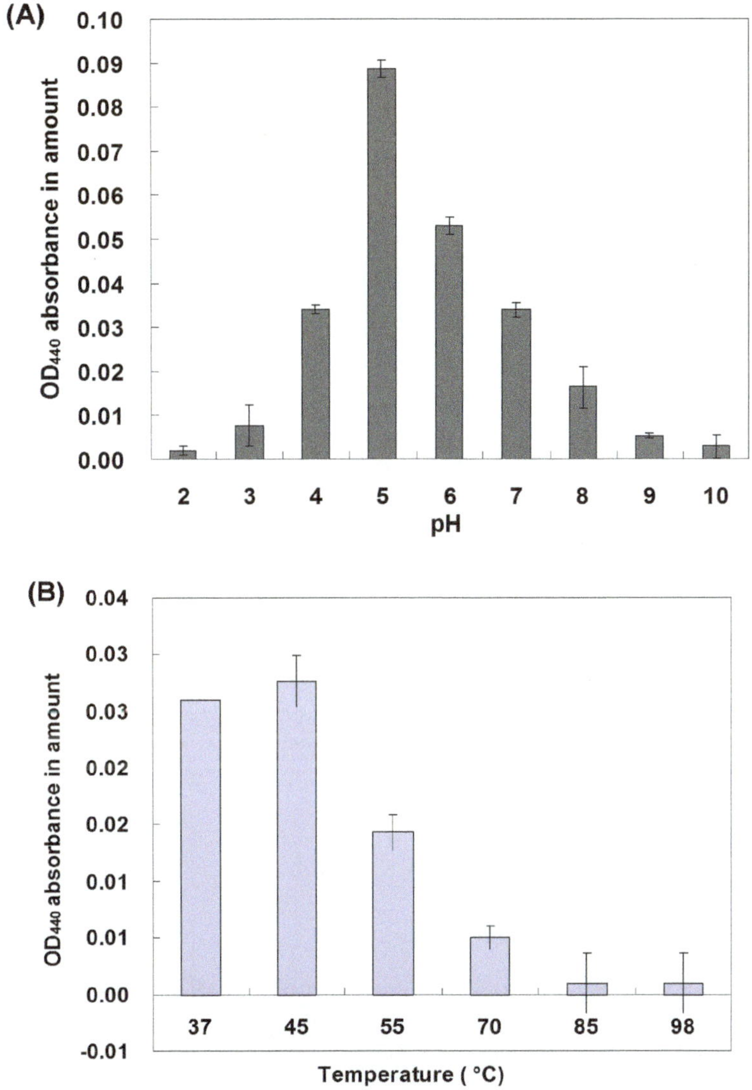

Fig. 3. The effects of pH and temperature on protease activity of fibrinolytic enzyme from *S. commune* (A) pH effects; (B) Temperature effects.

Three replicates of fibrinolytic enzyme were mixed with $HgCl_2$, $C_4H_6O_4Zn \cdot 2H_2O$, $CuSO_4$, $MgCl_2$, $CoCl_2$, $CaCl_2$, $Pb(NO_3)_2$, ddH_2O (control) and protease inhibitors of PMSF (serine protease inhibitor), EDTA (metalloprotease inhibitor), benzamidine hydrochloride hydrate (trypsin, trypsin-like protease inhibitor), pepstatin A (aspartyl peptidases inhibitor), aprotinin (serine protease) and phosphoramidon (endopeptidase inhibitor) respectively. Protease activity of mixed solution was then examined by azocasein assay. Figure 4 demonstrated the relative protease activity to the control treatment (100%=1) (Fig. 4A & 4B). The result indicated that the protease shows Mg^{2+}-dependent activity, and inhibits by EDTA.

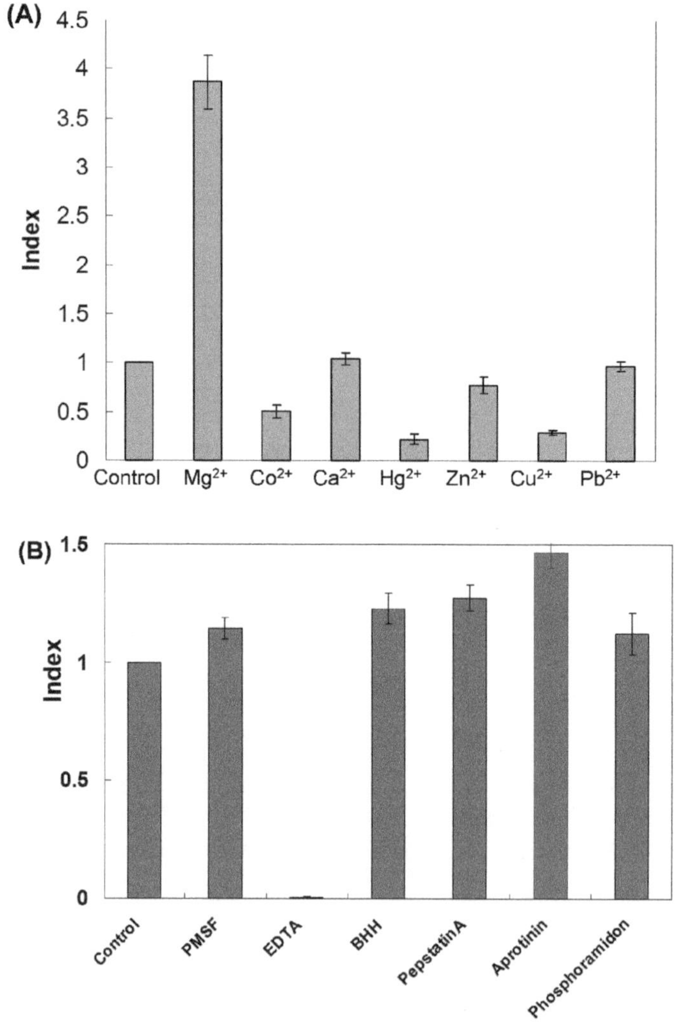

Fig. 4. The effects of divalent cations and protein inhibitors on protease activity of fibrinolytic enzyme from *S. commune* (A) Divalent cations effects; (B) Protein inhibitors effects.

N-terminal sequence of protease is another character for protein identification. However, fibrinolytic enzymes from microorganisms show low homology except the fibrinolytic enzyme from substilisin group, such as substilisin NAT (Nakamura et al., 1992) , substilisin E (Wong et al., 1984), substilisin DFE (Peng et al., 2003). Table 3 illustrated the results of N-terminal sequence comparison; it demonstrated the distinctive feature of the fibrinolytic enzyme from *S. commune* to other mushroom.

Enzyme[1)]	N-terminal amino acid sequence								
NK & substilisin DFE	A	Q	S	V	P	Y	G	I	S
GFMEP		T	Y	N	G	C	S	S	S
POMEP	A	T	F	V	G	C	S	A	
AMMEP	X[2)]	X	Y	N	G	X	T	X	
TSMEP	A	L	Y	V	G	X	S	P	
Fibrinolytic enzyme from *S. commune*	A	S	Y	N	G	X	S	S	

(1) GFMEP, POMEP, AMMEP and TSMEP are metalloendopeptidases from *Grifola frondosa, Pleurotus ostreatu, Armillariella mellea* and *Tricholoma saponaceum*. (2) X means amino acid undetermined.

Table 3. Comparison of N-terminal amino acid sequence of mushroom fibrinoytic enzyme

6. Antithrombotic effect of mushroom fibrinolytic enzymes

Current clinical thrombolytic agents are used to convert plasminogen to active enzyme, plasmin, which degrades fibrin and process antithrombotic effect. It's an effective way to decompose harmful thrombus in circulation system, but the side effect, the excess stimulation of plasmin that may cause haemorrhage within patients. Newly thrombolytic agents now are developed for fibrin-specific property, acting on the surface of thrombus that avoids excessive induction of systemic fibrinolytic system. Fibrinolytic enzymes from medicinal mushroom are novel proteases with fibrin and fibrinogen degradation activity, which directly break the clot and interfere the clotting system. In order to identify the clinical functions and risks of medicinal mushroom fibrinolytic enzymes; mushroom fibrinolytic enzymes were examined by various antithrombotic examinations, such as fibrinolytic assay, fibrinogenolytic assay, plasminogen activation assay, and the dynamic tracking of blood clot formation.

6.1 Fibrinolytic activity assay and clot degradation test

Fibrinolytic activity was determined by artificial fibrin plate assay, which synthesized by the method described by Astrup and Mullertz (Astrup & Mullertz, 1952). Fibrin plate was made at room temperature in a petri dish containing 1.5% agarose, 0.2% human fibrinogen, and 10U thrombin. Purified protease were loaded on a fibrin plate, and incubated for 24 hours. Plasmin from human plasma (3U/mg protein) was applied as a positive control. The result showed that the purified *S. commune* fibrinolytic enzyme display stronger fibrinolytic activity than commercial plasmin (Fig. 5).

Rat blood was withdrawn without anticoagulant and stayed for clot formation in a tube. A piece of clot was placed on Petri dish, followed by dropping 0.2μg and 0.5μg fibrinolytic enzyme on clot disks and incubated in 20°C for 8 hours to observe the clot degradation effects. Phosphate buffer was dropped on clot disk as the control treatment and another clot disk as the blank without any treatment. Figure 6 demonstrated that *S. commune* fibrinolytic enzyme digests the blood clot dominantly (Fig.6).

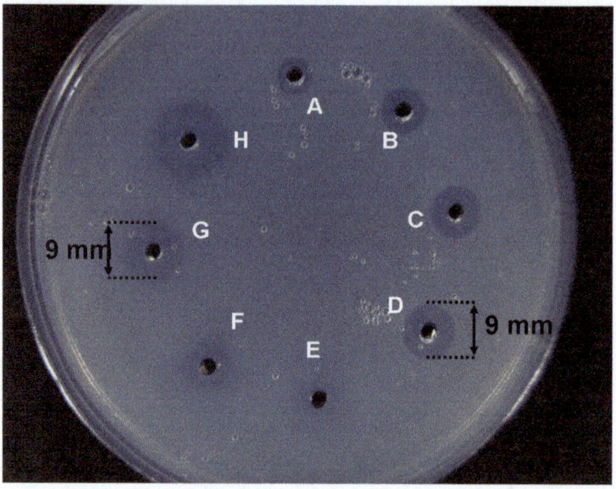

Fig. 5. Fibrinolytic activity of the fibrinolytic enzyme from *S. commune* Fibrin digestion zones A, B, C, and D: 0.2, 0.5, 0.8, and 1.0μg of human plasmin, respectively. Fibrin digestion zones E, F, G, and H: 0.1, 0.25, 0.5, and 1μg of fibrinolytic enzyme *S. commune*, respectively. Treatment of 1.0μg human plasmin (zones D) showed the equal digestion effects to 0.5μg fibrinolytic enzyme from *S. commune* (zone G).

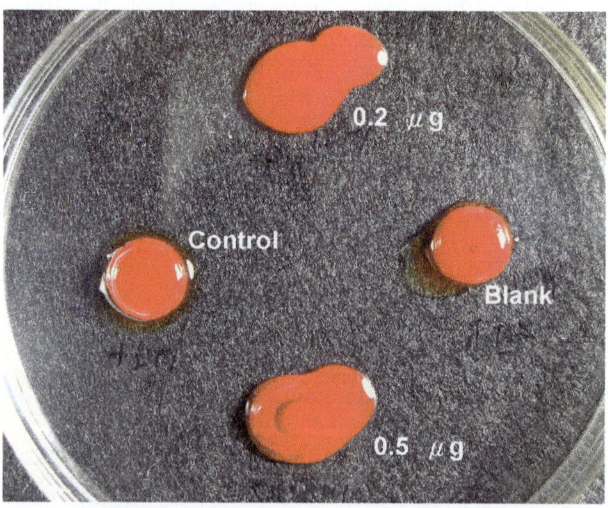

Fig. 6. Clot degradation assay of fibrinolytic enzyme from *S. commune* 0.2μg and 0.5μg: blood clot disks treated with 0.2μg and 0.5μg fibrinolytic enzyme; Control: clot disk treated with PBS; Blank: blank blood clot disk. Treatment of 0.2μg and 0.5μg fibrinolytic enzyme digested the clot disks, but opposite results were observed in control and blank.

6.2 Fibrinogenolytic activity assay

A fibrinogenolytic degradation assay was performed using a 1% fibrinogen solution mixed with the protease from *S. commune* at 37°C. At different time intervals, the reaction mixture was removed and SDS-PAGE electrophoresis was performed. Fibrinogen is composed of peptides α-, β- and γ-chains, and shows three bands on the SDS-PAGE. The fibrinogenolytic activity assay was carried out by the incubation of fibrinolytic enzyme with fibrinogen. Significant degradation of the fibrinogen α-chain and β-chain occurred within 0.5 hour after the reaction. Degradation of the γ-chain occurred after 6 hours of incubation. Most of the γ-chain was digested after 22 hours of incubation. Afterward, the fibrinolytic enzyme completely digested all of peptide chains in 30 hours. Fibrinolytic enzyme from *S. commune* displayed a higher activity in digesting the α-chain and β-chain, and it was less efficient in digesting the γ-chain (Fig. 7).

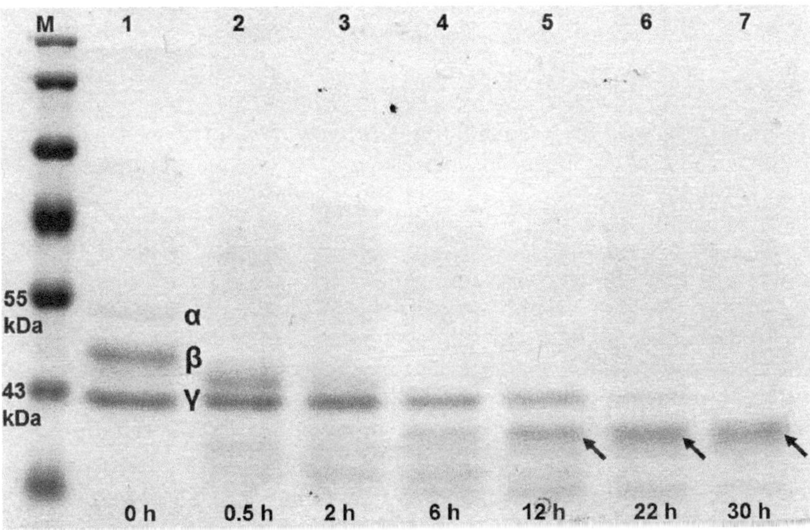

Fig. 7. Fibrinogenolytic assay of the fibrinolytic enzyme from *S. commune* α, β and γ means : α, β, and γ-chains of fibrinogen. M: Protein marker. Lane 1~7: Time course sampling in reaction. Black arrows: Fibrinogen degradation products.

6.3 Plasminogen activation assay

Plassminogen activation assay was performed by incubating 12.5μg purified enzyme with plasminogen (0.1UN) and a 10mM plasmin-specific substrate S2251 at 37°C. Substrate S2251 can be cleaved by plasmin and generated measurable p-nitroaniline (p-NA). In this assay, human plasmin and urokinase were used as positive control for S2251 digestion and plasminogen activation. In this assay, fibrinolytic enzyme from *S. commune* was mixed with plasminogen, plasmin substrate S2251and without fibrin. As the results in Fig. 8, the presence of urokinase, which is known to activate plasminogen, resulted in the digestion of S2251 and an increase in absorbance. The purified fibrinolytic enzyme from *S. commune* failed to digest the S2251 directly and without plasminogen activation activity.

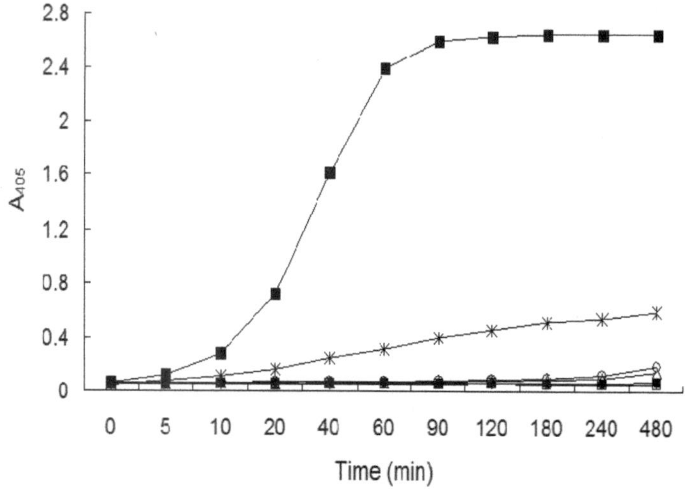

Fig. 8. Plasminogen activation assay of fibrinolytic enzyme from *S. commune* ○ : 1 μg purified enzyme and plasminogen at 0.0001 UN. ■ : 1 μg urokinase and plasminogen at 0.0001 UN. □ : 1 μg purified protease. ∗ : 1 μg human plasmin (≧3 units/mg protein). Δ : plasminogen at 0.0001 UN. • : 20 μg urokinase.

In order to test the plasminogen activation while fibrin present, fibrinolytic enzyme from *S. commune* was applied on a plasminogen-rich fibrin plate containing 0.1UN plasminogen. The protease-digested zone in the plasminogen-rich fibrin plate was then compared to that of the plasminogen-free fibrin plate. In the presence of fibrin, digest representation were equal on the plasminogen-rich and plasminogen-free plates (data not shown). Fibrinolytic enzyme from *S. commune* is not a serine protease for S2251 and does not activate plasminogen *in vitro*, regardless of whether fibrin exists or not.

6.4 Coagulation effects of protease from *S. commune*

In previous study, *G. lucidum* fibrinolytic enzyme was examined in human plasma by activated partial thromboplastin time (APTT) and thrombin time (TT). The anticoagulant activity was performed by inhibition of thrombin and hydrolyzation of fibrin and fibrinogen (Choi & Sa, 2000). In our research, *S. commune* fibrinolytic enzyme applied in rat citrated blood, and the coagulation processes were monitored by thromboelastrography (TEG®) analysis. Four TEG parameters, including the reaction time for clot initiation (R), the time to reach a 20-mm level of clot formation (K), the slope angle from R to K (α value), and the maximum vertical amplitude of the developed clot (MA) were measured. Fibrinolytic enzyme from *S. commune* was mixed with $CaCl_2$ and citrated rat blood to initiate recalcification. Fibrinolysis after MA was determined by measuring the loss of clot strength. Urokinase was used as positive treatment to activate plasminogen and resulted in fibrinolysis after reaching of MA (Fig. 9).

The result of TEG reaction time (R) indicated that the protease did not affect the clotting aggregation by the platelets, since clotting initiation completed normally. Fibrinolytic enzyme of *S. commune* suppressed blood coagulation without excessive fibrinolysis. Clotting time prolongation, clotting velocity depresses and significant decrease of clotting strength were occurred (Table 4). Coagulation suppression effect here we predicate may resulted from fibrinolysis and platelet-medicated clot retraction.

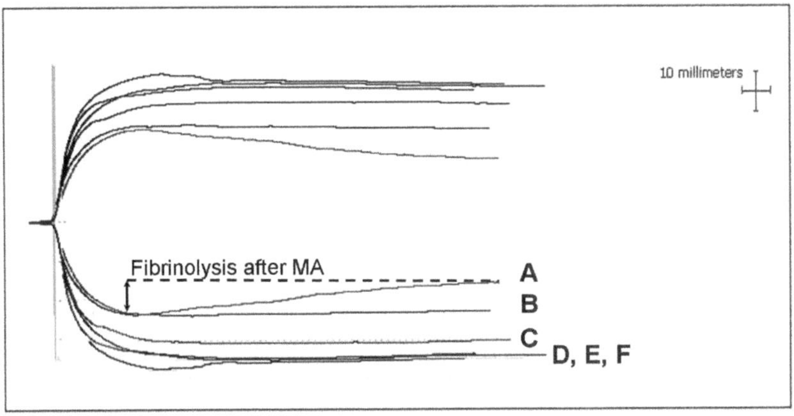

Fig. 9. Thromboelastography tracings of citrated blood added fibrinolytic enzyme from *S. commune* and human urokinase Curve A: Human urokinase 0.5μg added. Curve B, C, D, E and F: Purified protease from *S. commune* 1.5, 1.2, 0.6, 0.3, 0 (control)μg added respectively.

TEG parameters	Range of values	Amount of fibrinolytic enzyme in TEG				
		0 μg	0.3 μg	0.6 μg	1.2 μg	1.5 μg
Reaction time (R)	2.2-4.7 min	100[a]	92.07	86.43	88.14	81.21
Coagulation time (K)	0.8-3.3 min	100	96.67	97.96	127.22	165.17[b]
Velocity of clot (α)	50.2-81.1°	100	99.69	101.44	98.76	92.74[b]
Maximum amplitude (MA)	75.8-24.5 mm	100	94.14	90.00	73.51[b]	59.07[b]

[a] TEG Parameters with the addition of protease are presented as the relative index obtained by normalized to the values in control sample (0μg) which are set to 100%. (n=8).
[b] Statistically significant compared to control samples, $P<0.05$

Table 4. Effects of various concentrations of fibrinolytic enzyme from *S. commune* to citrated blood by thromboelastography analysis

In an experiment of coexist with Mg^{2+} ion, citrated blood treated by *S. commune* fibrinolytic enzyme and Mg^{2+} ion. The supplementation of Mg^{2+} ion stimulated the depression of blood clot amplitude (Fig. 10). Magnesium therapy in coronary heart disease has recently been proposed and documented in clinical trials (Shechter et al., 1999; Whiss & Andersson, 2002). Additionally, magnesium ion stimulate activity of fibrinolytic enzyme from *S. commune* was proved in study. Thus, a regulatory manipulation of *S. commune* fibrinolytic enzyme activity by magnesium supplementation can be expected in the future. The cost reducing effect by magnesium supplementation is also an innovative ideal for the currently used fibrin-specific antithrombotic agents.

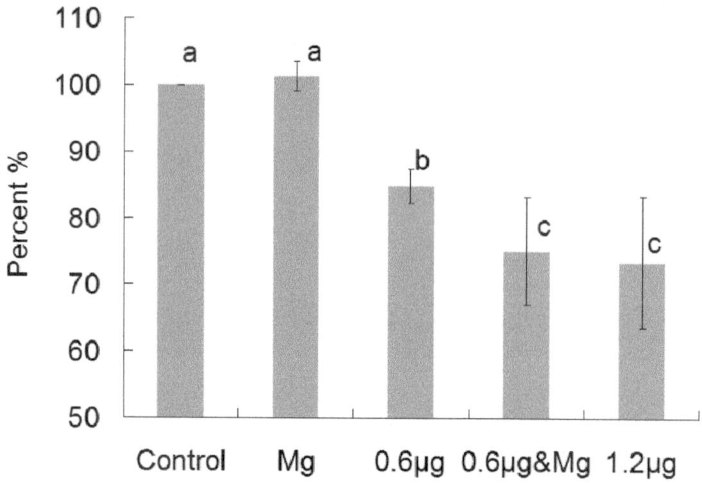

Fig. 10. Blood clot amplitude analysis of fibrinolytic enzyme coexist to Mg^{2+} Control: citrated blood mixed with PBS; Mg: citrated blood mixed with $MgCl_2$ (2.5mM); 0.6 / 1.2μg: citrated blood mixed with 0.6 / 1.2μg fibrinolytic enzyme; 0.6μg&Mg: citrated blood mixed with 0.6μg fibrinolytic enzyme and 2.5mM $MgCl_2$. Alphabetical lowercase means significant variation between other treatments, a difference was considered statistically significant at $p < 0.05$. TEG values were expressed relative to the control sample (at 100%).

7. Conclusion

In recent decades, pharmacologic intervention of an established thrombus has become an ideal therapeutic approach for thrombotic occlusive disease. Clinical application of several plasminogen activators results in activation of circulating plasminogen, which successfully digest abnormal blood clot occluded in arteries or vein; however, risk accompanied with this treatment option may include a life-threatening haemorrhage caused by the systemic activation of fibrinolytic mechanism. While scientists and clinicians are in the search for a better, safer therapeutic agent with fewer side effects, the discovery of the fibrinolytic enzymes from medicinal mushroom may shed the light for the modern thrombolytic therapies.

So far, the methods for purification and characterization of medicinal mushroom fibrinolytic enzyme were carried out by mushroom cultivation, crude protein extraction, protein chromatography, fibrin(ogen)olytic identification and characterization. Purification and characterization of fibrinolytic enzyme from *S. commune* were illustrated in this chapter, and might be a representative model for purification of medicinal mushroom fibrinolytic enzyme.

Fibrin(ogen)olytic enzymes were discovered from mushrooms, such as *P. ostreatus, A. mellea, T. saponaceum, C. militaris, G. lucidum, P. fraxinea, F. fraxinea, C. sinensis, F. velutipes, Fusarium* sp. BLB and *S. commune*. Most of them have been universal dietetic additives and traditional fork medicine for a long time. The low antigenicity and allergic affections of these mushroom fibrinolytic enzymes when human taken orally may be expected definitely. In previous studies, many mushroom fibrinolytic enzymes were investigated which perform superior fibrinolytic activity than known thrombolytic agents and inhibit thrombus formation. As fibrinolytic effect is not the only solution for thrombus occurrence and decomposition, the therapeutic effect of mushroom fibrinolytic enzyme were further examined in citrated blood on thromboelastography, which monitored the clotting and fibrinolysis action in blood. Fibrinolytic enzymes interfere in antithrombotic interaction may occur amongst plasma proteins, such as platelets, prothrombin, thrombin, plasminogen, plasmin, fibrinogen, and fibrin. As results have shown, using *S. commune* as the model, mushroom fibrinolytic enzyme suppressed blood coagulation without excessive fibrinolysis by prolonging clotting time and decreasing clotting velocity. Consequently, the result is a significant decrease in clot strength. Most notably, the initiation of clotting system and plasminogen activation occurrences were maintained at a normal level. Mushroom fibrinolytic enzyme did not offset haemostasis, but rather reduced the likelihood of thrombus formation without increasing the risk of haemorrhage. These data indicated that the safer antithrombotic effect of mushroom fibrinolytic enzyme and the board application may use as thrombolytic agents clinically.

Metal ion therapy in coronary heart disease has recently been proposed. Although the therapeutic results are still ambivalent, the researches and therapeutic development are never finished to search a more effective therapy for such coronary patients. Review of characters from medicinal mushroom fibrinolytic enzymes, these enzymes revealed significant metal ion dependent activity. The inhibition or the stimulation of enzyme activity by metal ion may provide a pharmaceutical conservation and a critical control manipulation in clinical uses. Although the description in this chapter showed the investigation of mushroom fibrinolytic enzyme is currently at stage of *in vitro* and *ex-vivo* experiment. However, the effectiveness of medicinal mushroom fibrinolytic enzyme in antithrombotic assays indicated the possible clinical application of mushroom fibrinolytic enzyme in the future.

8. Acknowledgement

The authors would like to thank Dr. Feei Sun (Council of Agriculture, Taiwan Agricultural Chemicals and Toxic Substance Research Institute) and Dr. Kurt M. Lin (Division of Medical Engineering Research, National Health Research Institutes, Taiwan) for critical reading of manuscript. We also thank Dr. Han Yin-Yi (Trauma Department, National Taiwan University Hospital) for equipment and technical assistance in TEG® analysis.

9. References

Archer, D. B., Mackenzie, D. A. & Ridout, M. J. (1995). Heterologous protein secretion by *Aspergillus niger* growing in submerged culture as dispersed or aggregated mycelia. *Applied Microbiology Biotechnology*, Vol. 44, No.1-2, pp. 157-160.

Archer, D. B. & Peberdy, J. F. (1997). The molecular biology of secreted enzyme production by fungi. *Critical Reviews Biotechnology*, Vol.17, No.4, pp. 273-306.

Astrup, T. & Mullertz, S. (1952). The fibrin plate method for estimating fibrinolytic activity. *Archives Biochemistry Biophysics*, Vol.40, No.2, pp. 346-351.

Avilan, L., Yarzabal, A., Jurgensen, C., Bastidas, M., Cruz, J. & Puig, J. (1997). Cloning, expression and purification of recombinant streptokinase: partial characterization of the protein expressed in *Escherichia coli*. *Brazilian Journal of Medical Biological Research*, Vol.30, No.12, pp. 1427-1430.

Bobowicz-Lassociska, T. & Grajek, W. (1995). Changes in protein secretion of *Aspergillus niger* caused by the reduction of the water activity by potassium chloride. *Acta Biotechnologica,* Vol.15, No.3, pp. 277-287.

Chen, H. M., Guan, A. L. & Markland, F. S., Jr. (1991). Immunological properties of the fibrinolytic enzyme (fibrolase) from southern copperhead (*Agkistrodon contortrix contortrix*) venom and its purification by immunoaffinity chromatography. *Toxicon,* Vol.29, No.6, pp. 683-694.

Cheng, M. B., Wang, J. C., Li, Y. H., Liu, X. Y., Zhang, X., Chen, D. W., Zhou, S. F. & Zhang, Q. (2008). Characterization of water-in-oil microemulsion for oral delivery of earthworm fibrinolytic enzyme. *Journal of Control Release,* Vol.129, No.1, 41-48.

Cheung, P. C. K. (1996). The hypocholesterolemic effect of two edible mushrooms: *Auricularia auricula* (tree-ear) and *Tremella fuciformis* (white jelly-leaf) in hypercholesterolemic rats. *Nutrition Research,* Vol.16, No.10, pp. 1721-1725.

Choi, H. S & Sa, Y.-S. (2000). Fibrinolytic and antithrombotic protease from *Ganoderma lucidum*. *Mycologia,* Vol.92, No.3, pp. 545-552.

Choi, H. S. & Shin, H. H. (1998). Purification and characterization of cysteine protease from *Pleurotus ostreatus*. *Bioscience Biotechnology Biochemistry,* Vol.62, No.7, pp. 1416-1418.

Choi, N. S., Chang, K. T., Jae Maeng, P. & Kim, S. H. (2004). Cloning, expression, and fibrin(ogen)olytic properties of a subtilisin DJ-4 gene from Bacillus sp. DJ-4. *FEMS Microbiolgy Letters* Vol.236, No.2, pp. 325-331.

Clemetson, K. J., Lu, Q. & Clemetson, J. M. (2007). Snake venom proteins affecting platelets and their applications to anti-thrombotic research. *Current Pharmaceutical Design,* Vol.13, No.28, pp. 2887-2892.

Collen, D. (1999). The plasminogen (fibrinolytic) system. *Thrombosis and Haemostasis,* Vol. 82, No.2, pp. 259-270.

Collen, D. & Lijnen, H. R. (1994). Staphylokinase, a fibrin-specific plasminogen activator with therapeutic potential? *Blood,* Vol.84, No.3, pp. 680-686.

Das, S. K., Masuda, M., Sakurai, A. & Sakakibara, M. (2010). Medicinal uses of the mushroom *Cordyceps militaris*: current state and prospects. *Fitoterapia,* Vol.81, No.8, pp. 961-968.

Desrochers, M., Jurasek, L. & Paice, M. G. (1981). High Production of beta-Glucosidase in *Schizophyllum commune*: Isolation of the Enzyme and Effect of the Culture Filtrate on Cellulose Hydrolysis. *Applied and Environmental Microbiology,* Vol.41, No.1, pp. 222-228.

Fan, Q., Wu, C., Li, L., Fan, R., Hou, Q. & He, R. (2001). Some features of intestinal absorption of intact fibrinolytic enzyme III-1 from *Lumbricus rubellus*. *Biochimica et Biophysica Acta,* Vol.1526, No.3, pp. 286-292.

Fiore, M. M. & Kakkar, V. V. (2003). Platelet factor 4 neutralizes heparan sulfate-enhanced antithrombin inactivation of factor Xa by preventing interaction(s) of enzyme with polysaccharide. *Biochemical and Biophysical Research Communications,* Vol.311, No.1, pp. 71-76.

Fujita, M., Hong, K., Ito, Y., Fujii, R., Kariya, K. & Nishimuro, S. (1995). Thrombolytic effect of nattokinase on a chemically induced thrombosis model in rat. *Biological and Pharmaceutical Bulletin,* Vol.18, No.10, pp. 1387-1391.

Fujita, M., Nomura, K., Hong, K., Ito, Y., Asada, A. & Nishimuro, S. (1993). Purification and characterization of a strong fibrinolytic enzyme (nattokinase) in the vegetable cheese natto, a popular soybean fermented food in Japan. *Biophysical Research Communications,* Vol.197, No.3, pp. 1340-1347.

Fukushima, Y., Itoh, H., Fukase, T. & Motai, H. (1991). Stimulation of protease production by *Aspergillus-Oryzae* with oils in continuous culture. *Applied Microbiology and Biotechnology,* Vol.34, No.5, pp. 586-590.

Gasmi, A., Chabchoub, A., Guermazi, S., Karoui, H., Elayeb, M. & Dellagi, K. (1997). Further characterization and thrombolytic activity in a rat model of a fibrinogenase from *Vipera lebetina* venom. *Thrombosis Research,* Vol.86, No.3, pp. 233-242.

Gentry, P. A. (2004). Comparative aspects of blood coagulation. *Veterinary Journal,* Vol.168, No.3, pp. 238-251.

Guillamon, E., Garcia-Lafuente, A., Lozano, M., D'arrigo, M., Rostagno, M. A., Villares, A. & Alfredo Martinez, J. (2010). Edible mushrooms: Role in the prevention of cardiovascular diseases. *Fitoterapia,* Vol.81, No.7, pp. 715-723.

Hahn, B. S., Chang, I. M. & Kim, Y. S. (1995). Purification and characterization of piscivorase I and II, the fibrinolytic enzymes from eastern cottonmouth moccasin venom (*Agkistrodon piscivorus piscivorus*). *Toxicon,* 33, No.7, pp. 929-941.

Heemskerk, J. W., Bevers, E. M. & Lindhout, T. (2002). Platelet activation and blood coagulation. *Thrombosis Haemostasis,* Vol.88, No.2, pp. 186-193.

Kadimaliev, D. A., Nadezhina, O. S., Atykian, N. A., Revin, V. V., Parshin, A. A., Lavrova, A. I. & Dukhovskis, P. V. (2008). Increased secretion of lignolytic enzymes by the *Lentinus tigrinus* fungus after addition of butanol and toluene in submerged cultivation. *Prikladnaia Biokhimiia Mikrobiologiia,* Vol.44, No.5, pp. 582-588.

Kim, J.-S., Kim, J.-E., Choi, B.-S., Park, S.-E., Sapkota, K., Kim, S., Lee, H.-H., Kim, C.-S., Park, Y., Kim, M.-K., Kim, Y.-S. & Kim, S.-J. (2008). Purification and characterization of fibrinolytic metalloprotease from *Perenniporia fraxinea* mycelia. *Mycological Research,* Vol.112, Part 8, pp. 990-998.

Kim, J. H. & Kim, Y. S. (1999). A fibrinolytic metalloprotease from the fruiting bodies of an edible mushroom, *Armillariella mellea. Bioscience, Biotechnology, and Biochemistry,* Vol.63, No.12, 2130-2136.

Kim, J. H. & Kim, Y. S. (2001). Characterization of a metalloenzyme from a wild mushroom, *Tricholoma saponaceum. Bioscience, Biotechnology, and Biochemistry,* Vol.65, No.2, pp. 356-362.

Kim, J. S., Sapkota, K., Park, S. E., Choi, B. S., Kim, S., Nguyen, T. H., Kim, C. S., Choi, H. S., Kim, M. K., Chun, H. S., Park, Y. & Kim, S. J. (2006). A fibrinolytic enzyme from the medicinal mushroom *Cordyceps militaris. Journal of Microbiology,* Vol.44, No.6, pp. 622-631.

Kim, S. H. & Choi, N. S. (2000). Purification and characterization of subtilisin DJ-4 secreted by Bacillus sp. strain DJ-4 screened from Doen-Jang. *Bioscience, Biotechnology, and Biochemistry,* Vol.64, No.8, pp. 1722-1725.

Kim, S. W., Xu, C. P., Hwang, H. J., Choi, J. W., Kim, C. W. & Yun, J. W. (2003). Production and characterization of exopolysaccharides from an enthomopathogenic fungus *Cordyceps militaris* NG3. *Biotechnology Progress,* Vol.19, No.2, pp. 428-435.

Kim, W., Choi, K., Kim, Y., Park, H., Choi, J., Lee, Y., Oh, H., Kwon, I. & Lee, S. (1996). Purification and characterization of a fibrinolytic enzyme produced from Bacillus sp. strain CK 11-4 screened from Chungkook-Jang. *Applied and Environmental Microbiology,* Vol.62, No.7, pp. 2482-2488.

Kino, K., Yamashita, A., Yamaoka, K., Watanabe, J., Tanaka, S., Ko, K., Shimizu, K. & Tsunoo, H. (1989). Isolation and characterization of a new immunomodulatory protein, ling zhi-8 (LZ-8), from *Ganoderma lucidium. The Journal of Biological Chemistry,* Vol.264, No.1, pp. 472-478.

Ko, J. H., Yan, J. P., Zhu, L. & Qi, Y. P. (2004). Identification of two novel fibrinolytic enzymes from *Bacillus subtilis* QK02. *Comparative Biochemistry and Physiology. C Toxicology Pharmacology,* Vol.137, No.1, pp. 65-74.

Kwok, Y., Ng, K. F., Li, C. C., Lam, C. C. & Man, R. Y. (2005). A prospective, randomized, double-blind, placebo-controlled study of the platelet and global hemostatic effects of *Ganoderma lucidum* (Ling-Zhi) in healthy volunteers. *Anesthesia and Analgesia,* Vol.101, No.2, pp. 423-426, table of contents.

Lee, J.-S., Baik, H.-S. & Park, S.-S. (2006). Purification and characterization of two novel fibrinolytic proteases from mushroom, *Fomitella fraxinea. Journal of Microbiology and Biotechnology,* Vol.16, No.2, pp. 264-271.

Lee, S.-Y., Kim, J.-S., Kim, J.-E., Sapkota, K., Shen, M.-H., Kim, S., Chun, H., Yoo, J.-C., Choi, H.-S., Kim, M.-K. & Kim, S.-J. (2005). Purification and characterization of fibrinolytic enzyme from cultured mycelia of *Armillaria mellea. Protein Expression and Purification,* Vol.43, No.1, pp. 10-17.

Li, H. P., Hu, Z., Yuan, J. L., Fan, H. D., Chen, W., Wang, S. J., Zheng, S. S., Zheng, Z. L. & Zou, G. L. (2007). A novel extracellular protease with fibrinolytic activity from the culture supernatant of *Cordyceps sinensis*: purification and characterization. *Phytotherapy Research,* Vol.21, No.12, pp. 1234-1241.

Liu, J. G., Yao, Y. C., Xu, R., Xu, W. W., Zhang, W., Kuang, R. G. & Gao, M. (2005). [Study on early fibrinolytic therapy to avoid acute myocardial infarction]. *Zhonghua Xin Xue Guan Bing Za Zhi,* Vol.33, No.9, pp. 782-784.

Lu, C. L. & Chen, S. N. (2010). Magnesium enhanced fibrinolytic activity of protease from *Schizophyllum commune. Taiwania,* Vol.55, No.2, pp. 117-122.

Lu, C. L., Chen, S. & Chen, S. N. (2010a). Purification and characterization of a novel fibrinolytic protease from *Schizophyllum commune.Journal of Food and Drug Analysis,* Vol.18, No.2, pp. 69-76.

Lu, C. L., Wang, J. P. & Chen, S. N. (2010b). Protease purified from *Schizophyllum commune* culture broth digests fibrins without activating plasminogen. *The American Journal of Chinese Medicine,* Vol.38, No.6, pp. 1223-1231.

Mao, X.-B., Eksriwong, T., Chauvatcharin, S. & Zhong, J.-J. (2005). Optimization of carbon source and carbon/nitrogen ratio for cordycepin production by submerged cultivation of medicinal mushroom *Cordyceps militaris. Process Biochemistry,* Vol.40, No.5, pp. 1667-1672.

Marsh, N. A. & Fyffe, T. L. (1996). Practical applications of snake venom toxins in haemostasis. *Bollettino della Società Italiana Biologia Sperimentale,* Vol.72, No.9-10, pp. 263-278.

Medved, L. & Nieuwenhuizen, W. (2003). Molecular mechanisms of initiation of fibrinolysis by fibrin. *Thrombosis Haemostasis,* Vol.89, No.3, pp. 409-419.

Mihara, H., Sumi, H., Yoneta, T., Mizumoto, H., Ikeda, R., Seiki, M. & Maruyama, M. (1991). A novel fibrinolytic enzyme extracted from the earthworm, *Lumbricus rubellus. The Japanese Journal of Physiology,* Vol.41, No.3, pp. 461-472.

Miles, P. G. & Chang, S.-T. (2004). Mushrooms: Cultivation, Nutritional Value, Medicinal Effect, and Environmental Impact. *Boca Raton, Florida: CRC Press. ISBN 0-8493-1043-1.*

Mine, Y., Kwan Wong, A. H. & Jiang, B. (2005). Fibrinolytic enzymes in Asian traditional fermented foods. *Food Research International,* Vol.38, No.3, pp. 243-250.

Moise, M. A. & Kashyap, V. S. (2008). Alfimeprase for the treatment of acute peripheral arterial occlusion. *Expert Opinion on Biological Therapy,* Vol.8, No.5, pp. 683-689.

Morozova, E. N., Falina, N. N., Denisova, N. P., Barkova, L. V. & Psurtseva, N. V. (1982). Analysis of the component constitution and substrate specificity of a fibrinolytic preparation from the fungus *Flammulina velutipes. Biokhimiia,* Vol.47, No.7, pp. 1181-1185.

Nakajima, N., Mihara, H. & Sumi, H. (1993). Characterization of potent fibrinolytic enzymes in earthworm, *Lumbricus rubellus. Bioscience, Biotechnology, and Biochemistry,* Vol.57, No.10, pp. 1726-1730.

Nakamura, T., Yamagata, Y. & Ichishima, E. (1992). Nucleotide sequence of the subtilisin NAT gene, aprN, of *Bacillus subtilis* (natto). *Bioscience, Biotechnology, and Biochemistry,* Vol.56, No.11, pp. 1869-1871.

Norris, L. A. (2003). Blood coagulation. *Best Practice & Research. Clinical Obstetrics & Gynaecology,* Vol.17, No.3, pp. 369-383.

Okamura-Matsui, T., Takemura, K., Sera, M., Takeno, T., Noda, H., Fukuda, S. & Ohsugi, M. (2001). Characteristics of a cheese-like food produced by fermentation of the mushroom *Schizophyllum commune. Journal of Bioscience and Bioengineering,* Vol.92, No.1, pp. 30-32.

Omura, K., Hitosugi, M., Zhu, X., Ikeda, M., Maeda, H. & Tokudome, S. (2005). A newly derived protein from *Bacillus subtilis* natto with both antithrombotic and fibrinolytic effects. *Journal of Pharmacological Sciences,* Vol.99, No.3, pp. 247-251.

Papagianni, M. (2004). Fungal morphology and metabolite production in submerged mycelial processes. *Biotechnology Advances,* Vol.22, No.3, pp. 189-259.

Park, J. P., Kim, S. W., Hwang, H. J. & Yun, J. W. (2001). Optimization of submerged culture conditions for the mycelial growth and exo-biopolymer production by *Cordyceps militaris. Letters in Applied Microbiology,* Vol.33, No.1, pp. 76-81.

Park, S. E., Li, M. H., Kim, J. S., Sapkota, K., Kim, J. E., Choi, B. S., Yoon, Y. H., Lee, J. C., Lee, H. H., Kim, C. S. & Kim, S. J. (2007). Purification and characterization of a fibrinolytic protease from a culture supernatant of *Flammulina velutipes* mycelia. *Bioscience, Biotechnology, and Biochemistry,* Vol.71, No.9, pp. 2214-2222.

Peng, Y., Huang, Q., Zhang, R. H. & Zhang, Y. Z. (2003). Purification and characterization of a fibrinolytic enzyme produced by *Bacillus amyloliquefaciens* DC-4 screened from douchi, a traditional Chinese soybean food. *Comparative Biochemistry and Physiology. Part B, Biochemistry & Molecular Biology,* Vol.134, No.1, pp. 45-52.

Peng, Y., Yang, X. & Zhang, Y. (2005). Microbial fibrinolytic enzymes: an overview of source, production, properties, and thrombolytic activity *in vivo. Applied Microbiology & Biotechnolgy,* Vol.69, No.2, pp. 126-132.

Peng, Y., Yang, X. J., Xiao, L. & Zhang, Y. Z. (2004). Cloning and expression of a fibrinolytic enzyme (subtilisin DFE) gene from *Bacillus amyloliquefaciens* DC-4 in Bacillus subtilis. *Research in Microbiology,* Vol.155, No.3, pp. 167-173.

Shechter, M., Merz, C. N., Paul-Labrador, M., Meisel, S. R., Rude, R. K., Molloy, M. D., Dwyer, J. H., Shah, P. K. & Kaul, S. (1999). Oral magnesium supplementation inhibits platelet-dependent thrombosis in patients with coronary artery disease. *The American Journal of Cardiology,* Vol.84, No.2, pp. 152-156.

Sugimoto, S., Fujii, T., Morimiya, T., Johdo, O. & Nakamura, T. (2007). The fibrinolytic activity of a novel protease derived from a tempeh producing fungus, *Fusarium* sp BLB.. *Bioscience Biotechnology and Biochemistry,* Vol.71, No.9, pp. 2184-2189.

Sumi, H., Hamada, H., Nakanishi, K. & Hiratani, H. (1990). Enhancement of the fibrinolytic activity in plasma by oral administration of nattokinase. *Acta Haematologica,* Vol.84, No.3, pp. 139-143.

Sumi, H., Hamada, H., Tsushima, H., Mihara, H. & Muraki, H. (1987). A novel fibrinolytic enzyme (nattokinase) in the vegetable cheese Natto; a typical and popular soybean food in the Japanese diet. *Experientia,* Vol.43, No.10, pp. 1110-1111.

Sumi, H., Nakajima, N. & Yatagai, C. (1995). A unique strong fibrinolytic enzyme (katsuwokinase) in skipjack "Shiokara," a Japanese traditional fermented food. *Comparative Biochemistry and Physiology. Part B, Biochemistry & Molecular Biology,* Vol.112, No.3, pp. 543-547.

Takada, A., Takada, Y. & Urano, T. (1994). The physiological aspects of fibrinolysis. *Thrombosis Research,* Vol.76, No.1, pp. 1-31.

Tang, Y., Liang, D., Jiang, T., Zhang, J., Gui, L. & Chang, W. (2002). Crystal structure of earthworm fibrinolytic enzyme component a: revealing the structural determinants of its dual fibrinolytic activity. *Journal of Molecular Biology,* Vol.321, No.1, pp. 57-68.

Toombs, C. F. (2001). New directions in thrombolytic therapy. *Current Opinion of Pharmacology,* Vol.1, No.2, pp. 164-168.

Ueda, M., Kubo, T., Miyatake, K. & Nakamura, T. (2007). Purification and characterization of fibrinolytic alkaline protease from *Fusarium* sp. BLB. *Applied Microbiology & Biotechnology,* Vol.74, No.2, pp. 331-338.

Wang, H. X., Liu, W. K., Ng, T. B., Ooi, V. E. & Chang, S. T. (1995). Immunomodulatory and antitumor activities of a polysaccharide-peptide complex from a mycelial culture of *Tricholoma* sp., a local edible mushroom. *Life Science,* Vol.57, No.3, pp. 269-281.

Whiss, P. A. & Andersson, R. G. (2002). Divalent cations and the protein surface co-ordinate the intensity of human platelet adhesion and P-selectin surface expression. *Blood Coagulation Fibrinolysis,* Vol.13, No.5, pp. 407-416.

Wolberg, A. S. (2007). Thrombin generation and fibrin clot structure. *Blood Reviews,* Vol.21, No.3, pp. 131-142.

Wong, A. H. & Mine, Y. (2004). Novel fibrinolytic enzyme in fermented shrimp paste, a traditional Asian fermented seasoning. *Journal of Agricultural and Food Chemistry,* Vol.52, No.4, pp. 980-986.

Wong, K.-H., Lai, C. K. M. & Cheung, P. C. K. (2011). Immunomodulatory activities of mushroom sclerotial polysaccharides. *Food Hydrocolloids,* Vol.25, No.2, pp. 150-158.

Wong, S. L., Price, C. W., Goldfarb, D. S. & Doi, R. H. (1984). The subtilisin E gene of *Bacillus subtilis* is transcribed from a sigma 37 promoter *in vivo*. *Proceedings of the National Academy of Sciences of the United States of America,* Vol.81, No.4, pp. 1184-1188.

Wu, D.-M., Duan, W.-Q., Liu, Y. & Cen, Y. (2010). Anti-inflammatory effect of the polysaccharides of Golden needle mushroom in burned rats. *International Journal of Biological Macromolecules,* Vol.46, No.1, pp. 100-103.

Yamamoto, J., Yamada, K., Naemura, A., Yamashita, T. & Arai, R. (2005). Testing various herbs for antithrombotic effect. *Nutrition,* Vol.21, No.5, pp. 580-587.

Yoon, S.-J., Yu, M.-A., Pyun, Y.-R., Hwang, J.-K., Chu, D.-C., Juneja, L. R. & Mourao, P. A. S. (2003). The nontoxic mushroom *Auricularia auricula* contains a polysaccharide with anticoagulant activity mediated by antithrombin. *Thrombosis Research,* Vol.112, No.3, pp. 151-158.

Zhang, X.-W., Sun, T., Huang, X.-N., Liu, X., Gu, D.-X. & Tang, Z.-Q. (1999). Recombinant streptokinase production by fed-batch cultivation of *Escherichia coli*. *Enzyme and Microbial Technology,* Vol.24, No.10, pp. 647-650.

16

Prediction and Rational Design of Antimicrobial Peptides

William F. Porto, Osmar N. Silva and Octávio L. Franco
Universidade Católica de Brasília,
Centro de Análises Proteômicas e Bioquímicas,
Brazil

1. Introduction

In recent decades the activity of conventional antibiotics against pathogenic bacteria has decreased due to the development of resistance. This phenomenon has generated the so-called 'superbugs', which are multi-resistant bacteria. In this context, antimicrobial peptides (AMP) appear as an alternative to control them. AMPs have been found in several sources, including animals, plants and fungi, constituting the first line of host defence against pathogens. However, the use of AMPs as therapeutic agents has some limitations, such as stability, cytotoxicity and mainly their amino acid length, since amino acids are expensive building blocks. Despite these limitations they have compensatory properties, including secondary activities such as immunomodulation or antitumor ones. Several methods have been applied since the 1990s for rational AMPs design, in order to generate analogues with improved activity, looking to reduce limitations and increase advantages. Computer-aided identification and design of AMPs play a crucial role in this area. The discovery of AMP properties, through the first rational design studies, will allow the development of methods for prediction of AMPs, which in turn, should lead to identification prior to synthesis of novel analogues. Thus, this chapter will be dedicated to describing important techniques in prediction and rational design of AMPs and their applications for drug development.

1.1 Multi-resistant bacteria: The 'superbugs'

A number of lethal infections became tractable and curable after the discovery and subsequent use of antimicrobial agents in clinical therapy, as the case of syphilis, rheumatic fever and cellulitis. However, this success has dimmed over the course of time due to the uncontrolled and inappropriate use of antibiotics, including the administration of under or overestimated doses, the insufficient duration of treatment and mistakes in the choice of drugs. Currently various microorganisms are resistant to antimicrobials, leading to the emergence and spread of so-called 'superbugs' resistant to virtually all available antibiotics on the market (Breidenstein *et al.*, 2011).

Among a variety of mechanisms of bacterial resistance, the production of β-lactamases is the main resistance factor of Gram-negative bacteria to β-lactam antibiotics.

Indeed, resistance to β-lactam antibiotics has increased in recent years, being mediated by a variety of mechanisms, most commonly the cleavage of β-lactam ring, antibiotics efflux and/or reduced drug uptake due to loss of outer membrane porin proteins (Pfeifer *et al.*, 2010). The large number of bacteria resistant to multiple antibiotics represents a challenge in the treatment of infections, since the rate of obtaining new antibiotics today cannot match the increasingly large number of resistant strains. Our next step must include the careful use of antibiotics in clinical and agricultural fields as well as the search for novel drugs.

1.2 Antimicrobial peptides

AMPs have emerged as an alternative strategy for the treatment of infections caused by resistant bacteria. These peptides are evolutionarily ancient molecules that have been isolated from microorganisms, plants, invertebrates, fish, amphibians, birds and mammals, including humans. They play an important role in the innate immune system and are the first line of defence to protect internal and external surfaces of the host (reviewed in Silva *et al.*, 2011). The AMPs may have a broad spectrum of antibacterial and antifungal activities. Moreover, in some cases, antiviral, antiparasitic and antitumor activities have also been observed (Nijnik & Hancock, 2009). Despite the enormous diversity in their sequences and structures, the majority of AMPs show a positive charge (+2 to +9), 12-100 amino acid residues and variable three-dimensional structures. Among them are included α-helices (*e.g.*, magainin, cecropin and cathelicidin), β-sheets (*e.g.*, hepcidin and human α-defensin 1), a combination of α-helices and β-sheets (*e.g.*, human β-defensin 1 and plant defensins), head-to-tail cyclized fold (*e.g.*, cyclotides), as well as extended and flexible loops (*e.g.*, indolicidins) (reviewed in Silva *et al.*, 2011). In addition to their action against microorganisms, AMPs have activities related to innate and adaptive immunity (immunomodulatory activity) that include the induction or modulation of proinflammatory cytokines and chemokines production, chemotaxis, apoptosis, inhibition of inflammatory response, recruitment and stimulation of proliferation of macrophages, neutrophils, eosinophils and T lymphocytes (Nijnik & Hancock, 2009).

The AMPs have a wide variety of mechanisms, showing that they clearly act bound to the lipid bilayer, using it as a primary target and leading to a membrane disruption (reviewed in Silva *et al.*, 2011). It was at first believed that the initial AMP mechanism of action was solely on the cell membrane. However, AMPs can also perform their functions through interactions with intracellular targets or by disturbing cellular processes, as well as causing synthesis inhibition of the cell wall, nucleic acids or proteins (Brogden, 2005).

The AMPs are molecules of great relevance to the pharmaceutical, biotechnology and food industries. The structural diversity and chemical nature displayed by these molecules is a condition that has led researchers to consider them as natural antibiotics, an innovative alternative to conventional antibiotics as a new class of drugs to prevent and treat systemic and topical infections (Gordon *et al.*, 2005). Due to these facts, some AMPs are already utilized with clinical and commercial purposes, including ambicin (nisin), polymixin B and gramicidin S (Bradshaw, 2003). However a restriction on the use of AMPs for therapeutic use is their limited stability (especially when composed of L-amino acids), toxicity against eukaryotic cells, susceptibility to proteolytic degradation and development of allergies. Thus, the rational design of AMPs emerges as an important tool that aims to develop AMPs with maximum performance against resistant bacteria.

1.3 Computer-aided identification and design of AMPs

Rational design of AMPs is a modern approach to antibiotic development, nevertheless, a more detailed target characterization is needed. Indeed, a target with sufficient differences between the host and the pathogen is necessary, in order to reduce or abolish adverse effects, according to the principle of selective toxicity. The principal barrier to the use of AMPs as antibiotics lies in their cytotoxicity for mammalian cells. This is perhaps not surprising since AMP activity is mostly dependent on membrane-peptide interaction. However, for AMPs become useful as broad-spectrum antibiotics it would be necessary to dissociate toxicity to mammalian cells from antimicrobial activity, which could be reached by increasing antimicrobial activity or reducing haemolytic activity, or both (Chen *et al.*, 2005). Another obstacle to the use of AMPs as antibiotics is their susceptibility to proteolysis, since peptides formed by L-amino acid are sensitive to degradation and clearance of serum components. These problems can be solved through amino acid substitutions, including replacement of L-amino acids to D-amino acids. These substitutions may promote alterations in amphipathicity/hydrophobicity, leading to a reduction in the cytotoxicity of the peptides to mammalian cells, without changing the antimicrobial activity, besides leaving the AMPs less susceptible to proteolytic degradation (Chen *et al.*, 2005; Pag *et al.*, 2004).

The first studies of rational design of AMPs generated several analogues of known AMPs (*e.g.*, cathelicidins, defensins, magainins and cecropins). Nevertheless, many of them were less active than the original prototype. In fact, these studies played a critical role in identifying the AMP properties involved in antimicrobial activity. These properties served as the basis for developing approaches for antimicrobial activity prediction, through several methods, such as support vector machine (SVM, Lata *et al.*, 2007; Porto *et al.*, 2010; Thomas *et al.*, 2010), artificial neural network (ANN, Fjell *et al.*, 2009; Torrent *et al.*, 2011) and quantitative structure-activity relationship (QSAR, Jenssen *et al.*, 2007) as will be further detailed. By using machine learning methods, this field became more scientific than descriptive. Nevertheless, the AMP mode of action is still an open subject since there are no definite models of prediction or rational design, and so novel methods tend to appear. Certainly, the AMPs emerge as a promising class of therapeutics, despite their limitations. Methods of prediction and rational design play a crucial role in improving AMP performance against resistant bacteria. Therefore, designing novel AMPs requires progress in methods for identifying the best candidate peptides prior to synthesis and then testing them against bacteria. Methods of rational design and prediction of AMPs emerged from early 90s and 2000s, respectively, and they will be reviewed in the next sections.

2. Methods of rational design

Rational design methods aim to create novel peptides with improved antimicrobial activity, lower toxicity to human cells and reduced size. In other words, it is much more specific in creating a pharmaceutical with higher specificity to microorganisms, avoiding side effects. This review classifies the rational design methods into three major classes: physicochemical, template-based and *de novo* methods. The first two methods use a previously known AMP as the basis for designing studies. While physicochemical approaches generate several analogues with different physicochemical properties, the template-based methods search for size reduction, adding selectivity and/or killer activity to known sequences. Furthermore, *de novo* methods that generate AMP without a template sequence, using only frequencies or patterns. Essentially, these three classes define the rational design methods, but there are also hybrid methods.

2.1 Physicochemical methods of rational design of AMPs

The first rational design methods were based on the most commonly proposed AMP mechanism of action, which is membrane disruption. This process is first mediated by electrostatic interactions among positive charged residues and negatively charged lipid heads, and then by insertions of hydrophobic residues into the membrane. The majority of physicochemical methods use α-helical peptides as the basis for study. Since α-helical peptides present wide distribution and the broadest activities spectrum, their physicochemical properties can be easily measured. In addition to charge and hydrophobicity, another property that can be easily measured is the hydrophobic moment, given by Eisenberg's equation (Eisenberg *et al.*, 1982):

$$\mu_H = \sqrt[2]{[\sum_i H_i \cos(\delta i)]^2 + [\sum_i H_i \sin(\delta i)]^2} \qquad (1)$$

Where δ is the angle separating side chains along the backbone (100° for α-helix); *i* is the number of residues and H_i is the hydrophobicity of amino acid *i* in a determined hydrophobicity scale, such as Eisenberg's (Eisenberg *et al.*, 1982) or Kite-Doolittle's (Kite & Doolittle, 1982). In fact, it is more common to use a normalized hydrophobic moment, dividing it by the total amino acid residues. These physicochemical properties are, apparently, directly involved in interactions between α-helical AMPs and bacterial membranes, by some "rules". First, increasing hydrophobicity boosts the lipid's affinity. Second, enhancing the hydrophobic moment may favour the α-helix peptide fold, and third, increasing the net charge could lead to a higher interaction with anionic membranes (Drin and Antonny, 2010).

Using this approach, Dathe *et al.* (1997) developed several magainin 2 analogues and an 18-residue model peptide with KLA repetitions, modulating their activity by changing only hydrophobicity, hydrophobic moment and the angle of positively charged face helix (Figure 1). Moreover other features were conserved, such as helix propensity and total charge. This showed that when the hydrophobicity and hydrophobic moment increase, the antimicrobial and haemolytic activities from those peptides also increase (Dathe *et al.*, 1997). It also showed that the angle of positively charged face has little influence on antimicrobial activity. Haemolytic activity increases if the angle is more obtuse than the original. Nonetheless, it varies according to peptide. For example, an angle of 120° applied to KLA model peptide increases haemolytic activity, but the same angle applied to magainin 2 does not affect its activity (Dathe *et al.*, 1997). On the other hand, very low hydrophobicity abolishes the antimicrobial activity of those peptides, which can be compensated by increasing the hydrophobic moment. Therefore, while increasing those parameters the peptide becomes unspecific, a selective peptide may be reached with moderated hydrophobicity, increasing the hydrophobic moment and keeping the angle of charged face small. Further, the same group would show that changes in net charge of magainin 2 also modulate its activity (Dathe *et al.*, 2001). This study designed six magainin 2 analogues, keeping helix propensity, hydrophobicity, hydrophobic moment and the angle of charged residues, and changing only the net charge. So for each charge modification, one or more amino acid substitutions were required to keep the other parameters (*i.e.*, the MK6 analogue has a charge of +7, while magainin 2 has +4; the identity between them is only 39%, but the other properties were very similar). A charge threshold was observed to develop an analogue with specificity to bacteria, and increasing the charge from +3 to +5 made the peptide more active against bacteria and less toxic to erythrocytes. Nevertheless increasing its charges to +6 or +7 could generate a very haemolytic analogue. The relation between the angle of charged face and haemolytic activity has a bias: the net charge must be great, but not too great. In the case of magainin 2 this threshold is +5. Therefore, increasing the peptide charge makes the peptide lose its specificity to bacterial membranes. Perhaps, when the charge is too positive, the neutral membrane ends up interacting with the peptide as an acidic membrane.

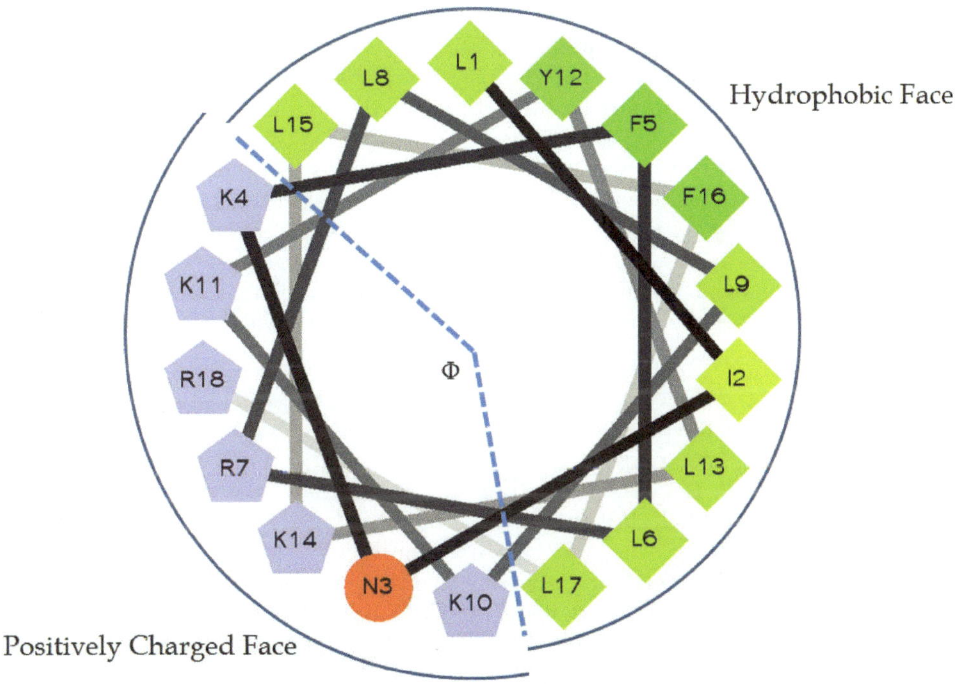

Fig. 1. Schematic representation of an α-helical amphipathic peptide. The angle of polar face is indicated by Φ. Positive charged residues are represented in pentagons, polar ones as circles and nonpolar as diamonds.

Giangaspero *et al.* (2001) also observed similar results in their study about α-helical peptides with non-proteinogenic amino acids. In fact, this work employs a hybrid method, first using a *de novo* technique and, subsequently, a physicochemical one. *De novo* design uses a model developed through amino acid frequencies by type of amino acid (structure determining, hydrophobic, hydrophilic, positively charged, negatively charged and polar uncharged). Then the model was filled up with a restricted set of amino acids: norleucine to hydrophobic positions and ornithine, glutamine or glutamic acid to hydrophilic ones. These residues were chosen since they have the same side-chain length, ensuring a homogeneous cross-section to the helix. This step generates two peptides, P19 (5) and P19 (6). In the next step, 18 novel sequences were derived from the initial model in order to verify the effects on activity by charge, helicity, amphipathicity, hydrophobicity and size reduction. Four sequences were developed with different charges: +1, +3, +8 and +9. As observed by Dathe *et al.* (2001), charge reduction leads to a decrease in antimicrobial activity. However, Giangaspero *et al.* (2001) propose that the activity is independent of positioning of charged residues within the helical domain. The addition of two ornithine or glutamic acid residues to the N-terminal of those peptides can increase or decrease the activity, respectively: adding two ornithine residues to the analogue with charge +1, its charge became +3 and it became active, with a similar spectrum of analogue with charge +3.

The amphipathicity and hydrophobicity were tested by developing a shuffled peptide version of P19 (6). This peptide has a moderate, and restricted to Gram-negative, antimicrobial activity when compared to P19 (6), even with the same amino acid composition, charge and hydrophobicity, showing that amphipathic arrangement is important to activity (Giangaspero *et al.*, 2001). Size reduction also was tested, by deleting either N- or C-terminal from the most active peptide. This reduces or abolishes the activity, however, switching polar to nonpolar residues, resulting in activity recovery, being similar to or better than the original peptide. These data indicate that in small peptides, there must be equilibrium among charges, helix formers and hydrophobic residues. Helicity modifications were also measured, showing that an increase in the helix propensity also increases the antimicrobial potency. However, it has little additional effect on peptides that have a high helix propensity. On the other hand, decreasing the helix propensity, by proline or D-amino acid insertions could clearly decrease the antimicrobial activity.

Nonetheless, Chen *et al.* (2005) observed a different relationship between helicity and antimicrobial activity in their study about analogues of V_{681}, a designed amphipathic α-helix antimicrobial peptide. Its nonpolar face comprises 12 amino acid residues, while the polar face shows 14 of these (Figure 1). The central residue from each face was chosen for substitutions, Ser^{11} and Val^{13} for polar and nonpolar faces, respectively. Amino acid substitutions were made by increasing or decreasing the peptide's hydrophobicity and/or amphipathicity. Each analogue were generated by only one amino acid substitution, being divided in two groups, the ones with alterations in the polar face named S11X, where 'S' is replaced by 'X'; and the second group with alterations in the nonpolar face named V13X, with the same logic as S11X analogues. Five L-amino acids (Leu, Val, Ala, Ser and Lys) plus glycine were selected to replace the central residues, representing a wide range of hydrophobicity, on a decreasing scale in the following order: Leu > Val > Ala > Gly > Ser > Lys. Moreover, D-enantiomers of each selected L-amino acid were also incorporated in the same positions in order to disrupt helical structures generating a total of 20 analogues. It was observed that some D-amino acid analogues were stronger than their L-amino acid equivalents. Probably, D-amino acids analogues overcome the helix disruption through other properties, such as hydrophobicity or amphipathicity. Moreover, they also observed that changes in the hydrophilic face of V_{681} does not reduce peptide activity against Gram-positive or -negative bacteria or human erythrocytes, in contrast to changes in hydrophobic face. A similar result was observed by Blondelle *et al.* (1996).

The D-enantiomer of analogue V13K was used by Jiang *et al.* (2011) as the basis of another physicochemical study. The all-D-enantiomer analogues were developed in order to create peptides with specificity to Gram-negative bacteria. Five analogues (D11, D14, D15, D16 and D22) were designed to investigate the influence of charge (analogue D11), hydrophobicity (analogue D22), insertions of charged residues into nonpolar face (analogue D14) and composition of the nonpolar face (analogues D15 and D16). They observed that when charge and hydrophobicity increase (comparing V13K and D11), antimicrobial activity also increases, as observed by Dathe *et al.* (2001) and Giangaspero *et al.* (2001). By increasing hydrophobicity (comparing D11 to D22), haemolytic activity increases, confirming the data proposed by Dathe *et al.* (1997). However, by introducing a second lysine in the nonpolar face (comparing D22 to D14), hydrophobicity can be kept higher and haemolytic activity can decrease. Finally, the composition of the nonpolar face (comparing D11 and D14 to D15 and D16, respectively), D15 and D16 were generated by switching all large side-chain hydrophobic residues for leucine residues. Those changes increase hydrophobicity and antimicrobial activity, but they have different effects on haemolytic activity, while D15 becomes more haemolytic and D16 becomes less haemolytic, probably due to the presence of second lysine residue in the polar face of D16. These data show that the same physicochemical rules can be applied to D- or L-enantiomers.

Few studies with another kind of folding have been reported. In 1999, Wu & Hancock carried out a study based on bactenecin, a 12-amino acid residue peptide that adopts a β-turn structure cyclized via disulphide bond. Linear analogues of bactenecin show its activity depleted. However, C-terminal amidation partially restores the activities. In this study, several changes in both forms (linear and cyclic) of bactenecins were evaluated. Several analogues were designed to test the importance of ring size (numbers of amino acids between the cysteine residues), charge, and amphipathicity. The results are similar to α-helical peptides, in which it was observed that increasing the charge leads to an improvement in antimicrobial activity. Moreover, the same study showed that the positions of charged residues are more important than the number of positive charged residues. Increasing the ring size by insertion of a tryptophan in the middle of the ring increases the activities, while a proline residue insertion was able to abolish the activity. Additionally, the cyclic analogues also have agglutination activities, in contrast to linear versions. The linear analogue Bac2A-NH2 was the most desirable candidate generated in this study, due to its broad spectrum of activity and absence of agglutination activity. Further, several analogues of Bac2A were developed through point substitutions, scrambling, and deletions in sequence; IDR1018 is the most promising of all Bac2A analogues. Besides bactericidal activity, IDR1018 also displays chemokine induction activity and suppresses pro-inflammatory responses to Gram-negative bacteria (Wieczorek *et al.*, 2010).

Conversely, this kind of analysis, considering the minimum inhibitory concentration (MIC) as a consequence of structural and physicochemical properties, leads us to false conclusions. MIC values can be very similar for peptides with different properties. This is easily observed when peptides KLA12 and KLA7 (Dathe *et al.*, 1997) are compared. They have similar MICs; nevertheless, their hydrophobicity and hydrophobic moment are different. The hydrophobicity of KLA7 is a half of KLA12, while its hydrophobic moment is 1.15 times higher than KLA12. Moreover, this kind of study is almost completely restricted to α-helical peptides. The lack of study of other varieties of folding might bring novel information about the relationship between physicochemical properties and antimicrobial activities.

2.2 Sequence template methods of rational design of AMP

Sequence template methods involve generating novel AMPs based on a known sequence, whether of an active or an inactive peptide. These approaches can seek to reduce size, add selectivity and/or increase the activity. In several cases, the information generated by physicochemical methods can be used to reach these objectives, by switching residues, changing the net charge or pursuing minor peptides with the same properties, without performing a physicochemical study itself.

In 1996, Thennarasu & Nagaraj developed three analogues of pardaxin by switching some amino acid residues for others with different properties. Pardaxin is a toxic peptide secreted by the sole fish from the genus *Pardachirus*. At low concentrations pardaxin is able to form ion channel-like structures and at high concentrations that causes cell membrane disruption. This toxin can also induce neurotransmitter release from neurons. Firstly, the authors identified the probable region responsible for membrane permeation activity. Preliminary studies have shown that the C-terminal region did not have this activity, since the positive charges were concentrated at N-terminal. Then, the first designed analogue was the N-terminal 18 residue segment, named 18P. The second analogue, 18A, was designed by switching the residue Pro^7 to an alanine residue, since proline residues cause structural distortions to helix backbone. The last analogue, 18Q, was developed switching the two lysines (Lys^8 and Lys^{16}) for glutamines in the 18A sequence, since glutamine residues play an important role in channel formation by peptides with neutral charges. Having designed these analogues, their activities were examined against *Escherichia coli*, *Staphylococcus aureus* and human erythrocytes. 18P analogue showed activity only against *E. coli*, while 18A showed haemolytic activity in addition to antimicrobial activity against *E. coli*. On the other hand, 18Q showed only haemolytic activity. No activities against *S. aureus* were observed. Although the minimum identity among the sequences was 83.3% (18P and 18Q) the activities were different. While 18P showed simply antimicrobial activity, 18Q had only haemolytic one. These differences can be explained by their intrinsic structures. Circular dichroism (CD) analysis showed that 18P had a low propensity to occur in helical conformation when compared to 18A, even though both have a typical helical CD spectrum, while 18Q adopted a clear β-structure, probably forming an amphipathic β-sheet, even in ~65% of 2,2,2-trifluoroethanol, indicating the importance of structure-activity relationship.

Ueno *et al.* (2011) developed a strategy that does not generate great conformational changes in relation to original sequences. This strategy is based on acid-amide substitutions by switching aspartic acids and glutamic acids to asparagine and glutamine, respectively. Since these substitutions are conservative, the structure has few changes and if the original peptide has basic residues, there will be an increased charge in the novel peptides. This strategy was successfully applied to three pro-regions of nematode cecropins. The pro-regions are inactive against bacteria and human erythrocytes. After modifications, the sequences became antimicrobial peptides with slight haemolytic activity. CD spectra reveal that the structure of original peptides and its analogues are similar.

Likewise, Ahn *et al.* (2006) developed an AMP based on the 11-residue α-helical domain from tenecin 1, an insect defensin isolated from *Tenebrio molitor*. This α-helical domain, named L1, shows no antimicrobial activity. In defensins, the activity is related to the γ-core motif, comprised of a β-hairpin (Yount & Yeaman, 2004). However, L1 shows physicochemical properties similar to well-known AMPs, except the net charge. L1 has a net charge of +2, while AMPs have a charge of +4 or +5. Three analogues of L1 were developed (L2, L3 and L4) by switching some residues. L4 was the most active analogue, showing even greater activity than tenecin 1. L4 was developed by switching an aspartic acid and a histidine for lysine residues, increasing the charge to +3. Thereafter, L4 showed activity against bacteria and fungi, including *E. coli, Pseudomonas aeruginosa* and *Candida albicans,* besides activity against *S. aureus* and *Micrococcus luteus,* while tenecin 1 had no activity toward these pathogens.

Another work involving defensins was developed in 2008 by Landon *et al.* In this study, 70 chimeric defensins were designed by combining conserved regions of *Anopheles gambiae* defensin and variable regions of other insect defensins. From these, 45 were expressed in yeast *Saccharomyces cerevisiae.* Five of them were selected for study. These five hybrid defensins originated from combinations of *A. gambiae* defensin (DEF-AAA) with defensins from *Belostoma gigas, T. molitor, Acrocinus longimanus* and/or *Drosophila melanogaster.* All hybrid defensins have the same structural scaffold of a cysteine-stabilized αβ motif. On the other hand, their activities against *S. aureus* multi-resistant strains were different. Two analogues were more effective *in vitro* against *S. aureus* (DEF-AcAA and DEF-DAA). DEF-DAA was toxic to mice models, with a lethal dose of 30 $mg{\cdot}kg^{-1}$, while DEF-AcAA showed a lethal dose higher than 100 $mg{\cdot}kg^{-1}$. Indeed, since the active site of defensins consists of the γ-core, these two hybrid defensins have an identical γ-core sequence, indicating that the N-terminal loop of defensins may also contribute to the activity. So the *in vivo* activity of DEF-AcAA and DEF-AAA was evaluated against *S. aureus* peritonitis model on mice. The results showed that both defensins have the same efficacy for *S. aureus* multi-sensitive strain, with a dose of 3 $mg{\cdot}kg^{-1}$. Nevertheless, on the same model with a multi-resistant strain DEF-AcAA was shown to be the most effective with a dose of 3 $mg{\cdot}kg^{-1}$, while DEF-AAA needed a dose of 10 $mg{\cdot}kg^{-1}$. These results demonstrate that these AMPs are more efficient than vancomycin, which requires a dose ranging from 10 to 30 $mg.kg^{-1}$ for treatment on the same model.

Also focusing on development of peptides with potential for systemic use, Sigurdardottir *et al.* (2006) identified a 21-amino acid fragment of human cathelicidin antimicrobial peptide LL-37 with similar or stronger activities than the complete peptide. LL-37 is an attractive candidate for treatment of sepsis, due to its broad spectrum of antimicrobial activity, the immune system's cell chemotactic abilities and also abilities to bind and neutralize bacterial lipopolysaccharides. However, LL-37 also has cytotoxic activity against eukaryotic cells. Therefore, using the helical propensity prediction of AGADIR (Lacroix, 1998) and the amino acid preference for α-helix terminals, they identified a fragment starting from Gly^{14} going up to Arg^{34}, named GKE. For comparisons, two other 21-amino acid fragments were derived, one from the N-terminal (LLG) and other from C-terminal (FKR). GKE was more active than LL-37 against bacteria and fungi. Moreover, GKE and LL-37 showed similar chemotaxis and inhibition of nitric oxide production activities. The same patterns were not observed for LLG and FKR fragments. Interestingly, all fragments showed 100% of identity to LL-37. However, they differed in helix propensity, although GKE and FKR do not present much difference in helical content and have 85.7% of identity. The amino acid preference for α-helix terminals might explain those differences in activity.

Another successful strategy is the use of synthetic combinatory libraries (Blondelle *et al.*, 1996). A synthetic combinatory library is composed of a set of mixtures of peptides generated from a template. In this approach, several positions of template sequence were chosen for amino acid substitutions. One of them was chosen for individual defined substitutions, where each modification is controlled, generating one subset for each modification. Thus it can generate 20 subsets, by using the proteinogenic amino acids. The other selected positions are randomly filled up in each subset. Following this strategy, Blondelle *et al.* (1996) used the sequence 'YKLLKKLLKKLKKLLKKL-NH_2' as a template for design of two synthetic combinatory libraries, the first one with changes in hydrophobic face and the second one in hydrophilic face. In hydrophobic face, Leu^4 was chosen for individual defined substitutions (represented by "O"), while Leu^7, Leu^{11} and Leu^{14} were chosen for random filling (represented by "X"); this library was represented by the sequence YKLO^{4}KKX^{7}LKKX^{11}KKX^{14}LKKL-NH_2. The same logic was applied to hydrophilic face, being represented by the sequence YKLLKO^{6}LLX^{9}KLKX^{13}LLX^{16}KL-NH_2. The assays showed that substitutions in hydrophobic face depleted the activity of the original peptide. During the second stage of design, the library with changes in polar face was redefined by changing the residue used for individual defined substitutions (positions 6, 9, 13 and 16). In this way, the best residues for each position could be selected. Thus leucine was selected at position 6 as a representative hydrophobic residue, proline at position 9, proline and glycine at position 13, and phenylalanine, isoleucine and proline at position 16, six peptides being generated by combining those selected residues ($1 \times 1 \times 2 \times 3 = 6$). Five of the six designed peptides showed a 10-fold improvement in activity.

Indeed, sequence template methods can generate novel AMPs with enhanced activity against bacteria and lower effectiveness toward mammalian cells than original sequences. However, there are no directives for using this kind of method. Which residues are important for activities? What must be substituted to obtain a higher deleterious activity and lower side effects? Are the physicochemical properties useful for this kind of study? Can the methodology applied to peptide A be applied to peptide B? For these methods, it is necessary to identify the governing principle that allows the enhancement of antimicrobial activity. In fact, there must be something beyond switching residues or development of chimeric proteins. Identifying this principle will be helpful for developing novel methods of rational design.

2.3 *De novo* methods of rational design of AMP

De novo methods are very interesting in terms of achieving a yield from multiple AMPs with little amino acid conservation. Instead of using one pivotal sequence to develop analogues, *de novo* methods can use amino acid patterns or amino acid frequencies and positioning preferences, generating several sequences with no clear relation.

Tossi *et al.* (1997) developed a *de novo* method that considers the length of the peptide, its cationicity, its amphipathicity and its helicity, in addition to sequence patterns. First, 20 residues from the N-terminal of 85 natural α-helical AMPs were aligned without gap insertions or any attempt to improve the alignment. Based on this alignment, frequencies and kinds of amino acids were extracted, in order to create a novel pattern. Despite the simplicity of this method, a well-defined pattern of residue distribution was developed. Next, the pattern was filled up with the most frequent amino acids in each position, generating an AMP with 20 amino acid residues. The same method was applied to mammalian cathelecidin with some modifications, using a reduced number of sequences and adding gaps in the alignment, creating three novel patterns ranging from 18 to 22 amino acids (Tossi *et al.*, 1997). Thus, the patterns were filled up with the most frequent amino acids. The helicity was evaluated by using secondary structure predictions and helical wheel diagrams. The four designed peptides showed a potent and broad-spectrum activity against Gram-positive and Gram-negative bacteria.

Some years later, Loose *et al.* (2006) developed a similar but more sophisticated *de novo* method, the linguistic model. According to this model, AMPs seem to be a formal language with grammar composed of several rules (patterns) and a vocabulary (amino acids). Instead of using alignments to define the patterns, the TEIRESIAS algorithm was used for pattern discovery (Rigoutsos & Floratos, 1998). Thus, ~700 grammars sequences were established, and then all possible grammar sequences with 20 amino acid residues were written out. Sequences with at least 60% of identity with natural AMPs were removed, resulting in 12 million remaining sequences. Next, by removing sequences with at least 70% of identity, 41 candidates were obtained. From these, one peptide was insoluble, but 18 had MIC at maximum of 256 $\mu g{\cdot}ml^{-1}$ and the remaining peptides showed no activity against *E. coli* and *Bacillus cereus*. Through this method, novel antimicrobial peptides were designed without any information about their structures (Loose *et al.*, 2006). In fact, as well as generating novel AMPs, this work was of great importance in that it explained some results of physicochemical methods of rational design. For each grammatical peptide, a non-grammatical peptide was designed with the same amino acid composition, by shuffling the sequence. Giangaspero *et al.* (2001) had already used this strategy and the shuffled peptide showed a reduced activity. However, Loose *et al.* (2006) generated shuffled peptides with no grammars, expecting that they had no activity because they were non-grammatical peptides. As result, only two shuffled peptides were active. From this, it could be seen that there are no direct relations between scalar physicochemical properties and antimicrobial activity, because shuffled and grammatical peptides have the same charge, hydrophobicity, size and molecular mass (Loose *et al.*, 2006). This explains why the conclusions of physicochemical methods were not completely correct, and in some cases, controversial (*i.e.*, equal MICs but different physicochemical properties and *vice-versa*). The scalar physicochemical properties had led researchers to false conclusions because they have a secondary role in activity.

On the other hand, a property widely used in physicochemical methods does change when the sequence is shuffled, and that is the hydrophobic moment, a vector property. Loose *et al.* (2006) also used the hydrophobic moment. In a second step of rational design, the best designed sequence was submitted to a redesign process to increase its activity using a heuristic approach, and one of proposals of this redesign process was to "improve the segregation of positive and hydrophobic residues based on a helical projection" or, in other words, to improve the hydrophobic moment. The fact that no structural information is needed is certainly an advantage, but this method has some limitations, such as the difficulty in designing larger proteins with complex structures. So this method is restricted to generating AMPs similar to those that are deposited in the main data set, *i.e.*, the two most active peptides obtained through this method have 50 and 60% of identity to natural AMPs.

All these methods have been helpful, in their time, in reaching a better understanding of relationships between sequence, structure, physicochemical properties and antimicrobial activity. Overall, these methods have been effective in designing potent AMPs able to kill bacteria at low concentrations. Furthermore, they have also been helpful in the development of antimicrobial prediction tools, as will be seen in the next section.

3. Methods to predict antimicrobial activity

The understanding of antimicrobial peptides' behaviour led some groups to propose different approaches to predict antimicrobial activity, and this field saw much progress in last years. Several methods of antimicrobial activity prediction emerged from studies of rational design, mainly the physicochemical and *de novo* methods. The rules extracted from rational design methods can be extrapolated to other sequences with good reliability by computer-aided predictions. As a result, several tools have been developed, such as prediction tools from Collection of Antimicrobial Peptides (CAMP, Thomas *et al.*, 2010) and AntiBP Server (Lata *et al.*, 2010). Overall, there are two main strategies for predicting AMPs, the empirical methods and the supervised machine learning ones.

3.1 Empirical methods of AMP prediction

The empirical methods are qualitative, being based only on characters of AMPs without taking into account peptides without antimicrobial activity. In fact, these models are based on rules or patterns correlated to antimicrobial activity. However, the methods cannot be extrapolated to other classes of AMP, being restricted to the class that generated the model. Moreover, they have no standard accuracy measurement, since there is no larger set of non-antimicrobial sequences to test them. In fact, there is no accuracy value, making it complicated to compare the methods, which are summarized in Table 1.

The most simple prediction method is that employed by the Antimicrobial Peptides Database prediction tool (Wang & Wang, 2004). In this case no artificial intelligence was used. It is based only on logical questions about the sequence. It returns a positive prediction whenever a sequence is less than 50 amino acid residues in length, has hydrophobicity below 75% and a cationic net charge. Moreover, the prediction is also going to be positive if the sequences present an even number of cysteines. This method seems to be merely based on rules extracted from physicochemical studies. However, this method neglects some AMPs (*e.g.*, anionic and hydrophilic antimicrobial peptides).

Despite these clear limitations the APD prediction tool was used by Nagarajan *et al.* (2006) for validating their prediction method. This method was developed in order to mine protein data sets, being based on Fourier transformations and Euclidian distances. Comparisons were made with a power spectrum generated by the Fourier transformation of five indices. The indices were based on hydrophobicity, charge, polarity, cysteine content and amino acid distribution. For analysis, the method was applied to six antimicrobial peptides with 16 amino acid residues. In all cases, the power spectrum shows a peak at period 5, and the major contribution to power spectrum is given by hydrophobicity index. Those power spectra were used to generate a reference power spectrum used in further comparisons. A set of 10,000 random peptides with 16 amino acid residues were generated by PERL scripts. The power spectrum of each random sequence was obtained and compared to the reference through Euclidian distance. From 10,000 random sequences, only three hits were obtained. Two of three hits had positive charge and were predicted as AMPs by the APD prediction tool. However, the three hits showed at least 30% of identity to a known AMP.

Similarly, Fernandes *et al.* (2009) developed a classification method based on fuzzy modelling, also focusing on data set mining. This approach is based on the linguistic model developed by Loose *et al.* (2006) and also on the peptide's amphipathicity. It made each screening into a data set, searching for sequences with a defined pattern. The found sequences were then classified by fuzzy modelling. This consists of a surface plot generated by two membership functions: a triangular function relating the ratio of polar to charged residues and a Gaussian hydrophobicity membership. The best candidates fall into a region between 2:1 and 1:1 polar to charged residues and the regions of moderated hydrophobicity in Gaussian membership, identifying the amphipathic sequences. Assuming hydrophobicity to be low or the ratio to be lower than 1, the sequence is a weak AMP; if hydrophobicity is medium and the ratio is adequate the peptide is a specific AMP; and if hydrophobicity or the ratio is high, the peptide is non-specific. The system was tested in NCBI's non-redundant protein data set (NR) and the seed sequence was Cn-AMP1 (Mandal *et al.*, 2009). Through this, three sequences were obtained from a total of 7,153,872 sequences in NR.

Another method that involves patterns is the multidimensional signatures developed by Yount & Yeaman (2004). This method is based on recognition of sequence patterns and motifs in three dimensional structures to correlate them to antimicrobial activity. It was successfully applied to cysteine-stabilized peptides. In this work, a γ-core motif was recognized by the patterns "X[1,3]GXCX[3,9]C", "CX[3,9]CXGX[1,3]" and "CX[3,9]GXCX[1,3]", where X corresponds to any natural amino acid and the numbers between brackets represent sequence variations (*i.e.*, X[3,9] represents an extension of three to nine residues, being composed of any natural amino acid residue). Based on a data set of 500 antimicrobial peptides with length of up to 75 amino acids residues and cysteine content, prototypic sequences were chosen as representative of their classes. The conserved motif GXC was identified by visual inspection of multiple alignments. However, in some sequences this motif was inverted, so the three patterns of γ-core motif were proposed. Structurally the three patterns are absolutely conserved, corresponding to an antiparallel β-sheet composed of two strands. In order to validate the model, two peptides without previously reported antimicrobial activity were selected, the sweet-tasting protein brazzein and the toxin charybdotoxin, both containing the γ-core motif in their 3D structures. These two peptides exerted direct antimicrobial activity against bacteria and fungi. The method was also validated by identification of γ-core sequence into well-known antimicrobial peptides without known 3D structure. The γ-core motif was identified in tachyplesins before its structure became available. Its three-dimensional structure really exhibits the motif of two antiparallel β-strands.

Jenssen *et al.* (2007) also developed and tested their model *in vitro*. They constructed a mathematical model for prediction based on the statistical methods, principal component analysis (PCA) and partial least squares (PLS). This model was filled up with three major classes of descriptors, (i) amino acid (charge, hydrophobicity and size); (ii) a series of contact energy for each pair of amino acids and (iii) 78 biophysical inductive and conventional quantitative QSAR descriptors. These data were extracted from a single-substitution Bac2a-library containing 228 peptides. This model was capable of predicting 84% of tested peptides.

Method	How is the evaluation done?	References
APD Prediction Tool	Well-known AMP properties	Wang & Wang, 2004
Fourier Transformation	Physicochemical properties	Nagarajan *et al.*, 2006
Fuzzy Modelling	Sequence motifs and physicochemical properties	Fernandes *et al.*, 2009
PCA/PLS	Amino acid descriptors, pairs of amino acid descriptors and QSAR descriptors	Jenssen *et al.*, 2007
Multidimensional Signatures	Sequence and structure motifs	Yount & Yeaman, 2004

Table 1. Empirical antimicrobial prediction methods.

In fact, the last two methods are the most important among the methods discussed so far, mainly due to the *in vitro* validation of predictions. Without this kind of validation, these methods become only good hypotheses, without contributing much knowledge. However, they can achieve a more accurate prediction when they are more restricted to some class of AMP, without a generalization model.

3.2 Supervised machine learning methods of AMP prediction

Supervised machine learning methods for predicting antimicrobial activities have a well-established validation procedure, allowing these methods to be compared. The reliability of these methods is evaluated by several parameters, the main three being calculated as follows:

$$Sensitivity = \frac{TP}{TP+FN} \times 100 \tag{2}$$

$$Specificity = \frac{TN}{TN+FP} \times 100 \tag{3}$$

$$Accuracy = \frac{TP+TN}{TP+TN+FN+FP} \times 100 \tag{4}$$

TP corresponds to the number of true positives; FN, the false negatives; TN, the true negative; and FP, the false positives. However, the evaluation of precision on positive predictions can be done by calculating the positive predictive value (PPV), given by the following equation:

$$PPV = \frac{Sensitivity}{Sensitivity+(100-Specificity)} \times 100 \therefore \frac{TP}{TP+FP} \times 100 \quad (5)$$

Here, comparisons among the methods are going to be made based on PPV and accuracy values (Table 2), since for discovering novel AMPs it is more important that the probability of a positive prediction be true. Overall, these methods require two data sets, the training set and the blind set. The training set is composed of two subsets, the positive data set (the AMPs) and the negative data set (the non-AMPs). Through these sets the algorithm is trained and then, tested against the blind set. From the results against the blind set, the true positives, negative and false positives and negatives are estimated and then the parameters (*e.g.*, accuracy) are calculated.

There are two major challenges in the usage of supervised machine learning to predict antimicrobial activity: the AMPs' size variation, and the absence of a dataset for non-antimicrobial peptides (Lata *et al.*, 2007). There are at least two choices of positive set, APD (Wang & Wang, 2004) and CAMP (Thomas *et al.*, 2010). Nevertheless, there are no non-antimicrobial data bases to use as a negative set. Another difficulty is the variation in size of AMPs, since the machine learning techniques need fixed length input vectors. Several strategies have been developed in order to overcome these problems.

Lata *et al.* (2007) developed the first supervised machine learning methods for prediction of antimicrobial activity. In this pioneer work, three algorithms were tested: SVM, ANN and quantitative matrices (QM). The positive data set was composed of 436 AMPs from APD and the negative set was composed of an equal number of non-secretory proteins randomly selected from SwissProt. Initially, an SVM model using amino acid composition of whole sequence was built with 20 inputs, one for each amino acid. This model achieves the highest accuracy of all generated models in 5-fold cross validation (89.04%). However, the authors proposed that is impossible to utilize this approach to search for AMPs in genomes or proteins due to the enormous size variation. Thus, it was decided that a fixed length would be used, using binary patterns, where each amino acid is represented by one binary pattern. SVM models with the 5, 10, 15 or 20 first N-terminal residues were constructed. The best accuracy observed in 5-fold cross validation was in SVM model with 15 residues (87.85%). Therefore, another two approaches using SVM were developed, the C-terminal approach (with 15 C-terminal residues) and the N+C-terminal approach (with 30 residues, 15 from N-terminal and 15 from C-terminal). The C-terminal approach achieves an accuracy of 85.16 %, while the N+C-terminal one achieves 92.11% in 5-fold cross validation. The three approaches were applied to QM and ANN. In both cases, the N+C-Terminal approach achieved the best accuracies in 5-fold cross validation, 90.37% and 88.17%, respectively. In a blind data set composed of 24 mature sequences extracted from SwissProt, the N+C-terminal approach had the higher performance in all algorithms, achieving a PPV of 91.66% for all algorithms.

In 2010, this system was improved, but only the SVM was used (Lata *et al.*, 2010). In this new version, the positive data set was composed of 999 AMPs and the negative data set was constructed with an equal number of non-secretory proteins extracted from SwissProt. The blind set was composed of 466 AMPs from SwissProt, none of which were present in the positive set. The N+C-terminal approach continued to show higher accuracy (91.64%). Despite the drop in precision (92.11 to 91.64%), the improved version was more reliable because the number of sequences used in training and testing were higher than the previous version.

Thomas *et al.* (2009) used two other methods in addition to SVM. Random forest (RF) and discriminant analysis (DA) were implemented. RF showed the finest accuracy (93.2%), followed by SVM (91.5%) and DA (87.5%). The positive data set was composed of 2578 AMPs and 4011 sequences derived from SwissProt or randomly generated sequences. 70% of each set was used for training the machines and the other 30% composed the blind set. The algorithms were trained with 275 features, including composition, physicochemical properties and structural characteristics of each amino acid. In contrast to that of Lata *et al.* 2007 and 2010, the method developed by Thomas *et al.* (2009) was able to predict antimicrobial activity for sequences with variable size.

Method	Positive Set	Negative Set	Accuracy (%)	PPV (%)	References
SVM	APD	SwissProt	92.11	92.11	Lata *et al.*, 2007
QM	APD	SwissProt	90.37	90.65	Lata *et al.*, 2007
ANN	APD	SwissProt	88.17	88.17	Lata *et al.*, 2007
SVM	APD	SwissProt	91.64	90.43	Lata *et al.*, 2010
RF	CAMP	SwissProt and Random Sequences	93.20	95.13	Thomas *et al.*, 2010
SVM	CAMP	SwissProt and Random Sequences	91.50	93.42	Thomas *et al.*, 2010
DA	CAMP	SwissProt and Random Sequences	87.50	87.45	Thomas *et al.*, 2010
SVM	APD	Random Transmembrane Sequences	83.02	96.54	Porto *et al.*, 2010
ANN	CAMP	SwissProt	89.20	88.60	Torrent *et al.*, 2011
ANN	AMP Library	Non-AMP Library	86.50	92.94	Fjell *et al.*, 2009

Table 2. Supervised machine learning methods of antimicrobial prediction.

Our group has developed an SVM model based on physicochemical properties for prediction of peptides with cysteine knot motifs (Porto *et al.*, 2010). Despite the absence of direct correlation between antimicrobial activity and physicochemical properties, their use solved the problem of size variation. However, it generated another problem, which is that shuffled sequences have the same scalar properties, since they are simple averages and the order of residues does not imply average modifications. That problem is avoided by including the hydrophobic moment, since the modification of sequence clearly modifies the hydrophobic moment. For the second challenge, a set of predicted transmembrane proteins was used as the set of non-antimicrobial peptides, since the transmembranes are non-secretory proteins. Through this approach, an overall accuracy of 83% was observed in a blind dataset. This model can be helpful to predict antimicrobial activity of a wide number of cysteine-stabilized peptides, such as conotoxins, proteinase inhibitors, metallothioneins, defensins and cyclotides. The only requirement is the presence of disulphide bonds in the peptide structure.

Also using physicochemical properties, Torrent *et al.* (2011) developed an ANN with eight properties: isoelectric point (pI), peptide length, α-helix, β-sheet and turn structure propensity, *in vivo* and *in vitro* aggregation propensity and hydrophobicity. The main data set was composed of 1157 AMPs from CAMP and 991 non-AMPs from SwissProt. The training set was composed of 1074 peptides, while the testing and validation sets contained 537 peptides. The system achieved an accuracy of 90%. Indeed, the aggregation propensity was seen to be crucial for this method in much the same way as the hydrophobic moment in the method developed by Porto *et al.* (2010). The aggregation propensity changes if the sequence is shuffled, but the other six properties do not. When the aggregation parameter is removed, the system's reliability decreases.

The methods discussed so far show that AMP size variation problem is easy to solve, by using fixed sizes or physicochemical properties. Both strategies achieve similar accuracies. On the other hand, the non-AMP data set only seems to be easier to solve by using random proteins or proteins from SwissProt. However, comparing AMPs to randomly selected proteins from SwissProt is almost the same as comparing oranges to strawberries; it is relatively easy to distinguish each, generating high accuracies. Moreover, as shown in section 2, two peptides with high identities and subsequently similar properties can have different activities, as is the case of peptides derived from pardaxin (Thennarasu & Nagaraj, 1996) and LL-37 (Sigurdardottir *et al.*, 2006).

Lately, a combined approach of QSAR and machine learning techniques has been developed (Fjell *et al.*, 2009). Through 44 QSAR descriptors, an ANN was built based on 1433 random nine-mer peptides. The ANN was trained to predict sequence activity in relation to the control peptide Bac2A. For model evaluation, a library of nine-mer peptides composed of approximately 100,000 sequences was screened *in silico*. These sequences were divided into four classes: (I) most likely to be more active than the control; (II) likely to be more active than the control; (III) likely to be less active than the control; and (IV) most likely to be less active than the control. The topmost 50 positions of each class were synthesized and tested. For class I, an accuracy of 94% was observed, although the overall accuracy was around 85%.

The methods discussed here show that the great difficulties in antimicrobial activity prediction are the absence of a non-antimicrobial database and the enormous variation in sequence size. The greatest challenge for prediction methods is perhaps the heterogeneity of AMPs, which are part of a group with different sequences, structures and mechanisms of action.

4. Conclusions and prospects

In the future, novel treatments against resistant bacteria should be developed, including strategies that use unnatural AMPs as their basis. The development of unnatural AMPs can be carried out by various methods, including those discussed here. This kind of study brings new knowledge and also generates novel AMPs, in turn boosting the development of prediction methods that can help evaluate rational designed AMPs. A more accurate prediction model may be developed when the patterns of the linguistic model can be used to train machine learning techniques. In addition, a more efficient approach to pattern recognition is needed, since a single sequence is insufficient for patterns identification. In this view TEIRESIAS could be used once two or more sequences were need by this approach. This methodology will be helpful not only for novel AMPs development, but also for other protein chemistry fields. The development of this new methodology is a real challenge and could reduce current limitations, leading us to develop novel and more potent antimicrobial peptide agents.

5. References

Ahn, H.S.; Cho, W.; Kang, S.H.; Ko, S.S.; Park, M.S.; Cho, H. & Lee, K.H. (2006). Design and synthesis of novel antimicrobial peptides on the basis of a helical domain of Tenecin 1, an insect defensin protein, and structure-activity relationship study. *Peptides*. Vol.27 ,No.1, pp. 640-648

Blondelle, S.E.; Takahashi, E.; Houghten, R.A. & Pérez-Payá, E. (1996). Rapid identification of compounds with enhanced antimicrobial activity by using conformationally defined combinatorial libraries. *Biochem J*. Vol.313, pp. 141-148

Bradshaw, J. (2003). Cationic antimicrobial peptides: issues for potential clinical use. *BioDrugs*. Vol.17, No.4, pp.233-240

Breidenstein, E.B.; de la Fuente-Núñez, C. & Hancock, R.E. (2011). *Pseudomonas aeruginosa*: all roads lead to resistance. *Trends Microbiol*. Vol.19, No.8, pp. 419-426

Brogden, K.A. (2005). Antimicrobial peptides: pore formers or metabolic inhibitors in bacteria? *Nat Rev Microbiol*. Vol.3, No.3, pp. 238-250

Chen, Y.; Mant, C.T.; Farmer, S.W.; Hancock, R.E.W.; Vasil, M.L. & Hodges, R.S. (2005). Rational design of α-helical antimicrobial peptides with enhanced activities and specificity/therapeutic index. *J Biol Chem*. Vol.280, No.13, pp. 12316-12329

Dathe, M.; Nikolenko, H.; Meyer, J.; Beyermann, M. & Bienert, M. (2001). Optimization of the antimicrobial activity of magainin peptides by modification of charge. *FEBS Lett*. Vol.541, pp. 146-150

Dathe, M.; Wieprecht, T.; Nikolenko, H.; Handel, L.; Maloy, W.L.; MacDonald, D.L.; Beyermann, M. & Bienert M. (1997). Hydrophobicity, hydrophobic moment and angle subtended by charged residues modulate antibacterial and haemolytic activity of amphipathic helical peptides. *FEBS Lett*. Vol.403, No.2, pp. 208-212

Drin, G. & Antonny, B. (2010). Amphipathic helices and membrane curvature. *FEBS Lett*. Vol.584, pp. 1840-1847.

Eisenberg, D.; Weiss, R.M.; Terwilliger, T.C. & Wilcox, W. (1982). Hydrophobic moments and protein structure. *Faraday Symp Chem Soc*. Vol.17, pp. 109-120

Fernandes, F.C.; Porto, W.F. & Franco, O.L. (2009). A wide antimicrobial peptides search method using fuzzy modeling. *LNCS*. Vol.5676, pp. 147-150

Fjell, C.D.; Jenssen, H.; Hilpert, K.; Cheung, W.A.; Panté, N.; Hancock, R.E.W. & Cherkasov, A. (2009). Identification of novel antibacterial peptides by chemoinformatics and machine learning. *Journal of Medicinal Chemistry*. Vol.52, No.1, pp. 2006-2015

Giangaspero, A.; Sandri, L. & Tossi, A. (2001). Amphipathic α helical antimicrobial peptides, a systemic study of the effects of structural and physical properties on biological activity. *Eur J Biochem*. Vol.268, pp. 5589-5600

Gordon, Y.J., Romanowski, E.G. & McDermott, A.M. (2005). A review of antimicrobial peptides and their therapeutic potential as anti-infective drugs. *Curr Eye Res*. Vol.30, No.7, pp. 505-515

Jenssen, H.; Lejon, T.; Hilpert, K.; Fjell, C.D.; Cherkasov, A. & Hancock, R.E.W. (2007). Evaluating Different Descriptors for Model Design of Antimicrobial Peptides with Enhanced Activity Toward *P. aeruginosa*. *Chem Biol Drug Des*. Vol.70, pp. 134-142

Jiang, Z.; Vasil, A.I.; Gera, L.; Vasil, M.L. & Hodges, R.S. (2011). Rational design of α-helical antimicrobial peptides to target gram-negative pathogens, *Acinetobacter baumannii* and *Pseudomonas aeruginosa*: utilization of charge, 'specificity determinants,' total hydrophobicity, hydrophobe type and location as design parameters to improve the therapeutic ratio. *Chem Biol Drug Des*. Vol.77, pp. 225-240

Kite, J. & Doolittle, R.F. (1982). A simple method for displaying the hydropathic character of a protein. *J Mol Biol*. Vol.157, pp. 105-132

Lacroix, E.; Viguera A.R. & Serrano, L. (1998). Elucidating the folding problem of α-helices: Local motifs, long-range electrostatics, ionic strength dependence and prediction of NMR parameters. *J Mol Biol*. Vol.284, pp. 173-191

Landon, C.; Barbault, F.; Legrain, M.; Guenneugues, M. & Vovelle, F. (2008). Rational design of peptides active against the gram positive bacteria *Staphylococcus aureus*. *Proteins*. Vol.72, pp. 229-239

Lata, S.; Mishra, N.K. & Raghava, G.P.S. (2010). AntiBP2: improved version of antibacterial petides prediction. *BMC Bioinformatics*. Vol.11

Lata, S.; Sharma, B.K. & Raghava, G.P.S. (2007). Analysis and prediction of antibacterial peptides. *BMC Bioinformatics*. Vol.8, No.236

Loose, C.; Jensen, K.; Rigoutsos, I. & Stephanopoulos, G. (2006). A linguistic model for the rational design of antimicrobial peptides. *Nature*. Vol.443, No.1, pp. 867-869

Mandal, S.M.; Dey, S.; Mandal, M.; Maria-Neto, S. & Franco, O.L. (2009). Identification and structural insights of three novel antimicrobial peptides isolated from green coconut water. *Peptides*. Vol.30, No.4, pp. 633-637

Nagarajan, V.; Kaushik, N.; Murali, B.; Zhang, C.; Lakhera, S.; Elasri, M.O. & Deng, Y. (2006). A Fourier Transformation based Method to Mine Peptide Space for Antimicrobial Activity. *BMC Bioinformatics*. 7(Suppl 2):S2.

Nijnik, A. & Hancock, R.E.W. (2009). Host defence peptides: antimicrobial and immunomodulatory activity and potential applications for tackling antibiotic-resistant infections. *Emerging Health Threats Journal*. Vol.2, No.e1, pp. 1-7

Pfeifer, Y.; Cullik, A. & Witte, W. (2010). Resistance to cephalosporins and carbapenems in Gram-negative bacterial pathogens. *Int J Med Microbiol*. Vol.300, No.6, pp. 371-379

Poirel, L.; Naas, T. & Nordmann, P. (2010). Diversity, epidemiology, and genetics of class D beta-lactamases. *Antimicrob Agents Chemother*. Vol.54, No. 1, pp. 24-38

Porto, W.F.; Fernandes, F.C. & Franco, O.L. (2010). An SVM model based on physicochemical properties to predict antimicrobial activity from protein sequences with cysteine knot motifs. *LNCS*. Vol.6268, pp. 59-62

Rigoutsos, L. & Floratos, A. (1998). Combinatorial pattern discovery in biological sequences: The TEIRESIAS algorithm. *Bioinformatics*. Vol.14, pp. 55-67

Sigurdardottir, T.; Andersson, P.; Davoudi, M.; Malmsten, M.; Schmidtchen, A. & Bodelsson, M. (2006). *In silico* identification and biological evaluation of antimicrobial peptides based on human cathelicidin LL-37. *Antimicrob Agents Chemother*. Vol.50, No.9, pp. 2983-2989

Silva, O.N.; Mulder, K.C.; Barbosa, A.A.; Otero-Gonzalez, A.J.; López-Abarrategui, C.; Dias, S.C.; Rezende, T.M. & Franco, O.L. (2011). Exploring the pharmacological potential of promiscuous host-defense peptides: from natural screenings to biotechnological applications. *Front Microbiol.* Vol.2

Thennarasu, S. & Nagaraj, R. (1996). Specific antimicrobial and hemolytic activities of 18-residue peptides derived from the amino terminal region of the toxin pardaxin. *Protein Eng.* Vol.9, No.12, pp. 1219-1224

Thomas, S.; Karnik, S.; Barai, R.S.; Jayaraman, V.K. & Idicula-Thomas, S. (2010). CAMP: a useful resource for research on antimicrobial peptides. *Nucl Acid Res.* Vol.38, Database issue.

Torrent, M.; Andreu, D.; Nogués, V.M. & Boix, E. (2011). Connecting peptide physicochemical and antimicrobial properties by a rational prediction model. *PLoS ONE.* Vol.6, No.2

Tossi, A.; Tarantino, C. & Romeo, D. (1997). Design of synthetic antimicrobial peptides based on sequence analogy and amphipathicity. *Eur J Biochem.* Vol.250, pp. 549-558

Ueno, S.; Minaba, M.; Nishiuchi, Y.; Taichi, M.; Tamada, Y.; Yamazaki, T. & Kato, Y. (2011). Generation of novel cationic antimicrobial peptides from natural non-antimicrobial sequences by acid-amide substitution. *Ann Clin Microbiol Antimicrob.* Vol.10, No.11

Wang, Z. & Wang, G. (2004). APD: the Antimicrobial Peptide Database. *Nucl Acid Res.* Vol.32, No.1, pp. D590-D592

Wieczorek, M.; Jenssen, H.; Kindrachuk, J.; Scott, W.R.P.; Elliott, M.; Hilpert, K.; Cheng, J.T.J.; Hancock, R.E.W. & Straus, S.K. (2010). Structural studies of a peptide with immune modulating and direct antimicrobial activity. *Chemistry & Biology.* Vol.17, No.1, pp. 970-980

Wu, M. & Hancock, R.E.W. (1999). Improved Derivatives of Bactenecin, a Cyclic Dodecameric Antimicrobial Cationic Peptide. *Antimicrob Agents Chemother.* Vol.43, No.5, pp. 1274-1276

Yount, N.Y. & Yeaman, M.R. (2004). Multidimensional signatures in antimicrobial peptides. *PNAS.* Vol.101, No.1, pp. 7363-7368

Zhang, X. (2010). Human in check: new threat from superbugs equipped with NDM-1. *Protein Cell.* Vol.1, No.12, pp.1051-1052.

17

Phospholipases A_2 Protein Structure and Natural Products Interactions in Development of New Pharmaceuticals

Marcos Toyama[1], Selma D. Rodrigues[1],
Daneila O. Toyama[2], Veronica C.G. Soares[3],
Camila Ap Cotrim[4], Rafael Ximenes[5] and Marcelo L. Santos[6]
*[1]Universidade Júlio de Mesquita Filho,
Campus Experimental do Litoral Paulista,
[2]Universidade Presbiteriana Mackenzie,
[3]Instituto de Biologia, UNICAMP,
[4]Institut für Biochemie und Biotechnologie,
Martin-Luther-Universität Halle-Wittenberg,
[5]Departamento de Fisiologia e Farmacologia UFC Fortaleza,
[6]Universidade Federal de Sergipe,
[1,2,3,5,6]Brasil,
[4]Germany*

1. Introduction

Phospholipases are enzymes that hydrolyze phospholipids into fatty acids and other lipophilic substances. There are four main classes, named A, B, C and D, which are distinguished by the type of reaction they catalyze. The phospholipase A2 (PLA2), EC 3.1.1.4, are enzymes that release fatty acids from the second carbon group of glycerol. These enzymes are common in many living organisms and can be found both intra and extracellularly, which are referred to as secretory PLA2 (sPLA2) and cytosolic PLA2 (cPLA2), respectively (Burke & Dennis, 2009; Dennis, 1997; Diz Filho et al., 2009; Schaloske & Dennis, 2006). The enzymatic action of cPLA2 on lipids, allows the formation of a compound that acts as mediators of inflammatory diseases (Chakraborti, 2003; Dennis, 1994). Furthermore sPLA2 is divided in 5 different groups among them group I (sPLA2 from mammalian pancreas) and group II (sPLA2 from venom snake) are the most studied since they are also involved in pathological processes (Fonteh et al., 2000). The focus of this chapter is on the structure and interaction of snake venom sPLA2 with natural compounds. These sPLA2 also present pharmacological activities such as: neurotoxicity, cardiotoxicity, myotoxicity, edema formation and hemorrhagic effects (Kini Chan, 1999; Valentin & Lambeau, 2000). Due to the implication of these enzymes in the inflammatory process, several studies have been proposed in order to find new compounds able to inhibit the action of sPLA2.

2. Structural features and biological activities of sPLA2

Enzymatic reactions promoted by sPLA2 happen in a lipid-aqueous interface and the phospholipase activity is more efficient on substrates such as monolayers, bilayers, micelles, membranes and vesicles with monomolecular dispersed soluble substrates. This phenomenon has been termed "interfacial activation" and includes "inter-facial-binding" enzyme and 'activation' steps (Berg et al., 2001; Chakraborti, 2003; Burke & Dennis, 2009). The proposed mechanism of sPLA2 involves the side chain of Asp99 and His48 and a molecule of water (Scott, 1994). According to the mechanism, a proton in the position 3 of the imidazole ring from His48 is bound to the carboxylic group of Asp99, and rotation is prevented, this way the nitrogen (position 1) is able to attack the molecule of water releasing a molecule of hydroxyl. The hydroxyl through a nucleophilic attack, to tie up the carbon from ester group substrate and an intermediate is formed. Afterwards, the oxygen form ester group attacks the proton (position 1) of imidazole ring to produce a lysophospholipid alcohol and a double bond between carbon and oxygen of ester group is remade and then occurs the hydrolysis of acyl-ester sn-2 releasing two products: a phospholipid and a fatty acid (Verheij et al., 1980).

They are known several isoforms of sPLA2 all of then have a conserved His/Asp catalytic dyad and a Ca^{2+}-binding loop. sPLA2-IB, a pancreatic sPLA2, is characterized by a N-terminal pro-peptide whose proteolytic removal gives rise to a functional enzyme, besides the presence of a Cys11-Cys77 disulfide bonds (group I-specific disulfide), which is a unique bond of pancreas sPLA2s. The group II subfamily (IIA, IIC, IID, IIE and IIF) is characterized by the absence of the pro-peptide and the presence of Cys 49 in the C-terminal extension (group II-specific disulfide). sPLA2-IIF has a long C-terminal extension, which is pro-rich. sPLA2-V is evolutionarily close to the group II subfamily, but has no group II-specific disulfide and no C-5 terminal extension. sPLA2-X has properties of both groups I and II, since it has an N-terminal pro-peptide and both groups I and II-specific disulphide bonds. sPLA2-III is unique considering the central domain sPLA2, which is more similar to bee venom PLA2 than to group I / II / V / X sPLA2s, that is flanked by unique and highly cationic N-terminal and C-terminal domains. C-terminal domain is removed to produce a unique domain, the mature sPLA2 form. The group XII contains two isoforms collection, XIIa and XIIb, whose general structures (except for the catalytic domain and Ca^{2+}-binding site) do not show any homology with other sPLA2s. The catalytic site is replaced by Leu in sPLA2-XIIb, indicating that this enzyme has no catalytic active (Murakami et al., 2010; Murakami et al., 2011).

The sPLA2 compose a superfamily of hydrolases, which can be divided into calcium depended and independent and can be classified mainly based on sequence homology and the position of disulfide bonds. Recently, a large number of new enzymes that hydrolysis glycerophospholipids in the sn-2 position have been characterized and classified into three main groups and several subgroups were extended. There is structural similarity between sPLA2 and that its catalytic domain is characterized by a three-tier architecture employing a being saved the Asp catalytic dyad instead of the classical catalytic triad, these similarities and other structural and biochemical evidences suggest that sPLA2 evolved a common ancestral gene. The information gathered on the physiological action, structure and functioning of these enzymes suggest that in addition to its complex distribution, these enzymes may have different regulatory mechanisms and functional roles (Schaloske, 2006; Murakami et al., 2011).

The sPLA2 molecules are well characterized and established as important enzymes. Evidences show that some of these sPLA2 are involved in the release of arachidonic acid from cellular phospholipids for the biosynthesis of eicosanoids in the inflammatory process (Lambeau, 2008; Lättig et al., 2007; Murakami et al., 1997; Oliveira et al., 2008; Toyama et al., 2011; ValentinLambeau, 2000). Although several experimental data show that sPLA2 may also modulate the activity of cPLA2 and that this modulation may be mediated through specific cell receptors for sPLA2, there is still a discrete mechanism elucidation of the molecular events generated by the interaction of sPLA2 with their receptors (Fonteh et al., 2000; Boilard et al., 2010).

Experimental evidence shows that certain sPLA2 Group IIA, V, X can have many different pro-atherogenic properties in the blood vessel walls. This pro-atherogenic activity depends on the generation of pro-inflammatory lipid mediators such as prostaglandins, thromboxanes, leukotrienes, and lysophospholipids; hydrolysis of low-density lipoprotein (LDL) and conversion of them into more pro-atherogenic particles, besides promote multiple inflammatory processes in various artery wall cells (Boyanovsky et al., 2010; (Divchev, 2008; Hanasaki, 2002; Jayaraman et al., 2011; Murakami, 2003; Oestvang & Johansen, 2006; Rosengren et al., 2006; Webb, 2005). Scientific evidence supports the importance of sPLA2 in physiological and pathological processes.

Due to the large variability of the structure and function of sPLA2, they can be considered the most important class of phospholipases. Therefore this important class of enzymes is promising as a therapeutic target for the development of new drugs sPLA2 inhibitors which may be used in the treatment and control of cellular processes, such as acute inflammation.

2.1 Natural compounds as inhibitors of sPLA2

Due to the role of sPLA2 in the inflammatory process, there is interest in pharmacological inhibitors of sPLA2 from natural products isolated from plants, algae and other sources. Polyphenols constitute one of the most abundant and widely distributed plant secondary metabolites, present in plants that are commonly consumed in the diet including various grains, vegetables, fruits, extra virgin olive oil, red wine and tea (Garcia Salas et al., 2010; Yang et al., 2006).

Polyphenols have been reported to have a wide range of biological activities. Many studies show that the beneficial effects of phenolic compounds and suggests its role as a promising therapeutic tools in several acute and chronic diseases. They are extensively metabolized *in vivo* and several studies have focused on the interaction of polyphenols with intracellular proteins involved in vital signalling pathways to cell function. One important anti-inflammatory mechanism played by polyphenols is the inhibition of eicosanoid generating enzymes, including PLA2, cyclooxygenase and lipoxygenase, thereby reducing the concentration of prostanoids and leukotrienes. Arachidonic acid (AA) is released from membrane phospholipids by PLA2 cleavage, which can be metabolized by cyclooxygenases (COX) into prostaglandins (PGs) and thromboxane A2 (TXA2), or by lipoxygenases (LOX) to hydroperoxyeicosatetraenoic acids (HPETE), hydroxyeicosatetraenoic acids (HETES) and leukotrienes (Lts) (Blanchard et al., 1998; Kim et al., 2008; Tanaka, 1995; Terracciano et al., 2006).

Among the polyphenolic compounds, flavonoids are the most common in human diet. Studies have shown that the ability of sPLA2 inhibition by flavonoids is related to structural features, such as: the 5-hydroxyl group and the double bond and the double bonded oxygen in the ring oxana, and that the groups in the 3'-hydroxyl and 4 ' -position are necessary for the selective inhibition of sPLA2 (Lindahl, 1997). However, the exact mechanism by which flavonoids inhibit sPLA2 remains unclear. Iglesias et al. (2005) showed that morin modifies the secondary structure of sPLA2 from *Crotalus durissus cascavella* venom, but did not significantly affect its pharmacological activity.

Flavonoids and other natural compounds isolated from plants have shown promising results in the development of anti-inflammatory candidates, because they may interact with key enzymes such as PLA2, COX and LOX. In this chapter is presented an overview of general techniques that can be used to measure the effectiveness of new sPLA2 inhibitory compounds, including techniques for analyzing proteins in tanden using HPLC detectors in combination with circular dichroism, fluorescence, UV-Vis and other techniques such as mass spectrometry and molecular docking. The sPLA2 isolated from *Crotalus durissus* and *Bothrops jararacussu* are the two main models used by our research group as molecular targets to study mechanisms of different anti-inflammatory compounds, because they are abundant in snake venom and structurally well characterized with atomic coordinates deposited in protein data banks (Chioato & Ward, 2003 Oliveira, Fonseca, Antunes et al., 2008; Lomonte et al., 2003; Nunes et al., 2009; Soares et al., 2003).

Furthermore, integrated biochemical-pharmacological techniques such as paw and skin edema, mast cell degranulation and myotoxicity are shown here. Finally, techniques for measuring enzyme activity using specific substrates for sPLA2 are also explained.

2.2. Experimental procedures to evaluate interactions between sPLA2 and natural compounds

2.2.1 Snake venom sPLA2: a model for molecular target of NC

sPLA2 used for assessing potential inhibitors from natural compounds in our research group are usually isolated from *Crotalus durissus terrificus* and *Bothrops jararacussu* venoms. The first sPLA2 was purified by crystallization in 1938 by Slotta and Fraenkel-Conrat from the venom of *Crotalus terrificus* (*Crotalus durissus terrificus*) and has been one of the catalytically active sPLA2 molecules better characterized in terms of cellular interaction and an example of cooperation between different venom components, because when sPLA2 is associated with crotapotin, the pharmacological and biochemical properties change (Nunes, Zychar, Della-Casa et al., 2009). The sPLA2 from *Bothrops jararacussu* venom has been studied since the 80's and the great interest in this poison is the presence of a non-catalytically active sPLA2 (BthTX-I), which represents approximately 28% of dried whole venom. In addition, the whole venom of *Bothrops jararacussu* has approximately 7 to 8% of catalytically active sPLA2 (BthTX-II) (Lomonte, Angulo, Santamaria, 2003). These two sPLA2 have already been well characterized in terms of their pharmacological, physiological, biochemical and structural properties, highlighting good crystallographic data deposited in proteins data banks. The fractioning of these two venoms have been determined and maximized to obtain the maximum achievement of sPLA2 (Chioato, 2003).

For the isolation and purification of sPLA2 from the venom of *Crotalus durissus terrificus*, two chromatographic steps are done. First, venom of *Crotalus durissus terrificus* is fractionated on a molecular exclusion HPLC column, which allows the purification of the major toxin groups: convulxin (CLC, ~ 85kDa), gyroxin (Gyr, ~ 50 kDa), crotoxin (Crtx, ~ 35kDa) and crotamine (Crot, ~ 10kDa). The crotoxin fraction is then subjected to a new chromatographic run on a reversed-phase HPLC column, isolating the catalytically active sPLA2.

Bothrops jararacussu venom has two types of sPLA2, the catalytically active (D49) and catalytically inactive (K49) sPLA2s. Both proteins account for approximately 35% of dried venom. Again, two chromatographic steps are performed to obtain these sPLA2s. The initial step comprises the fractioning of whole venom on a column of Cation Exchange Chromatography (CEC), which allows to obtain both catalytically active sPLA2 (D49, BhtTx-II) and catalytically inactive sPLA2 (K49, BthTX-II). These two fractions are then subjected to a new chromatographic run using a reverse phase HPLC column. All sPLA2 obtained from both venoms of *Crotalus durissus terrificus* and *Bothrops jararacussu* are examined to characterize its enzymatic activity and molecular weight by MALDI-TOF mass spectrometry (Fig. 1).

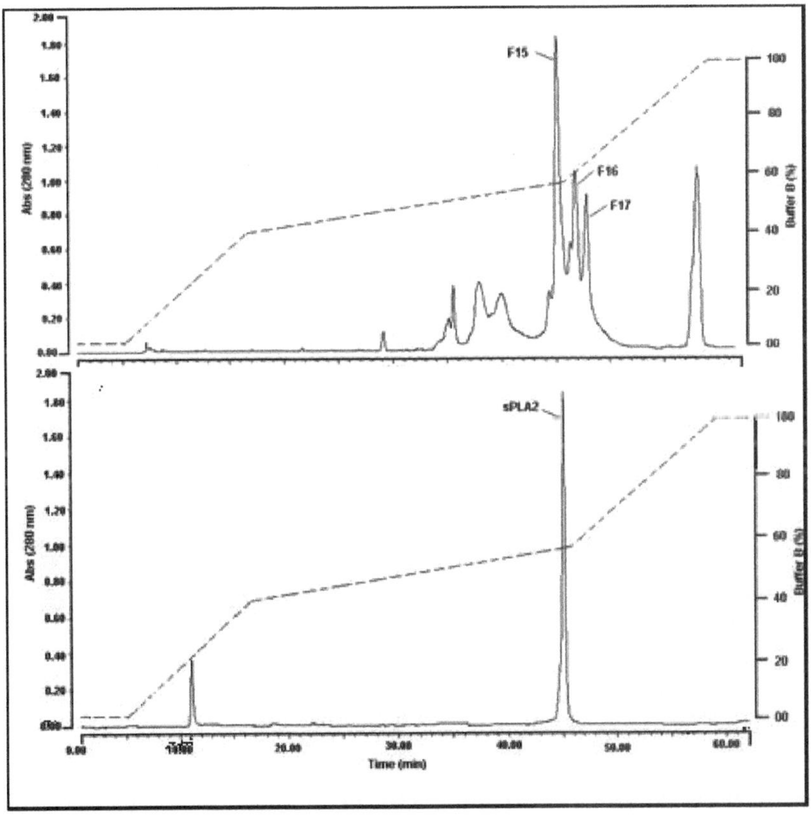

Fig. 1. Profile of purification of the whole venom of *Crotalus durissus terrificus* (Cotrim et al., 2010).

2.2.2 Natural compounds (bioassay-guided fractioning of secondary metabolites)

Isolation of natural compounds (NC) with specific pharmacological activities requires a bioassay-guided fractioning which usually uses crude extracts and polarity/chromatographic fractioning to find active fractions for the specific pharmacological effects wanted (Militao et al., 2007). Herein, sPLA2 and specific substrate 4-nitro-3-octanoyloxy-benzoic acid (4NOB3A) were used to screening samples with inhibitory properties (Cotrim, De Oliveira, Diz Filho et al., 2010). Briefly, crude extract is tested and if it inhibited the substrate cleavage by the enzyme, this extract is subjected to different extractions with increasing polarity solvents. Then, these new fractions are tested and those with inhibitory properties subjected to chromatographic techniques to isolate secondary metabolites according to their chemical classes as described by (Proença Da Cunha Roque, 2009).

2.2.3 Treatment of sPLA2 with NC

The development of this kind of protocol is important to create an adequate environment for molecular interaction between sPLA2 and NC, and allow the isolation of sPLA2 chemically treated, without any kind of interference. In some cases, it is possible to obtain also the modified NC. The incubation of sPLA2 with NC (mol:mol), followed procedures described by (Iglesias, Aparicio, Rodrigues-Simioni et al., 2005). Natural compounds were dissolved in dimethyl sulphoxide (DMSO) at maximum concentration of 1%. Purified sPLA2 (1.5 mg, 100 nmol/mL) was dissolved in 1000 μl of water and after complete homogenization, 10 μl of NC solution (100 nmol) was added and the samples were warmed for 30 minutes in a water bath at 37 °C. Samples of 200 μl of this mixture were loaded onto a preparative reverse phase HPLC column to separate the modified sPLA2 (sPLA2:NC) from NC. Samples were eluted using a discontinuous buffer gradient (66.6% of acetonitrile in TFA 0.1%) at a constant flow rate of 2.0 mL/min. The chromatographic run was monitored at Λ280 nm.

After purification of chemically treated sPLA2 with NC (sPLA2: NC), assessments on the ability of NC to inhibit sPLA2 are carried out. Most of the NC are coloured substances that can interfere with the measurements of inhibition of enzymatic activity, since the reaction product, of sPLA2 with 4N3OBA, absorbs at 405 - 425 nm. This substrate was tested and optimized for simple microplate assay for PLA2 in human serum. Using this substrate, PLA2 activity levels were similar to that measured with the previously characterized chromogenic phospholipid substrate 1,2-bis heptanoylthioglycerophosphocholine, which is one of the substrates routinely used to determine the activity enzymatic of PLA2. The use of NOB substrate introduces some technical advantages over the use of 1,2-bis heptanoylthioglycerophosphocholine, in terms of one step product quantification as well as its stability. The evaluation of residual enzyme activity of sPLA2 is a standard procedure to check the ability of NC to reduce the enzymatic activity of these enzymes. The protocol for measuring the activity of the enzyme following protocols described by (Rigden et al., 2003) and modified by (Toyama et al., 2000) or the 96-well plate, using 4N3OBA as substrate is follow described.

Enzymatic activity is expressed as the initial velocity of the reaction (Vo) and is calculated from the increase in absorbance after 30 min of experiment. All assays are performed using n = 12 and absorbance at 425 nm are measured using a SpectraMax 340 multi-wells plate reader (Molecular Devices, Sunnyvale, CA). After the addition of sPLA2 native or treated with NC (20 mg), the reaction mixture is incubated for 30 min at 37°C and absorbance is read at 10 min intervals. Moreover, using this procedure is possible to estimate the minimum inhibitory concentration of NC on the enzymatic activity of sPLA2 in this way can be assessed in a comparative inhibitory activity of different NC.

In addition, sPLA2 treated with NC allows us to estimate structural changes induced by their interaction. The detection of fluorescence from either specific molecular probes or the inherent fluorescence of molecules allows the investigation of the interactions between of sPLA2 and NC. Monitoring the intrinsic fluorescence of tryptophan (Trp) is used to estimate protein unfolding, but this technique is not efficient when the NC absorbs the same wavelength that the Trp.

Circular Dichroism (CD) spectroscopy is a powerful technique that is more sensitive to evaluate the molecular interactions between NC and sPLA2. CD spectroscopy is an extremely valuable technique for the study of the secondary structural components of proteins, such as α-helix and β-sheet. In this technique, circularly polarized light interacts differently with chiral centers and is absorbed. CD is characterized as a very fast way for obtaining information about protein structural integrity, conformational changes and folding-unfolding process (Corrêa, 2009). Recently, such method has been used for the evaluation of the sPLA2 treatment with different NC, mainly flavonoids. (Cotrim, De Oliveira, Diz Filho et al., 2010; De Oliveira et al., 2009; Iglesias, Aparicio, Rodrigues-Simioni et al., 2005; Santos et al., 2011; Toyama, Diz Filho, Cavada et al., 2011). In these studies native and modified PLA2 with different flavonoids (naringin, quercetin and morin) were prepared in 10 mM sodium phosphate buffer pH 7.4. Samples were transferred to quartz cuvettes with an optical path length of 1 mm.CD spectra in the wavelength range 185-300 nm were acquired in-house on a J720 spectropolarimeter (Jasco Corp., Japan) using a bandwidth of 1 nm and a response between 1-4 s. Scans were accumulated for each sample and all spectra were corrected by subtraction of buffer blanks. Comparison of spectra was performed after concentration and optical path length normalization. Results from CD revealed significant changes in the secondary structure composition of sPLA2 after treatment with flavonoids. In order to correlate CD spectral information and sPLA2 structure, homology modeling and secondary structure prediction using SWISS-MODEL (Schwede et al., 2003) and PSIPRED (Mcguffin et al., 2000) web servers, respectively, were performed. This evaluation suggested that CD signal alteration was consequence of modifications of mainly helical components, which in sPLA2 are responsible for forming the hydrophobic channel (Arni, 1996).

2.2.4 Mass spectroscopy

Mass Spectrometry (MS) can also be used for the evaluation of sPLA2 modifications induced by NC treatment, as was previously done to naringin and quercetin (Cotrim, De Oliveira, Diz Filho et al., 2010; Santos, Toyama, Oliveira et al., 2011). This technique can be applied to measure the mass-to-charge ratio of native and NC-treated PLA2. In these studies, mass measurements are performed through matrix-assisted laser desorption/ionization-time-of-flight (MALDI-TOF) using a Voyager-DE PRO MALDI-TOF mass spectrometer (Applied Biosystems). Samples of one microliter are mixed with 2 μL of the matrix α-cyano-4-hydroxycinnamic acid, 50% acetonitrile, and 0.1% TFA (v/v). The matrix is prepared with

30% acetonitrile and 0.1% TFA (v/v). Ion masses are determined under the following conditions: accelerate voltage 25 kV, the laser operated at 2890 μJ/com2, delay of 300 nanoseconds, and in the linear analysis mode. Results from MALDI-TOF mass spectrometry displayed an increasing in sPLA2 mass-to-charge ratio after NC treatment in comparison with the native protein if chemical stable complexes are formed. Our data shows that chemical shift observed for naringin treatment was discrete; therefore, modifications on the PLA2 structure as consequence of this flavonoid action did not indicate an attachment of such specie, similarly to the results found in a previous study of sPLA2 treated with umbelliferone (Toyama, Diz Filho, Cavada et al., 2011).

2.2.5 Molecular docking

When the target protein has its crystal structure elucidated and deposited in databases online, these data can be used to perform a virtual screening of possibly active compounds which bind in specific regions of protein surface. For this purpose, the structural optimization of the natural compounds are initially achieved using the quantum chemical AM1 method (Tang et al., 1999) implemented in the BioMedCache program with default values for the convergence criteria. Docking calculations are performed with the GOLD 4.0 program (Jones et al., 1997) to obtain the in silica affinity of natural compounds to the target proteins, herein, sPLA2 from snake venoms. These PLA2 structures were taken from the RCSB Protein Data Bank [PDB]. Usually, if the chains are homologous and the active sites (docking regions) are not close in the dimeric form, only one chain is chosen for calculations. Docking calculations are performed to consider the flexibility of the natural compound ligand in such a way that torsions were considered active during the calculation. The active site is defined as all atoms within a 10 °A radius from His48, an important residue according to the literature (Scott, 1994; Scott et al., 1990). When the docking is performed in Asp49 PLA2s, calculation maintaining the Ca2+ ion and the coordination water molecule is also done (Fig. 2; Fig. 3).

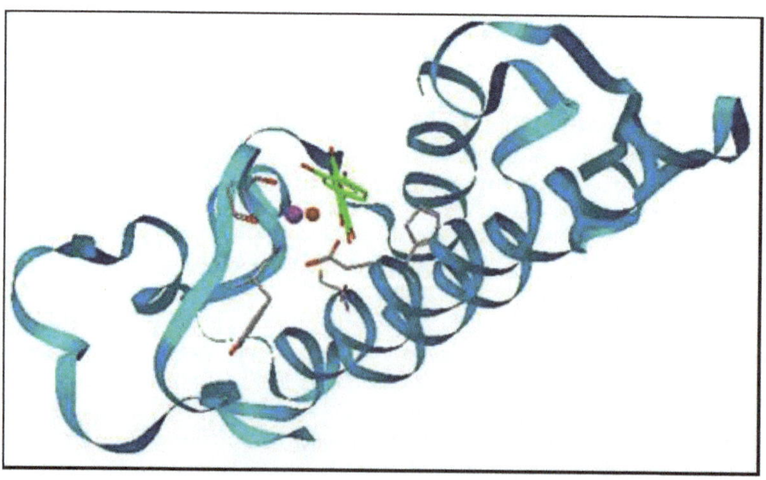

Fig. 2. Analysis of the interaction of a natural compound with phospholipase A2 by docking technique (Cotrim, De Oliveira, Diz Filho et al., 2010).

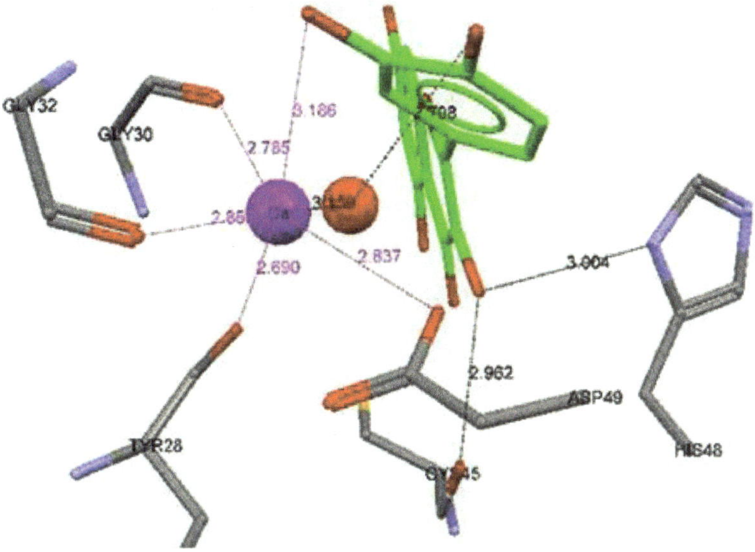

Fig. 3. Analysis of the interaction of a natural compound with phospholipase A2 by docking technique (Cotrim, De Oliveira, Diz Filho et al., 2010).

2.2.6 Small Angle X-ray Scattering (SAXS)

Small Angle X-ray Scattering (SAXS) is a very useful technique for analysing shape and size of macromolecules in solution (Svergun, Koch, 2002). Using SAXS, structural information about tertiary and quaternary structures of proteins at the nanometer level can be recovered from the elastic scattering of X-rays by the sample. Such methodology has been also employed in the study of structural modifications induced in PLA2 by its treatment with different flavonoids, especially naringin (Santos, Toyama, Oliveira et al., 2011). SAXS measurements were carried out at the D02A-SAXS2 beamline of the Brazilian Synchrotron Light Laboratory (LNLS) (Kellermann et al., 1997). Data collection was performed at a sample-to-detector distance of 985.7 mm using a MAR CCD detector (MAR Research) with X-rays at a wavelength of 1.488 Å, to cover q ($q = 4\pi\sin\theta/\lambda$) values ranging from 0.013 to 0.34 Å-1, with samples in different concentrations. X-ray scattering intensities were reduced to a 1D profile using the Fit2D program (Hammersley et al., 1996). The several frames collected for each sample (native and naringin-treated PLA2) were inspected and averaged with the program PRIMUS (Konarev et al., 2003). GNOM program (Svergun, 1992) was used to calculate distance distribution functions, p(r), and radius of gyration, Rg, by the indirect Fourier transform method. Rg values were also obtained employing Guinier analysis (Guinier, 1955). The degree of protein globularization was evaluated by the so-called Kratky plot (Iq2 x q) (Putnam et al., 2007). Molecular weight of proteins was estimated using the web tool "SAXS MoW" (Fischer et al., 2010). Low resolution models were recovered from the scattering curves through the ab initio procedure embedded in the program GASBOR (Svergun et al., 2001). Final envelopes were obtained by spatial average of 50 independent models with the program DAMAVER (Konarev, Volkov, Sokolova et al., 2003). Interpretation of native and naringin-tretated PLA2 SAXS models were performed by the superposition of crystallographic models of several PLA2 found in the Protein Data Bank (PDB) using the program SUPCOMB (Kozin, 2001). The smallest discrepancy (Chi) calculated by the program CRYSOL (Svergun et al., 1995) between experimental data and PLA2 PDB models was achieved to the dimer composed of chains A and E of agkistrodotoxin (PDB ID 1BJJ) (Tang, Zhou, Lin, 1999).

According to our results, SAXS models displayed a dimeric enlogated shape, but completely different from two previous reports published regarding PLA2 SAXS studies (Arni et al., 1999) (Murakami et al., 2007). The most remarkable result derived from our work corresponds to a conformational change observed for the PLA2 dimer after naringin treatment. SAXS envelope obtained for naringin-treated PLA2 exhibited a clear bending in comparison with the models for native PLA2 that has been attributed to the action of naringin in the dimerization interface of this protein (Santos, Toyama, Oliveira et al., 2011).

3. Conclusion

One of the biggest challenges for pharmacology is to find molecules that are able to interact with specific sPLA2 because the side effects are often related to low-affinity inhibitors of the proteins. The screening of inhibitors by in vitro enzyme activity inhibition enables a quick approach to molecules of interest, which will be assessed for their efficiency by techniques of molecular interaction as SAX and molecular docking.

4. Acknowledgment

We gratefully acknowledge the National Laboratory of Synchrotron Light (LNLS) for the use of the Small Angle X-ray Scattering beamlines under proposal D11A-SAXS1-5366. We are also grateful to Daniel Razzo for helping with CD data collection. This work was supported by CNPq, FAPESP and FAEPEX/PRP/UNICAMP.

5. References

Arni, R. K.; Fontes, M. R.; Barberato, C. et al., 1999. Crystal structure of myotoxin II, a monomeric Lys49-phospholipase A2 homologue isolated from the venom of Cerrophidion (Bothrops) godmani. *Arch Biochem Biophys*, v.366, n.2, Jun 15, p.177-82.

Arni, R.; Ward, K.R. J. 1996 Phospholipase A2--a structural review. *Toxicon*, v.34, n.8, Aug, p.827-41.

Berg, O. G.; Gelb, M. H.; Tsai, D. et al., 2001 Interfacial enzymology: the secreted phospholipase A(2)-paradigm. *Chem Rev*, v.101, n.9, Sep, p.2613-54.

Blanchard, S. G.; Andrews, R. C., Brown, P. J., et al. 1998 Discovery of bioavailable inhibitors of secretory phospholipase A2. *Pharm Biotechnol*, v.11, p.445-63.

Boilard, E.; Lai, Y.; Larabee, K. et al. 2010 A novel anti-inflammatory role for secretory phospholipase A2 in immune complex-mediated arthritis. *EMBO Mol Med*, v.2, n.5, May, p.172-87.

Boyanovsky, B. B.; Li. X.; Shridas, P. et al. 2010 Bioactive products generated by group V sPLA(2) hydrolysis of LDL activate macrophages to secrete pro-inflammatory cytokines. *Cytokine*, v.50, n.1, Apr, p.50-7.

Burke, J.E.E. & Dennis, A. 2009 Phospholipase A2 structure/function, mechanism, and signaling. *J Lipid Res*, v.50 Suppl, Apr, p.S237-42.

Chakraborti, S. 2003 Phospholipase A(2) isoforms: a perspective. *Cell Signal*, v.15, n.7, Jul, p.637-65.

Chioato, L.R. & Ward. J. 2003 Mapping structural determinants of biological activities in snake venom phospholipases A2 by sequence analysis and site directed mutagenesis. *Toxicon*, v.42, n.8, Dec 15, p.869-83.

Corrêa, D. & Ramos, C. 2009 The use of circular dichroism spectroscopy to study protein folding, form and function. *African Journal of Biochemistry Research*, v.3, p.164-173.

Cotrim, C. A.; De Oliveira, S. C.; Diz Filho, E. B. et al. 2010. Quercetin as an inhibitor of snake venom secretory phospholipase A2. *Chem Biol Interact*, v.189, n.1-2, Jan 15, p.9-16.

De Oliveira, D.; Murakami, M.; Cintra, A., et al. 2009 Functional and structural analysis of two fibrinogen-activating enzymes isolated from the venoms of Crotalus durissus terrificus and Crotalus durissus collilineatus. *Acta Biochim Biophys Sin (Shanghai)*, v.41, n.1, Jan, p.21-9.

Dennis, E. A. 1994 Diversity of group types, regulation, and function of phospholipase A2. *J Biol Chem*, v.269, n.18, May, p.13057-60.

Dennis, E. A. 1997 The growing phospholipase A2 superfamily of signal transduction enzymes. *Trends Biochem Sci*, v.22, n.1, Jan, p.1-2.

Divchev, D. & Schieffer, B. 2008 The secretory phospholipase A2 group IIA: a missing link between inflammation, activated renin-angiotensin system, and atherogenesis? *Vasc Health Risk Manag*, v.4, n.3, p.597-604.

Diz Filho, E. B.; Marangoni, S.; Toyama, D. O. et al. 2009 Enzymatic and structural characterization of new PLA2 isoform isolated from white venom of Crotalus durissus ruruima. *Toxicon*, v.53, n.1, Jan, p.104-14.

Fischer, H.; M.D.O.N.; Napolitano, H. B. et al. 2010 Determination of the molecular weight of proteins in solution from a single small-angle x-ray scattering measurement on a relative scale. *J. Appl. Cryst.* , p.43-9.

Fonteh, A. N.; Atsumi, G.; Laporte, T. et al. 2000 Secretory phospholipase A2 receptor-mediated activation of cytosolic phospholipase A2 in murine bone marrow-derived mast cells. *J Immunol*, v.165, n.5, Sep 1, p.2773-82.

Garcia-Salas, P.; Morales-Soto, A.; Segura-Carretero, A. et al. 2010 Phenolic-compound-extraction systems for fruit and vegetable samples. *Molecules*, v.15, n.12, p.8813-26.

Guinier, A.G. Fournet. 1955. *Small-Angle Scattering of X-rays*. New York: John Wiley & Sons.

Hammersley, A.; Svensson, S.; Hanfland, M. et al. 1996 Two-dimensional detector software: From real detector to idealised image or two-theta scan. *High. Press. Res*, p.235-248.

Hanasaki, K.H. Arita. 2002 Phospholipase A2 receptor: a regulator of biological functions of secretory phospholipase A2. *Prostaglandins Other Lipid Mediat*, v.68-69, Aug, p.71-82.

Iglesias, C. V.; Aparicio, R.; Rodrigues-Simioni, L. et al. 2005 Effects of morin on snake venom phospholipase A2 (PLA2). *Toxicon*, v.46, n.7, Dec, p.751-8.

Jayaraman, S.; Gantz, D. L.; Gursky, O. 2011 Effects of phospholipase A(2) and its products on structural stability of human LDL: relevance to formation of LDL-derived lipid droplets. *J Lipid Res*, v.52, n.3, Mar, p.549-57.

Jones, G.; Willett, P.; Glen, R. C. et al. 1997 Development and validation of a genetic algorithm for flexible docking. *J Mol Biol*, v.267, n.3, Apr 4, p.727-48.

Kellermann, G.; Vicentin, F.; Tamura, E. et al. 1997 The small-angle x-ray scattering beamline of the brazilian synchrotron light laboratory. . 1997, 30, 4. *J. Appl. Cryst*, v.4, p.30.

Kim, H. J.; Park, S. H.; Park, S. Y. et al. 2008 Epigallocatechin-3-gallate inhibits interleukin-1beta-induced MUC5AC gene expression and MUC5AC secretion in normal human nasal epithelial cells. *J Nutr Biochem*, v.19, n.8, Aug, p.536-44.

Kini, R. M.; Chan, Y. M. 1999 Accelerated evolution and molecular surface of venom phospholipase A2 enzymes. *J Mol Evol*, v.48, n.2, Feb, p.125-32.

Konarev, P.; Volkov, V.; Sokolova, A. et al. 2003 Primus: A windows pc-based system for small-angle scattering data analysis. . *J. Appl. Cryst.* , p.1277-1282.

Kozin, M. & Svergun, D. 2001. Automated matching of high- and low-resolution structural models. . *J. Appl. Cryst.*, p.33-41.

Lambeau, G.M. & Gelb, H. 2008. Biochemistry and physiology of mammalian secreted phospholipases A2. *Annu Rev Biochem*, v.77, p.495-520.

Lättig, J.; Böhl, M.; Fischer, P. et al. 2007 Mechanism of inhibition of human secretory phospholipase A2 by flavonoids: rationale for lead design. *J Comput Aided Mol Des*, v.21, n.8, Aug, p.473-83.

Lindahl, M.C. & Tagesson. 1997 Flavonoids as phospholipase A2 inhibitors: importance of their structure for selective inhibition of group II phospholipase A2. *Inflammation*, v.21, n.3, Jun, p.347-56.

Lomonte, B.; Angulo, Y.; Santamaria, C. 2003 Comparative study of synthetic peptides corresponding to region 115-129 in Lys49 myotoxic phospholipases A2 from snake venoms. *Toxicon*, v.42, n.3, Sep, p.307-12.

Mcguffin, L. J.; Bryson, K.; Jones, D. T. 2000 The PSIPRED protein structure prediction server. *Bioinformatics*, v.16, n.4, Apr, p.404-5.

Militao, G. C.; Pinheiro, S. M.; Dantas, I. N. et al. 2007 Bioassay-guided fractionation of pterocarpans from roots of Harpalyce brasiliana Benth. *Bioorg Med Chem*, v.15, n.21, Nov 1, p.6687-91.

Murakami, M.I. Kudo. 2003 New phospholipase A(2) isozymes with a potential role in atherosclerosis. *Curr Opin Lipidol*, v.14, n.5, Oct, p.431-6.

Murakami, M.; Nakatani, Y.; Atsumi, G. et al. 1997 Regulatory functions of phospholipase A2. *Crit Rev Immunol*, v.17, n.3-4, p.225-83.

Murakami, M.; Taketomi, Y.; Miki, Y. et al. 2010 Recent progress in phospholipase A research: from cells to animals to humans. *Prog Lipid Res*, v.50, n.2, Apr, p.152-92.

Murakami, M.; Taketomi, Y.; Sato, H. et al. 2011 Secreted phospholipase A2 revisited. *J Biochem*, v.150, n.3, Sep, p.233-55.

Murakami, M.; Viçoti, M.; Abrego, J. et al. 2007 Interfacial surface charge and free accessibility to the PLA2-active site-like region are essential requirements for the activity of lys49 PLA2 homologues. *Toxicon*, v.49, p.378-387.

Nunes, F. P.; Zychar, B. C.; Della-Casa, M. S. et al. 2009 Crotoxin is responsible for the long-lasting anti-inflammatory effect of Crotalus durissus terrificus snake venom: involvement of formyl peptide receptors. *Toxicon*, v.55, n.6, Jun 1, p.1100-6.

Oestvang, J. & Johansen, B. 2006 PhospholipaseA2: a key regulator of inflammatory signalling and a connector to fibrosis development in atherosclerosis. *Biochim Biophys Acta*, v.1761, n.11, Nov, p.1309-16.

Oliveira, S. C.; Fonseca, F. V.; Antunes, E. et al. 2008 Modulation of the pharmacological effects of enzymatically-active PLA2 by BTL-2, an isolectin isolated from the Bryothamnion triquetrum red alga. *BMC Biochem*, v.9, p.16.

Proença Da Cunha, A. O. R. R. Obtenção de moléculas com actividade farmacológica a partir de material vegetal e sua transformação em medicamento. In: A. Proença (Ed.). *Farmacognosia e Fitoquímica*. Lisboa: CALOUSTE GULBENKIAN, v.1, 2009. Obtenção de moléculas com actividade farmacológica a partir de material vegetal e sua transformação em medicamento., p.670

Putnam, C. D.; Hammel, M.; Hura, G. L. et al. 2007 X-ray solution scattering (SAXS) combined with crystallography and computation: defining accurate macromolecular structures, conformations and assemblies in solution. *Q Rev Biophys*, v.40, n.3, Aug, p.191-285.

Rigden, D. J.; Hwa, L. W.; Marangoni, S. et al. 2003 The structure of the D49 phospholipase A2 piratoxin III from Bothrops pirajai reveals unprecedented structural displacement of the calcium-binding loop: possiblerelationship to cooperative substrate binding. *Acta Crystallogr D Biol Crystallogr*, v.59, n.Pt 2, Feb, p.255-62.

Rosengren, B.; Jonsson-Rylander, A. C.; Peilot, H. et al. 2006 Distinctiveness of secretory phospholipase A2 group IIA and V suggesting unique roles in atherosclerosis. *Biochim Biophys Acta*, v.1761, n.11, Nov, p.1301-8.

Santos, M. L.; Toyama, D. O.; Oliveira, S. C. et al. 2011 Modulation of the pharmacological activities of secretory phospholipase A2 from Crotalus durissus cascavella induced by naringin. *Molecules*, v.16, n.1, p.738-61.

Schaloske, R. H.; Dennis, E. A. 2006 The phospholipase A2 superfamily and its group numbering system. *Biochim Biophys Acta*, v.1761, n.11, Nov, p.1246-59.

Schwede, T.; Kopp, J.; Guex, N. et al. 2003 SWISS-MODEL: An automated protein homology-modeling server. *Nucleic Acids Res*, v.31, n.13, Jul 1, p.3381-5.

Scott, D. L. P.; Sigler, B. 1994 Structure and catalytic mechanism of secretory phospholipases A2. *Adv Protein Chem*, v.45, p.53-88.

Scott, D. L.; White, S. P.; Otwinowski, Z. et al. 1990 Interfacial catalysis: the mechanism of phospholipase A2. *Science*, v.250, n.4987, Dec 14, p.1541-6.

Soares, A. M.; Marcussi, S.; Stabeli, R. G. et al. 2003 Structural and functional analysis of BmjMIP, a phospholipase A2 myotoxin inhibitor protein from Bothrops moojeni snake plasma. *Biochem Biophys Res Commun*, v.302, n.2, Mar 7, p.193-200.

Svergun, D. 1992 Determination of the regularization parameter in indirect-transform methods using perceptual criteria.. *J. Appl. Cryst.*, p.495-503.

Svergun, D.; Barberato, C.; Koch, M. H. J. 1995 Crysol - a program to evaluate x-ray solution scattering of biological macromolecules from atomic coordinates. *J. Appl. Cryst.*, v.28, p.6.

Svergun, D. I.; Koch, M. H. 2002 Advances in structure analysis using small-angle scattering in solution. *Curr Opin Struct Biol*, v.12, n.5, Oct, p.654-60.

Svergun, D. I.; Petoukhov, M. V.; Koch, M. H. 2001 Determination of domain structure of proteins from X-ray solution scattering. *Biophys J*, v.80, n.6, Jun, p.2946-53.

Tanaka, K.& Arita, H. 1995 Secretory phospholipase A2 inhibitors. Possible new anti-inflammatory agents. *Agents Actions Suppl*, v.46, p.51-64.

Tang, L.; Zhou, Y. C.; Lin, Z. J. 1999 Structure of agkistrodotoxin in an orthorhombic crystal form with six molecules per asymmetric unit. *Acta Crystallogr D Biol Crystallogr,* v.55, n.Pt 12, Dec, p.1986-96.

Terracciano, S.; Aquino, M.; Rodriquez, M. et al. 2006 Chemistry and biology of anti-inflammatory marine natural products: molecules interfering with cyclooxygenase, NF-kappaB and other unidentified targets. *Curr Med Chem,* v.13, n.16, p.1947-69.

Toyama, D. D. O.&E. B. Diz Filho&B. S. Cavada, et al. 2011 Umbelliferone induces changes in the structure and pharmacological activities of Bn IV, a phospholipase A(2) isoform isolated from Bothrops neuwiedi. *Toxicon,* v.57, n.6, May, p.851-60.

Toyama, M.; Carneiro, E.; Marangoni, S. et al. 2000 Biochemical characterization of two crotamine isoforms isolated by a single step RP-HPLC from Crotalus durissus terrificus (South American rattlesnake) venom and their action on insulin secretion by pancreatic islets. *Biochim Biophys Acta,* v.1474, n.1, Mar, p.56-60.

Valentin, E..; Lambeau, G. 2000 Increasing molecular diversity of secreted phospholipases A(2) and their receptors and binding proteins. *Biochim Biophys Acta,* v.1488, n.1-2, Oct, p.59-70.

Verheij, H. M.; Volwerk, J. J.; Jansen, E. H. et al. 1980 Methylation of histidine-48 in pancreatic phospholipase A2. Role of histidine and calcium ion in the catalytic mechanism. *Biochemistry,* v.19, n.4, Feb 19, p.743-50.

Webb, N. R. 2005 Secretory phospholipase A2 enzymes in atherogenesis. *Curr Opin Lipidol,* v.16, n.3, Jun, p.341-4.

Yang, C. S.; Lambert, J. D.; Hou, Z. et al. 2006 Molecular targets for the cancer preventive activity of tea polyphenols. *Mol Carcinog,* v.45, n.6, Jun, p.431-5.

Permissions

The contributors of this book come from diverse backgrounds, making this book a truly international effort. This book will bring forth new frontiers with its revolutionizing research information and detailed analysis of the nascent developments around the world.

We would like to thank Dr. Eshel Faraggi, for lending his expertise to make the book truly unique. He has played a crucial role in the development of this book. Without his invaluable contribution this book wouldn't have been possible. He has made vital efforts to compile up to date information on the varied aspects of this subject to make this book a valuable addition to the collection of many professionals and students.

This book was conceptualized with the vision of imparting up-to-date information and advanced data in this field. To ensure the same, a matchless editorial board was set up. Every individual on the board went through rigorous rounds of assessment to prove their worth. After which they invested a large part of their time researching and compiling the most relevant data for our readers. Conferences and sessions were held from time to time between the editorial board and the contributing authors to present the data in the most comprehensible form. The editorial team has worked tirelessly to provide valuable and valid information to help people across the globe.

Every chapter published in this book has been scrutinized by our experts. Their significance has been extensively debated. The topics covered herein carry significant findings which will fuel the growth of the discipline. They may even be implemented as practical applications or may be referred to as a beginning point for another development. Chapters in this book were first published by InTech; hereby published with permission under the Creative Commons Attribution License or equivalent.

The editorial board has been involved in producing this book since its inception. They have spent rigorous hours researching and exploring the diverse topics which have resulted in the successful publishing of this book. They have passed on their knowledge of decades through this book. To expedite this challenging task, the publisher supported the team at every step. A small team of assistant editors was also appointed to further simplify the editing procedure and attain best results for the readers.

Our editorial team has been hand-picked from every corner of the world. Their multi-ethnicity adds dynamic inputs to the discussions which result in innovative outcomes. These outcomes are then further discussed with the researchers and contributors who give their valuable feedback and opinion regarding the same. The feedback is then collaborated with the researches and they are edited in a comprehensive manner to aid the understanding of the subject.

Apart from the editorial board, the designing team has also invested a significant amount of their time in understanding the subject and creating the most relevant covers. They scrutinized every image to scout for the most suitable representation of the subject and create an appropriate cover for the book.

The publishing team has been involved in this book since its early stages. They were actively engaged in every process, be it collecting the data, connecting with the contributors or procuring relevant information. The team has been an ardent support to the editorial, designing and production team. Their endless efforts to recruit the best for this project, has resulted in the accomplishment of this book. They are a veteran in the field of academics and their pool of knowledge is as vast as their experience in printing. Their expertise and guidance has proved useful at every step. Their uncompromising quality standards have made this book an exceptional effort. Their encouragement from time to time has been an inspiration for everyone.

The publisher and the editorial board hope that this book will prove to be a valuable piece of knowledge for researchers, students, practitioners and scholars across the globe.

List of Contributors

Azhari Aziz
Cell Biology and Molecular Pathogenesis Laboratory, Australia

Paul R. Fisher
Molecular Cell Biology Laboratory, Department of Microbiology, Australia

Naomi E. Bishop
Cell Biology and Molecular Pathogenesis Laboratory, Australia
Olga Tennison Autism Research Centre, La Trobe University, Melbourne, Victoria, Australia

Jasmina Ilievska
Cell Biology and Molecular Pathogenesis Laboratory, Australia
Molecular Cell Biology Laboratory, Department of Microbiology, Australia

Helmut Durchschlag
Institute of Biophysics and Physical Biochemistry, University of Regensburg, Regensburg, Germany

Peter Zipper
Physical Chemistry, Institute of Chemistry, University of Graz, Graz, Austria

Homayoun Valafar and Stephanie J. Irausquin
University of South Carolina, Columbia, SC, USA

Eric Paquet and Herna L. Viktor
National Research Council & University of Ottawa, Canada

Jürgen M. Schmidt
School of Biosciences, University of Kent, Canterbury, Kent, United Kingdom

Frank Löhr
Institute of Biophysical Chemistry, Center for Biomolecular Magnetic Resonance, Goethe-University, Frankfurt am Main, Germany

Jiaan Yang
MicrotechNano LLC, USA

Wei-Hua Lee
Department of Pediatrics and Anatomy, Indiana University School of Medicine, Indianapolis, USA

Kiattisak Lugsanangarm
Department of Chemistry, Faculty of Science, Chulalongkorn University, Thailand

Nadtanet Nunthaboot
Department of Chemistry, Faculty of Science, Mahasarakham University, Mahasarakham, Thailand

Somsak Pianwanit and Sirirat Kokpol
Center of Excellence for Petroleum, Petrochemicals, and Advanced Materials, Chulalongkorn University, Bangkok, Thailand
Department of Chemistry, Faculty of Science, Chulalongkorn University, Thailand

Fumio Tanaka
Department of Chemistry, Faculty of Science, Chulalongkorn University, Thailand
Laser Biochemistry Division, Institute for Laser Technology, Osaka, Japan

Yu-Tung Chien and Shao-Wei Huang
Department of Medical Informatics, Tzu Chi University, Hualien, Taiwan

Jenn-Kang Hwang
Institute of Bioinformatics and Systems Biology, National Chiao Tung University, Hsinchu, Taiwan, R.O.C.

Liina Kangur and Arvi Freiberg
University of Tartu, Estonia

John D. Olsen and C. Neil Hunter
University of Sheffield, United Kingdom

Takahiro Takekiyo and Yukihiro Yoshimura
Department of Applied Chemistry, National Defense Academy, Yokosuka, Japan

Minoru Kato
Department of Pharmacy, Ritsumeikan University, Kusatsu, Japan

Azumi Hirata and Kazufumi Takano
Department of Biomolecular Chemistry, Kyoto Prefectural University, Japan
Department of Material and Life Science, Osaka University, Japan

Aya Sato, Takashi Tadokoro, Yuichi Koga and Shigenori Kanaya
Department of Material and Life Science, Osaka University, Japan

Takashi Sugiyama and Tohru Yoshioka
Kaohsiung Medical University, Taiwan

Olga Stepanenko, Alexander Fonin, Olesya Stepanenko, Irina Kuznetsova and Konstantin Turoverov
Institute of Cytology RAS, Laboratory of Structural Dynamics, Stability and Folding of Proteins, Russia

Ilya M. Mukovozov and Lisa A. Robinson
The University of Toronto, The Hospital for Sick Children, Canada

Chung-Lun Lu and Shiu-Nan Chen
College of Life Science, National Taiwan University, Taiwan

William F. Porto, Osmar N. Silva and Octávio L. Franco
Universidade Católica de Brasília, Centro de Análises Proteômicas e Bioquímicas, Brazil

Marcos Toyama and Selma D. Rodrigues
Universidade Júlio de Mesquita Filho, Campus Experimental do Litoral Paulista, Brasil

Daneila O. Toyama
Universidade Presbiteriana Mackenzie, Brasil

Veronica C.G. Soares
Instituto de Biologia, UNICAMP, Brasil

Camila Ap Cotrim
Institut für Biochemie und Biotechnologie, Martin-Luther-Universität Halle-Wittenberg, Germany

Rafael Ximenes
Departamento de Fisiologia e Farmacologia UFC Fortaleza, Brasil

Marcelo L. Santos
Universidade Federal de Sergipe, Brasil